Processing and Application of Weather Radar Data

Processing and Application of Weather Radar Data

Guest Editors

Youcun Qi
Zhe Zhang
Zhanfeng Zhao
Bong-Chul Seo
Huiqi Li

MDPI

Basel • Beijing • Wuhan • Barcelona • Belgrade • Novi Sad • Cluj • Manchester

Guest Editors

Youcun Qi
Key Laboratory of Water Cycle and Related Land Surface Processes
Institute of Geographic Sciences and Natural Resources Research
Chinese Academy of Sciences
Beijing
China

Zhe Zhang
Shenzhen National Climate Observatory
Shenzhen
China

Zhanfeng Zhao
Key Laboratory of Water Cycle and Related Land Surface Processes
Institute of Geographic Sciences and Natural Resources Research
Chinese Academy of Sciences
Beijing
China

Bong-Chul Seo
Department of Civil, Architectural, and Environmental Engineering
Missouri University of Science and Technology
Rolla
USA

Huiqi Li
Guangzhou Institute of Tropical and Marine Meteorology
Guangzhou
China

Editorial Office

MDPI AG
Grosspeteranlage 5
4052 Basel, Switzerland

This is a reprint of the Special Issue, published open access by the journal *Remote Sensing* (ISSN 2072-4292), freely accessible at: https://www.mdpi.com/journal/remotesensing/special_issues/PSMBQE4ZKV.

For citation purposes, cite each article independently as indicated on the article page online and as indicated below:

Lastname, A.A.; Lastname, B.B. Article Title. *Journal Name* **Year**, *Volume Number*, Page Range.

ISBN 978-3-7258-4489-0 (Hbk)
ISBN 978-3-7258-4490-6 (PDF)
https://doi.org/10.3390/books978-3-7258-4490-6

Contents

remote sensing

Editorial

An Editorial for the Special Issue "Processing and Application of Weather Radar Data"

Youcun Qi [1,*], Zhe Zhang [2], Zhanfeng Zhao [1], Bong-Chul Seo [3] and Huiqi Li [4]

1 Key Laboratory of Water Cycle and Related Land Surface Processes, Institute of Geographic Sciences and Natural Resources Research, Chinese Academy of Sciences, Beijing 100101, China
2 Shenzhen National Climate Observatory, Shenzhen 518040, China
3 Department of Civil, Architectural, and Environmental Engineering, Missouri University of Science and Technology, 229 Butler-Carlton Hall, 1401 N. Pine St., Rolla, MO 65409, USA; bongchul.seo@mst.edu
4 Guangzhou Institute of Tropical and Marine Meteorology, China Meteorological Administration, Guangzhou 510641, China
* Correspondence: youcun.qi@igsnrr.ac.cn

In 2019, the World Meteorological Organization (WMO) pointed out the following based on the statistics from 2007 to 2019: in natural disasters, 90% of losses are related to meteorology, of which heavy storms and floods account for more than 70%. Heavy precipitation plays a very important role in the early warning of meteorological, hydrological, and geological disasters. Therefore, accurate monitoring and early warning and forecasting of heavy rainfall induced by strong convection are the basis for improving our ability to prevent natural disasters, such as floods, landslides, and mudslides.

Currently, the most powerful technique for monitoring natural hazards induced by heavy rainstorms is to use weather radars (e.g., ground-based radars, profiling radars, and space-born radars) [1,2]. Dual-polarization or dual-frequency radar data are used to derive water mixing ratios and number concentrations as well as to improve the capability of the convection-permitting numerical weather prediction (NWP) models to forecast severe storms at scales varying from a few hundred meters to kilometers. Advanced quantitative precipitation forecast (QPF) products are of great assistance in short-term weather and hydrological forecasting. Associated surface in situ observations, such as from rain gauges, runoff gauges, and disdrometers, are also required for calibrating radar observational variables and products [3,4].

In this Special Issue, studies covered several important topics, involving the development of radar signal processing methods; characterization of errors/uncertainties in remote sensing precipitation products and retrieval algorithm functions of different conditions; new sensing techniques as well as attenuation correction and calibration techniques; applications of radar data in data assimilation to improve the performance of NWP models; development of new analysis methods (e.g., machine learning and data assimilation) to maximize the benefits of using extensive datasets, multi-scale remote sensing data, and in situ data fusion; and application of radar data in disastrous weather (e.g., heavy rain, hail, and tornado) analysis and radar observations of hydrometeorological extremes. All of which improve the skills of QPF, radar signal processing methods, etc.

Routinely, operational weather radars could suffer from many difficulties that limit their data quality and applications. Efforts are made in proposing new bin-by-bin approximation methods employing the European Centre for Medium-Range Weather Forecasts (ECMWF) re-analysis data trying to address the attenuation caused by atmospheric gases and stratiform clouds [5], training the dilated and self-attentional UNet model to improve the completion of weather radar missing data [6], developing novel optimization strategy to mitigate the effects of sidelobes in strong convection weather process [7], and developing techniques for noise cancelation and recovery of radial velocity to improve the quality of three-dimensional radar wind fields [8].

Understanding the characteristics of the raindrop size distribution (DSD) is crucial to improve our knowledge of the microphysical processes of precipitation and to improve the accuracy of radar quantitative precipitation estimation (QPE). Topics in this Special Issue presented the spatial variability of DSD in different geographic regions and its influence on radar QPE [3], performance differences in the QPE relationship of dual-polarization radars under different schemes, radar wavelengths, and rainfall rates R classes in typical arid areas of China [9]. A new uniformity index for the axis ratios derived from dual-polarization weather radar data was proposed for raindrop area identification and analysis [10]. Microphysics schemes were also tested to depict the contrast between convective and stratiform regions in terms of the DSD [11]. Aircraft observations were analyzed to gain insights in the vertical distribution of cloud microphysical properties in different parts of stratocumulus clouds [12].

Weather forecasting plays a pivotal role in modern society, aiding individuals and decision makers in making informed choices and preparations. Recent advancements in weather radar technology and Blending forecast have propelled meteorological research and forecasting capabilities to new heights. For example, the prediction abilities of the Radar Extrapolation Forecast (REF), Wuhan Rapid Update Cycle (WHRUC), GRAPES_3 km, and Blending are compared and analyzed. It is shown that Blending is obviously better than the single forecast, especially in the heavy precipitation echo forecast, and plays a positive role in the convective forecast [13]. Another notable area of progress lies in the field of nowcasting and QPF. Traditionally, radar echo extrapolation methods have been used for nowcasting, but they often suffer from spatial inaccuracies. However, recent studies have showcased the efficacy of deep learning techniques in improving nowcasting performance. Despite their success, current deep learning-based models face challenges in accurately representing spatial variability, leading to a "blurry" effect in forecasts. To address this issue, researchers have proposed novel approaches, such as the Spatial Variability Representation Enhancement (SVRE) loss function and the Attentional Generative Adversarial Network (AGAN) model [14]. These innovations offer promising solutions for achieving high-precision radar nowcasting applications.

Moreover, the integration of multiscale representations (MSRs) of the atmosphere holds immense potential for advancing model–data fusion techniques in weather forecasting [15]. By reconstructing radar echoes from weather model simulations and satellite products, researchers have unveiled stratified features within the atmosphere, providing valuable insights into small-scale patterns and larger-scale information. This holistic understanding of atmospheric dynamics paves the way for more accurate and reliable forecasts, transcending conventional limitations in nowcasting.

In parallel, advancements in precipitation forecasting have been significant, particularly in addressing the challenges on sub-seasonal to seasonal scales. The development of the Quantile Mapping of Matching Precipitation Threshold by Time Series (MPTT-QM) method represents a breakthrough in precipitation bias correction, offering improved spatial distribution and temporal consistency in forecasts [16]. Furthermore, innovations, such as the nonlinear grid transformation (NGT) method, show promise in enhancing convective echo extrapolation prediction, thereby refining precipitation forecasts and mitigating potential inaccuracies [17].

Beyond forecasting techniques, studies on weather systems, such as mesovortices (MVs) during the rainy season, provide valuable insights into their spatiotemporal distributions and environmental influence [18]. Additionally, the utilization of machine learning algorithms for retrieving temperature and relative humidity profiles demonstrates the interdisciplinary nature of meteorological research [19], offering new avenues for enhancing data accuracy and reliability. Despite the significant progress in radar technology, challenges persist in radar processing and applications. Efforts are under way to improve data quality and maximize the utility of operational weather radars.

In summary, the continuous advancements in weather radar technology hold immense promise for revolutionizing meteorological forecasting and disaster preparedness. As we navigate an increasingly volatile climate, investing in these innovations is essential for safeguarding lives and livelihoods. By harnessing the full potential of weather forecasting technology, one can better anticipate and mitigate the impact of extreme weather events, ensuring a safer and more resilient future for us all.

Author Contributions: This editorial was prepared by Z.Z. (Zhe Zhang), Z.Z. (Zhanfeng Zhao) and H.L. and reviewed by Y.Q. and B.-C.S. All authors have read and agreed to the published version of the manuscript.

Funding: The Guest Editors of this Special Issue Z.Z. (Zhe Zhang) was funded by the Shenzhen Science and Technology Program (JCYJ20230807154400002) and the Key Laboratory of Atmospheric Sounding of China Meteorological Administration (2022KLAS06M). Z.Z. (Zhanfeng Zhao) was funded by the National Key Research and Development Project (2022YFC3002904), the Hainan Key Research and Development Project (ZDYF2023SHFZ125), the Anyang Key Research and Development Project (2022A02SF005), and the National Natural Science Foundation of China (42105031).

Acknowledgments: The Guest Editors would like to thank the authors who contributed to this Special Issue and the reviewers who helped to improve the quality of the Special Issue by providing constructive feedback to the authors.

Conflicts of Interest: The authors declare no conflicts of interest.

References

1. Qi, Y.; Zhang, J.; Zhang, P. A Real-Time Automated Convective and Stratiform Precipitation Segregation Algorithm in Native Radar Coordinates. *Q. J. R. Meteorol. Soc.* **2013**, *139*, 2233–2240. [CrossRef]
2. Yang, Z.; Qi, Y.; Zhang, Z.; Li, D. Can CINRAD Radar With VCP-21 Mode Capture the Accumulated Rainfall Pattern and Intensity of Fast-Moving Storms? *IEEE Trans. Geosci. Remote Sens.* **2024**, *62*, 4100813. [CrossRef]
3. Zhang, Z.; Li, H.; Li, D.; Qi, Y. Spatial Variability of Raindrop Size Distribution at Beijing City Scale and Its Implications for Polarimetric Radar QPE. *Remote Sens.* **2023**, *15*, 3964. [CrossRef]
4. Zhu, Z.; Qi, Y.; Cao, Q.; Li, D.; Zhang, Z.; Cao, J.; Xue, M. Particle Size Distribution Characteristics Within Different Regions of Mature Squall-Line Based on the Analysis of Global Precipitation Measurement Dual-Frequency Precipitation Radar Retrieval. *IEEE Geosci. Remote Sens. Lett.* **2020**, *19*, 3500205. [CrossRef]
5. Chang, Y.; Chen, H.; Huang, X.; Bi, Y.; Duan, S.; Wang, P.; Liu, J. Correction for the Attenuation Due to Atmospheric Gas and Stratiform Clouds in Triple-Frequency Radar Observations of the Microphysical Properties of Snowfall. *Remote Sens.* **2023**, *15*, 4843. [CrossRef]
6. Gong, A.; Chen, H.; Ni, G. Improving the Completion of Weather Radar Missing Data with Deep Learning. *Remote Sens.* **2019**, *29*, 4568. [CrossRef]
7. Hu, J.; Dong, X.; Tian, W.; Hu, C.; Feng, K.; Lu, J. A Novel Optimization Strategy of Sidelobe Suppression for Pulse Compression Weather Radar. *Remote Sens.* **2023**, *15*, 3188. [CrossRef]
8. Park, S.Y.; Jung, S.H.; Kim, K.H. Correction of Dual-PRF Velocity for Operational S-Band Doppler Weather Radar. *Remote Sens.* **2023**, *15*, 1920. [CrossRef]
9. Zeng, Y.; Li, J.; Yang, L.; Li, H.; Li, X.; Tong, Z.; Jiang, Y.; Liu, J.; Zhang, J.; Zhou, Y. Microphysical Characteristics of Raindrop Size Distribution and Implications for Dual-Polarization Radar Quantitative Precipitation Estimations in the Tianshan Mountains, China. *Remote Sens.* **2023**, *15*, 2668. [CrossRef]
10. Sun, Y.; Xiao, H.; Yang, H.; Chen, H.; Feng, L.; Shu, W.; Yao, H. A Uniformity Index for Precipitation Particle Axis Ratios Derived from Radar Polarimetric Parameters for the Identification and Analysis of Raindrop Areas. *Remote Sens.* **2023**, *15*, 534. [CrossRef]
11. Sun, Y.; Zhou, Z.; Gao, Q.; Li, H.; Wang, M. Evaluating Simulated Microphysics of Stratiform and Convective Precipitation in a Squall Line Event Using Polarimetric Radar Observations. *Remote Sens.* **2023**, *15*, 1507. [CrossRef]
12. Xiong, J.; Liu, X.; Wang, J. Study on the Vertical Structure and the Evolution of Precipitation Particle Spectrum Parameters of Stratocumulus Clouds over North China Based on Aircraft Observation. *Remote Sens.* **2023**, *15*, 2168. [CrossRef]
13. Wang, J.; Wang, Z.; Ye, J.; Lai, A.; Ma, H.; Zhang, W. Technical Evaluation of Precipitation Forecast by Blending Weather Radar Based on New Spatial Test Method. *Remote Sens.* **2023**, *15*, 3134. [CrossRef]
14. Gong, A.; Li, R.; Pan, B.; Chen, H.; Ni, G.; Chen, M. Enhancing Spatial Variability Representation of Radar Nowcasting with Generative Adversarial Networks. *Remote Sens.* **2023**, *15*, 3306. [CrossRef]
15. Zhu, M.; Liao, Q.; Wu, L.; Zhang, S.; Wang, Z.; Pan, X.; Wu, Q.; Wang, Y.; Su, D. Multiscale Representation of Radar Echo Data Retrieved through Deep Learning from Numerical Model Simulations and Satellite Images. *Remote Sens.* **2023**, *15*, 3466. [CrossRef]
16. Li, X.; Wu, H.; Nanding, N.; Chen, S.; Hu, Y.; Li, L. Statistical Bias Correction of Precipitation Forecasts Based on Quantile Mapping on the Sub-Seasonal to Seasonal Scale. *Remote Sens.* **2023**, *15*, 1743. [CrossRef]

17. Sun, Y.; Xiao, H.; Tian, Y.; Yang, H. A Nonlinear Grid Transformation Method for Extrapolating and Predicting the Convective Echo of Weather Radar. *Remote Sens.* **2023**, *15*, 1406. [CrossRef]
18. Tang, Y.; Xu, X.; Ju, Y.; Wu, Z.; Zhang, S.; Chen, X.; Xu, Q. Statistical Analysis of Mesovortices during the First Rainy Season in Guangdong, South China. *Remote Sens.* **2023**, *15*, 2176. [CrossRef]
19. Luo, Y.; Wu, H.; Gu, T.; Wang, Z.; Yue, H.; Wu, G.; Zhu, L.; Pu, D.; Tang, P.; Jiang, M. Machine Learning Model-Based Retrieval of Temperature and Relative Humidity Profiles Measured by Microwave Radiometer. *Remote Sens.* **2023**, *15*, 3838. [CrossRef]

remote sensing

MDPI

Article

Spatial Variability of Raindrop Size Distribution at Beijing City Scale and Its Implications for Polarimetric Radar QPE

Zhe Zhang [1], Huiqi Li [2], Donghuan Li [3] and Youcun Qi [3,*]

1 Shenzhen National Climate Observatory, Shenzhen 518040, China
2 Guangzhou Institute of Tropical and Marine Meteorology, China Meteorological Administration/China Meteorological Administration Tornado Key Laboratory, Guangzhou 510640, China
3 Key Laboratory of Water Cycle and Related Land Surface Processes, Institute of Geographic Sciences and Natural Resources Research, Chinese Academy of Sciences, Beijing 100101, China
* Correspondence: youcun.qi@igsnrr.ac.cn

Abstract: Understanding the characteristics of the raindrop size distribution (DSD) is crucial to improve our knowledge of the microphysical processes of precipitation and to improve the accuracy of radar quantitative precipitation estimation (QPE). In this study, the spatial variability of DSD in different regions of Beijing and its influence on radar QPE are analyzed using 11 disdrometers. The DSD data are categorized into three regions: Urban, suburban, and mountainous according to their locations. The DSD exhibits evidently different characteristics in the urban, suburban, and mountain regions of Beijing. The average raindrop diameter is smaller in the urban region compared to the suburban region. The average rain rate and raindrop number concentration are lower in the mountainous region compared to both urban and suburban regions. The difference in DSD between urban and suburban regions is due to the difference in DSD for the same precipitation types, while the difference in DSD between mountain and plains (i.e., urban and suburban regions) is the combined effect of the convection/stratiform ratio and the difference of DSD for the same precipitation types. Three DSD-based polarimetric radar QPE estimators were retrieved and estimated. Among these three QPE estimators, $R(Z_H)$, $R(K_{dp})$, and $R(K_{dp}, Z_{DR})$, $R(K_{dp}, Z_{DR})$ performs best, followed by $R(K_{dp})$, and $R(Z_H)$ performs worst. $R(K_{dp})$ is more sensitive to the representative parameters, while $R(Z_H)$ and $R(K_{dp}, Z_{DR})$ are more sensitive to observational error and systematic bias (i.e., calibration).

Keywords: raindrop size distribution (DSD); polarimetric radar; quantitative precipitation estimation (QPE)

1. Introduction

Raindrop size distribution (DSD) represents the combined effect of dynamic, thermodynamic, and microphysical processes in precipitation systems. Therefore, analyzing DSD is crucial for the development or validation of microphysical parameterization schemes in numerical weather prediction models [1–3], as well as for understanding the microphysical characteristics in precipitation systems [4–7], which is of great help in improving weather forecasts. DSD modeling and retrieval are also useful for improving the radar quantitative precipitation estimation (QPE) [8–10], which is critical for meteorological and hydrological applications. In addition, DSD is closely related to the kinetic energy of rain, which is critical in understanding the erosive and runoff processes of soil and the subsequent hydrological hazards [11,12].

DSD is affected by various factors, including environmental conditions (temperature, pressure, humidity, wind, aerosol, etc.), evaporation, drop sorting, clustering and breakup, and so on [5,13–15]. As a result, DSD exhibits significant variation across different climatic regimes, seasons, and precipitation types. Numerous studies have extensively investigated the characteristics of DSD worldwide, utilizing both in situ and remote sensing instruments such as radars. Bringi et al. [16] (hereinafter BR03) analyzed the DSD of convection and stratiform in different climate regimes and showed that the DSD of convection can be classified as "maritime" and "continental". Tang et al. [17] showed significant differences in DSD between the northern and southern regions of China. Other studies revealed the characteristics of DSD in different regions of China [6,18,19]. It is also found that the convection exhibits a larger mass-weighted mean diameter (D_m) and normalized intercept parameter (N_w) as compared to the stratiform [4,7,20,21]. Zeng et al. [22–24] found that both the number concentration and the drop size are larger at the top of Mt. Tianshan than at its foot.

Beijing, the capital of China with a population of over 21 million, has experienced rapid infrastructure development in recent decades. Extreme precipitation events and subsequent floods have caused great losses to the city every year. The impact of urbanization on precipitation has been widely recognized, and several possible mechanisms have been identified. These include the destabilization and perturbation of the boundary layer due to the urban heat island effect [25–27], enhanced convergence due to the large roughness in urban areas [28–30], and increased cloud condensation nuclei due to the high aerosol concentration in urban areas [31–33]. The climatological and statistical characteristics of precipitation and the mechanism of extreme precipitation systems in Beijing have been studied using rain gauges, radars, and models [34–39]. However, only a few studies have focused on the DSD characteristics in Beijing due to the lack of DSD measurements. Tang et al. [17] compared the difference in DSD characteristics between Beijing and southern China. Ji et al. [40] analyzed the DSD in Beijing based on 14-month DSD observations from one disdrometer. Ma et al. [41] studied the statistical characteristics of DSD during the rainy seasons in Beijing urban areas. These above works were only based on the observation of a single disdrometer and cannot represent the spatial variation of DSD characteristics in the whole Beijing region. As pointed out by Jaffrain et al. [42], DSD could vary even on the kilometer scale. Is there any DSD variability in different areas (urban, suburban, and mountains) of Beijing? What is the cause of the variability? These questions have not been well answered yet.

In recent years, quite a few second-generation OTT Parsivel (hereafter Parsivel^2) laser-optical disdrometers have been deployed in different areas of Beijing, providing a good opportunity to investigate the spatial variability of DSD. In this paper, we aim to reveal the DSD variability in Beijing as well as its impact on radar QPE, which would enhance our understanding of the microphysical characteristics of precipitation and improve the accuracy of radar QPE. This paper is organized as follows: Section 2 describes the data used, as well as the method for quality control, precipitation type classification, and analysis. The spatial characteristics of DSD in Beijing and its implication for polarimetric radar QPE are analyzed in Section 3. Sections 4 and 5 provide the discussion and conclusion, respectively.

2. Data and Methodology

2.1. Dataset

In this study, DSDs were collected using 11 Parsivel^2 disdrometers. The locations of these disdrometers are shown in Figure 1. In brief, Parsivel^2 is a laser-optical disdrometer that can simultaneously measure the size and falling velocity of particles. The sampling area of the Parsivel^2 is 54 cm^2 (18 cm in length and 3 cm in width). The measured size and falling velocity are divided into 32 bins. These bins are non-uniform, ranging from 0.062 to 24.5 mm for size and 0.05 to 20.8 m s^{-1} for falling velocity. Tokay et al. [43] evaluated the performance of Parsivel^2. Compared to its predecessor Parsivel, Parsivel^2 performs much better in measuring particle size, as Parsivel tends to underestimate the size of small particles and overestimate the size of large particles.

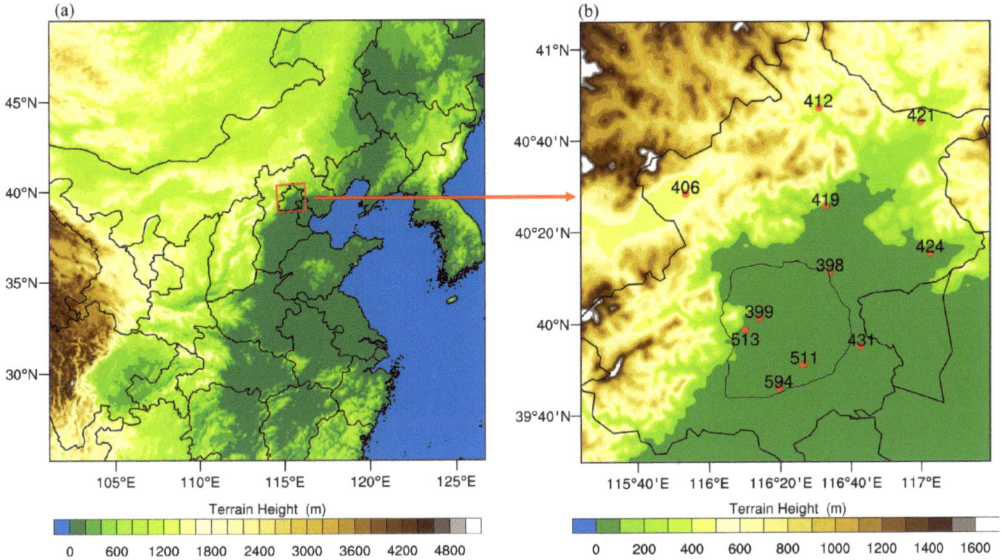

Figure 1. (**a**) Location of Beijing in China and (**b**) topography of Beijing and locations of the disdrometers used in this study. The thin black lines in (**b**) denote the 6th Ring Road of Beijing.

All the disdrometers were configured to measure DSD with a 1 min temporal resolution. The DSD data were collected from May to September 2017. Rainfall during these months accounts for 90% of the annual rainfall in Beijing.

2.2. Quality Control of DSD Dataset

Various sources affect the observational quality of the disdrometer, such as splashing of raindrops, wind effect, and margin fallers [5,13,43–46]. Therefore, quality control must be applied before using the data for analysis. The quality control procedure used here is similar to that proposed by Tokay et al. [46]. For the dataset of 1 min each, if the total drop number is less than 10, or the rain rate is less than 0.1 mm h^{-1}, this 1-min DSD is considered as noise and discarded. Drops exceeding $\pm$ 50% of their theoretical terminal falling velocity are also discarded from the DSD spectrum because such an observation may be due to splashing or wind effects. The theoretical terminal falling speed used here is based on the result of Brandes et al. [47]. The drop sizes that exceed 8 mm are also eliminated [17,41] because the largest raindrop recorded in nature are around 8 mm [48], and drops larger than 8 mm are unlikely to be raindrops.

A total number of 124,647 1-min DSD observations from 11 disdrometers passed the quality control and were used for analysis.

2.3. Separation of Precipitation Types Based on DSD Data

Previous studies have demonstrated that the characteristic DSD parameters are related to precipitation types. Therefore, it is necessary to separate different precipitation types when analyzing DSD characteristics. Numerous methods have been proposed to separate the precipitation type into convection and stratiform based on disdrometer observation [16,20,49,50]. The criteria of these methods are different, but the principles are similar: Convection usually exhibits heavier rainfall that may vary from time to time, while stratiform generally has a weaker but steadier rainfall. Therefore, when the rain rate and its variation are large, the precipitation can be classified as convection. Otherwise, it can be classified as stratiform. In this study, the precipitation separation method is similar to that of BR03. To be clear, for a 1-min DSD observation at time t, if its rain rate is greater than

5 mm h^{-1} and the standard deviation from t− Δt to t+ Δt is larger than 1.5 mm h^{-1}, the DSD of this minute is classified as convection. Otherwise, it is stratiform. Δt is set to 5 min.

2.4. Raindrop Size Distribution

The direct measurement of the disdrometer provides the number of drops in each bin (i.e., i size bins and j falling velocity bins). The mid-value of each bin is taken as the representative size of the bin. The following parameters are calculated to represent the characteristics of DSD, including the total number concentration N_t, mass-weighted diameter D_m, normalized intercept parameter N_w, and rain rate R:

$$N_t = \int_{D_0}^{D_{max}} N(D)dD \tag{1}$$

$$D_m = \int_{D_0}^{D_{max}} D^4 N(D)dD / \int_{D_0}^{D_{max}} D^3 N(D)dD \tag{2}$$

$$N_w = \frac{4^4}{6} \left(\int_{D_0}^{D_{max}} D^3 N(D)dD \right) / D_m^4 \tag{3}$$

$$R = \frac{\pi}{6} \int_{D_0}^{D_{max}} D^3 V(D) N(D)dD \tag{4}$$

where *N(D)* is the normalized number of drops in each size bin:

$$N(D) = \sum_{j=1}^{32} \frac{n_j}{A\Delta t V_j \Delta D} \tag{5}$$

where Δt, V_j, and ΔD are the measuring time, the falling velocity at a given size bin, and the size bin width, respectively. A is the effective sampling area [46]:

$$A = L(W - D_i/2) \tag{6}$$

where L and W are the length and width of the sampling area, respectively.

2.5. DSD-Based Polarimetric Radar QPE Estimators

To simulate the radar QPE of operational X-band polarimetric radars in Beijing, polarimetric radar variables are calculated from DSD data using the T-matrix method [51], including horizontal (vertical) reflectivity $Z_{H(V)}$ (mm^6m^{-3}), differential reflectivity Z_{DR}, and specific differential phase K_{dp} ($^{\circ}km^{-1}$):

$$Z_{H,V} = \frac{4\lambda^4}{\pi^4 |K_w|^2} \sum_{i=1}^{32} |f_{hh,vv}(180, D_i)|^2 N(D_i)\Delta D_i \tag{7}$$

$$Z_{DR} = Z_H / Z_V \tag{8}$$

$$K_{dp} = \frac{180}{\pi} \sum_{i=1}^{32} \mathrm{Re}|f_{hh}(0, D_i) - f_{vv}(0, D_i)| N(D_i)\Delta D_i \tag{9}$$

where $f_{hh,vv}(180, D_i)$ is the back scattering amplitude of horizontal and vertical polarization for a drop; $f_{hh}(0, D_i)$ and $f_{vv}(0, D_i)$ are the forward scattering amplitudes of horizontal and vertical polarization, respectively; K_w is the dielectric factor of water (0.9639) and λ is radar wavelength (32 mm in this study).

Three widely used radar estimators are applied for radar QPE:

$$R(Z_{\mathrm{H}}) = aZ_{\mathrm{H}}^{b} \tag{10}$$

$$R(K_{\mathrm{dp}}) = aK_{\mathrm{dp}}^{b} \tag{11}$$

$$R(K_{\mathrm{dp}}, Z_{\mathrm{DR}}) = aK_{\mathrm{dp}}^{b}Z_{\mathrm{DR}}^{c} \tag{12}$$

where a, b, and c are parameters.

To quantitatively evaluate the performance of different QPE estimators, 3 statistical scores are used, including the correlation coefficient (CC), root mean square error (RMSE), and relative mean bias (RMB):

$$CC = \frac{\sum_{i=1}^{n} (R(es)_i - \overline{R(es)})(R(\mathrm{d})_i - \overline{R(\mathrm{d})})}{\sqrt{\sum_{i=1}^{n} (R(es)_i - \overline{R(es)})^2 \sum_{i=1}^{n} (R(\mathrm{d})_i - \overline{R(\mathrm{d})})^2}} \tag{13}$$

$$RMSE = [\frac{1}{n}\sum_{i=1}^{n} (R(es)_i - R(\mathrm{d})_i)^2]^{1/2} \tag{14}$$

$$RMB = \sum_{i=1}^{n} (R(es)_i - R(\mathrm{d})_i) / \sum_{i=1}^{n} (R(\mathrm{d})_i) \tag{15}$$

where $R(es)_i$ is the rain rate using one of the radar estimators (i.e., Equations (10)–(12)) with radar variables simulated using DSD data with the T-matrix method (i.e., Equations (7)–(9)), and $R(\mathrm{d})_i$ is the rain rate calculated directly from the DSD data using Equation (4).

3. Results

3.1. DSD Variability in Different Areas of Beijing

The areas within the 6th Ring Road have dense infrastructure, heavy traffic, and frequent human activities. Over 90% of Beijing's population lives and works in the areas inside the 6th Ring Road, while outside the 6th Ring Road, there are mostly farms, forests, and wastelands. The topography of Beijing is characterized by plains in the center and southeast and mountains in the west and north (Figure 1). Accordingly, the 11 disdrometers used in this study were categorized into three groups based on their location: Urban stations (stations located within the 6th Ring Road, i.e., stations 399, 511, 513, and 594); suburban stations (stations located outside the 6th Ring Road with an elevation of less than 200 m above sea level, i.e., stations 398, 419, 424, and 431), and mountain stations (stations above 200 m above sea level, i.e., stations 406, 412, and 421).

The average DSD characteristics derived with all these 11 disdrometers are shown in Figure 2. Figure 2a shows the density scatter plot of D_m versus R, superimposed with the power–law relationship obtained using the least-square fit method. D_m increases with the increase in R (positive exponent in power–law relationship). As shown in the figure, D_m increases rapidly when R is less than 50 mm h^{-1}. This is because both the raindrop size and number concentration effectively increase within this rain rate range [52]. The increase in D_m becomes much slower (around 2.2~2.5 mm) when R is greater than 50 mm h^{-1}. Apparently, the increase in the rain rate mainly relies on the increase in raindrop concentration rather than raindrop size. This fact implies that the accurate estimation of particle number concentration in numerical models is crucial for better forecasting of extreme precipitation events. In addition, the spread of D_m becomes narrower with the increase in R. Such a fact suggests that when the rain rate is small, the breakup and coalescence processes of raindrops may be unbalanced, resulting in a wide spread of D_m. At a high rain rate, the breakup and coalescence are likely to reach a more balanced state. This result also explains why retrieving the rain rate using Equation (10) (traditional approach of QPE for single-polarimetric radar) is not accurate for small rain rate cases. Since parameters a and b in Equation (10) depend on DSD and there is a wide spread of DSD parameters with a small rain rate, if a fixed combination of a and b is used for QPE (the common approach for operational QPE), large errors will appear. Figure 2b is the scatter plot of D_m versus N_w for convection and stratiform in Beijing, superimposed with BR03's results. There is a clear boundary between convection and stratiform with some overlap of samples. For convection, only 4.23% and 13.1% of the samples lie in "maritime" and "continent" clusters in BR03, respectively. The mean value point of D_m-N_w lies between these two clusters, suggesting that the characteristics of convection in Beijing be different from those places recorded in BR03. As for the stratiform, the mean value point of D_m-N_w and 90% of samples lie on the left side of the least square fitting line of stratiform in BR03, indicating that the stratiform in Beijing has a smaller raindrop size and concentration. It is notable that DSD studies in other locations of China (Nanjing, eastern China; Guangzhou and Yangjiang, southern China; Naqu, Tibet Plateau) also suggest a lower raindrop size and concentration in the stratiform as compared to BR03 [6,7,17,18].

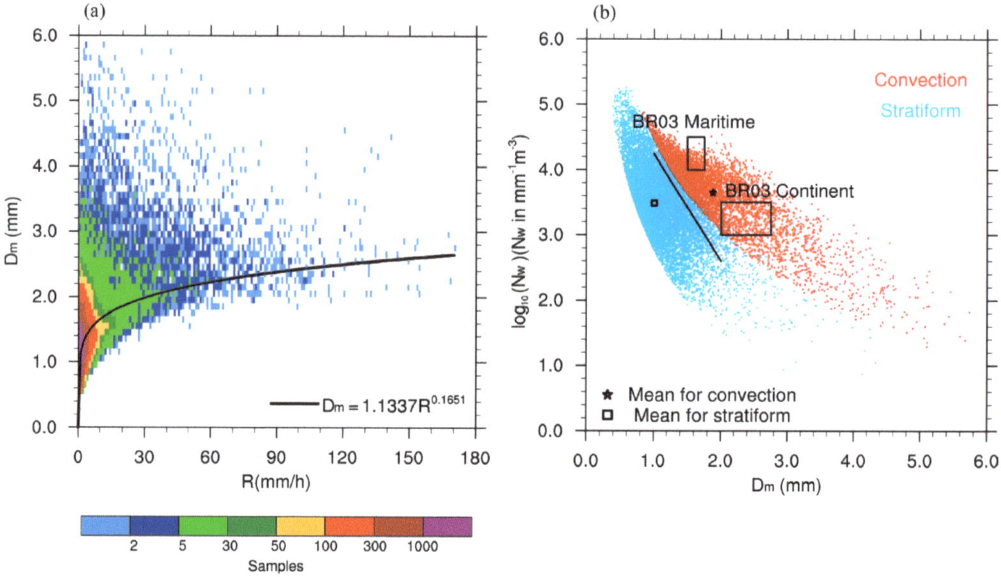

Figure 2. (**a**) Scatter density plot for R versus D_m, superimposed with the power–law relationship obtained using the least-square fit method and (**b**) scatter plot for D_m versus N_w. Red (blue) dots represent convection (stratiform). The star and square symbols represent the mean values for convection and stratiform, respectively. The black line is the log10(N_w)–D_m relationship for stratiform in BR03. Two rectangles indicate the maritime and continental convective clusters in BR03.

Figure 3 shows the variations of mean number concentration versus raindrop size in different areas of Beijing. The number concentration in mountain areas is lower than in the plains (i.e., urban and suburban areas) from the smallest raindrop size to up to 5 mm in diameter. The differences are most pronounced at the smallest sizes and around 3 mm. The urban and suburban curves are similar, but urban areas have a higher number concentration of raindrops less than 1 mm, and a lower number concentration of raindrops greater than 1 mm, indicating a smaller mean raindrop size.

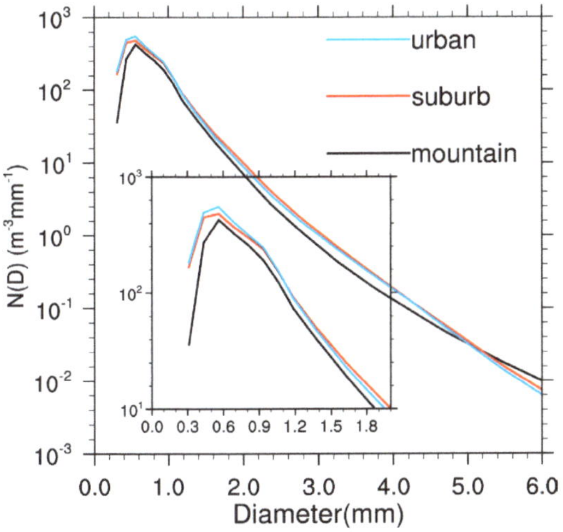

Figure 3. Average raindrop spectra for different areas of Beijing.

Table 1 shows the mean values of DSD parameters in different areas of Beijing. Among these three areas, the urban average drop size (D_m) is the smallest, and this value is comparable with previous studies based on urban disdrometer data in Beijing [40,41]. The average drop size in the suburban areas is the largest and between the two in the mountain areas. In terms of the average rain rate, the mountain areas have the smallest average rain rates, the suburban areas have the largest average rain rates, and the urban areas are in the middle. The distribution of R is consistent with previous works on the precipitation in Beijing based on rain gauge measurements [38,53,54], which found that the average hourly precipitation intensity in the mountains is smaller than that in the plains, but the total precipitation hours are larger in the mountains, mainly because light rain occurs more frequently in the mountains of Beijing. This phenomenon might be related to the specific geographical location of Beijing. The southeast flow coming from the sea is the main moisture source for precipitation systems in Beijing. The mountain areas are located in the northwest part of Beijing, which means that the mountain areas of Beijing generally receive less moisture than the plain areas of Beijing. However, although the mountain areas receive less moisture, light precipitation can easily occur when the southeast flow is elevated by the mountains. In terms of the number concentration (N_t), the mountain areas have the smallest average number concentration, the urban areas have the largest average number concentration, and the suburban areas are in the middle. In the mountain areas of Beijing, the smallest average number concentration may also be related to the frequent occurrence of light rain, as light rain is usually associated with fewer drop numbers. The comparison of these parameters in urban and suburban areas reveals that the urban environment modifies the precipitation microphysics, such that the drop size is suppressed while a greater number of drops are produced. This phenomenon may be related to the high aerosol emission in urban areas of Beijing due to human activities such as traffic. A high aerosol concentration tends to reduce the average drop size and increase the number concentration by providing more cloud condensation nuclei (CCN) [15,31,55,56].

Table 1. Mean values of DSD parameters in different areas of Beijing.

Location	Samples	D_m(mm)	R (mm h^{-1})	N_t (m^{-3})
Urban	47325	1.17	2.69	328.6
Suburb	44429	1.26	2.83	306.2
Mountain	32893	1.20	2.15	233.1

Figure 4 shows the probability distribution functions (PDF) of D_m, R, and N_t in different areas. D_m in these areas all peak around 0.9 mm, with suburban areas having the highest frequency around the peak and urban areas having the lowest (Figure 3a). The distribution of N_t in the mountain areas is sharper and more symmetrical compared to those in the plains (i.e., urban and suburban areas). Both the urban and suburban areas have a broader distribution around the peak, and the frequency decreases faster toward the higher N_t end than toward the lower N_t end. In addition, urban areas have a higher distribution for N_t larger than 10^3 and a lower distribution for N_t from 10^2 to $10^{2.5}$. As for the PDF for R, R less than $10^{0.5}$ mm h^{-1} is mainly responsible for the differences in different areas. There is a sharper and higher peak in mountain areas at the lower end of R, indicating that light rain occurs more frequently in mountain areas than in plains. There is a higher frequency of rain rate from 10^0 to $10^{0.5}$ and larger than $10^{1.0}$ mm h^{-1} in plains than in mountain areas, while the rain rate from $10^{0.5}$–$10^{1.0}$ is quite close together. This result suggests that the mountains in Beijing may play a role in modifying precipitation microphysics mainly for precipitation with a rain rate less than $10^{0.5}$ or larger than $10^{1.0}$ mm h^{-1}. Light rain occurs more frequently in mountain areas because mountain areas receive less moisture than the plains in Beijing as the southeast wet flow travels further to reach the mountain areas in the western part of Beijing [38,53,54]. As for the differences between urban and suburban areas, suburban areas have a high frequency of rain rate of less than $10^{0.3}$ mm h^{-1} and more than $10^{1.3}$ mm h^{-1}, with a lower frequency in the middle.

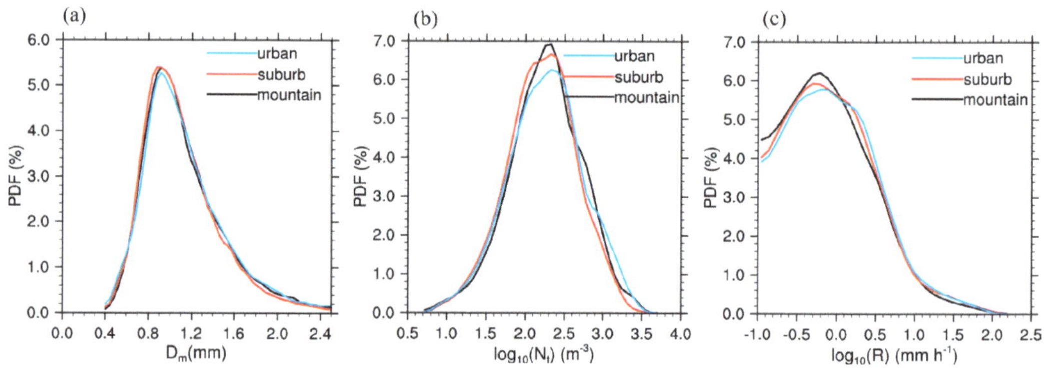

Figure 4. The probability distribution functions (PDF) of (**a**) D_m, (**b**) N_t, and (**c**) R for different areas of Beijing.

Numerous studies have shown that D_m, R, and N_t are larger in convection than in stratiform [16,17,20], and other studies suggested that the terrain or urban environments can modify the microphysics processes in precipitation systems and change the DSD characteristics [24,57]. Consequently, there might be three causes responsible for the variation in DSD characteristics in different regions of Beijing: (1) The ratio of convection/stratiform might be different in different regions and the higher frequency of convection might lead to larger D_m, R, and N_t; (2) for the same precipitation type, DSD characteristics in different regions might be different due to the terrain or urban effect; or (3) the combination of (1) and (2).

Accordingly, DSD observations were classified into convection and stratiform for further analysis. The DSD parameters, number of samples, and percentage of convection and stratiform in different areas of Beijing are shown in Table 2. First, the difference between urban and suburban areas was analyzed. Although the average D_m is larger in suburban areas than that in urban areas, it is surprising to see that the percentage of convection in suburban areas is almost identical to that in urban areas (8.58% versus 8.60%). Consequently, the differences in D_m between urban and suburban areas are not likely due to differences in convection/stratiform ratios but rather are more likely due to differences in DSD characteristics for the same precipitation type. For both convection and stratiform, D_m is smaller in urban areas than in suburban areas, perhaps due to the high aerosol concentration in urban areas of Beijing. High aerosol concentration usually causes smaller raindrop sizes by providing high CCN [15]. A study on the spatial distribution of PM2.5 (particulate matter with aerodynamic diameters of less than 2.5 μm) in Beijing shows that the PM2.5 concentration is much higher in urban areas than in suburbs during rainy seasons [58]. On the other hand, smaller raindrops tend to evaporate more quickly after falling out of the cloud, which may further lead to smaller raindrops in urban areas. Urban areas have smaller R and N_t for convection than suburban areas, but larger ones for stratiform. The result suggests urbanization affects convection and stratiform differently, whereby the urban environment tends to reduce the intensity of rain and the number concentration of raindrops in convection while positively influencing them in stratiform. It appears that the differences in DSD between urban areas and suburban areas are not due to differences in convection/stratiform ratios, but rather due to differences in DSD characteristics for the same precipitation types.

Table 2. Mean values of DSD parameters, number of samples, and percentage of convection and stratiform in different areas of Beijing.

Location	Precipitation Type	Samples	Percentage (%)	D_m(mm)	R (mm h^{-1})	N_t (m^{-3})
Urban	convection	4070	8.60	1.85	18.25	863.4
	stratiform	43,255	91.40	0.97	1.22	278.3
Suburb	convection	3810	8.58	1.92	20.29	949.8
	stratiform	40,619	91.42	1.02	1.20	245.7
Mountain	convection	2149	6.53	1.85	16.93	681.3
	stratiform	30,744	93.47	1.05	1.11	201.7

The convection/stratiform ratio in mountain areas is lower than that in the plains, and only 6.53% of the total precipitation is convection. For convection, D_m in mountain areas is almost the same as that in urban areas and smaller than that in suburban areas. However, for stratiform, D_m in mountain areas is larger than that in the plains. R and N_t for both convection and stratiform are smaller in mountain areas than in the plains. Therefore, smaller R and N_t values in mountain areas are the combined result of a smaller convection/stratiform ratio and smaller R and N_t values for the same precipitation types. Such a result may be related to the moisture conditions in Beijing. Beijing typically receives its moisture from the east (from the ocean), which travels hundreds of kilometers before reaching Beijing (the nearest ocean is 160 km away). The mountain areas on the west side of Beijing receive less moisture than the plain areas on the east, thereby reducing convection frequency, rain intensity, and number concentration.

3.2. Implication for QPE of Polarimetric Radar

Several X-band polarimetric radars (λ = 3.2 cm) have been deployed in Beijing in recent years, aiming at providing better QPE products to meet the needs of meteorological and hydrological applications. These radars all operate in VCP 21 mode, which completes a volume scan in 3 min with radial and azimuth resolutions of 75 m and 0.95°, respectively. To study the X-band radar QPE using DSD data, the polarimetric radar variables of Z_H, Z_{DR}, and K_{dp} were calculated from 1 min DSD observations. The parameters of a, b, and c in Equations (10)–(12) were then derived using the nonlinear least square fitting. The fitted parameters using DSD data collected in all locations, namely, urban, suburban, and mountain areas, are listed in Table 3. As Table 3 shows, these parameters vary in different regions of Beijing due to the DSD variability.

Table 3. The fitted parameters for QPE estimators.

Estimator	Location	a	b	c
$R(Z_H)$	Entire	0.1232	0.4758	\
	Urban	0.1243	0.4756	\
	Suburb	0.107	0.4927	\
	Mountain	0.1203	0.4646	\
$R(K_{dp})$	Entire	15.83	0.7727	\
	Urban	15.87	0.7721	\
	Suburb	15.97	0.8078	\
	Mountain	14.99	0.7277	\
$R(K_{dp}, Z_{DR})$	Entire	30.31	0.9676	−1.409
	Urban	29.17	0.9554	−1.309
	Suburb	29.78	0.9856	−1.38
	Mountain	30.04	0.9324	−1.431

Figure 5 illustrates the scatter density plots of R estimated from four estimators (Table 3) versus R calculated directly from the 1 min DSD data. The statistical values of CC, RMSE and RMB are also shown. As shown in Figure 5, the estimator $R(Z_H)$ performs the worst (Figure 5a) with the smallest CC and largest RMSE and RMB. The uncertainty of QPE increases greatly with the intensity of R. The difference between $R(es)_i$ and $R(\mathrm{d})_i$ can be up to approximately 4 times (e.g., 30 mm h^{-1} of $R(es)_i$ versus 120 mm h^{-1} of $R(\mathrm{d})_i$). When polarimetric variables of Z_{DR} and K_{dp} are introduced, the accuracy of QPE is much better (Figure 5b–c). Among these three estimators, $R(K_{dp}, Z_{DR})$ performs the best, providing the most accurate estimation for light rain to heavy rain.

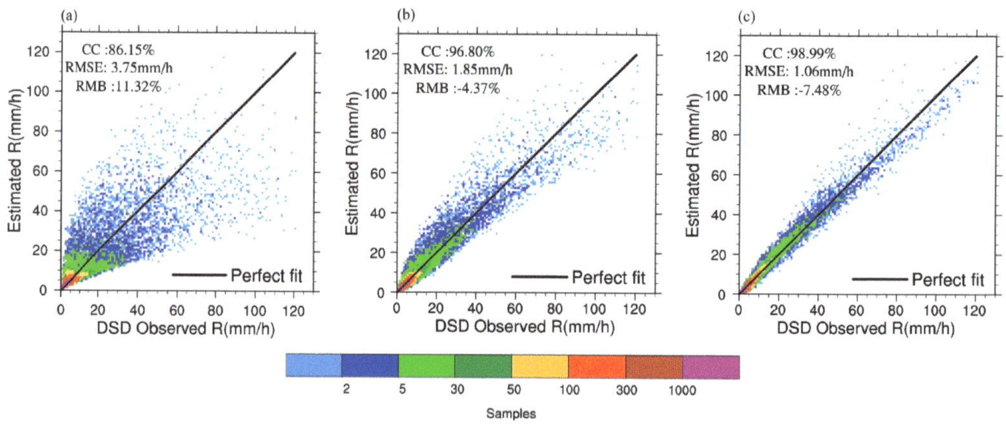

Figure 5. Scatter density plots of R from all 11 disdrometer observations and R estimated using estimators: (**a**) $R(Z_H)$, (**b**) $R(K_{dp})$, and (**c**) $R(K_{dp}, Z_{DR})$. The black line in each panel is the perfect fit line (i.e., y = x). Statistical scores of CC, RMSE, and RMB are superimposed.

The results shown in Figure 5 can be regarded as the theoretical upper limit of the performances of the estimators. When performing QPE estimators into operational radar, three aspects below affect the accuracy of the QPE result: Random observational errors of the radar variables, the systematic bias of the radar variables due to miscalibration, and the variability of the parameters of the QPE estimators due to DSD variability. It is well-known that the DSD variability in different climate regions significantly affects QPE accuracy. However, for city scales such as Beijing, it is unclear how much the DSD variability affects the QPE accuracy compared to radar variables measurement errors and bias. What is the dominant source of error for QPE? To find the answer, a series of experiments were performed using these DSD data.

The actual distribution of radar observational errors can be very complicated. However, for ideal experiments using DSD data, let us assume that the errors conform to the most general type of error distribution, the normal distribution $N(\mu, \sigma^2)$, where μ and σ are the mean value and standard deviation. Approximately 68% and 95% of the total samples lie between $\mu - \sigma$ to $\mu + \sigma$ and $\mu - 2\sigma$ to $\mu + 2\sigma$, respectively. Assuming that the radar observational variables follow the normal distribution, then the observational variables can be perturbed by multiplying $N(\mu, \sigma^2)$ to simulate measurement errors and systematic bias. For example, $Z_h' = Z_h \bullet N(1, 0.15^2)$ means measurement errors exist in Z_h, while approximately 95% of the measured Z_h are between 0.7 and 1.3 times the theoretical Z_h; $Z_h' = Z_h \bullet N(1.05, 0.15^2)$ means both measurement errors and systematic bias exist in Z_h, the mean observational Z_h is stronger for 5% than the theoretical Z_h, and approximately 95% of the measured Z_h are between 0.75 to 1.35 time of the theoretical Z_h.

Table 4 shows the experiment design of $R(Z_H)$. All the DSD data collected by these 11 disdrometers are used in these experiments. In the control experiment, the rain rate is estimated using parameters obtained for the whole region of Beijing; it is the theoretical upper limit capability of applying $R(Z_H)$ to perform QPE. In the DSD variability experiment, the rain rate is estimated using parameters obtained for the mountain region of Beijing. The purpose of this experiment is to find out how much the DSD variability can affect the accuracy of QPE when the parameters for specific regions (e.g., mountains) are used to estimate the rain rate for the whole region of Beijing. In the measurement error experiment, the Z_h is perturbed by multiplying $N(1, 0.05^2)$. This experiment aims to find out if there are measurement errors of Z_h between operational radar and disdrometer and how much the error can affect the accuracy of QPE. Furthermore, the systematic bias experiment is designed to find out how much the error and systematic bias (i.e., calibration issues) can affect the accuracy of QPE, and which of these above issues affect the accuracy of QPE the most.

Table 4. Experiment design of $R(Z_H)$ estimator.

Name	Description
Control experiment	Perform $R(Z_H)$ to estimate rain rate using all the DSD data with parameters for the whole region of Beijing (i.e., a = 0.1232 and b = 0.4758)
DSD variability experiment	Perform $R(Z_H)$ to estimate rain rate using all the DSD data with parameters for the mountain region of Beijing (i.e., a = 0.1202 and b = 0.4646)
Measurement error experiment	Perturb Z_h by multiplying $N(1, 0.05^2)$
Systematic bias experiment	Perturb Z_h by multiplying $N(1.05, 0.05^2)$

Figure 6 shows the results of the $R(Z_H)$ experiment. If inappropriate parameter values (DSD variability experiment, Figure 6b) are used in QPE, such as using the parameters obtained in the mountain area to estimate the rain rate for the entire region of Beijing, it will lead to systematic bias in QPE. In this case, the rain rate is underestimated, as can be seen in Figure 6b, where more dots appear in the lower right part. The RMSE does not change much, with the RMSE increasing from 3.75 mm h^{-1} to 3.88 mm h^{-1}, and an even higher CC. The measurement errors affect the accuracy of QPE more significantly, as shown in Figure 6c; even if 95% of the observational Z_h are within 10% measurement errors, the QPE accuracy drops significantly, especially for heavy precipitation. With a rain rate larger than 50 mm h^{-1}, the dots become much more scattered than in Figure 6a, and RMSE rises to 4.23 mm h^{-1}. When both measurement errors and systematic bias of Z_h coexist, as shown in Figure 6d, the QPE accuracy decreases even more. The QPE overestimates the rain rate by 38.82%, with more dots appearing in the upper right part and becoming more scattered, and the RMSE increases significantly to 5.26 mm h^{-1}. This result suggests that for a city-scale region such as Beijing, when R(Z_H) is used for QPE, the variability of DSD certainly affects the QPE accuracy, but the main influencing factors on QPE accuracy are the measurement errors and calibration of reflectivity, and they affect the QPE accuracy to a greater extent than the influence of the variability of DSD. Therefore, we should focus more on improving the quality of the reflectivity when utilizing R(Z_H) in operation.

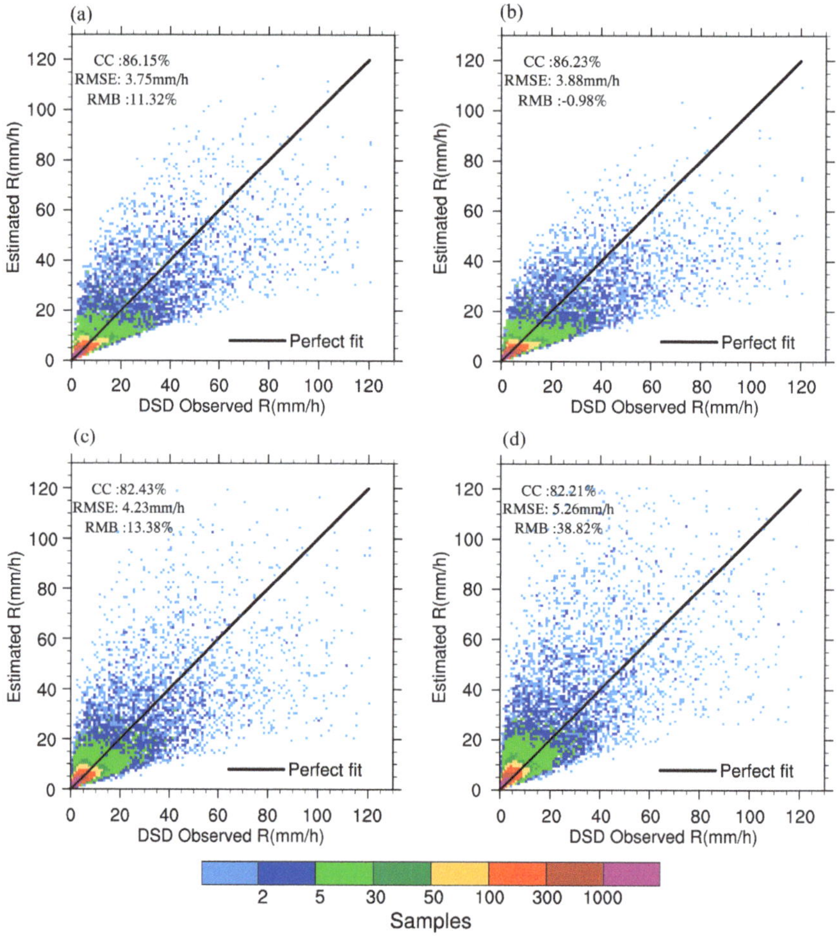

Figure 6. Scatter density plots of R in the whole region of Beijing from 11 disdrometer observations and R estimated using estimator R(Z_H): (**a**) Control experiment, (**b**) DSD variability experiment, (**c**) measurement error experiment, and (**d**) systematic bias experiment as described in Table 4. The black line in each panel is the perfect fit line (i.e., y = x). Statistical scores of CC, RMSE, and RMB are superimposed.

Similarly, the experiment design of R(K_{dp}) is shown in Table 5. Since K_{dp} is immune to calibration, the systematic bias experiment was discarded, and an additional measurement error experiment was added. The results are shown in Figure 7. Some previous works have suggested that the R(K_{dp}) estimator is relatively insensitive to the variability of DSD compared to R(Z_H) [47,59,60], but by comparing Figure 7a,b, it is clear that the variability of DSD does affect the accuracy of QPE using R(K_{dp}), at least for heavy precipitation. In this case, using parameters for the mountain region to estimate the rain rate for the entire region of Beijing results in the underestimation of heavy precipitation, as shown in Figure 7b. More dots with a rain rate larger than 50 mm h^{-1} appear in the lower right flank of the perfect line. The measurement errors, on the other hand, do not significantly affect the accuracy of QPE. Perturbing K_{dp} by multiplying $N(1, 0.05^2)$ does not degrade the performance much (Figure 7c), with CC, RMSE, and RMB quite close to the control experiment. Even when perturbing K_{dp} by multiplying $N(1, 0.15^2)$, which means assuming large measurement errors for K_{dp} (approximately 32% of the K_{dp} observation errors are larger than 15%), the QPE accuracy does not deteriorate significantly (Figure 7d), and it is comparable to the result of Figure 7b. This series of experiments on R(K_{dp}) suggest that the variability of DSD even at the city scale could lead to systematic bias in QPE, especially for heavy precipitation. The variability of DSD may affect the accuracy of QPE even more than K_{dp} measurement errors. Therefore, when utilizing R(K_{dp}) to perform QPE in operational usage, special attention should be paid to obtaining appropriate parameters.

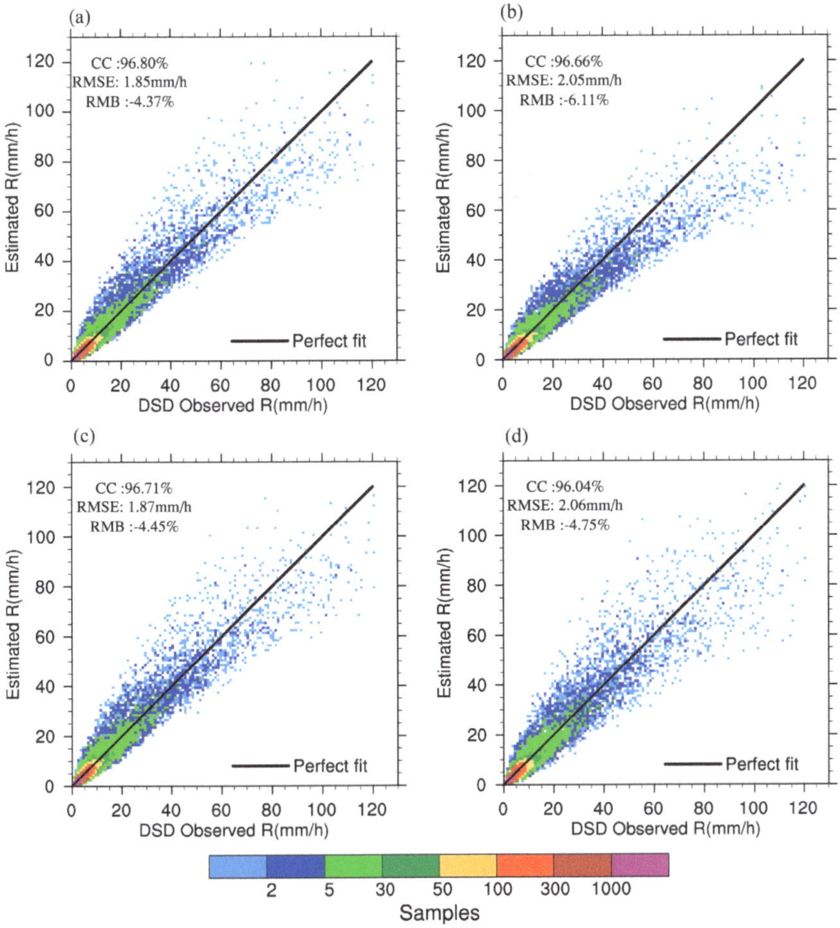

Figure 7. Scatter density plots of R in the whole region of Beijing from 11 disdrometer observations and R estimated using estimator R(K_{dp}): (**a**) Control experiment, (**b**) DSD variability experiment, (**c**) measurement error experiment 1, and (**d**) measurement error experiment 2 as described in Table 5. The black line in each panel is the perfect fit line (i.e., y = x). Statistical scores of CC, RMSE, and RMB are superimposed.

Table 5. Experiment design of R(K_{dp}) estimator.

Name	Description
Control experiment	Perform R(K_{dp}) to estimate rain rate using all the DSD data with parameters for the whole region of Beijing (i.e., a = 15.83 and b = 0.7727)
DSD variability experiment	Perform R(K_{dp}) to estimate rain rate using all the DSD data with parameters for the mountain region of Beijing (i.e., a = 14.99 and b = 0.7727)
Measurement error experiment 1	Perturb K_{dp} by multiplying $N(1, 0.05^2)$
Measurement error experiment 2	Perturb K_{dp} by multiplying $N(1, 0.15^2)$

Experiments on R(K_{dp}, Z_{DR}) are also performed using the design outlined in Table 6, and the results are shown in Figure 8. Similar to the R(K_{dp}) experiment, although R(K_{dp}, Z_{DR}) is relatively insensitive to the variability of DSD, the variability of DSD does affect the accuracy of R(K_{dp}, Z_{DR}), at least for heavy precipitation above 50 mm h^{-1}. As shown in Figure 8b, R(K_{dp}, Z_{DR}) underestimates heavy precipitation above 50 mm h^{-1} when inappropriate parameters are used. When Z_{DR} is assumed to have observational errors (Figure 8c), the accuracy of QPE drops significantly, especially for heavy precipitation above 50 mm h^{-1}, resulting in more scattered dots. When both observational errors and systematic bias coexist (Figure 8d), the accuracy of QPE becomes worse. In this case, the QPE systematically overestimates the rain rate, with more dots appearing in the upper left flank of the perfect line, and the dots become more scattered. These results suggest that the accuracy of R(K_{dp}, Z_{DR}) may be more sensitive to observational errors and systematic bias rather than the representative parameters. It could be due to the negative parameter c, which puts ZDR in the denominator. Given that Z_{DR} is small in rain (generally less than 3 dB), a small fluctuation or deviation of Z_{DR} may lead to significant errors in QPE. Therefore, accurate Z_{DR} observation is crucial to the QPE accuracy for R(K_{dp}, Z_{DR}) estimator. Therefore, accurate and well-calibrated Z_{DR} observations are crucial to ensure the accuracy of QPE using the R(K_{dp}, Z_{DR}) estimator. Introducing Z_{DR} into QPE may not necessarily have a positive impact, but rather a negative impact on QPE accuracy if Z_{DR} is not measured accurately and well-calibrated.

Table 6. Experiment design of R(K_{dp}, Z_{DR}) estimator.

Name	Description
Control experiment	Perform R(K_{dp}, Z_{DR}) to estimate rain rate using all the DSD data with parameters for the whole region of Beijing(i.e., a = 30.31, b = 0.9676, and c = −1.409)
DSD variability experiment	Perform R(K_{dp}, Z_{DR}) estimate rain rate using all the DSD data with parameters for the mountain region of Beijing(i.e., a = 30.04, b = 0.9324, and c = −1.431)
Measurement error experiment	Perturb Z_{DR} by multiplying $N(1, 0.10^2)$
Systematic bias experiment	Perturb Z_{DR} multiplying $N(0.95, 0.1^2)$

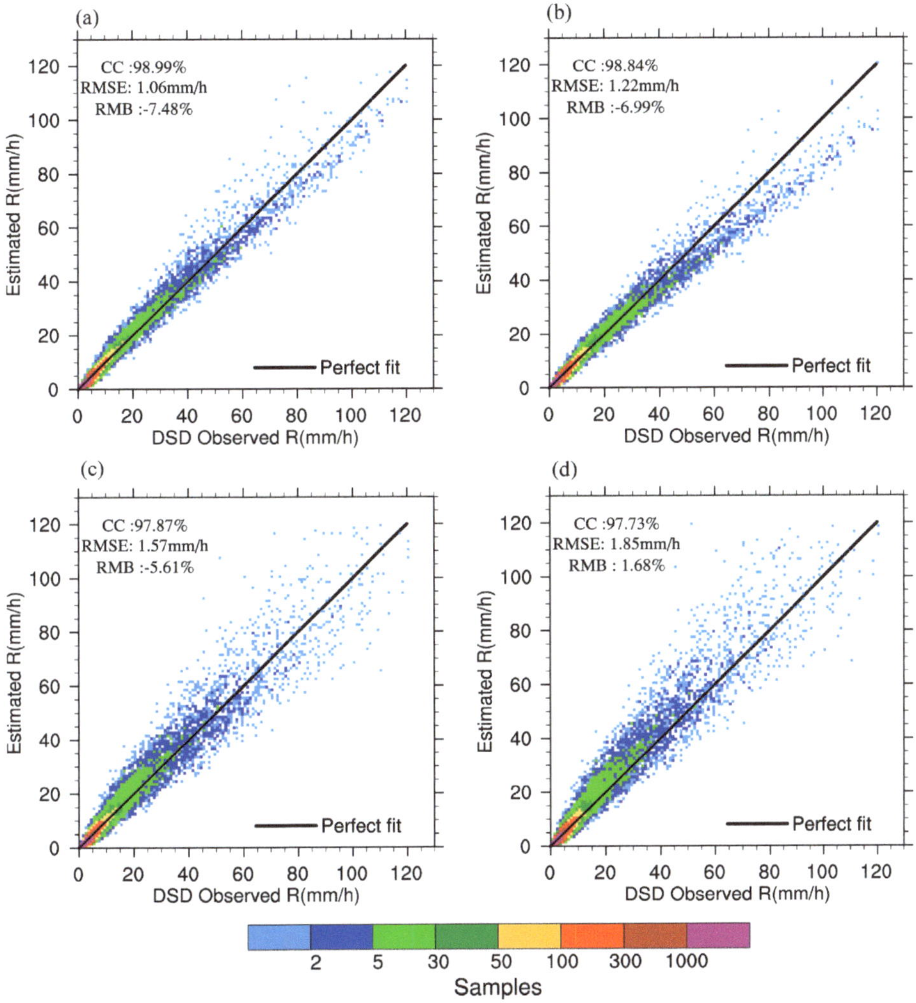

Figure 8. Scatter density plots of R in the whole region of Beijing from 11 disdrometer observations and R estimated using estimator $R(K_{dp}, Z_{dr})$: (**a**) Control experiment, (**b**) DSD variability experiment, (**c**) measurement error experiment, and (**d**) systematic bias experiment as described in Table 6. The black line in each panel is the perfect fit line (i.e., y = x). Statistical scores of CC, RMSE, and RMB are superimposed.

4. Discussion

In this study, disdrometer data collected from 11 sites in Beijing in 2017 are analyzed to reveal the city-scale spatial variability of DSD, and to investigate its impact on radar QPE. We found that the average precipitation intensity is smaller in the mountain areas of Beijing and more light rain occurs. It should be pointed out that this phenomenon is only specific to the mountain areas of Beijing, not worldwide. This phenomenon is possible due to the specific location of Beijing, that is, the mountain areas are further away from the sea and receive less moisture than the plains in Beijing. As for other places, some previous studies suggested that the mountains tend to enhance the precipitation [24,61]. A series of sensitivity experiments were conducted to investigate the effect of DSD variability on radar QPE. However, it should be pointed out that these experiments are ideal experiments based on DSD data and no real radar data are used yet. Utilizing real radar data to perform QPE is more complicated, and the errors of radar observation may not simply conform to a Gaussian distribution. In addition, the disdrometer measures DSD information at ground level while the weather radar measures microwave electromagnetic scattering of precipitation particles in the air. These two types of instruments do not measure precipitation at the same location, and the sampling volumes of these two types of instruments are also different, with the sampling volume of weather radar being much larger. This work provides insights into the relative importance of the factors that affect the accuracy of QPE with sensitivity experiments, and more work needs to be performed when applying these results to the operational usage of weather radar.

5. Conclusions

In this study, disdrometer data collected from 11 sites in Beijing in 2017 are analyzed to reveal the variability of DSD and to investigate its impact on radar QPE. The main conclusions are summarized as follows:

DSD exhibits evidently different characteristics in urban, suburban, and mountain areas of Beijing. Specifically, the average raindrop diameter is smaller in the urban area compared to the suburban area. Additionally, the average rain rate and raindrop number concentration are lower in mountain areas compared to both urban and suburban areas.

The convection/stratiform ratio is almost the same in urban and suburban areas, indicating that the difference in DSD between urban and suburban areas is due to the difference in DSD within the same precipitation types. In the urban area, both convection and stratiform exhibit smaller average raindrop diameters compared to the suburban area. This difference may be attributed to higher aerosol concentrations in the urban area.

The lower average rain rate and raindrop number concentration in mountain areas is the combined effect of the convection/stratiform ratio and the DSD difference for the same precipitation types. Convection occurs less frequently in mountain areas, and the rain rate and raindrop number concentration are also smaller both for convection and stratiform, resulting in the smaller average rain rate and raindrop number concentration in mountain areas.

Among the three QPE estimators of $R(Z_H)$, $R(K_{dp})$, and $R(K_{dp}, Z_{DR})$, $R(K_{dp}, Z_{DR})$ performs best, followed by $R(K_{dp})$, and $R(Z_H)$ performs worst. The $R(K_{dp})$ is more sensitive to the representative parameters while $R(Z_H)$ and $R(K_{dp}, Z_{DR})$ are more sensitive to the observational error and systematic bias (i.e., calibration).

Our conclusions suggest that when performing QPE at the city scale using different QPE estimators, special attention should be paid to different aspects to improve the accuracy of QPE. However, these results are based on DSD data and sensitivity experiments, and it should be noted that this conclusion needs to be further confirmed by using polarimetric radar data in the future. In addition, there are also other factors that influence the operational radar QPE, such as beam blockage by terrains, the undersampling of the disdrometer, and the variation in DSD when raindrops are falling in the air (where the radar samples them) and on the ground (where the disdrometer samples them). Moreover, for other regions such as Southwest China where there are many mountains, how these mountains affect the DSD variability is worth studying. All of these issues are crucial to obtaining accurate QPE and will be further studied in future work.

Author Contributions: Conceptualization, Z.Z. and H.L.; data curation, D.L. and Y.Q.; formal analysis, Z.Z., H.L. and Y.Q.; methodology, Z.Z. and H.L. supervision, D.L. and Y.Q.; writing—original draft, Z.Z.; writing— review and editing, Z.Z., H.L. and Y.Q. All authors have read and agreed to the published version of the manuscript.

Funding: This research was funded by the National Key Research and Development Program of China (2022YFC3003902 and 2022YFC3002904), and National Natural Science Foundation of China (41905047).

Data Availability Statement: Not applicable.

Acknowledgments: The authors would like to thank the editors and the reviewers for their thorough comments, which really helped improve the manuscript.

Conflicts of Interest: The authors declare no conflict of interest.

References

1. Milbrandt, J.; Yau, M. A multimoment bulk microphysics parameterization. Part I: Analysis of the role of the spectral shape parameter. *J. Atmos. Sci.* **2005**, *62*, 3051–3064. [CrossRef]
2. Morrison, H.; Curry, J.; Khvorostyanov, V. A new double-moment microphysics parameterization for application in cloud and climate models. Part I: Description. *J. Atmos. Sci.* **2005**, *62*, 1665–1677. [CrossRef]
3. Zhang, G.; Sun, J.; Brandes, E.A. Improving parameterization of rain microphysics with disdrometer and radar observations. *J. Atmos. Sci.* **2006**, *63*, 1273–1290. [CrossRef]
4. Maki, M.; Keenan, T.D.; Sasaki, Y.; Nakamura, K. Characteristics of the Raindrop Size Distribution in Tropical Continental Squall Lines Observed in Darwin, Australia. *J. Appl. Meteorol.* **2001**, *40*, 1393–1412. [CrossRef]
5. Cao, Q.; Zhang, G.; Brandes, E.; Schuur, T.; Ryzhkov, A.; Ikeda, K. Analysis of video disdrometer and polarimetric radar data to characterize rain microphysics in Oklahoma. *J. Appl. Meteorol. Climatol.* **2008**, *47*, 2238–2255. [CrossRef]
6. Wen, L.; Zhao, K.; Zhang, G.; Xue, M.; Zhou, B.; Liu, S.; Chen, X. Statistical characteristics of raindrop size distributions observed in East China during the Asian summer monsoon season using 2-D video disdrometer and Micro Rain Radar data. *J. Geophys. Res. Atmos.* **2016**, *121*, 2265–2282. [CrossRef]
7. Wang, H.; Kong, F.; Wu, N.; Lan, H.; Yin, J. An investigation into microphysical structure of a squall line in South China observed with a polarimetric radar and a disdrometer. *Atmos. Res.* **2019**, *226*, 171–180. [CrossRef]
8. Zhang, G.; Vivekanandan, J.; Brandes, E. A method for estimating rain rate and drop size distribution from polarimetric radar measurements. *IEEE Trans. Geosci. Remote Sens.* **2001**, *39*, 830–841. [CrossRef]
9. Cao, Q.; Zhang, G.; Brandes, E.A.; Schuur, T.J. Polarimetric radar rain estimation through retrieval of drop size distribution using a Bayesian approach. *J. Appl. Meteorol. Climatol.* **2010**, *49*, 973–990. [CrossRef]
10. Chen, G.; Zhao, K.; Zhang, G.; Huang, H.; Liu, S.; Wen, L.; Yang, Z.; Yang, Z.; Xu, L.; Zhu, W. Improving Polarimetric C-Band Radar Rainfall Estimation with Two-Dimensional Video Disdrometer Observations in Eastern China. *J. Hydrometeorol.* **2017**, *18*, 1375–1391. [CrossRef]
11. Angulo-Martínez, M.; Barros, A. Measurement uncertainty in rainfall kinetic energy and intensity relationships for soil erosion studies: An evaluation using PARSIVEL disdrometers in the Southern Appalachian Mountains. *Geomorphology* **2015**, *228*, 28–40. [CrossRef]
12. Caracciolo, C.; Napoli, M.; Porcù, F.; Prodi, F.; Dietrich, S.; Zanchi, C.; Orlandini, S. Raindrop size distribution and soil erosion. *J. Irrig. Drain. Eng.* **2012**, *138*, 461–469. [CrossRef]
13. Friedrich, K.; Higgins, S.; Masters, F.J.; Lopez, C.R. Articulating and stationary PARSIVEL disdrometer measurements in conditions with strong winds and heavy rainfall. *J. Atmos. Ocean. Technol.* **2013**, *30*, 2063–2080. [CrossRef]
14. May, P.T.; Bringi, V.N.; Thurai, M. Do We Observe Aerosol Impacts on DSDs in Strongly Forced Tropical Thunderstorms. *J. Atmos. Sci.* **2011**, *68*, 1902–1910. [CrossRef]
15. Tao, W.; Chen, J.; Li, Z.; Wang, C.; Zhang, C. Impact of aerosols on convective clouds and precipitation. *Rev. Geophys.* **2012**, *50*, RG2001. [CrossRef]
16. Bringi, V.; Chandrasekar, V.; Hubbert, J.; Gorgucci, E.; Randeu, W.; Schoenhuber, M. Raindrop size distribution in different climatic regimes from disdrometer and dual-polarized radar analysis. *J. Atmos. Sci.* **2003**, *60*, 354–365. [CrossRef]
17. Tang, Q.; Xiao, H.; Guo, C.; Feng, L. Characteristics of the raindrop size distributions and their retrieved polarimetric radar parameters in northern and southern China. *Atmos. Res.* **2014**, *135*, 59–75. [CrossRef]
18. Chen, B.; Hu, Z.; Liu, L.; Zhang, G. Raindrop Size Distribution Measurements at 4500 m on the Tibetan Plateau During TIPEX-III. *J. Geophys. Res. Atmos.* **2017**, *122*, 11092–11106. [CrossRef]
19. Niu, S.; Jia, X.; Sang, J.; Liu, X.; Lu, C.; Liu, Y. Distributions of raindrop sizes and fall velocities in a semiarid plateau climate: Convective versus stratiform rains. *J. Appl. Meteorol. Climatol.* **2010**, *49*, 632–645. [CrossRef]
20. Chen, B.; Yang, J.; Pu, J. Statistical characteristics of raindrop size distribution in the Meiyu season observed in eastern China. *J. Meteorol. Soc. Japan. Ser. II* **2013**, *91*, 215–227. [CrossRef]
21. Luo, L.; Xiao, H.; Yang, H.; Chen, H.; Guo, J.; Sun, Y.; Feng, L. Raindrop size distribution and microphysical characteristics of a great rainstorm in 2016 in Beijing, China. *Atmos. Res.* **2020**, *239*, 104895. [CrossRef]
22. Zeng, Y.; Tong, Z.; Jiang, Y.; Zhou, Y. Microphysical characteristics of seasonal rainfall observed by a Parsivel disdrometer in the Tianshan Mountains, China. *Atmos. Res.* **2022**, *280*, 106459. [CrossRef]
23. Zeng, Y.; Yang, L.; Tong, Z.; Jiang, Y.; Chen, P.; Zhou, Y. Characteristics and Applications of Summer Season Raindrop Size Distributions Based on a PARSIVEL2 Disdrometer in the Western Tianshan Mountains (China). *Remote Sens.* **2022**, *14*, 3988. [CrossRef]
24. Zeng, Y.; Yang, L.; Zhou, Y.; Tong, Z.; Jiang, Y.; Chen, P. Characteristics of orographic raindrop size distribution in the Tianshan Mountains, China. *Atmos. Res.* **2022**, *278*, 106332. [CrossRef]

25. Baik, J.; Kim, Y.; Chun, H. Dry and Moist Convection Forced by an Urban Heat Island. *J. Appl. Meteorol.* **2001**, *40*, 1462–147 [CrossRef]
26. Shepherd, J.M.; Burian, S.J. Detection of Urban-Induced Rainfall Anomalies in a Major Coastal City. *Earth Interact.* **2003**, *7*, 1–1 [CrossRef]
27. Zhong, S.; Qian, Y.; Zhao, C.; Leung, R.; Wang, H.; Yang, B.; Fan, J.; Yan, H.; Yang, X.; Liu, D. Urbanization-induced urban hea island and aerosol effects on climate extremes in the Yangtze River Delta region of China. *Atmos. Chem. Phys.* **2016**, *17*, 5439–545 [CrossRef]
28. Bornstein, R.; Lin, Q. Urban heat islands and summertime convective thunderstorms in Atlanta: Three case studies. *Atmos Environ.* **2000**, *34*, 507–516. [CrossRef]
29. Guo, X.; Fu, D.; Wang, J. Mesoscale convective precipitation system modified by urbanization in Beijing City. *Atmos. Res.* **2006**, *8* 112–126. [CrossRef]
30. Han, J.-Y.; Baik, J.-J.; Lee, H. Urban impacts on precipitation. *Asia-Pac. J. Atmos. Sci.* **2014**, *50*, 17–30. [CrossRef]
31. Rosenfeld, D.; Lohmann, U.; Raga, G.B.; O'Dowd, C.D.; Kulmala, M.; Fuzzi, S.; Reissell, A.; Andreae, M.O. Flood or drough How do aerosols affect precipitation? *Science* **2008**, *321*, 1309–1313. [CrossRef]
32. Van Den Heever, S.C.; Cotton, W.R. Urban aerosol impacts on downwind convective storms. *J. Appl. Meteorol. Climatol.* **2007**, *4* 828–850. [CrossRef]
33. Zhong, S.; Qian, Y.; Zhao, C.; Leung, R.; Yang, X.Q. A case study of urbanization impact on summer precipitation in the Greate Beijing Metropolitan Area: Urban heat island versus aerosol effects. *J. Geophys. Res. Atmos.* **2015**, *120*, 10903–10914. [CrossRef]
34. Chen, M.; Wang, Y.; Gao, F.; Xiao, X. Diurnal variations in convective storm activity over contiguous North China during th warm season based on radar mosaic climatology. *J. Geophys. Res.* **2012**, *117*, D20115. [CrossRef]
35. Li, H.; Cui, X.; Zhang, D.-L. On the initiation of an isolated heavy-rain-producing storm near the central urban area of Beijin metropolitan region. *Mon. Weather Rev.* **2017**, *145*, 181–197. [CrossRef]
36. Li, H.; Cui, X.; Zhang, D.L. A statistical analysis of hourly heavy rainfall events over the Beijing metropolitan region during th warm seasons of 2007–2014. *Int. J. Climatol.* **2017**, *37*, 4027–4042. [CrossRef]
37. Liu, L.; Ran, L.; Sun, X. Analysis of the structure and propagation of a simulated squall line on 14 June 2009. *Adv. Atmos. Sc* **2015**, *32*, 1049. [CrossRef]
38. Yin, S.; Li, W.; Chen, D.; Jeong, J.; Guo, W. Diurnal variations of summer precipitation in the Beijing area and the possible effect c topography and urbanization. *Adv. Atmos. Sci.* **2011**, *28*, 725–734. [CrossRef]
39. Zhang, D.L.; Lin, Y.; Zhao, P.; Yu, X.; Wang, S.; Kang, H.; Ding, Y. The Beijing extreme rainfall of 21 July 2012:"Right results" bu for wrong reasons. *Geophys. Res. Lett.* **2013**, *40*, 1426–1431. [CrossRef]
40. Ji, L.; Chen, H.; Li, L.; Chen, B.; Xiao, X.; Chen, M.; Zhang, G. Raindrop Size Distributions and Rain Characteristics Observed by PARSIVEL Disdrometer in Beijing, Northern China. *Remote Sens.* **2019**, *11*, 1479. [CrossRef]
41. Ma, Y.; Ni, G.; Chandra, C.V.; Tian, F.; Chen, H. Statistical characteristics of raindrop size distribution during rainy seasons in th Beijing urban area and implications for radar rainfall estimation. *Hydrol. Earth Syst. Sci.* **2019**, *23*, 4153–4170. [CrossRef]
42. Jaffrain, J.; Studzinski, A.; Berne, A. A network of disdrometers to quantify the small-scale variability of the raindrop siz distribution. *Water Resour. Res.* **2011**, *47*, W00H06. [CrossRef]
43. Tokay, A.; Wolff, D.B.; Petersen, W.A. Evaluation of the new version of the laser-optical disdrometer, OTT Parsivel2. *J. Atmo Ocean. Technol.* **2014**, *31*, 1276–1288. [CrossRef]
44. Krajewski, W.F.; Kruger, A.; Caracciolo, C.; Golé, P.; Barthes, L.; Creutin, J.-D.; Delahaye, J.-Y.; Nikolopoulos, E.I.; Ogden, Vinson, J.-P. DEVEX-disdrometer evaluation experiment: Basic results and implications for hydrologic studies. *Adv. Water Resou* **2006**, *29*, 311–325. [CrossRef]
45. Jaffrain, J.; Berne, A. Experimental Quantification of the Sampling Uncertainty Associated with Measurements from PARSIVE Disdrometers. *J. Hydrometeorol.* **2011**, *12*, 352–370. [CrossRef]
46. Tokay, A.; Petersen, W.A.; Gatlin, P.; Wingo, M. Comparison of raindrop size distribution measurements by collocated disdrom ters. *J. Atmos. Ocean. Technol.* **2013**, *30*, 1672–1690. [CrossRef]
47. Brandes, E.A.; Zhang, G.; Vivekanandan, J. Experiments in Rainfall Estimation with a Polarimetric Radar in a Subtropic Environment. *J. Appl. Meteorol.* **2002**, *41*, 674–685. [CrossRef]
48. Beard, K.V.; Johnson, D.B.; Baumgardner, D. Aircraft observations of large raindrops in warm, shallow, convective cloud *Geophys. Res. Lett.* **1986**, *13*, 991–994. [CrossRef]
49. Marzano, F.S.; Cimini, D.; Montopoli, M. Investigating precipitation microphysics using ground-based microwave remote senso and disdrometer data. *Atmos. Res.* **2010**, *97*, 583–600. [CrossRef]
50. Testud, J.; Oury, S.; Black, R.A.; Amayenc, P.; Dou, X. The Concept of "Normalized" Distribution to Describe Raindrop Spectra: Tool for Cloud Physics and Cloud Remote Sensing. *J. Appl. Meteorol.* **2001**, *40*, 1118–1140. [CrossRef]
51. Waterman, P. Matrix formulation of electromagnetic scattering. *Proc. IEEE* **1965**, *53*, 805–812. [CrossRef]
52. Bringi, V.N.; Chandrasekar, V. *Polarimetric Doppler Weather Radar: Principles and Applications*; Cambridge University Pres Cambridge, UK, 2001.
53. Liu, W.; You, H.; Ren, G.; Yang, P.; Zhang, B. Subtle Precipitation Characteristics in Beijing Area. *Clim. Environ. Res.* **2014**, *1* 61–68. (In Chinese)

54. Yang, P.; Xiao, Z.; Shi, W. Fine-Scale Characteristics of Rainfall in Beijing Urban Area Based on a High-Density Autonomous Weather Stations (AWS) Dataset. *Chin. J. Atmos. Sci.* **2017**, *41*, 475–489. (In Chinese)
55. Zhang, D.-L. Rapid urbanization and more extreme rainfall events. *Sci. Bull.* **2020**, *65*, 516–518. [CrossRef]
56. Rosenfeld, D. Suppression of rain and snow by urban and industrial air pollution. *Science* **2000**, *287*, 1793–1796. [CrossRef]
57. Zhong, S.; Yang, X.-Q. Mechanism of urbanization impact on a summer cold-frontal rainfall process in the greater Beijing metropolitan area. *J. Appl. Meteorol. Climatol.* **2015**, *54*, 1234–1247. [CrossRef]
58. Ji, W.; Wang, Y.; Zhuang, D. Spatial distribution differences in PM2.5 concentration between heating and non-heating seasons in Beijing, China. *Environ. Pollut.* **2019**, *248*, 574–583. [CrossRef]
59. Ryzhkov, A.; Zrnić, D. Comparison of dual-polarization radar estimators of rain. *J. Atmos. Ocean. Technol.* **1995**, *12*, 249–256. [CrossRef]
60. Matrosov, S.Y. Evaluating polarimetric X-band radar rainfall estimators during HMT. *J. Atmos. Ocean. Technol.* **2010**, *27*, 122–134. [CrossRef]
61. Napoli, A.; Crespi, A.; Ragone, F.; Maugeri, M.; Pasquero, C. Variability of orographic enhancement of precipitation in the Alpine region. *Sci. Rep.* **2019**, *9*, 13352. [CrossRef]

remote sensing

MDPI

Article

Microphysical Characteristics of Raindrop Size Distribution and Implications for Dual-Polarization Radar Quantitative Precipitation Estimations in the Tianshan Mountains, China

Yong Zeng [1,2,3], Jiangang Li [1,2,3], Lianmei Yang [1,2,3,*], Haoyang Li [1,2,3], Xiaomeng Li [1,2,3], Zepeng Tong [1,2,3], Yufei Jiang [1,2,3], Jing Liu [1,2,3], Jinru Zhang [1,2,3] and Yushu Zhou [4,5]

1 Institute of Desert Meteorology, China Meteorological Administration, Urumqi 830002, China; zengyong@idm.cn (Y.Z.); lijg@idm.cn (J.L.); lihy@idm.cn (H.L.); lixm@idm.cn (X.L.); tongzp@idm.cn (Z.T.); jiangyf@idm.cn (Y.J.); liujing@idm.cn (J.L.); zhangjr@idm.cn (J.Z.)

2 Field Scientific Observation Base of Cloud Precipitation Physics in West Tianshan Mountains, Urumqi 830002, China

3 Xinjiang Cloud Precipitation Physics and Cloud Water Resources Development Laboratory, Urumqi 830002, China

4 Laboratory of Cloud-Precipitation Physics and Severe Storms, Institute of Atmospheric Physics, Chinese Academy of Sciences, Beijing 100029, China; zys@mail.iap.ac.cn

5 College of Earth Science, University of Chinese Academy of Sciences, Beijing 100049, China

* Correspondence: yanglm@idm.cn; Tel.: +86-135-7990-3530

Abstract: In order to improve the understanding of the microphysical characteristics of raindrop size distribution (DSD) under different rainfall rates (R) classes, and broaden the knowledge of the impact of radar wavelengths and R classes on the QPE of dual-polarization radars in the Tianshan Mountains, a typical arid area in China, we investigated the microphysical characteristics of DSD across R classes and dual-polarimetric radar QPE relationships across radar wavelengths and R classes, based on the DSD data from a PARSIVEL^2 disdrometer at Zhaosu in the Tianshan Mountains during the summers of 2020 and 2021. As the R class increased, the DSD became wider and flatter. The mean value of the mass-weighted mean diameters (D_m) increased, while the mean value of logarithm normalized intercept parameters ($\log_{10} N_w$) decreased after increasing from C1 to C3, as the R class increased. The largest contributions to R and the radar reflectivity factor from large raindrops (diameter > 3 mm) accounted for approximately 50% and 97%, respectively, while 84% of the total raindrops were small raindrops (diameter < 1 mm). Dual-polarization radars—horizontal polarization reflectivity (Z_h), differential reflectivity (Z_{dr}), and specific differential phase (K_{dp})—were retrieved based on the DSD data using the T-matrix scattering method. The DSD-based polarimetric radar QPE relations of a single-parameter ($R(Z_h)$, $R(K_{dp})$), and double-parameters ($R(Z_h,Z_{dr})$, $R(K_{dp},Z_{dr})$) on the S-, C-, and X-bands were derived and evaluated. Overall, the performance of the $R(K_{dp})$ ($R(K_{dp},Z_{dr})$) scheme was better than that of $R(Z_h)$ ($R(Z_h,Z_{dr})$) for the QPE in the three bands. Furthermore, we have for the first time confirmed and quantified the performance differences in the QPE relationship of dual-polarization radars under different schemes, radar wavelengths, and R classes in typical arid areas of China. Therefore, selecting an appropriate dual-polarization radar band and QPE scheme for different R classes is necessary to improve the QPE ability compared with an independent scheme under all R classes.

Keywords: raindrop size distribution; dual-polarization radar; quantitative precipitation estimation; rain rate class; Tianshan Mountains

1. Introduction

The microphysical characteristics of raindrop size distribution (DSD) are important for understanding the dynamic processes of precipitation [1–3]. Furthermore, DSD has an important application value in improving the parameterization scheme of microphysical processes in numerical weather prediction models [4–6] and in enhancing the ability to estimate rainfall kinetic energy [7–9]. More importantly, DSD information is very helpful for improving the ability of quantitative precipitation estimation (QPE), whether using ground-based or spaceborne radars [10–12].

The microphysical characteristics of DSD vary with the climate region, terrain, rainfall type, season, and weather system [13–20]. China is a vast country with significant differences in climatic characteristics among its different regions. Researchers have conducted detailed surveys of DSD in the southern [21,22], eastern [23–26], northern [27–29], and Tibetan Plateau [30,31] regions of China directly affected by the monsoon system and obtained the characteristics of DSD in these regions. However, the research on Xinjiang, which accounts for one-sixth of China's total land area and is characterized by an arid climate, is insufficient. Xinjiang is not directly affected by the monsoon system and has an uneven distribution of precipitation [32]. Affected by the terrain, the Tianshan Mountains in central Xinjiang, China, and their adjacent areas are rich in precipitation, while the famous Taklimakan Desert and Gurbantunggut Desert form farther away from the Tianshan Mountains [33,34]. In recent years, several studies have partially revealed the characteristics of DSD on the Tianshan Mountains [35,36]. Several recent studies have also shown significant differences in DSD in different seasons (spring, summer, and fall), locations (western and central regions), and altitudes (foot and top stations) in the Tianshan Mountains [37–39]. However, further in-depth research is needed on the microphysical characteristics of DSD across the rainfall rate (R) classes in the Tianshan Mountains.

DSD information is of great significance for improving the accuracy of local ground-based radar QPEs by providing accurate microphysical characteristics of raindrops [14,20,40]. The QPE of single-polarization radars has been revealed in many studies by establishing a relationship between radar reflectivity factors and rain rates based on DSD data [41–48]. Single-polarization radar QPE relationships in the southern [22], eastern [25], northern [28], and Tibetan Plateau [31] of China have been established based on local DSD information. Similarly, based on the DSD data observed in the Tianshan Mountains, researchers have established single-polarization radar QPE relationships for different seasons, rainfall types, and altitudes [35–39]. However, the accuracy of the QPE for single-polarization radars is lower than that for dual-polarization radars [49–52]. Therefore, in recent years, the QPE relationships for dual-polarization radars have been established in different regions of China and have shown significantly better QPE capabilities than single-polarization radars [23,24,26–29,53,54]. More importantly, both the microphysical characteristics of the DSD and the QPE of the dual-polarization radar are closely related to the rain rate classes [17,18,28,55], and the QPE of the dual-polarization radar is very sensitive to the radar wavelength [24,53]. However, the previous studies mentioned above have mostly focused on the impact of one or two of the three factors, namely the R class, radar wavelength, and QPE scheme, on the QPE of dual-polarization radars. There has been relatively little comprehensive analysis of the impact of these three factors on the QPE of dual-polarization radars, and similarly, there is still a gap in these knowledge in the Tianshan Mountains, a typical arid area in China.

To reveal the microphysical characteristics of DSD under different rain rate classes and to explore the effects of radar wavelengths and rain rate classes on the QPE of dual-polarization radars in the Tianshan Mountains, we conducted this study using DSD data from the Tianshan Mountains and dual-polarization radar variables based on the T-matrix scattering method. The remainder of this paper is organized as follows: The data and methodology are presented in Section 2. Section 3 presents the microphysical characteristics of the DSD under different rain rate classes and the QPE of the dual-polarization radar with different wavelengths in the Tianshan Mountains. Section 4 provides the summary and conclusions of the study.

2. Data and Methodology

2.1. Study Area and Research Data

Tianshan Mountain is located in northwest China, as well as in Central Asia, and is not directly affected by the monsoon system. Summer is the most important rainfall period. In this study, the DSD measurements were collected at Zhaosu (1850.8 m ASL, 43.14°N, 81.13°E) over the Tianshan Mountains, China (Figure 1), during the 2020 and 2021 summer season, based on the second-generation OTT Particle Size Velocity (PARSIVEL^2) disdrometer [54]. The PARSIVEL^2 disdrometer obtains DSD information by simultaneously recording particle sizes (32 unequal intervals from 0.062 to 24.5 mm) and fall speeds (32 unequal intervals from 0.05 to 20.8 m s^{-1}) within a resolution of 1 min [56,57].

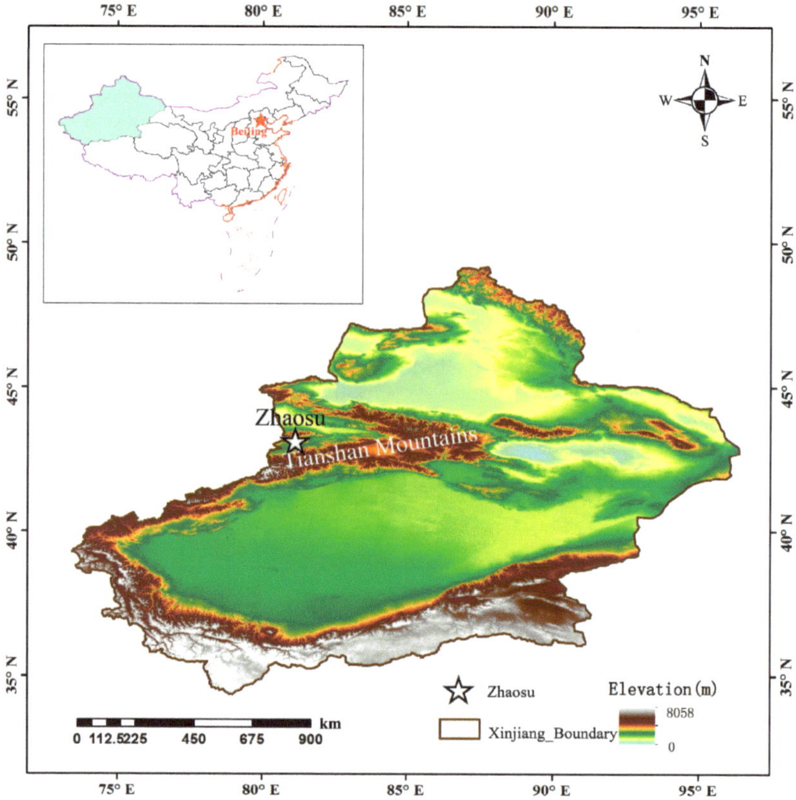

Figure 1. Location of Zhaosu (the black dot), with shading representing the topography (m) of the Tianshan Mountains.

2.2. Data Quality Control

The DSD data collected by the PARSIVEL^2 disdrometer were affected by many factors, such as the measurement accuracy and environmental conditions. Therefore, it is necessary to perform quality control before using DSD data for further analysis. In this study, the first two particle size bins were not considered because of their low signal-to-noise ratios [17,45]. Detected raindrops with very large diameters are likely to be generated by overlapping raindrops rather than by actual independent raindrops [58]; therefore, raindrops with a diameter of more than 8 mm were deleted. Furthermore, to reduce marginal effects [57], strong wind, and splashing effects [59] when measuring raindrops using a PARSIVEL^2 disdrometer, the theoretical raindrop fall speed–diameter relation, proposed by Atlas et al. [60] was used to constrain (within ±60%) the correlation between the raindrop fall speed and diameter during measurement, thereby eliminating unrealistic raindrops [59,61]. In this process, the theoretical raindrop fall speed–diameter relation [60] was adjusted by considering a correction factor (1.07) for air density adjustments related to the terrain height of Zhaosu [54]. In addition, samples with raindrop numbers of <10 or with rain rates of <0.1 mm h^{-1} were removed [14,62]. After data quality control, 14,609 DSD samples were used.

2.3. DSD and DSD-Based Polarimetric Radar QPE Relations

The DSD ($N(D_i)$, m^{-3} mm^{-1}) for raindrop per unit volume per unit diameter interval can be calculated according to Equation (1):

$$N(D_i) = \sum_{j=1}^{32} \frac{n_{ij}}{A_{eff}(D_i) \cdot \Delta t \cdot V(D_i) \cdot \Delta D_i} \tag{1}$$

where D_i (mm) represents the equivalent spherical raindrop diameter of the ith size class; n_{ij} is the number of drops within the ith size class and the ith velocity bin; Δt (s) represents the sampling time resolution (60 s in this study); ΔD_i (mm) is the diameter interval of the ith size class; and A_{eff} (D_i) (m^2) is the effective sampling area calculated according to Equation (2):

$$A_{eff}(D_i) = 180 \times 10^{-6} \cdot (30 - 0.5 \cdot D_i) \tag{2}$$

$V(D_i)$ (m s^{-1}) is the raindrop velocity at the ith size class [18,45,54,60], which can be expressed as

$$V(D_i) = (9.65 - 10.3 \cdot \exp(-0.6 \cdot D_i)) \cdot \delta(h) \tag{3}$$

where $\delta(h)$ represents the correction factor for air density adjustments (1.07), and h (m) is the terrain height of Zhaosu.

The rain rate R (mm h^{-1}), liquid water content LWC (g m^{-3}), total number concentration of raindrops N_t (m^{-3}), median volume diameter D_0 (mm), radar reflectivity factor Z (mm^6 m^{-3}), the normalized intercept parameter N_w (mm^{-1} m^{-3}), and the mass-weighted mean diameter D_m (mm) are expressed by Equations (4)–(10), respectively:

$$R = \frac{6\pi}{10^4} \cdot \sum_{i=1}^{32} N(D_i) \cdot D_i{}^3 \cdot V(D_i) \cdot \Delta D_i \tag{4}$$

$$LWC = \frac{\pi}{6000} \cdot \sum_{i=1}^{32} N(D_i) \cdot D_i^3 \cdot \Delta D_i \tag{5}$$

$$N_t = \sum_{i=1}^{32} N(D_i) \cdot \Delta D_i \tag{6}$$

$$\sum_{i=1}^{D_0} N(D_0) \cdot D_i^3 \cdot \Delta D_i = \sum_{i=D_0}^{32} N(D_0) \cdot D_i^3 \cdot \Delta D_i \tag{7}$$

$$Z = \sum_{i=1}^{32} N(D_i) \cdot D_i^6 \cdot \Delta D_i \tag{8}$$

$$N_w = \frac{4^4}{\pi \cdot \rho_w} \cdot \frac{10^3 \cdot W}{D_m^4} \tag{9}$$

$$D_m = \frac{\sum_{i=1}^{32} N(D_i) \cdot D_i^4 \cdot \Delta D_i}{\sum_{i=1}^{32} N(D_i) \cdot D_i^3 \cdot \Delta D_i} \tag{10}$$

The gamma model describing DSD [13] is given by Equation (11).

$$N(D) = N_0 \cdot D^{\mu} \cdot \exp(-\Lambda \cdot D) \tag{11}$$

where N_0 (mm$^{-1\text{-}\mu}$ m^{-3}), μ (−), and Λ (mm^{-1}) represent the intercept parameter, the shape factor, and the slope parameter of the gamma model, respectively [19]. These three parameters were calculated using the truncated moment method [63,64] with the third–fourth–sixth moments [13–15,18,45,54], where the nth order moment M_n (mm^{n} m^{-3}), N_0, μ, and Λ can be calculated according to the following Equation:

$$M_n = \int_0^{\infty} D^n \cdot N(D) \cdot dD \quad (12)$$

$$G = \frac{M_4^3}{M_3^2 M_6} \quad (13)$$

$$N_0 = \frac{M_3 \cdot \Lambda^{\mu+4}}{\Gamma(\mu+4)} \quad (14)$$

$$\mu = \frac{11 \cdot G - 8 + \sqrt{G \cdot (G+8)}}{2(1-G)} \quad (15)$$

$$\Lambda = (\mu+4) \cdot \frac{M_3}{M_4} \quad (16)$$

DSD-based QPE relations of dual-polarization radars have been proven to be very helpful in improving the accuracy of QPE [23,24,26–29,53,54], and these DSD-based QPE relations are established by the dual-polarization radar variables: radar reflectivity at horizontal or vertical polarization $Z_{h,v}$ (mm^6 m^{-3}), differential reflectivity Z_{dr} (dB), and the specific differential phase K_{dp} (° km^{-1}), which can be calculated using the observed DSD based on the method of T-matrix scattering [52,65–67] as follows:

$$Z_{h,v} = \left(\frac{4 \cdot \lambda^4}{\pi^4 \cdot |K_w|^2}\right) \cdot \int_{D_{min}}^{D_{max}} |f_{hh,vv}(D)|^2 \cdot N(D) \cdot dD \quad (17)$$

$$Z_{dr} = 10 \cdot \log_{10}\left(\frac{Z_h}{Z_v}\right) \quad (18)$$

$$K_{dp} = 10^{-3} \cdot \frac{180}{\pi} \cdot \lambda \cdot Re\left\{\int_{Dmin}^{D_{max}} [f_h(D) - f_v(D)] \cdot N(D) \cdot dD\right\} \quad (19)$$

where λ (mm) and K_w (−) represent the radar wavelength (for the S-, C-, and X-band, the values are 111.0 mm, 53.5 mm, and 33.3 mm, respectively), and the dielectric constant factor of water (here is 0.9639), respectively; $f_{hh,vv}(D)$ and $f_{h,v}(D)$ represent the backscattering amplitude and the forward scattering amplitude of a raindrop with horizontal and vertical polarizations, respectively. In addition, the raindrops followed the axis–ratio relationship proposed by Brandes [52].

The DSD-based QPE relationships, including $R(Z_h)$, $R(K_{dp})$, $R(Z_h,Z_{dr})$, and $R(K_{dp},Z_{dr})$ of the dual-polarization radar, were derived for the S-, C-, and X-bands. The dual-polarization radar QPE estimators are as follows:

$$R(Z_h) = \alpha \cdot Z_h^{\beta} \quad (20)$$

$$R(K_{dp}) = \alpha \cdot K_{dp}^{\beta} \quad (21)$$

$$R(Z_h, Z_{dr}) = \alpha \cdot Z_h^{\beta} \cdot 10^{\gamma \cdot Z_{dr}} \quad (22)$$

$$R(K_{dp}, Z_{dr}) = \alpha \cdot K_{dp}^{\beta} \cdot 10^{\gamma \cdot Z_{dr}} \quad (23)$$

where α, β, and γ are coefficients in the corresponding QPE estimator.

2.4. Assessing the Accuracy of QPE Estimators

The R value calculated using Equation (4), containing DSD information, was used to evaluate the performance of the QPE algorithms (Equations (20)–(23)) in the S-, C-, and X-band dual-polarization radars at Zhaosu. The correlation coefficient (CC), root mean square error (RMSE), and normalized mean absolute error (NMAE) were adopted for the evaluation of the QPE algorithms in this study, and are defined as

$$CC = \frac{\sum_{i=1}^{n} (R_i - R) \cdot (R_{e,i} - R_e)}{\sqrt{\sum_{i=1}^{n} (R_i - R)^2} \cdot \sqrt{\sum_{i=1}^{n} (R_{e,i} - R_e)^2}} \tag{24}$$

$$RMSE = \sqrt{\frac{1}{n} \cdot \sum_{i=1}^{n} (R_{e,i} - R_i)^2} \tag{25}$$

$$NMAE = \frac{\frac{1}{n} \cdot \sum_{i=1}^{n} |R_{e,i} - R_i|}{R} \tag{26}$$

where n represents the number of samples; R_i and R are the individual and mean R calculated from the DSD data, respectively; and $R_{e,i}$ and R_e represent the individual and mean R calculated from the QPE algorithms, respectively.

3. Results

3.1. DSD Characteristics under Different Rain Rate Classes

Many previous studies have revealed that the characteristics of DSD vary with rain rates (R) [17,18,24,30,53]. To reveal the microphysical characteristics of DSD under different rain rate classes, referring to previous classification standards [36,38], all the samples were classified into six classes on the basis of R: C1: 0.1–0.5 mm h^{-1}, C2: 0.5–1 mm h^{-1}, C3: 1–2 mm h^{-1}, C4: 2–5 mm h^{-1}, C5: 5–10 mm h^{-1}, and C6: $\geq$10 mm h^{-1}. The number of samples for each class is listed in Table 1. The accumulated rain duration (red histogram) and amount (blue line) for the six R classes in Zhaosu are shown in Figure 2. As the R class increased, its contribution to the total rainfall duration decreased. Specifically, the first two classes (C1 and C2) contributed the most to the total rain duration, accounting for 40.9% and 20.7%, respectively, whereas the last two classes (C5 and C6) contributed the least to the total rain duration, contributing less than 7%. The largest contributor to the total amount of rain was the fourth class (C4), followed by the last class (C6), accounting for 27.4% and 23.2%, respectively.

Table 1. Several important DSD parameters for the six R classes.

Parameters	No. of Samples	R (mm h^{-1})	D_0 (mm)	LWC (g m^{-3})	Z (dBZ)	N_t (m^{-3})
C1	5970	0.26	0.90	0.02	14.67	188
C2	3030	0.72	1.00	0.05	20.66	222
C3	2563	1.43	1.13	0.09	25.04	237
C4	2109	3.07	1.34	0.17	30.25	249
C5	612	6.91	1.72	0.32	36.21	254
C6	325	16.85	2.30	0.66	43.03	260

The DSD variations for the different R classes in Zhaosu are shown in Figure 3. As the R class increased, the raindrop spectrum widened, along with the increased concentrations of large raindrops (diameter > 3 mm), medium-size raindrops (1 $\leq$ diameter $\leq$ 3 mm), and small raindrops (diameter < 1 mm) [18,68], and reached the peak concentration at small raindrops around 0.6 mm in diameter for all classes (color lines) and all samples (black line). For small raindrops, the raindrop spectra of all the samples were between those of C1 and C2, whereas the raindrop spectrum of medium-size raindrops for all samples was similar to that of C3. For large raindrops, the raindrop spectra of all the samples were between C4 and C5. Box-and-whisker plots of the variations in the mass-weighted mean diameter (D_m) and normalized intercept parameter ($\log_{10}N_w$) for the six R classes are shown in Figure 4. The mean D_m value increased from 0.92 mm at C1 to 2.40 mm at C6 with an increasing R class, while the mean $\log_{10}N_w$ value first increased and then decreased with an increasing R class, reaching a maximum at C3 (3.63 mm^{-1} m^{-3}) and a minimum at C6 (3.27 mm^{-1} m^{-3}). Furthermore, several other important DSD parameters, such as R, median volume diameter (D_0), liquid water content (LWC), radar reflectivity factor (Z), and total number concentration of raindrops (N_t), for different R classes in Zhaosu are shown in Table 1. Their mean values increased with an increasing R class.

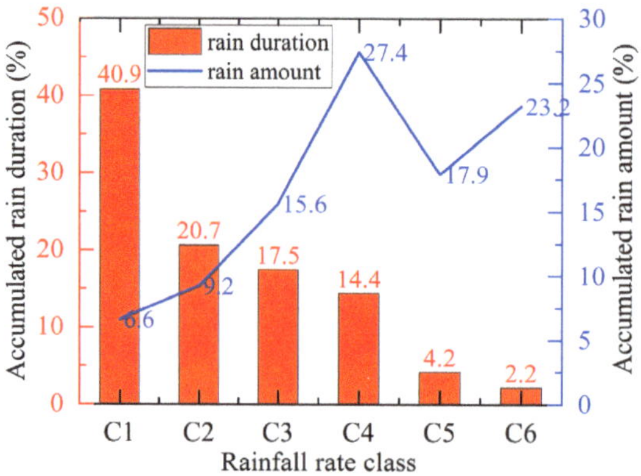

Figure 2. Accumulated rain duration (red histogram) and rain amount (blue line) for the six R classes in Zhaosu.

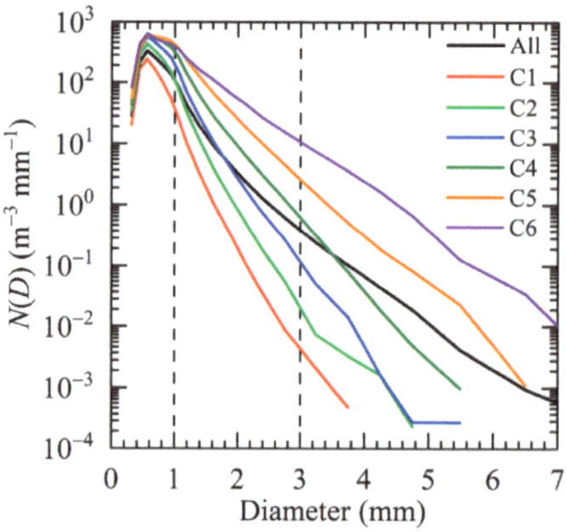

Figure 3. Mean DSD variations for different R classes (color lines) and all samples (black line) in Zhaosu. The two vertical dashed lines on the left and right distinguish the raindrop spectrum of small and medium-size raindrops, and the raindrop spectrum of medium-size and large raindrops, respectively.

The DSD was composed of raindrops of different sizes (large, medium, and small) and their corresponding concentrations. Therefore, discussing the contribution of raindrops of various sizes to the parameters is conducive to further understanding the DSD [37]. Figure 5 illustrates the contributions of small, medium-size, and large drops to R, Z, LWC, and N_t in Zhaosu. Large raindrops contributed nearly half of R; small raindrops contributed less than 14% to R; and medium-size raindrops contributed nearly 37% to R. The vast majority of the contributions to Z were from large raindrops (over 96%), whereas the contribution of small raindrops to Z was minimal (less than 0.2%). Most contributions to LWC came from medium-size raindrops (>40%), followed by large raindrops (>33%). The majority of the contribution to N_t was from small raindrops (over 96%), whereas the contribution of large raindrops to N_t was minimal (approximately 0.3%).

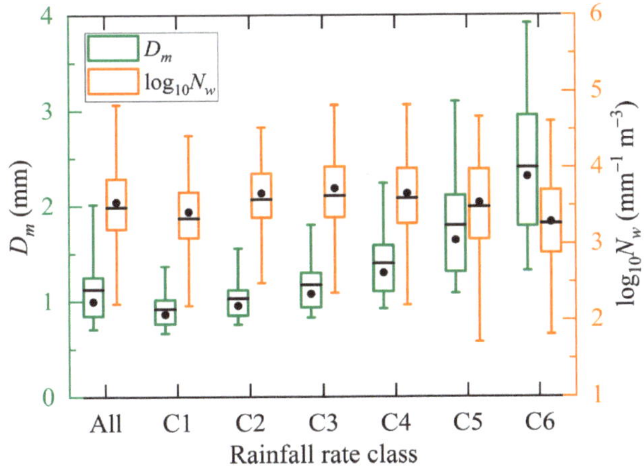

Figure 4. Variations of the D_m and the $\log_{10}N_w$ in Zhaosu for the six R classes. The line and dot of the box indicate the mean (black line) and median (black dot), respectively. The bottom (top) lines of the box indicate the 25th (75th) percentiles. The bottom (top) lines of the vertical lines out of the box indicate the 5th (95th) percentiles.

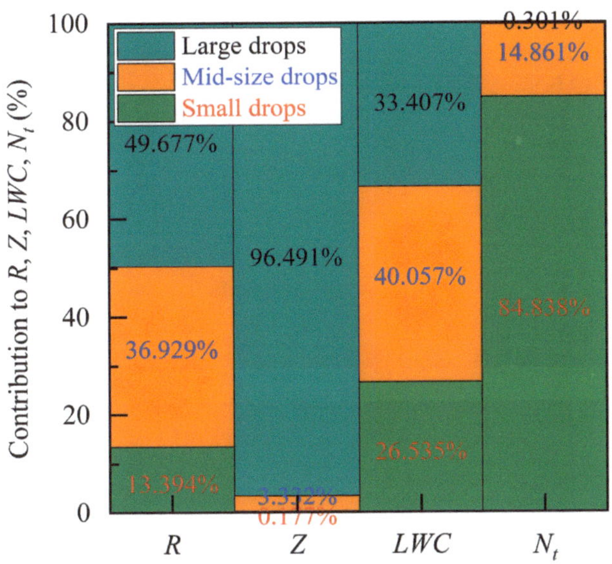

Figure 5. The contribution of small, medium, and large drops to R, Z, LWC, and N_t in Zhaosu.

3.2. Dual-Polarization Radar QPE Relations

Previous studies have revealed the advantages of using DSD information to retrieve dual-polarization radar variables for QPE based on the T-matrix scattering method [23,24,26–29,54]. The dual-polarization radar variables Z_h, Z_{dr}, and K_{dp} were calculated using Equations (17)–(19). Figure 6 shows the scatterplots of Z_{dr} versus Z_h and K_{dp} versus Z_h and the power–law fitting algorithms derived for Z_{dr}–Z_h and K_{dp}–Z_h on the S-, C-, and X-bands. For these three-band Z_{dr}–Z_h relations, the coefficient values ranged from 3.975×10^{-5} to 9.237×10^{-5}, and the exponent values varied between 2.595 and 2.842. Specifically, the Z_{dr}–Z_h relationship in the S-band (C-band) had the smallest coefficient (index) value and the largest index (coefficient) value. For the K_{dp}–Z_h relations on the S-, C-, and X-bands, the coefficient values ranged from 3.466×10^{-13} (on the X-band) to 9.261×10^{-13} (on the C-band), and the exponent values varied between 7.153 (on the C-band) and 7.541 (on the X-band). From the above results, it can be seen that the Z_{dr}–Z_h relation corresponding to different radar bands had obvious differences, as did the K_{dp}–Z_h relation, which further illustrates the necessity of studying dual-polarization radar variables and their relationships (including the QPE) at different radar bands.

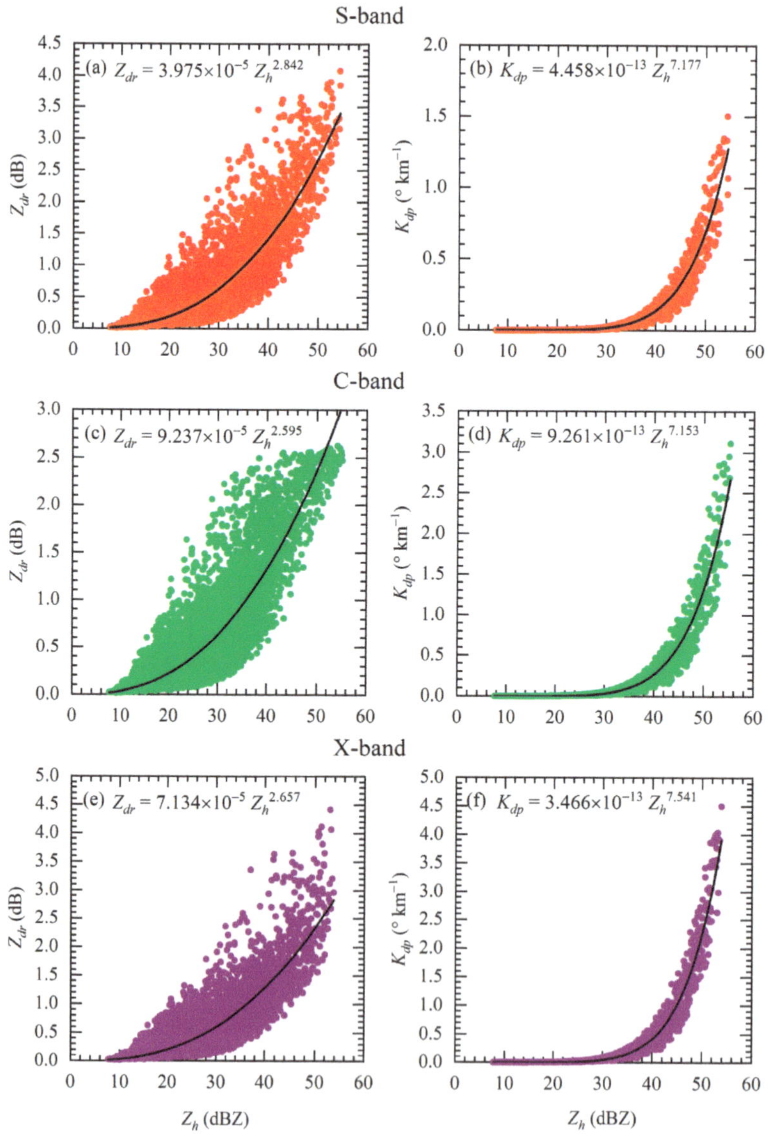

Figure 6. Scatterplots of Z_{dr} versus Z_h, and the Z_{dr}–Z_h relations represented by black line and equation on the (**a**) S-band, (**c**) C-band, and (**e**) X-band. Scatterplots of K_{dp} versus Z_h, and the K_{dp}–Z_h relations represented by black line and equation on the (**b**) S-band, (**d**) C-band, and (**f**) X-band.

The DSD-based dual-polarization radar QPE relationships ($R(Z_h)$, $R(K_{dp})$, $R(Z_h,Z_{dr})$, and $R(K_{dp},Z_{dr})$) for the S-, C-, and X-bands at Zhaosu were derived in this study and are listed in Table 2. Differences were observed in the QPE estimators for the different bands. Specifically, the two coefficients (α and β, as shown in Equation (20)) of the $R(Z_h)$ relation on the S- and C-bands were very similar, with a small difference compared to them on the X-band. For the $R(K_{dp})$ relations, the difference in the α coefficient was relatively large (13.053 to 27.831), while the difference in the β coefficient (as shown in Equation (21)) was relatively small (0.639 to 0.668) on these three bands. For the $R(Z_h,Z_{dr})$ relations, the differences in the α and γ coefficients were relatively large, while the difference in the β coefficient (as shown in Equation (22)) was relatively small. The α coefficient (as shown in Equation (23)) was 23.265 on the X-band, while the α coefficient on the C-band was about twice that, and the α coefficient on the S-band was about three times that at the $R(K_{dp},Z_{dr})$ relation.

Table 2. The DSD-based dual-polarization radar QPE relations for S-, C-, and X-band at Zhaosu.

Band	$R(Z_h)$	$R(K_{dp})$	$R(Z_h,Z_{dr})$	$R(K_{dp},Z_{dr})$
S	$R(Z_h) = 0.096\ Z_h^{0.468}$	$R(K_{dp}) = 27.831\ K_{dp}^{0.639}$	$R(Z_h,Z_{dr}) = 0.013\ Z_h^{0.824}10^{-0.352Zdr}$	$R(K_{dp},Z_{dr}) = 75.719\ K_{dp}^{0.845}10^{-0.172Zdr}$
C	$R(Z_h) = 0.098\ Z_h^{0.457}$	$R(K_{dp}) = 16.914\ K_{dp}^{0.641}$	$R(Z_h,Z_{dr}) = 0.010\ Z_h^{0.900}10^{-0.556Zdr}$	$R(K_{dp},Z_{dr}) = 51.816\ K_{dp}^{0.890}10^{-0.251Zdr}$
X	$R(Z_h) = 0.070\ Z_h^{0.497}$	$R(K_{dp}) = 13.053\ K_{dp}^{0.668}$	$R(Z_h,Z_{dr}) = 0.018\ Z_h^{0.744}10^{-0.294Zdr}$	$R(K_{dp},Z_{dr}) = 23.265\ K_{dp}^{0.816}10^{-0.147Zdr}$

It is important to evaluate the performance of various DSD-based dual-polarization radar QPE relations in QPE applications. R calculated from DSD (Equation (4)) was used to evaluate the QPE relations [27,28,53,54,69]. In this study, three evaluation indicators—the correlation coefficient (CC), root mean square error (RMSE), and normalized mean absolute error (NMAE)—were used to evaluate the different QPE relations for different bands [27,53]. Figures 7–9 show the scatterplots of R computed from the QPE relations and the DSD information on the S-, C-, and X-bands. The performances of the double-parameter schemes ($R(Z_h, Z_{dr})$ and $R(K_{dp}, Z_{dr})$) were superior to those of the single-parameter schemes ($R(Z_h)$ and $R(K_{dp})$) for all bands, characterized by a larger CC and smaller RMSE and NMAE. The performance of the $R(K_{dp})$ scheme is better than that of the $R(Z_h)$ scheme for single-parameter schemes. Similarly, the $R(K_{dp}, Z_{dr})$ scheme showed a relatively better performance than the $R(Z_h, Z_{dr})$ scheme in the double-parameter schemes. Moreover, both single-parameter schemes performed the best for the X-band, whereas both double-parameter schemes performed the best for the C-band.

3.3. QPE Relations under Different Rain Rate Classes

R classes have an important impact on the performance of dual-polarization radar QPE estimators [53,55,69] as well as radar bands [53,55,69]. To quantify the performance of QPE estimators under different R classes and different radar bands, we used CC, RMSE, and NMAE to evaluate the performance of these QPE estimators in detail. Before evaluating the performance of the QPE estimators, we first provided the distribution and average values of the dual-polarization radar variables required to establish these QPE estimators in Table 2 for different R classes and different radar bands, as shown in Figure 10 and Table 3. Z_h increased with an increasing R class for all bands, and the mean Z_h value in the X-band was the largest for all R classes except C6, compared to that in the S- and X-bands. The distribution of Z_h was narrowest in C2 and widest in C6 for all the bands. Similar to Z_h, Z_{dr} also increased with an increasing R class for all bands; however, in the first two R classes (C1 and C2), Z_{dr} was largest in the X-band, whereas in the middle two R classes (C3 and C4), Z_{dr} was largest in the C-band, and in the last two R classes (C5 and C6), Z_{dr} was largest in the S-band. The distribution of Z_{dr} was narrower in the first three R classes and widened in the last three R classes, particularly in the last R class (C6), where Z_{dr} had the widest distribution. Interestingly, during the process of increasing K_{dp} as the R class increased, the mean K_{dp} value of the next R class was about three times that of the previous R class for all bands (for example, 11.3×10^{-3} ° km^{-1} in C3 and 32.8×10^{-3} ° km^{-1} in C4 for S-band). The mean K_{dp} value in the C-band was about twice that in the S-band, and the mean K_{dp} value in the X-band was about three times that in the S-band for each R class (for example, 11.3×10^{-3} ° km^{-1} in the S-band, 24.2×10^{-3} ° km^{-1} in the C-band, and 39.8×10^{-3} ° km^{-1} in the X-band for C3).

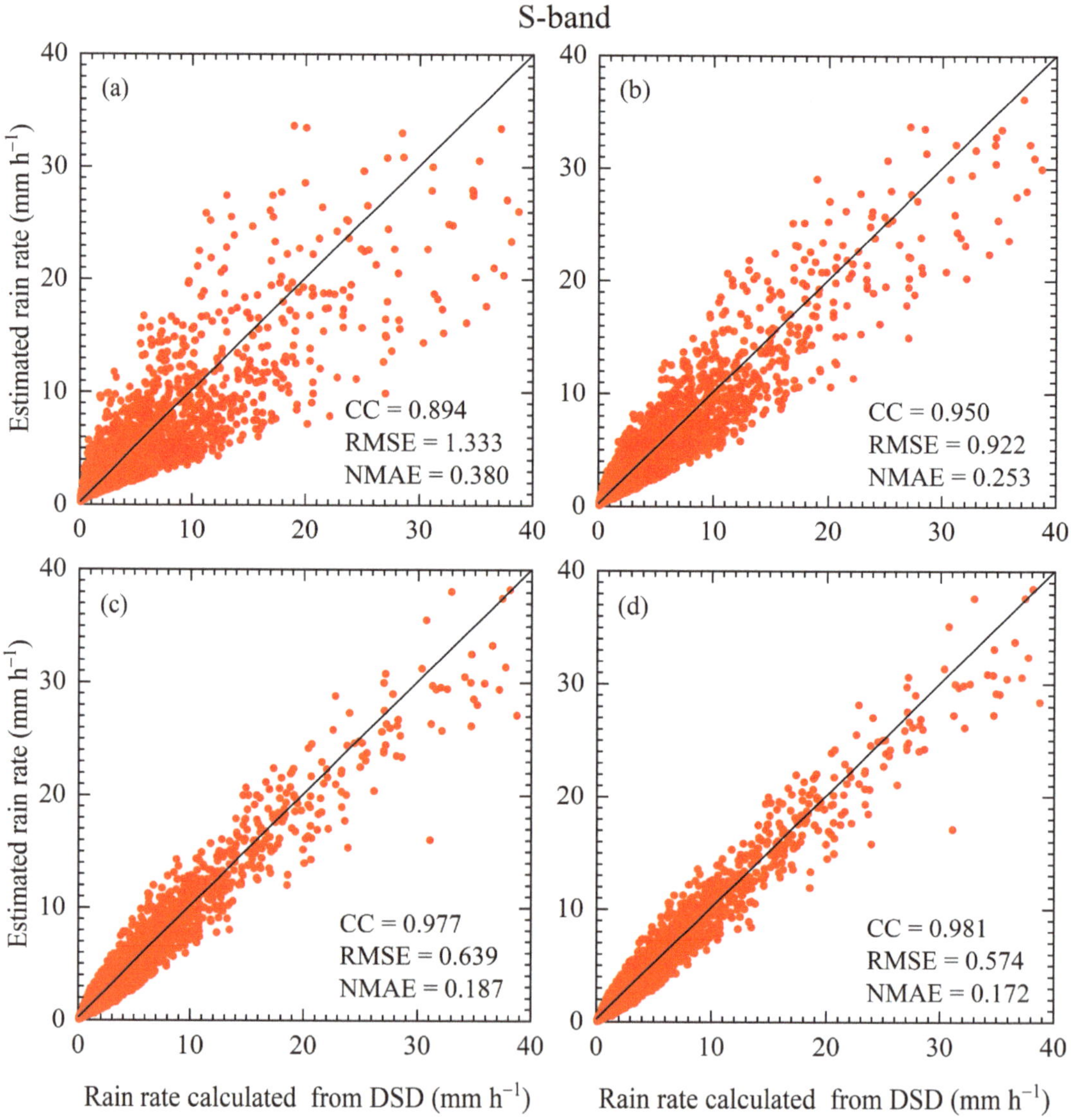

Figure 7. Scatterplot of R calculated from (**a**) $R(Z_h)$, (**b**) $R(K_{dp})$, (**c**) $R(Z_h,Z_{dr})$, and (**d**) $R(K_{dp},Z_{dr})$ relations versus the R computed from DSD for S-band in Zhaosu.

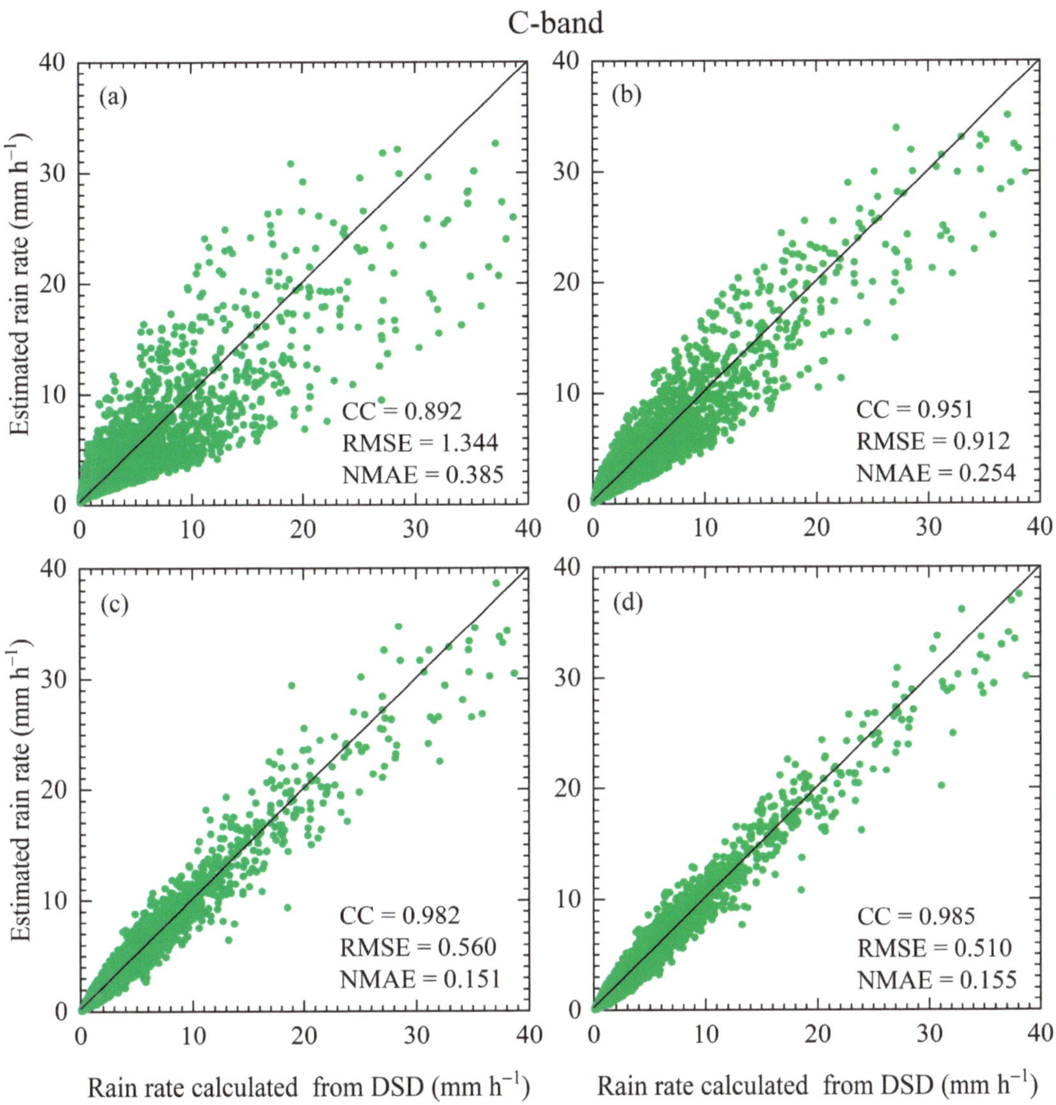

Figure 8. Scatterplot of R calculated from (**a**) $R(Z_h)$, (**b**) $R(K_{dp})$, (**c**) $R(Z_h,Z_{dr})$, and (**d**) $R(K_{dp},Z_{dr})$ relations versus the R computed from DSD for C-band in Zhaosu.

Table 3. The mean of Z_h, Z_{dr}, and K_{dp} on the S-, C-, and X-bands for the six R classes. Red font indicates the maximum value at the same R class.

Band	Z_h (dBZ)						Z_{dr} (10^{-1} dB)						K_{dp} (10^{-3} km^{-1})					
	C1	C2	C3	C4	C5	C6	C1	C2	C3	C4	C5	C6	C1	C2	C3	C4	C5	C6
S	15.82	21.73	26.08	31.27	37.47	44.69	10.47	10.71	11.05	11.71	12.93	15.17	1.4	4.5	11.3	32.8	110.1	426.7
C	15.93	21.89	26.31	31.64	38.03	45.51	10.48	10.72	11.09	11.79	12.90	14.71	2.9	9.5	24.2	71.5	241.7	933.6
X	16.09	22.10	26.58	31.94	38.18	45.22	10.49	10.73	11.07	11.70	12.75	14.69	4.7	15.5	39.8	116.2	379.9	1394.0

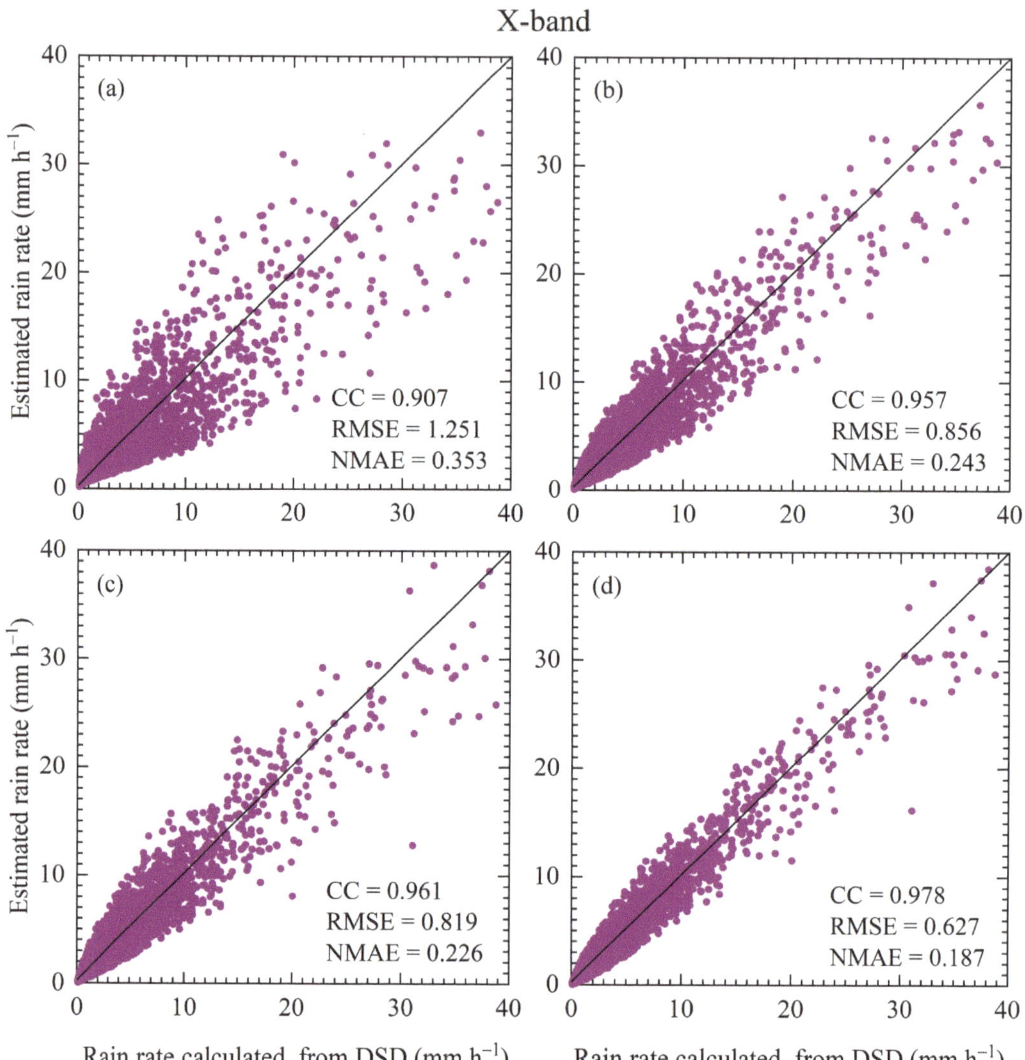

Figure 9. Scatterplot of R calculated from (**a**) $R(Z_h)$, (**b**) $R(K_{dp})$, (**c**) $R(Z_h,Z_{dr})$, and (**d**) $R(K_{dp},Z_{dr})$ relations versus the R computed from DSD for X-band in Zhaosu.

Figure 11 shows the CC, RMSE, and NMAE of R estimated from the dual-polarization radar QPE estimators against R calculated from the DSD under different R classes (C1–C6) and radar bands (the S-, C-, and X-bands). The performances of the four schemes for these three bands differed under different R classes. For the S-band radar, the $R(Z_h)$ estimator had the worst performance, characterized by a relatively lower CC and higher RMSE and NMAE for all R classes, followed by the $R(K_{dp})$ estimator. The $R(K_{dp},Z_{dr})$ estimator performed the best (highest CC and lowest RMSE and NMAE) for all R classes. The RMSE of all the estimators increased with an increasing R class, whereas the CC and NMAE did not monotonically increase or decrease during this process (Figure 11a–c). For the C-band radar, similar to the S-band radar, the performance of the $R(Z_h)$ estimator remained the worst, followed by that of the $R(K_{dp})$ estimator for all R classes. However, unlike in the

S-band radar, the performance of the $R(K_{dp},Z_{dr})$ estimator was not always the best for all the R classes in the C-band radar. Specifically, when the R class was between C1 and C4, the $R(Z_h,Z_{dr})$ estimator was slightly superior to the $R(K_{dp},Z_{dr})$ estimator, with a higher CC and lower RMSE and NMAE; however, the opposite was true when the R class was C5 and C6 (Figure 11d–f). For the X-band radar, similar to the S-band radar, the $R(Z_h)$ estimator exhibited the worst performance, whereas the $R(K_{dp},Z_{dr})$ estimator exhibited the best performance for all the R classes. However, the gap in performance between the $R(K_{dp})$ and $R(Z_h,Z_{dr})$ estimators for X-band radars narrowed compared to the S- and C-band radars, as reflected in the narrowing of the gap between the three evaluation parameters (CC, RMSE, and NMAE) (Figure 11d–f).

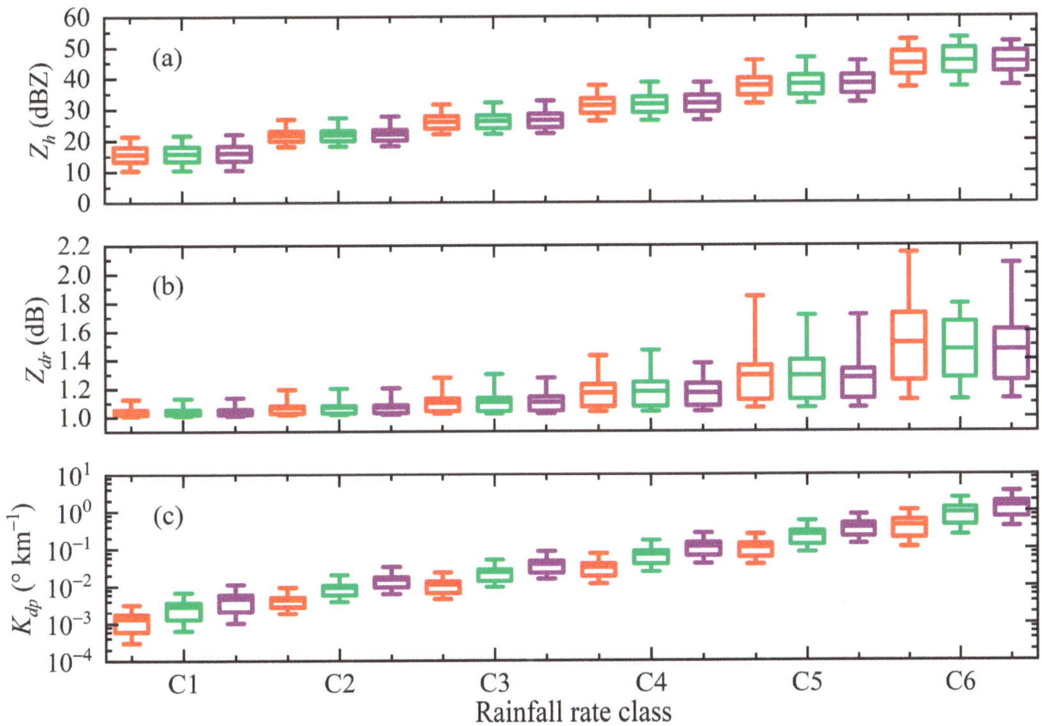

Figure 10. Variations of (**a**) Z_h, (**b**) Z_{dr}, and (**c**) K_{dp} on the S- (red), C- (green), and X-bands (purple) for the six R classes. The line of the box indicates the mean. The bottom (top) lines of the box indicate the 25th (75th) percentiles. The bottom (top) lines of the vertical lines out of the box indicate the 5th (95th) percentiles.

For the same type of QPE estimator and the same R class, the performances of the estimators in the different bands were different. Taking the $R(Z_h,Z_{dr})$ estimator at C3 as an example, for the $R(Z_h,Z_{dr})$ estimator at C3, the CC, RMSE, and NMAE were 0.669, 0.318 mm h^{-1}, and 0.73 mm h^{-1} in the C-band, respectively, while the CC, RMSE, and NMAE were 0.554 (0.471), 0.417 (0.501) mm h^{-1}, and 0.227 (0.258) mm h^{-1} in the S-band (X-band), respectively. Therefore, among the three bands, the C-band estimator performed the best, whereas the X-band estimator performed the worst for the $R(Z_h,Z_{dr})$ estimator at C3. Similarly, taking the $R(Z_h)$ estimator at C6 as an example again, for the $R(Z_h)$ estimator at C6, the CC, RMSE, and NMAE were 0.689, 6.032 mm h^{-1}, and 0.295 mm h^{-1} in the X-band, respectively, while the CC, RMSE, and NMAE were 0.634 (0.646), 6.749 (6.651) mm h^{-1}, and 0.332 (0.329) mm h^{-1} in the S-band (X-band), respectively. Therefore, among the three bands, the X-band estimator performed best, whereas the S-band estimator performed worst for the $R(Z_h)$ estimator at C6. Overall, the performances of the double-parameter schemes were significantly better than those of the single-parameter schemes for all bands and R classes. Furthermore, the performance of the dual-parameter estimators in the C-band was better than that in the S- and X-bands for all R classes, and the performance of the $R(Z_h,Z_{dr})$ estimator was better compared to the $R(K_{dp},Z_{dr})$ estimator at lower R classes (C1 to C4, R less than 5 mm h^{-1}), while the performance of the $R(K_{dp},Z_{dr})$ estimator was better compared to the $R(Z_h,Z_{dr})$ estimator at higher rainfall rates (C5 to C6, R greater than 5 mm h^{-1}) for the C-band. It is worth noting that previous studies have shown the importance of selecting suitable estimators for actual dual-polarization radar QPEs, and suitable estimators need to be provided for different regions and different band radars [55,69–71]. For a dual-polarization radar QPE estimator in the Tianshan Mountains, we plan to conduct the relevant research based on dual-polarization radar observational data in the future.

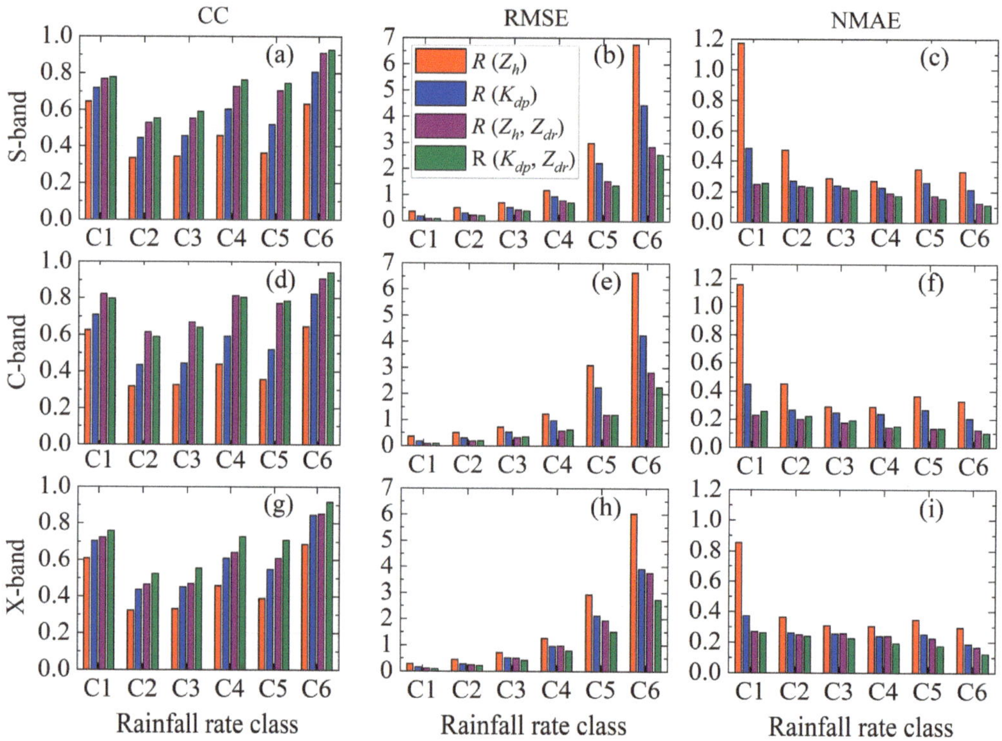

Figure 11. The (**a**,**d**,**g**) CC, (**b**,**e**,**h**) RMSE, and (**c**,**f**,**i**) NMAE of R estimated from the dual-polarization radar QPE estimators against R calculated from the DSD under different R classes and different radar bands, (**a**–**c**) for S-band, (**d**–**f**) for C-band, and (**g**–**i**) for X-band, respectively.

4. Summary and Conclusions

To reveal the microphysical characteristics of the raindrop size distribution (DSD) across rainfall rate (R) classes, and more importantly, to quantify the effects of radar wavelengths, QPE estimators, and R classes on the QPE of dual polarization radars in typical arid areas of China, DSD data from a PARSIVEL^2 disdrometer at Zhaosu in the Tianshan Mountains during summer 2020 and 2021 were used to investigate the microphysical characteristics of DSD for six rain rate (R) classes (C1: 0.1–0.5 mm h^{-1}, C2: 0.5–1 mm h^{-1}, C3: 1–2 mm h^{-1}, C4: 2–5 mm h^{-1}, C5: 5–10 mm h^{-1}, and C6: $\geq$10 mm h^{-1}) and DSD-based polarimetric radar quantitative precipitation estimation (QPE) relations on the S-, C-, and X-bands for different R classes. The analysis revealed that the first two R classes

(C1 and C2) contributed the most to the total rain duration, which accounted for 40.9% and 20.7%, respectively, and the largest contributors to the total rain amount were the fourth class (C4) and the last class (C6), which accounted for 27.4% and 23.2%, respectively. The raindrop spectrum widened, characterized by an increase in the concentrations of large raindrops (diameter > 3 mm), medium-size raindrops ($1 \leq$ diameter ≤ 3 mm), and small raindrops (diameter < 1 mm) with an increasing R class. The mean D_m value increased from 0.92 mm in C1 to 2.40 mm in C6, while the mean logarithm of $\log_{10}N_w$ value decreased after increasing from C1 to C3 with an increasing R class. In addition, the mean values of R, D_0, LWC, Z, and N_t increased with the R class. For the entire dataset, large raindrops had the largest contribution to R and Z, accounting for 50% and 97% of the total contribution, respectively, compared with small and medium-size raindrops, whereas small raindrops had the largest contribution to N_t, accounting for more than 84% of the total contribution.

Dual-polarization radar parameters including Z_h, Z_{dr}, and K_{dp} were retrieved based on the DSD data using the T-matrix scattering method. The Z_{dr}-Z_h and K_{dp}-Z_h relations were established in a power–law fitting form on the S-, C-, and X-bands. The Z_{dr}-Z_h relation corresponding to different radar bands had obvious differences as well as the K_{dp}-Z_h relation. The DSD-based dual-polarization radar QPE estimators ($R(Z_h)$, $R(K_{dp})$, $R(Z_h,Z_{dr})$, and $R(K_{dp},Z_{dr})$) for the S-, C-, and X-bands were derived. For the $R(Z_h)$ relations, the two coefficients (α and β) on the S- and C-bands were very similar, with a small difference compared to them on the X-band. For the $R(K_{dp})$ relations, the difference in the α coefficient was relatively large (13.053 to 27.831), while the difference in the β coefficient was relatively small (0.639 to 0.668) on these three bands. For the $R(Z_h,Z_{dr})$ relations, the differences in the α and γ coefficients were relatively large, while the difference in the β coefficient was relatively small. For the $R(K_{dp},Z_{dr})$ relations, the α coefficient was 23.265 on the X-band, while the α coefficient on the C-band was about twice that, and the α coefficient on the S-band was about three times that. The CC, RMSE, and NMAE of R estimated from the dual-polarization radar QPE estimators against R calculated from the DSD were used to evaluate the performance of these dual-polarization radar QPE estimators. The result revealed that the performance of double-parameter estimators ($R(Z_h,Z_{dr})$ and $R(K_{dp},Z_{dr})$) was superior to that of single-parameter estimators ($R(Z_h)$ and $R(K_{dp})$), and the performance of the $R(K_{dp})$ ($R(K_{dp},Z_{dr})$) estimator was superior to that of the $R(Z_h)$ ($R(Z_h,Z_{dr})$) estimator for all the bands. Overall, the single-parameter estimator performed the best for the X-band, whereas the double-parameter estimator performed the best for the C-band.

Furthermore, the distribution and mean values of the dual-polarization radar variables establishing these QPE estimators across R classes and radar wavelengths were determined, and the performance of these four types of estimators ($R(Z_h)$, $R(K_{dp})$, $R(Z_h,Z_{dr})$, and $R(K_{dp},Z_{dr})$) for the three bands (S-, C-, and X-bands) showed differences across the R classes. Generally, for all the R classes, the dual-parameter estimators had better performances in the C-band than the other two bands, and the performance of the $R(Z_h,Z_{dr})$ ($R(K_{dp},Z_{dr})$) estimator was better compared to the $R(K_{dp},Z_{dr})$ ($R(Z_h,Z_{dr})$) estimator at lower (higher) R classes for the C-band. Our conclusion emphasizes that when conducting dual-polarization radar QPE applications, it is necessary to consider both the appropriate radar wavelength and the type of estimator, as well as the impact of R classes on the accuracy of QPE. It should be noted that although this study reported promising findings, they need to be further confirmed using dual-polarization radar observations in the future.

Author Contributions: Conceptualization, Y.Z. (Yong Zeng) and J.L. (Jiangang Li); data curation, Y.Z. (Yong Zeng), Y.Z. (Yushu Zhou), J.L. (Jiangang Li), and L.Y.; formal analysis, Y.Z. (Yong Zeng); funding acquisition, L.Y. and Y.Z. (Yong Zeng); methodology, Y.Z. (Yong Zeng), J.L. (Jiangang Li), and Y.J.; project administration, X.L., H.L., Z.T. and Y.J.; resources, Y.Z. (Yong Zeng) and Z.T.; supervision, J.L. (Jing Liu), Y.Z. (Yushu Zhou), J.Z. and L.Y.; writing—original draft, Y.Z. (Yong Zeng); writing—review and editing, Y.Z. (Yong Zeng). All authors have read and agreed to the published version of the manuscript.

Funding: This research was funded by the Natural Science Foundation of Xinjiang Uygur Autonomous Region (Grant No. 2022D01B227), the National Natural Science Foundation of China

(U2003106), the Tianshan Mountains Talent Project (Grant No. 2021-32), the Natural Science Foundation of Xinjiang Uygur Autonomous Region (Grant No. 2022D01B75), the Tianshan Mountains Talent Project (Grant No. 2022TSYCLJ0003), the Science and Technology Innovation Development Fund Project of Xinjiang Meteorological Bureau (Grant No. MS202210), the S&T Development Fund of IDM (Grant No. KJFZ202303, KJFZ202301), and the Uygur Autonomous Region Tianchi Project for Introducing High-Level Talents (2019).

Data Availability Statement: Not applicable.

Acknowledgments: The authors would like to thank the Institute of Desert Meteorology, China, Meteorological Administration, Urumqi, for providing the data of Disdrometers. Thanks also goes to the reviewers for their thorough comments that really helped improve the manuscript.

Conflicts of Interest: The authors declare no conflict of interest.

References

1. Rosenfeld, D.; Ulbrich, C.W. Cloud microphysical properties, processes, and rainfall estimation opportunities. *Meteorol. Monogr* **2003**, *30*, 237–258. [CrossRef]
2. Zhang, G.; Sun, J.; Brandes, E.A. Improving parameterization of rain microphysics with disdrometer and radar observations *J. Atmos. Sci.* **2006**, *63*, 1273–1290. [CrossRef]
3. Mason, B.J. Physics of clouds and precipitation. *Nature* **1954**, *174*, 957–959. [CrossRef]
4. Milbrandt, J.A.; Yau, M.K. A multimoment bulk microphysics parameterization. Part I: Analysis of the role of the spectral shape parameter. *J. Atmos. Sci.* **2005**, *62*, 3051–3064. [CrossRef]
5. Wainwright, C.E.; Dawson, D.T.I.; Xue, M.; Zhang, G. Diagnosing the intercept parameters of the exponential drop size distributions in a single-moment microphysics scheme and impact on supercell storm simulations. *J. Appl. Meteorol. Climatol* **2014**, *53*, 2072–2090. [CrossRef]
6. McFarquhar, G.M.; Hsieh, T.-L.; Freer, M.; Mascio, J.; Jewett, B.F. The characterization of ice hydrometeor gamma size distributions asvolumes in N_0–λ–μ phase space: Implications for microphysical process modeling. *J. Atmos. Sci.* **2015**, *72*, 892–909. [CrossRef]
7. Kinnell, P.I.A. Rainfall Intensity-Kinetic Energy Relationships for Soil Loss Prediction1. *Soil Sci. Soc. Am. J.* **1981**, *45*, 153 [CrossRef]
8. Steiner, M.; Smith, J.A. Reflectivity, Rain Rate, and Kinetic Energy Flux Relationships Based on Raindrop Spectra. *J. Appl. Meteorol* **2000**, *39*, 1923–1940. [CrossRef]
9. Seela, B.K.; Janapati, J.; Kalath Unnikrishnan, C.; Lin, P.-L.; Le Loh, J.; Chang, W.-Y.; Kumar, U.; Reddy, K.K.; Lee, D.-I.; Venkatram Reddy, M. Raindrop Size Distributions of North Indian Ocean Tropical Cyclones Observed at the Coastal and Inland Stations in South India. *Remote Sens.* **2021**, *13*, 3178. [CrossRef]
10. Ryzhkov, A.V.; Zrnic, D.S. Comparison of dual polarization radar estimators of rain. *J. Atmos. Ocean. Technol.* **1995**, *12*, 249–256 [CrossRef]
11. Liao, L.; Meneghini, R.; Tokay, A. Uncertainties of GPM DPR rain estimates caused by DSD parameterizations. *J. Appl. Meteor Climatol.* **2014**, *53*, 2524–2537. [CrossRef]
12. Chen, B.; Yang, J.; Pu, J. Statistical characteristics of raindrop size distribution in the Meiyu season observed in eastern China *J. Meteorol. Soc. Jpn.* **2013**, *91*, 215–227. [CrossRef]
13. Ulbrich, C.W. Natural variations in the analytical form of the raindrop size distribution. *J. Clim. Appl. Meteorol.* **1983**, *22* 1764–1775. [CrossRef]
14. Tokay, A.; Short, D.A. Evidence from tropical raindrop spectra of the origin of rain from stratiform versus convective clouds *J. Appl. Meteorol.* **1996**, *35*, 355–371. [CrossRef]
15. Bringi, V.N.; Chandrasekar, V.; Hubbert, J.; Gorgucci, E.; Randeu, W.L.; Schoenhuber, M. Raindrop size distribution in differen climatic regimes from disdrometer and dual-polarized radar analysis. *J. Atmos. Sci.* **2003**, *60*, 354–365. [CrossRef]
16. Konwar, M.; Das, S.K.; Deshpande, S.M.; Chakravarty, K.; Goswami, B.N. Microphysics of clouds and rain over the Western Gha *J. Geophys. Res. Atmos.* **2014**, *119*, 6140–6159. [CrossRef]
17. Chen, B.; Hu, Z.; Liu, L.; Zhang, G. Raindrop Size Distribution Measurements at 4500 m on the Tibetan Plateau During TIPEX-II *J. Geophys. Res. Atmos.* **2017**, *12211*, 11092–12006. [CrossRef]
18. Seela, B.K.; Janapati, J.; Lin, P.-L.; Reddy, K.K.; Shirooka, R.; Wang, P.K. A Comparison Study of Summer Season Raindrop Size Distribution between Palau and Taiwan, Two Islands in Western Pacific. *J. Geophys. Res. Atmos.* **2017**, *122*, 11–787. [CrossRef]
19. Suh, S.-H.; You, C.-H.; Lee, D.-I. Climatological characteristics of raindrop size distributions in Busan, Republic of Korea. *Hydro Earth Syst. Sci.* **2016**, *20*, 193–207. [CrossRef]
20. Wen, L.; Zhao, K.; Chen, G. Drop size distribution characteristics of seven typhoons in China. *J. Geophys. Res. Atmos.* **2018**, *123* 6529–6548. [CrossRef]
21. Zheng, J.; Liu, L.; Chen, H. Characteristics of warm clouds and precipitation in South China during the pre-flood season using datasets from a cloud radar, a ceilometer, and a disdrometer. *Remote Sens.* **2019**, *11*, 3045. [CrossRef]

22. Zhang, A.; Hu, J.; Chen, S. Statistical characteristics of raindrop size distribution in the monsoon season observed in southern China. *Remote Sens.* **2019**, *11*, 432. [CrossRef]
23. Wen, L.; Zhao, K.; Zhang, G. Statistical characteristics of raindrop size distributions observed in East China during the Asian summer monsoon season using 2-D video disdrometer and Micro Rain Radar data. *J. Geophys. Res. Atmos.* **2016**, *121*, 2265–2282. [CrossRef]
24. Pu, K.; Liu, X.; Wu, Y.; Hu, S.; Liu, L.; Gao, T. A comparison study of raindrop size distribution among five sites at the urban scale during the East Asian rainy season. *J. Hydrol.* **2020**, *590*, 125500. [CrossRef]
25. Wen, L.; Zhao, K.; Wang, M. Seasonal variations of observed raindrop size distribution in East China. *Adv. Atmos. Sci.* **2019**, *36*, 346–362. [CrossRef]
26. Zhang, H.; Zhang, Y.; He, H. Comparison of raindrop size distributions in a midlatitude continental squall line during different stages as measured by Parsivel over East China. *J. Appl. Meteorol. Climatol.* **2017**, *56*, 2097–2111. [CrossRef]
27. Luo, L.; Guo, J.; Chen, H. Microphysical characteristics of rainfall observed by a 2DVD disdrometer during different seasons in Beijing, China. *Remote Sens.* **2021**, *13*, 2303. [CrossRef]
28. Ma, Y.; Ni, G.; Chandra, C.V. Statistical characteristics of raindrop size distribution during rainy seasons in the Beijing urban area and implications for radar rainfall estimation. *Hydrol. Earth Syst. Sci.* **2019**, *23*, 4153–4170. [CrossRef]
29. Ji, L.; Chen, H.N.; Li, L.; Chen, B.J.; Xiao, X.; Chen, M.; Zhang, G.F. Raindrop size distributions and rain characteristics observed by a PARSIVEL disdrometer in Beijing, Northern China. *Remote Sens.* **2019**, *11*, 1479. [CrossRef]
30. Wang, G.; Zhou, R.; Zhaxi, S.; Liu, S. Raindrop size distribution measurements on the Southeast Tibetan Plateau during the STEP project. *Atmos. Res.* **2021**, *249*, 105311. [CrossRef]
31. Wang, G.; Li, R.; Sun, J. Comparative analysis of the characteristics of rainy season raindrop size distributions in two typical regions of the Tibetan Plateau. *Adv. Atmos. Sci.* **2022**, *39*, 1062–1078. [CrossRef]
32. Zhang, J.B.; Deng, Z.F. *A Generality of Rainfall in Xinjiang*; Meteorological Press: Beijing, China, 1987; pp. 1–10. (In Chinese)
33. Yang, L.M.; Li, X.; Zhang, G.X. Some advances and problems in the study of heavy rain in Xinjiang. *Clim. Environ. Res.* **2011**, *16*, 188–198. (In Chinese)
34. Zeng, Y.; Yang, L. Triggering mechanism of an extreme rainstorm process near the Tianshan Mountains in Xinjiang, an arid region in China, based on a numerical simulation. *Adv. Meteorol.* **2020**, *2020*, 8828060. [CrossRef]
35. Zeng, Y.; Yang, L.; Zhang, Z.; Tong, Z.; Li, J.; Liu, F.; Zhang, J.; Jiang, Y. Characteristics of clouds and raindrop size distribution in Xinjiang, using cloud radar datasets and a disdrometer. *Atmosphere* **2020**, *11*, 1382. [CrossRef]
36. Zeng, Y.; Yang, L.; Tong, Z.; Jiang, Y.; Zhang, Z.; Zhang, J.; Zhou, Y.; Li, J.; Liu, F.; Liu, J. Statistical characteristics of raindrop size distribution during rainy seasons in Northwest China. *Adv. Meteorol.* **2021**, *2021*, 6667786. [CrossRef]
37. Zeng, Y.; Tong, Z.; Jiang, Y.; Zhou, Y. Microphysical characteristics of seasonal rainfall observed by a Parsivel disdrometer in the Tianshan Mountains, China. *Atmos. Res.* **2022**, *280*, 106459. [CrossRef]
38. Zeng, Y.; Yang, L.; Zhou, Y.; Tong, Z.; Jiang, Y. Statistical characteristics of summer season raindrop size distribution in the western and central Tianshan Mountains in China. *J. Meteorol. Soc. Jpn.* **2022**, *100*, 855–872. [CrossRef]
39. Zeng, Y.; Yang, L.; Zhou, Y.; Tong, Z.; Jiang, Y.; Chen, P. Characteristics of orographic raindrop size distribution in the Tianshan Mountains, China. *Atmos. Res.* **2022**, *278*, 106332. [CrossRef]
40. Fu, Z.; Dong, X.; Zhou, L.; Cui, W.; Wang, J.; Wan, R.; Leng, L.; Xi, B. Statistical characteristics of raindrop size distributions and parameters in Central China during the Meiyu seasons. *J. Geophys. Res. Atmos.* **2020**, *125*, e2019JD031954. [CrossRef]
41. Marshall, J.S.; Palmer, W.M. The distribution of raindrops with size. *J. Meteor.* **1948**, *5*, 165–166. [CrossRef]
42. Fulton, R.A.; Breidenbach, J.P.; Seo, D.-J. The WSR-88D Rainfall Algorithm. *Weather. Forecast.* **1998**, *13*, 377–395. [CrossRef]
43. Atlas, D.; Ulbrich, C.W.; Marks, F.D. Systematic variation of drop size and radar-rainfall relations. *J. Geophys. Res.* **1999**, *104*, 6155–6169. [CrossRef]
44. Ulbrich, C.W.; Atlas, D. Microphysics of raindrop size spectra: Tropical continental and maritime storms. *J. Appl. Meteor. Climatol.* **2007**, *46*, 1777–1791. [CrossRef]
45. Janapati, J.; Seela, B.K.; Lin, P.-L. Raindrop size distribution characteristics of Indian and Pacific Ocean tropical cyclones observed at India and Taiwan sites. *J. Meteorol. Soc. Jpn.* **2020**, *98*, 299–317. [CrossRef]
46. Janapati, J.; Seela, B.K.; Lin, P.-L.; Lee, M.-T.; Joseph, E. Microphysical features of typhoon and non-typhoon rainfall observed in Taiwan, an island in the northwestern Pacific. *Hydrol. Earth Syst. Sci.* **2021**, *25*, 4025–4040. [CrossRef]
47. Kim, H.-J.; Jung, W.; Suh, S.-H.; Lee, D.-I.; You, C.-H. The Characteristics of raindrop size distribution at windward and leeward side over mountain area. *Remote Sens.* **2022**, *14*, 2419. [CrossRef]
48. Li, R.; Wang, G.; Zhou, R.; Zhang, J.; Liu, L. Seasonal variation in microphysical characteristics of precipitation at the entrance of water vapor channel in Yarlung Zangbo Grand Canyon. *Remote Sens.* **2022**, *14*, 3149. [CrossRef]
49. Cifelli, R.; Chandrasekar, V.; Lim, S.; Kennedy, P.C.; Wang, Y.; Rutledge, S.A. A new dual-polarization radar rainfall algorithm: Application in Colorado precipitation events. *J. Atmos. Ocean. Technol.* **2011**, *28*, 352–364. [CrossRef]
50. You, C.-H.; Kang, M.; Lee, D.-I. Rainfall estimation by S-band polarimetric radar in Korea. Part I: Preprocessing and preliminary results. *Meteorol. Appl.* **2014**, *21*, 975–983. [CrossRef]
51. Cao, Q.; Zhang, G.; Brandes, E.A. Polarimetric radar rain estimation through retrieval of drop size distribution using a Bayesian approach. *J. Appl. Meteorol. Climatol.* **2010**, *49*, 973–990. [CrossRef]

52. Brandes, E.A.; Zhang, G.; Vivekanandan, J. Experiments in rainfall estimation with a polarimetric radar in a subtropical environment. *J. Appl. Meteorol.* **2002**, *41*, 674–685. [CrossRef]
53. Li, Q.; Wei, J.; Yin, J.; Qiao, Z.; Cao, J.; Shi, Y. Microphysical characteristics of raindrop size distribution and implications for radar rainfall estimation over the northeastern Tibetan Plateau. *J. Geophys. Res. Atmos.* **2022**, *127*, e2021JD035575. [CrossRef]
54. Zeng, Y.; Yang, L.; Tong, Z.; Jiang, Y.; Chen, P.; Zhou, Y. Characteristics and applications of summer season raindrop size distributions based on a PARSIVEL^2 disdrometer in the western Tianshan Mountains (China). *Remote Sens.* **2022**, *14*, 3988. [CrossRef]
55. You, C.-H.; Suh, S.-H.; Jung, W.; Kim, H.-J.; Lee, D.-I. Dual-Polarization Radar-Based Quantitative Precipitation Estimation of Mountain Terrain Using Multi-Disdrometer Data. *Remote Sens.* **2022**, *14*, 2290. [CrossRef]
56. Löffler-Mang, M.; Joss, J. An optical disdrometer for measuring size and velocity of hydrometeors. *J. Atmos. Ocean. Technol.* **2000**, *17*, 130–139. [CrossRef]
57. Yuter, S.E.; Kingsmill, D.E.; Nance, L.B.; Löffler-Mang, M. Observations of precipitation size and fall speed characteristics within coexisting rain and wet snow. *J. Appl. Meteorol.* **2006**, *45*, 1450–1464. [CrossRef]
58. Beard, K.V.; Johnson, D.B.; Baumgardner, D. Aircraft observations of large raindrops in warm, shallow, convective clouds. *Geophys. Res. Lett.* **1986**, *13*, 991–994. [CrossRef]
59. Friedrich, K.; Higgins, S.; Masters, F.J.; Lopez, C.R. Articulating and stationary PARSIVEL disdrometer measurements in conditions with strong Winds and heavy rainfall. *J. Atmos. Oceanic Technol.* **2013**, *30*, 2063–2080. [CrossRef]
60. Atlas, D.; Srivastava, R.C.; Sekhon, R.S. Doppler radar characteristics of precipitation at vertical incidence. *Rev. Geophys.* **1973**, *11*, 1–35. [CrossRef]
61. Jaffrain, J.; Berne, A. Experimental quantification of the sampling uncertainty associated with measurements from PARSIVEL disdrometers. *J. Hydrometeorol.* **2011**, *12*, 352–370. [CrossRef]
62. Tokay, A.; Petersen, W.A.; Gatlin, P.; Wingo, M. Comparison of raindrop size distribution measurements by collocated disdrometers. *J. Atmos. Ocean. Technol.* **2013**, *30*, 1672–1690. [CrossRef]
63. Ulbrich, C.W.; Atlas, D. Rainfall microphysics and radar properties: Analysis methods for drop size spectra. *J. Appl. Meteorol.* **1998**, *37*, 912–923. [CrossRef]
64. Zhang, G.; Vivekanandan, J.; Brandes, E.A.; Meneghini, R.; Kozu, T. The shape-slope relation in observed gamma raindrop size distributions: Statistical error or useful information? *J. Atmos. Ocean. Technol.* **2003**, *20*, 1106–1119. [CrossRef]
65. Kalogiros, J.; Anagnostou, M.N.; Anagnostou, E.N. Optimum estimation of rain microphysical parameters from X-band dual polarization radar observables. *IEEE Trans. Geosci. Remote Sens.* **2013**, *51*, 3063–3076. [CrossRef]
66. Leinonen, J. High-level interface to T-matrix scattering calculations: Architecture, capabilities and limitations. *Opt. Express* **2014**, *22*, 1655–1660. [CrossRef]
67. Waterman, P.C. Matrix formulation of electromagnetic scattering. *Proc. IEEE* **1965**, *53*, 805–812. [CrossRef]
68. Tokay, A.; Bashor, P.G.; Habib, E.; Kasparis, T. Raindrop size distribution measurements in tropical cyclones. *Mon. Weather. Rev.* **2008**, *136*, 1669–1685. [CrossRef]
69. Guo, Z.; Hu, S.; Liu, X.; Chen, X.; Zhang, H.; Qi, T.; Zeng, G. Improving S-band polarimetric radar monsoon rainfall estimation with two-dimensional video disdrometer observations in South China. *Atmosphere* **2021**, *12*, 831. [CrossRef]
70. Chen, H.; Chandrasekar, V. The quantitative precipitation estimation system for Dallas-Fort Worth (DFW) urban remote sensing network. *J. Hydrol.* **2015**, *531*, 259–271. [CrossRef]
71. Chen, H.; Chandrasekar, V.; Bechini, R. An improved dual-polarization radar rainfall algorithm (DROPS2.0): Application in NASA IFloodS field campaign. *J. Hydrometeorol.* **2017**, *18*, 917–937. [CrossRef]

remote sensing

MDPI

Article

A Uniformity Index for Precipitation Particle Axis Ratios Derived from Radar Polarimetric Parameters for the Identification and Analysis of Raindrop Areas

Yue Sun [1], Hui Xiao [1,2,*], Huiling Yang [1], Haonan Chen [3], Liang Feng [1], Weixi Shu [1,2] and Han Yao [4]

1 Key Laboratory of Cloud-Precipitation Physics and Severe Storms (LACS), Institute of Atmospheric Physics, Chinese Academy of Sciences, Beijing 100029, China
2 College of Earth Sciences, University of Chinese Academy of Sciences (UCAS), Beijing 100049, China
3 Department of Electrical and Computer, Colorado State University, Fort Collins, CO 80523, USA
4 Qingdao Meteorology Bureau, Qingdao 266003, China
* Correspondence: hxiao@mail.iap.ac.cn; Tel.: +86-10-82995318

Abstract: A uniformity index for the axis ratios (U_{ar}) derived from dual polarization weather radar data is proposed for raindrop area identification and analysis. The derivation of this new parameter is based on radar scattering simulations and assumptions. U_{ar} is between 0 and 1 and can be calculated from the differential reflectivity (Z_{DR}) and the copolar correlation coefficient (ρ_{hv}), which reflects the uniformity of the axis ratio (r) of the particle group. For raindrops, U_{ar} is close to 1 under ideal conditions, but is clearly different from that of ice particles whose value is close to 0. Studies conducted during two convective weather events observed by X-band and S-band radar are presented to show the U_{ar} features. In convective areas, high U_{ar} presents a U-shaped vertical structure. One branch corresponds to the Z_{DR} column, while the other branch is located at the rear of the convective cloud zone and is lower in altitude, representing the process of ice particles melting into raindrops and then being transported upward by a strong updraft. In stratiform cloud areas, a more than 95% overall identification ratio is obtained when the threshold of U_{ar} is set to 0.2~0.3 for discriminating rain layers.

Keywords: dual polarization weather radar; axis ratio; rain area identification

1. Introduction

The distribution and variation characteristics of the phase state (liquid, ice, mixed phase, etc.) of hydrometeors in clouds are extremely important issues in precipitation physics. Dual polarization weather radar obtains polarimetric variables, such as the horizontal/vertical reflectivity factor (Z_H/Z_V), differential reflectivity (Z_{DR}), copolar correlation coefficient (ρ_{hv}) and differential propagation phase shift (K_{DP}), which are closely related to the microphysical properties of hydrometeors in clouds [1–3]. For example, large raindrops show a flat shape and a corresponding high Z_{DR} value when under air resistance [4]. This is clearly different from the various shapes of tumbling hail and graupel, which make it possible to roughly distinguish liquid and solid particles in the high Z_H region [5]. Furthermore, lower Z_H appears in snow and ice crystals due to the lower dielectric constant, while lower ρ_{hv} appears in mixed phase particles and sometimes in ice phase particles due to the variation of dielectric constant, shape and orientation [3,6]. Since different kinds of particles are not easily and directly identified due to the overlapping of the range of polarization parameters, a hydrometeor classification algorithm (HCA) based on the fuzzy logic algorithm [6] is the most feasible solution to obtain a general qualitative result. This kind of algorithm has been developed and improved over the past 20 years [7–16]. However, the HCA still has limitations in terms of subjectivity and experience [16]. Thus, its results cannot be regarded as absolutely accurate or the only microphysical analysis results, nor can the HCA completely replace the analysis of the original observed variables and other analysis methods. In particular, an additional input temperature profile is needed in most of the methods above, which means that the original polarimetric observables still cannot completely identify the hydrometeor phase independently.

Another principle algorithm that can distinguish the phase state of hydrometeor particles in clouds involves identifying the melting layer (ML) so that the part below the ML is identified as the rain area. In stratiform mixed phase clouds, the ML exhibits a Z_H peak in the vertical direction [17,18], known as the bright band (BB) in radar meteorology. There is also a Z_{DR} peak and a valley of ρ_{hv} in the ML. These dual polarization weather radar signals are closely related to changes in the dielectric constant, particle density, size and shape during the falling of ice phase particles [3]. Weather radar mostly adopts the polar coordinate volume scanning mode, where automatic ML detection algorithms are mainly built based on radar image features, including the gradient and extreme of Z_H in a vertical profile of reflectivity (VPR) [18–20], the boundary of high ρ_{hv} [21], the thresholds or gradient of Z_H, Z_{DR} and ρ_{hv} [22–25], and the matching degree with the ideal model [26]. Automatic ML detection results can reduce contamination in radar quantitative precipitation estimation (QPE). Furthermore, ML is helpful for studying cloud and precipitation physical processes such as the ML sinking due to the riming or coalescence of snow [3,27,28]. However, such ML features mainly exist in large-scale stable stratiform precipitation and are difficult to be identified in convective clouds with severe temporal and spatial variability. Hence, the current algorithms above are not easily applied to the study of the melting or freezing processes within convective clouds. In addition, the accuracy of these algorithms mostly also depends on the additional input temperature profile, and it is usually necessary to summarize the thresholds of multiple variables. Therefore, it is still meaningful to find a more accurate and reliable method or some variables based on weather radar data to identify the hydrometeor phase.

In this study, a new parameter involving the microphysical characteristics of hydrometeor particles is proposed. The new parameter is derived from existing polarimetric radar observables and is found to reflect the uniformity of precipitation particles' axial ratio. By this parameter alone, a simpler method for identifying raindrop areas is presented by setting a threshold. The derivation and demonstration of the new parameter will aim at the S-band (wavelength 10 cm) and X-band (wavelength 3.2 cm), which are commonly used in weather radars. The S-band is the most common band of operational weather radar, which has little rain attenuation and a long detection distance. X-band radar often has a smaller antenna and is easy to deploy in mobile platforms, and it is sensitive to weak precipitation.

The process of establishing the new parameter is described in Section 2. Section 3 shows and discusses the typical vertical structure characteristics of the new parameter in terms of radial height indicator (RHI) data from X-band radar and the performance and simple application of the new parameter in S-band weather radar volume scan data. The limitations of the new parameter are discussed in Section 4. A summary is given in Section 5.

2. Axis Ratio Uniformity Index

2.1. Approximate Relationship between the Reflectivity Ratio, Dielectric Properties and Axial Ratio

In this section, an approximate relationship between the reflectivity ratio, dielectric properties and axial ratio is proposed for the derivation of the new parameter presented in Section 2.3. When the scattering amplitude of ellipsoidal particles is calculated using the Rayleigh–Gans formula [1], both the axis ratio of the particle (r) and the dielectric constant (ε) exist in a nonlinear form, and these two parameters are not easily separated to form independent product terms. In previous studies, dielectric parameters were often regarded as fixed values according to the phase state of the particles, and then the approximate relationship of other parameters was discussed. For example, K_{DP} can be approximated as the product of the rain content, mass-weighted axial ratio of the raindrop, and a constant term containing the given ε [29]. However, for mixed phase clouds, the phase state of particles needs to be considered as a variable since they are not fully known in advance.

The dielectric property parameters of particles are one of the key parameters that determine the scattering ability of particles. These parameters are usually considered to be related to the material of the object, incident wavelength and ambient temperature. There are two equivalent descriptions of dielectric properties: the complex dielectric constant $\varepsilon = \varepsilon_r + i\varepsilon_i$ and the complex refractive index $m = m_r + im_i$, where ε_r and ε_i denote the real and imaginary parts of ε, m_r and m_i denote the real and imaginary parts of m, respectively, and $i^2 = -1$. The conversion relationships between ε and m are as follows:

$$m_r^2 = 0.5\left(\sqrt{\varepsilon_r^2 + \varepsilon_i^2} + \varepsilon_r\right) \quad (1)$$

$$m_i^2 = 0.5\left(\sqrt{\varepsilon_r^2 + \varepsilon_i^2} - \varepsilon_r\right) \quad (2)$$

For different hydrometeor phase states, the Ray scheme [30] is used to calculate the dielectric parameters ε of pure water and pure ice. For an ice water mixture and spongy ice (mixture of ice and air), the overall dielectric constant is calculated according to the mass fraction method, and the Debye scheme [31] is selected as follows:

$$\frac{\varepsilon^{(mix)} - 1}{\varepsilon^{(mix)} + 2} = f\cdot\left(\frac{\varepsilon^{(1)} - 1}{\varepsilon^{(1)} + 2}\right) + (1 - f)\cdot\left(\frac{\varepsilon^{(2)} - 1}{\varepsilon^{(2)} + 2}\right) \quad (3)$$

where $\varepsilon^{(1)}$ and $\varepsilon^{(2)}$ are the complex dielectric constants of the two components, $\varepsilon^{(mix)}$ is the overall complex dielectric constant of the mixture, and f is the volume fraction of the first component. The comparison of different schemes [31] shows that although the scheme of Equation (3) is not the most accurate scheme, the difference is very small compared with the best Maxwell–Garnett scheme, and the constraint conditions are the least accurate.

Some typical phases of hydrometeor particles in clouds and their corresponding dielectric properties are listed to find a simplified representation. The dielectric properties of pure water (clouds and raindrops) and pure ice (solid graupel and hail) are set according to the corresponding material. Mixtures of ice and water with $f = 0.5$ are used to characterize particles that are melting or freezing. Mixtures of ice and air (aggregated snow and ice crystals) with $f = 0.1$ and 0.5 are used to characterize spongy ice particles. For pure water, ambient temperatures of 0, 10, and 20 °C are selected to reflect the effect of temperature change on the dielectric properties. Since the dielectric properties of ice vary little with temperature, pure ice and other mixtures are set to 0 °C. Another problem is that ε_r and ε_i may vary differently with temperature, which leads to two variables of comparable magnitudes that need to be discussed. Note that m_r is clearly larger than m_i (Equations (1) and (2)); thus, only m_r is taken as a dielectric characteristic parameter in the following attempts to characterize different phases.

Figure 1a gives m_r at different phases and temperatures. The m_r of water increases slightly with temperature in the X-band, while in the S-band, it decreases slightly or can be considered as undergoing little change. However, when water transitions to an ice/water mixture or ice, m_r decreases, which generally has both nonlinear and nonmonotonic characteristics that are not easily used to form a simple model. However, if the reciprocal of m_r is taken, it can be found that the ${m_r}^{-1}$ of spongy ice, pure ice, an ice/water mixture and pure water decrease somewhat linearly (Figure 1b). The different temperatures have little effect on ${m_r}^{-1}$ at this time. Therefore, ${m_r}^{-1}$ can be used as an available paramcter to characterize the dielectric properties of the different phases.

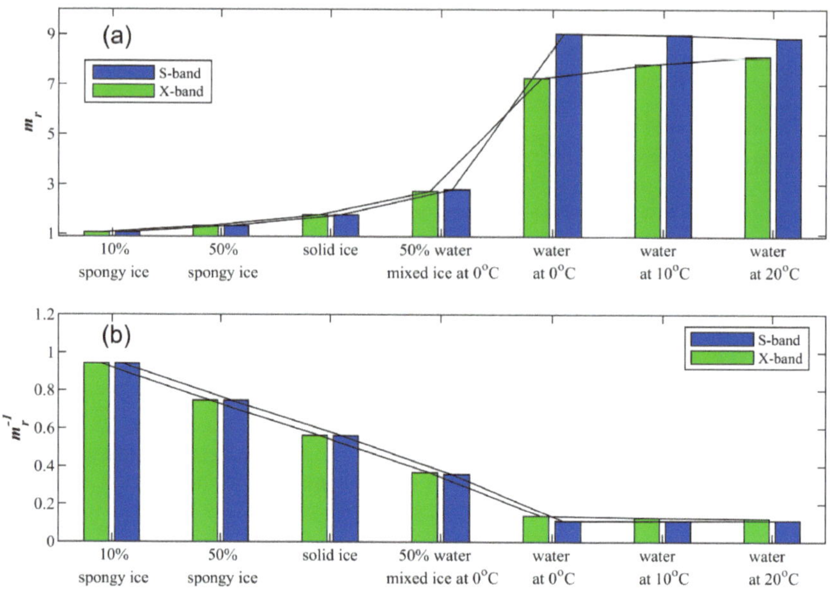

Figure 1. Dielectric property parameters: (**a**) m_r and (**b**) $m_r{}^{-1}$ of hydrometeors at different phases and temperatures.

The reflectivity variables are calculated using the T-matrix method [32]. For raindrops, the relevant theoretical calculation schemes of reflectivity variables [1] are as follows:

$$Z_{h,v} = \frac{4\lambda^4}{\pi^4|K_w|^2} \int_{D_{min}}^{D_{max}} |S_{hh,vv}(D)|^2 N(D) dD \tag{4}$$

$$Z_{H,V} = 10\log_{10} Z_{h,v} \tag{5}$$

$$Z_{dr} = \frac{Z_h}{Z_v} \tag{6}$$

$$Z_{DR} = 10\log_{10} Z_{dr} = Z_H - Z_V \tag{7}$$

where Z_h (or Z_v), with lowercase subscripts, have linear units (mm^6/m^3), Z_H (or Z_V), with uppercase subscripts, are in log units (dBZ), and H or V represent horizontal or vertical polarization, respectively. Z_{DR} is in log units (dB). Z_{dr} is the dimensionless reflectivity ratio. K_w is associated with the dielectric property ($K_w = (\varepsilon - 1)/(\varepsilon + 2)$). λ (unit: m) is the wavelength of the radar. $S_{hh,vv}$ is the backscattering amplitude of a single hydrometeor particle in a horizontal or vertical channel. D is the equivalent spherical diameter of a particle, and $N(D)$ is the particle number concentration density. D_{min} and D_{max} are the lower and upper limits of the drop size distribution, respectively.

A common problem in the simulation of particle scattering properties is that there are many dimensions that can be discussed, such as phase, shape, axis ratio and size distribution. Here, an individual particle is first discussed, trying to find some available laws that are less affected by particle size. To avoid confusion with the Z_{dr} of the particle group, the reflectivity ratio of a single particle is represented by the symbol η_{dr}. Taking an ellipsoidal particle with $r = 2$ as an example, the effect of different phases on η_{dr} is analyzed (Figure 2). Note that r here is defined as the ratio of the horizontal scale to the vertical scale of the particle relative to the polarization direction of the radar beam, which is contrary to the definition of the raindrop axis ratio used in previous studies [33–36]. Figure 2 shows that η_{dr} increases as the phase of the particles becomes closer to pure water. For D greater than approximately 5 mm (in the S-band) and 2.5 mm (in the X-band), the η_{dr} of the pure water particle shows large oscillations due to the effect of Mie scattering. For the mixed phase, η_{dr} increases slightly with D. However, the general rule is still that the phase corresponding to m_r^{-1} can amplify the value of η_{dr}. Therefore, we can use this relationship between the phase and η_{dr} of a single particle as an available approximation and the Mie scattering effect caused by the change in D as a potential error factor leading to the inaccuracy of this approximation. For example, based on this approximation, raindrops with a particle size greater than 6 mm in the S-band will introduce uncertainty, while particles of 3 to 4 mm in the X-band will lead to overestimation of the axial ratio. Considering that the actual radar detection variable is an integral of a group of particles, the above error factors will not always dominate.

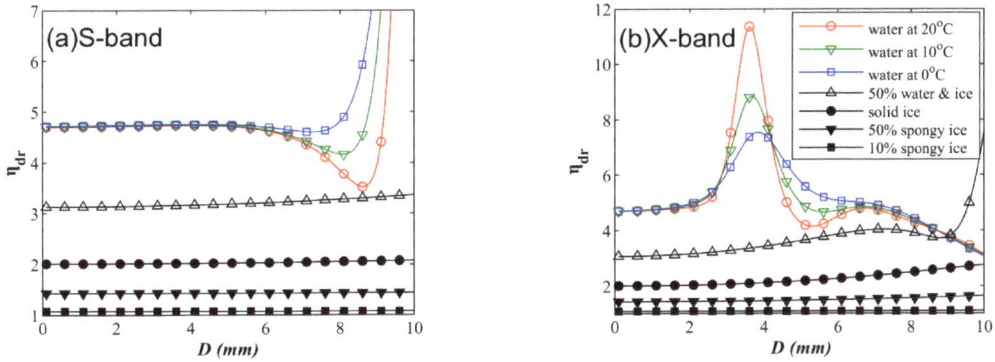

Figure 2. Variation of η_{dr} with the spherical equivalent diameter D for a single ellipsoid water particle with an axial ratio r = 2 under different phase conditions. (**a**): S-band, (**b**): X-band. η_{dr} represents the Z_{dr} of a single particle.

Then, if there is an available relationship with η_{dr}, phase and r can be discussed by ignoring the effect of D on η_{dr}, provided that the r is fixed, and D is fixed to 1 mm in the simulations. Figure 3 shows that the contribution of r also amplifies η_{dr}. However, this is not easily applied since η_{dr} changes along both r and phase. Hence, an approximate significant linear relationship is proposed here, taking $X = m_r^{-1}$ as an independent variable and $Y = (\eta_{dr}^{0.5})/(r-1)$ as a dependent variable to form the linear regression $Y = a_1X + a_0$ (Figure 4).

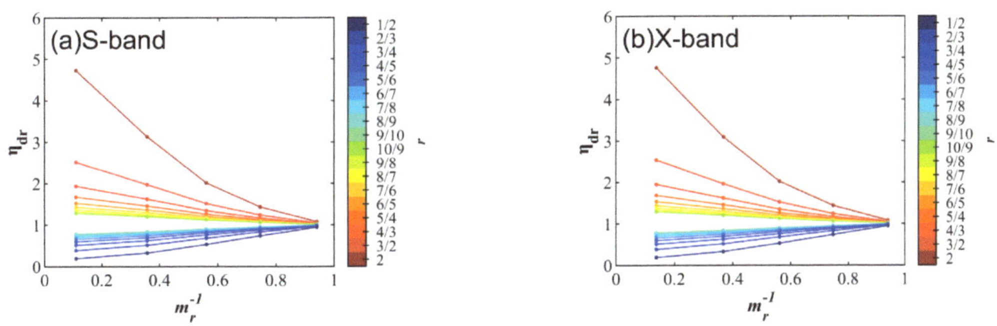

Figure 3. Variation of η_{dr} with m_r^{-1} for a single ellipsoidal water particle with different axial ratios (colors of the lines) when D = 1 mm. (**a**) S-band, (**b**) X-band.

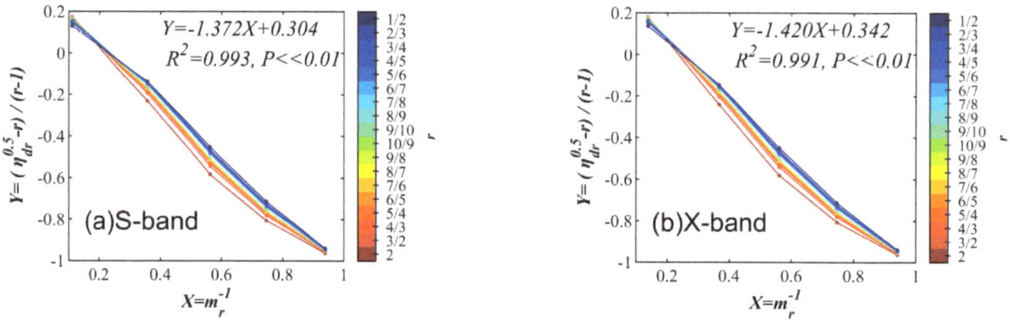

Figure 4. Linear approximation between m_r, η_{dr} and r of a single ellipsoidal particle. (R^2 is the goodness of fit, and P is the significance of the linear fit. (**a**) S-band, (**b**) X-band.

According to the slope a_1 and intercept a_0 obtained by the linear fit shown in Figure 4, a new particle phase parameter E is defined instead of m_r to characterize the phase state:

$$E = a_1 / m_r + a_0 \tag{8}$$

Furthermore, for a single ellipsoidal particle, a simple relationship between η_{dr}, E and r can be obtained, as shown in Equation (9):

$$\sqrt{\eta_{dr}} = E \cdot (r-1) + r \tag{9}$$

Three points also need to be noted:

(1) The approximate linear relationship in Figure 4 is not the most accurate approximation. If r is logarithmic, an approximation with less error can be constructed. However, the approximate linear relationship shown in Figure 4 and Equation (9) is now more readily available for deriving the new index presented below. Therefore, Equation (9) is still used.

(2) The process of eliminating E will be shown in Section 2.3; thus, the values of E, a_1, and a_0 in Equation (8) are no longer discussed in the following sections.

(3) The effects of radar scanning elevation, particle orientation and nonellipsoidal shape are not considered here. Therefore, r should be considered as the flattening or narrowing of the particle in the horizontal and vertical polarization directions as detected by the radar.

2.2. Approximate Relationship between ρ_{hv} and Reflectivity

In this section, an approximate relationship between ρ_{hv} and the reflectivity variables is proposed for the derivation of the new parameter presented in Section 2.3. According to the principles involved in dual polarization radar detection, ρ_{hv} itself reflects the uniformity of the particle shape in the detection volume, but it is also affected by the radar observation system and noise [37,38]. Referring to the basic definition of ρ_{hv}, the ideal ρ_{hv} formula [3] is shown in Equation (10):

$$\rho_{hv}^{(Ideal)} = \frac{\langle S_{hh}^{*} S_{vv} \rangle}{\sqrt{\langle |S_{hh}|^2 \rangle \langle |S_{vv}|^2 \rangle}} \tag{10}$$

where <...> represents the volumetric average. The $S_{hh,vv}$ values are complex, and the molecular component included in Equation (10) requires conjugate multiplication, which makes it difficult to establish numerical connections with known parameters. For this reason, an approximate ρ_{hv} is proposed here as Equation (11):

$$\rho_{hv}^{(Approx)} = \frac{\sum(\sqrt{b_h} \cdot \sqrt{b_v})}{\sqrt{\sum b_h \cdot \sum b_v}} \tag{11}$$

where b_h and b_v reflect the contribution of a single particle to Z_h and Z_v, respectively, as follows:

$$\sum b_h = Z_h, \sum b_v = Z_v \tag{12}$$

In this way, the relationship between ρ_{hv} and the reflectivity of a single particle is established.

To verify the hypothetical approximation of Equation (11), a particle size distribution is necessary since ρ_{hv} is based on a particle group and is calculated by an integral or volumetric average. The range of the axial ratio and size distribution of ice particles may be too large and random, which is not easily resolved in a representative enumeration study; thus, a simpler raindrop size distribution (RSD) model is taken. By enumerating some typical RSDs, the difference between the ideal and approximate ρ_{hv} in different cases can be compared. A common RSD model can be expressed by a gamma distribution with three parameters [1] as follows:

$$N(D) = N_T \frac{(3.67+\mu)^{\mu+1}}{\Gamma(\mu+1)D_0}\left(\frac{D}{D_0}\right)^{\mu} e^{\left[-(3.67+\mu)\frac{D}{D_0}\right]} \tag{13}$$

where N_T is the number concentration of particles, D_0 is the spherical equivalent volume median diameter, and μ is the shape parameter of the RSD. When considering the "shape size" (D with r) model of raindrops, since the trends of D with r are not much different in previous models using fixed parameters [33,35,36], the scheme presented in [36] is taken as a typical case. Another key issue is the setting of D_{max}. In common rainfall, raindrops larger than 6 mm are rare, but in severe convective rainfall, large raindrops of approximately 10 mm are often observed. Choosing a different D_{max} may result in a large difference in the variables after integration according to RSD, thus D_{max} values of 6 and 10 mm are both taken to represent common and typical severe rainfall cases, respectively. D is from 0.1 to D_{max} with a 0.1 mm interval.

When enumerating different sets of RSD parameters, N_T is not enumerated since ρ_{hv} does not involve the absolute number of particles. D_0 starts from 0.1 mm with a 0.1 interval, and its upper limit is determined according to the constraint relation $D_{max}/D_0 \geq 2.5$ [39] to limit the truncation error. μ is from -0.8 to 16 with a 0.2 interval to represent exponential shape ($\mu \leq 0$) and single peak shape (larger μ) distributions. By the combination of different D_{max}, D_0 and μ, RSD parameters in wide value ranges are enumerated to cover possible real conditions of raindrops.

The difference between the approximate ρ_{hv} and the ideal ρ_{hv} is evaluated by common statistics, including the correlation coefficient (R), mean absolute error (MAE) and mean relative error (MRE, see Appendix A for definitions). The results are shown in Figure 5. For general rainfall in the S-band (Figure 5a), the approximate ρ_{hv} can be considered to be consistent with the ideal ρ_{hv}. In other cases, the approximate ρ_h is larger than the ideal ρ_h, but the deviation is generally limited. The largest deviation appears in the case of severe rainfall in the X-band (Figure 5b), but the MRE is only 0.13%. Therefore, the approximate ρ_{hv} in Equation (11) is considered basically consistent with the ideal ρ_{hv} for rainfall cases.

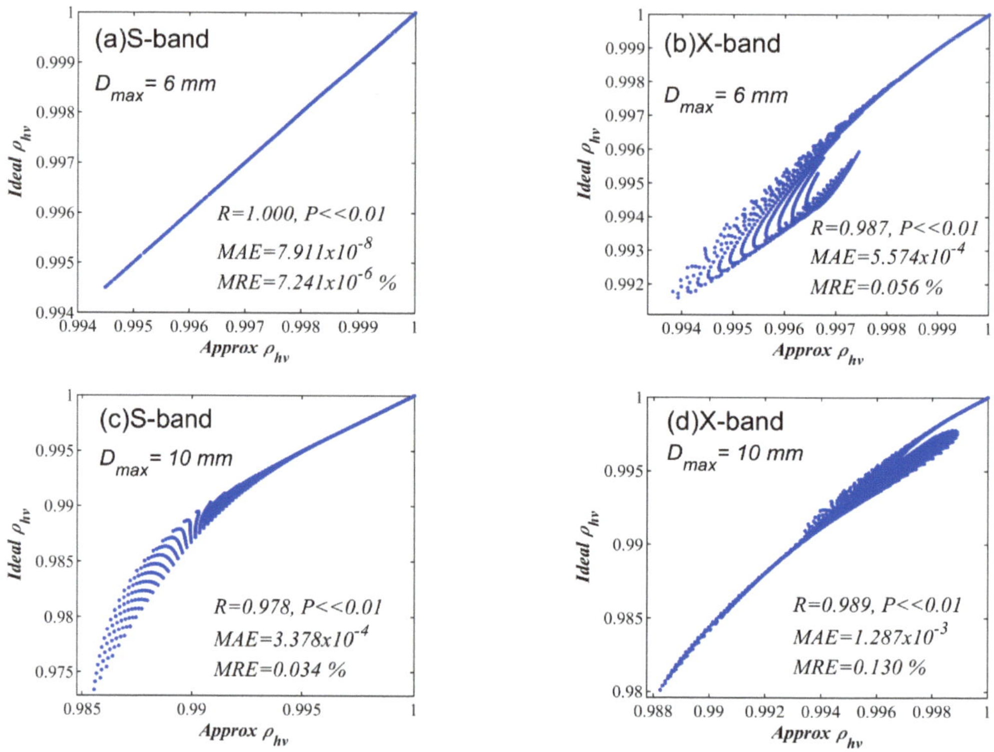

Figure 5. Relationships between approximate ρ_{hv} in this study and ideal ρ_{hv} in raindrops. (The values of D_0 are from 0.1 mm to D_{max} in 0.1 mm intervals. The values of μ range from −0.8 to 16 in intervals of 0.2. R is the correlation coefficient between the approximate ρ_{hv} and ideal ρ_{hv} enumerated samples, and *P* is the significance of R. MAE is the mean absolute error, and MRE is the mean relative error. See Appendix A). (**a**) S-band and D_{max} = 6 mm, (**b**) X-band and D_{max} = 6 mm, (**c**) S-band and D_{max} = 10 mm, (**d**) X-band and D_{max} = 10 mm.

2.3. Construction of the New Parameter

In this section, the derivation of the new parameter is proposed based on the approximations obtained in the previous two sections. First, according to the provisions of b_h and b_v (Equation (12)), Equation (9) can be transformed into Equations (14) and (15):

$$\frac{\sqrt{b_h}}{\sqrt{b_v}} = E\cdot(r-1)+r \tag{14}$$

$$\sqrt{b_h} = (E\cdot(r-1)+r)\cdot\sqrt{b_v} \tag{15}$$

The above still applies to a single particle. When considering an integrated particle group, a new weighted average axial ratio is defined as Equation (16):

$$\bar{r} = \frac{\sum(b_v\cdot r)}{\sum b_v} \tag{16}$$

The $\bar{r}$ here is actually the "vertical reflectivity weighted average axial ratio" and can be considered to reflect the overall average axial ratio of the particle group. Then, combining

Equations (9) and (14) and the definition mode of Equation (16), $\overline{r^2}$ is further defined as follows:

$$\overline{r^2} = \frac{\sum(b_v \cdot r^2)}{\sum b_v} \tag{17}$$

Then, the Z_{dr} of a particle group can be written as a relation of E, $\overline{r}$ and $\overline{r^2}$ as Equation (18):

$$Z_{dr} = \frac{\sum b_h}{\sum b_v} \tag{18}$$

$$= \frac{\sum\left[b_v \cdot (E \cdot r - E + r)^2\right]}{\sum b_v} \tag{19}$$

$$\approx (E+1)^2 \cdot \overline{r^2} - 2E \cdot (E+1) \cdot \overline{r} + E^2 \tag{20}$$

The E here is eventually moved outside the summation sign from Equations (19) to (20), where it represents the general phase of a particle group (which may be a mixture) and ignores the differences caused by the different phases of each particle. Furthermore, combining Equations (11), (14) and (18), Z_{dr} and ρ_{hv} can be written as the relation between E and $\overline{r}$, as shown in Equation (19):

$$\rho_{hv}\sqrt{Z_{dr}} = \frac{\sum[b_v \cdot (E \cdot r - E + r)]}{\sum b_v} \tag{21}$$

$$\approx E \cdot \overline{r} - E + \overline{r} \tag{22}$$

Thus far, there are two radar variables (Z_{dr} and ρ_{hv}) that are used, while there are three unknowns: E, $\overline{r}$ and $\overline{r^2}$. Although absolute quantities such as total concentration and water content are avoided, there is still no way to solve all unknowns. To this end, a solution is proposed here that eliminates E by a combination of Z_{dr} and ρ_{hv} to finally obtain a relationship between $\overline{r}$ and $\overline{r^2}$:

$$\frac{\left(\rho_{hv}\sqrt{Z_{dr}} - 1\right)^2}{Z_{dr} - 2\rho_{hv} \cdot \sqrt{Z_{dr}} + 1} = \frac{(E+1)^2 \cdot (\overline{r}-1)^2}{(E+1)^2 \cdot \overline{(r-1)^2}} \tag{23}$$

$$= \frac{(\overline{r}-1)^2}{\overline{(r-1)^2}} \tag{24}$$

Note that Equation (24) can reflect the uniformity of r relative to 1 for a particle group. For example, the shape and orientation of ice particles may differ greatly, resulting in a large denominator and small molecule component in Equation (24), which eventually leads to the value of Equation (24) being close to 0. However, for raindrops, r is greater than 1 for most particles. This results in a value of Equation (24) between 0 and 1 and close to 1. In addition, the elimination of E between Equations (23) and (24) is equivalent to the elimination of the impact of phase on amplifying η_{dr} or Z_{dr}. Finally, only one descriptive quantity for the axial ratio distribution uniformity of the particle group is obtained, and it is named the "axis ratio uniformity index" (U_{ar}):

$$U_{ar} = \frac{\left(\rho_{hv}\sqrt{Z_{dr}} - 1\right)^2}{Z_{dr} - 2\rho_{hv} \cdot \sqrt{Z_{dr}} + 1} \tag{25}$$

where the dimensionless Z_{dr} can be transformed by $Z_{dr} = 10^{ZDR}$, and Z_{DR} (in dB) is observed by radar. Therefore, U_{ar} is a variable that can be calculated from radar measurements Z_{DR} and ρ_{hv}.

The numerical distribution of U_{ar} is shown in Figure 6, which is based on the same enumerated ranges of RSD as in Section 2.2. In the S-band, U_{ar} values are mostly concentrated between 0.8 and 0.9, mostly above 0.7. In the X-band, the value distribution of U_{ar} is more dispersed, but most values are more than 0.6. Less than 10% of the data have U_{ar} values less than 0.1, and the relationship between U_{ar}, D_0, and μ is further examined (Figure 7). U_{ar} rapidly decreased to 0 when D_0 was less than 0.5 mm, i.e., indicating that U_{ar} does not have the ability to distinguish particles in the ice phase from raindrop groups with small particle diameters. However, since the case in which D_0 is less than 0.5 mm rarely appears in previous joint observation and retrieval studies based on weather radar and RSD [29,40,41], it can be considered that U_{ar} can show a value close to 1 for raindrops, which is obviously different from the value close to 0 for most ice phase particles.

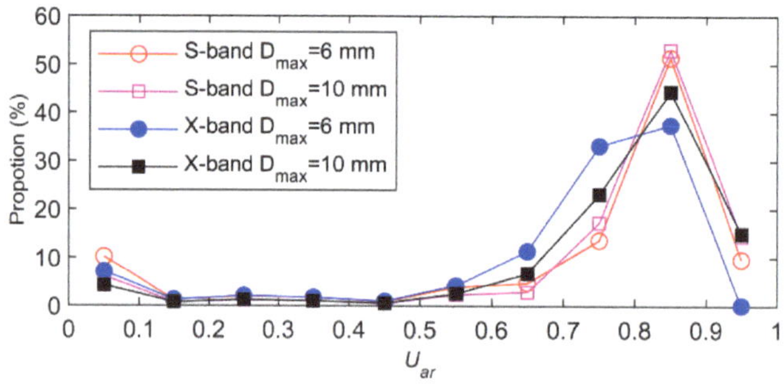

Figure 6. Distribution of U_{ar} within the enumeration range of the RSD parameters. (The values of D_0 and μ are the same as those in Figure 5).

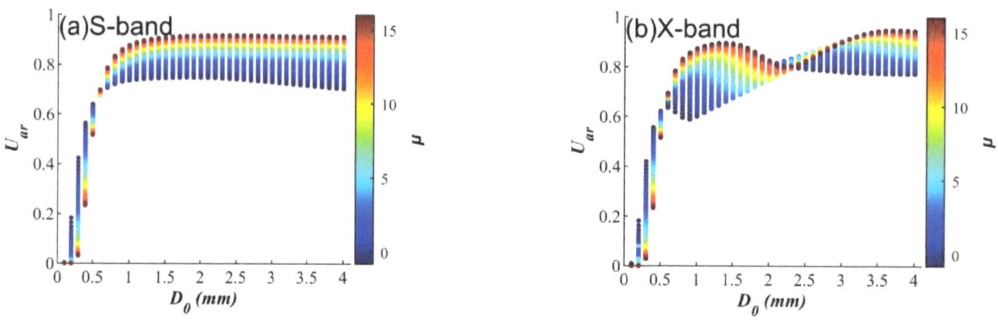

Figure 7. Relationships of U_{ar} versus D_0 and μ when D_{max}=10 mm. (D_{max} = 10 mm. The values of D_0 and μ are the same as those in Figure 5). (**a**) S-band, (**b**) X-band.

In summary, U_{ar} has the potential to identify rain areas. However, the U_{ar} probability distribution shown in Figure 6 is not the probability distribution considered for actual detection, and none of the above discussions include noise in ρ_{hv}. The practical application effect of U_{ar} will be discussed below.

3. Performance of U_{ar} on Real Observations

3.1. Typical Features of Vertical Structures of U_{ar} on X-Band RHI Radar Data

3.1.1. Overview of RHI Data during a Convective Event

In this section, the vertical distribution characteristics of U_{ar} are discussed using RHI data obtained by X-band dual polarization radar. The selected case is a convective event in Beijing that occurred during 18:00–19:30 (Local Standard Time (LST), GMT+8) on 7 September 2016. The radar is a 714XDP-A type X-band dual polarization mobile radar belonging to the Key Laboratory of Cloud-Precipitation Physics and Severe Storms (LACS), Institute of Atmospheric Physics (IAP), Chinese Academy of Sciences (CAS). That radar was deployed for field observations at the Beijing Olympic Water Park (116.68°E, 40.18°N) during the summer from 2015 to 2019. The main characteristics of this radar are listed in Table 1. Other information, including quality control and attenuation correction methods, can be seen in [42,43]. The vertical temperature profile used for the analysis was derived from the neighboring sounding station at 08:00 LST (station number: 54511, 116.28°E, 39.93°N). For the observation mode, plan position indicator (PPI) and RHI scans were switched manually in these observations. The PPI at 4° elevation was scanned first, followed by an RHI scan that aimed at the strong convection center, forming a cycle to track the evolution of the vertical structure of the severe convective cell. Figure 8 shows the convective system to the southwest of the radar moving southeastward.

Table 1. Characteristics of the X-band dual polarization radar used in this paper.

Attribute	Value	Attribute	Value
Antenna diameter	2.4 m	Linear dynamic range	>90 dB
Frequency	9.37 GHz	Beam width	1°
Antenna gain	41.6 dB	Radial resolution	150 m
Peak power	80 kW	Observation range	150 km
Polarization	Horizontal/Vertical	Elevation resolution in RHI mode	0.17°
Pulse width	0.5/1/2 μs		

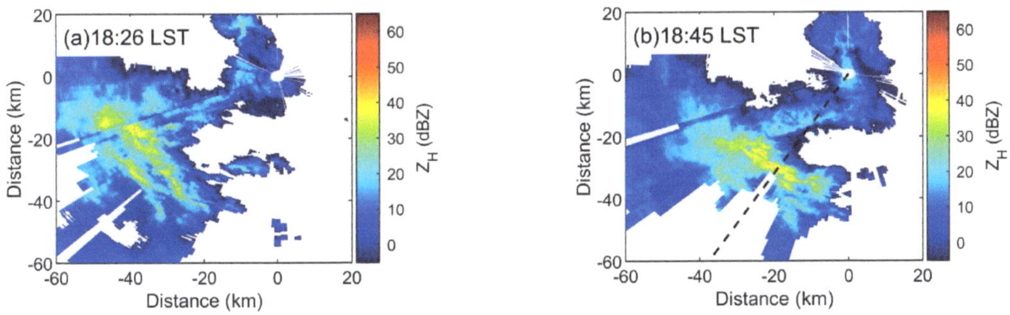

Figure 8. Sample PPI observations of Z_H. (The elevation of the PPI is 4°. The azimuth of the dashed line in Figure 8b is 212°. The data collection time is 7 September 2016). (**a**) 18:26 LST, (**b**) 18:45 LST.

Figure 9a–d show radar data at the RHI of 212° azimuth. The K_{DP} is not shown here since it is not discussed in this paper. There is a stratiform cloud area from 0 to 25 km in the horizontal direction, and the BB characteristics of the ML are found below the 0 °C layer among Z_H, Z_{DR} and ρ_{hv}. The ML, at a horizontal distance of approximately 20 km, has a sinking feature, which may be related to the rapid fall caused by the riming or coalescence of snow [3,27,28]. In the convective cloud region, the convective core is located at a horizontal distance of 35 km (Figure 9a), and there is a distinct Z_{DR} column [44] feature (Figure 9b). The value of ρ_{hv} in the lower layer and 40 km away is less than 0.4 (Figure 9c), which should be attributed to invalid observations caused by attenuation and noise. The corresponding Z_H and Z_{DR} values at such areas are automatically masked during the quality control process, despite some of the Z_{DR} values remaining abnormally low at the rear of the radar beams. The Doppler radial velocity (V_R, Figure 9d) shows convergence below 6 km at the convective core, while divergence appears at a height of approximately 10 km, indicating a strong updraft. Figure 9e shows the U_{ar} proposed in this paper. In addition, a hydrometeor classification (HC) result is shown in Figure 9f as a reference for the particle phase, using the scheme from the work of Feng et al. [45], which is an ensemble and improved version from previous studies [9–12].

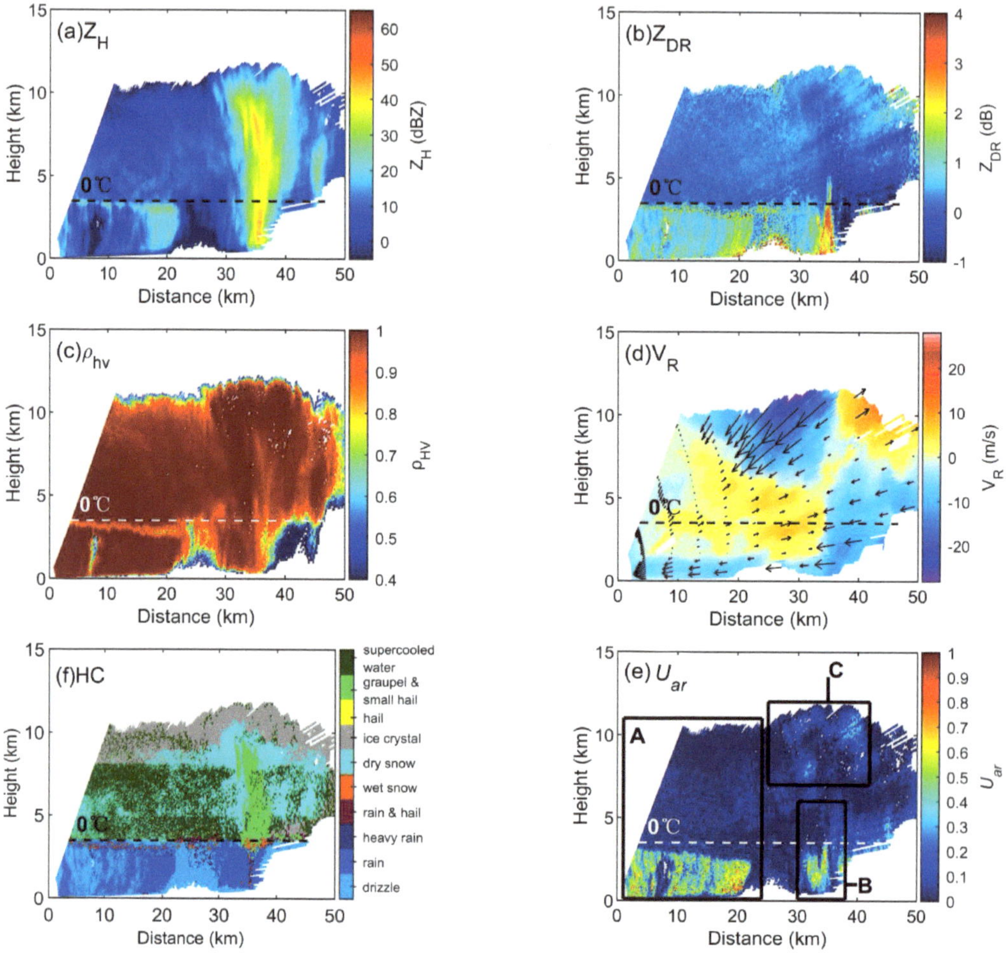

Figure 9. Sample RHI observations and retrievals of the convective system from the X-band dual polarization weather radar. (The time is 18:46:30 LST. The azimuth is the same as the dashed line in Figure 8b. The maximum elevation is 44°). (**a**) Z_H, (**b**) Z_{DR}, (**c**) ρ_{hv}, (**d**) V_R, (**f**) HC, (**e**) U_{ar}.

The following focuses on the three regions shown in Figure 9e, where region A is a stratiform cloud area, region B is the lower area of the convective cloud, and region C is the upper area of the convective cloud.

In region A, U_{ar} suddenly appears as a whole layer with values larger than approximately 0.5 from a certain distance below the 0 °C layer to the ground (Figure 9e), while the HC result shows wet snow corresponding to the ML, and then rain and drizzle below. These results show that HC and U_{ar} are basically consistent in terms of identifying rainy areas. Moreover, the top of the high-value area of U_{ar} exhibits a sinking feature consistent with ML.

In region B, there is an obvious difference between HC and U_{ar}. The high value areas of U_{ar} show a clear 'U'-shaped spatial distribution, while HC shows a very narrow range of heavy rain, grauples and hail mixing with rain in other locations.

In region C, U_{ar} has a high value area with the maximum value approaching 0.4 at the divergence area of V_R, corresponding to dry snow in the HC results. This phenomenon may be explained by the fact that strong horizontal winds contributed to the formation of high-level snow clustering and the maintenance of quasi-horizontal orientation. However, since the spatial scope of that feature is limited, the value of U_{ar} is not so large, and ice habits are not a focus of this study, there is no further discussion of it in the following sections. In Sections 3.1.2 and 3.1.3 below, only areas A and B are discussed.

3.1.2. Analysis of the Stratiform Cloud Area

Average vertical profiles of polarimetric variables in region A are counted and shown in Figure 10. There are notable ML features with extreme values and strong gradients, where Z_H has a peak and ρ_{hv} has a valley from the 0 °C level (3.48 km) down to approximately 9 °C (2.37 km). Z_{DR} also shows a small peak in this ML. However, U_{ar} is close to 0 in the cold cloud area and changes in the ML, mutating into an average of more than 0.4 below the ML. This result apparently shows that U_{ar} is a step-like mutation signal in a three-layered cold cloud, which is unlike conventional variables (such as Z_H, Z_{DR} and ρ_{hv}) that exhibit extreme values in ML.

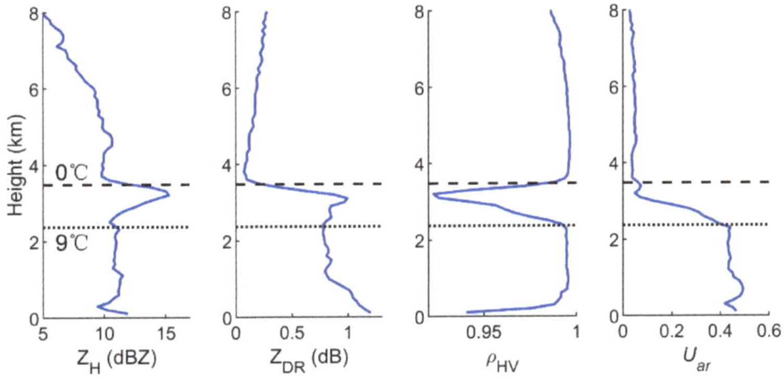

Figure 10. Mean vertical profile in area A. (Statistics by data points with Z_H > 0 dBZ).

The threshold value of U_{ar} used for determining the raindrop areas should be given according to the U_a probability distribution statistics below the ML. If the data below the ML where Z_H is larger than 0 dBZ are counted (Figure 11a), U_{ar} is concentrated between 0.4 and 0.5, but there is a certain distribution from 0 to 1 that makes it difficult to select a threshold. If the statistics are performed in areas with a slightly stronger reflectivity, such as 20 dBZ, which is generally considered to have obvious rainfall, U_{ar} values range, at most, from 0.6 to 0.7 (Figure 11b), and more than 90% are above 0.4. Therefore, it is possible to identify raindrop areas easily only by setting a U_{ar} threshold and without temperature input instead of determining the boundaries of ML first. Further quantitative analysis of the threshold is presented in Section 3.3.

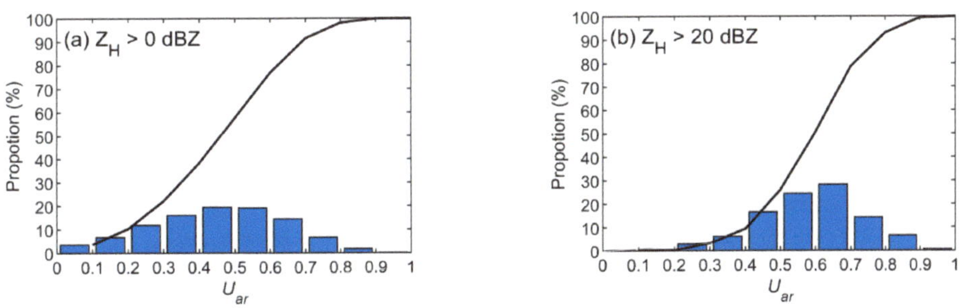

Figure 11. Proportions of U_{ar} values in the warm layer (below 2.37 km, warmer than 9 °C) of area A: (**a**) points with Z_H > 0 dBZ and (**b**) points with Z_H > 20 dBZ. The bars are the proportions within given ranges, and the black lines are the accumulated proportions.

3.1.3. Analysis of a Convective Cloud Area in the Lower Levels

Figure 12 shows the enlarged vertical structure of variables in region B. The divergence of V_R is added (Figure 12c) to better diagnose the distribution of vertical airflow in the cloud, which is defined as dV_R/ds, where s is the radial distance of a radar beam. Strong convergence ($dV_R/ds < 0$) extends from the ground to a height of 5 km and reaches a maximum above the 0 °C layer, which indicates a deep and strong updraft here that is consistent with the position of the Z_{DR} column shown in Figure 12b. The Z_{DR} column is a phenomenon in which the high Z_{DR} region extends above the 0 °C height [45] and is thought to be closely related to the transport of large raindrops and strong updrafts in the supercooled layer [46–50]. The right branch of the U-shape of U_{ar} corresponds to the Z_{DR} column and the strong convergence of V_R, which indicates that the corresponding raindrops are transported upward by the updraft. The upper bound of the right branch of that U-shape crosses the 0 °C layer, indicating that the raindrops freeze after being transported to the supercooled layer. The left branch of the U-shape is 500 m lower than the right branch and is deflected away from the strong convergence area, which can be inferred as raindrops formed by the falling and melting of high-level ice particles.

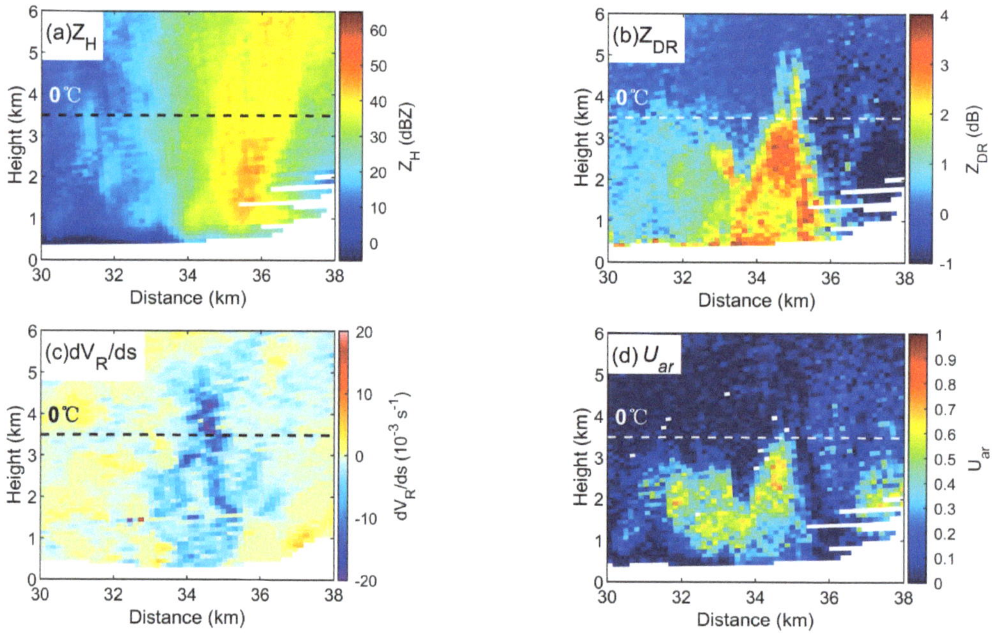

Figure 12. Variables in area B with the Z_{DR} column. (**a**) Z_H, (**b**) Z_{DR}, (**c**) dV_R/ds, (**d**) U_{ar}.

In the analysis of the above features that change with time, due to the large amount of low-level occlusion during the period preceding what is shown in Figures 9 and 12, a later time is selected to track the change in the convective cell. Figure 13 shows PPI data 7 min later and RHI data 9 min later, where the RHI is obtained by tracking the horizontal movement of the Z_H core. Figure 14 shows an enlarged view of the convective core at a low level, which is similar to Figure 12. The divergence of V_R ($dV_R/ds > 0$, Figure 14c) below 1 km indicates the dominating downdraft caused by rainfall. The Z_{DR} column no longer exists (Figure 14b), which may be the result of deep updrafts disappearing. At the same time, U_{ar} no longer displays a U-shape (Figure 14d). The upper bound of the large value of U_{ar} is approximately 1 km below the 0 °C layer, showing the characteristics of large-scale melting of ice particles into raindrops. In summary, it can be inferred that the left and right branches of the U-shaped U_{ar} in the lower troposphere correspond to rain formed by the melting of ice particles and rain transported upward by the updraft, respectively. This set of processes is very similar to Conway and Zrnic's explanation of the formation mechanism of the Z_{DR} column [46]. Therefore, the study presented here can not only be used as evidence to support previous studies but can also expand the means of future research on the Z_{DR} column.

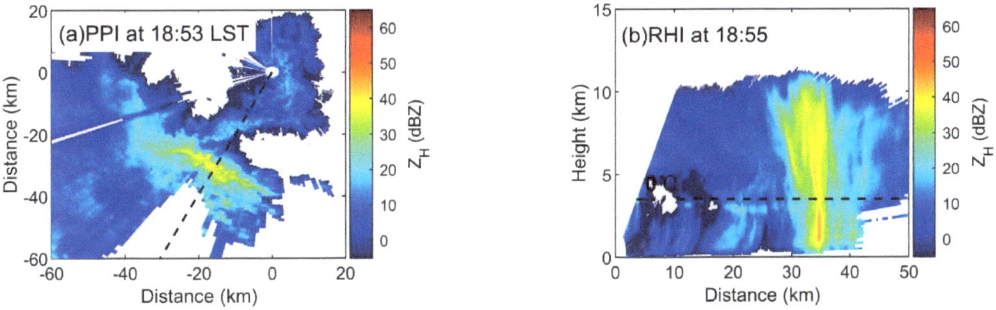

Figure 13. PPI and RHI of Z_H at subsequent time intervals. (**a**) PPI at 18:53 LST, (**b**) RHI at 18:55.

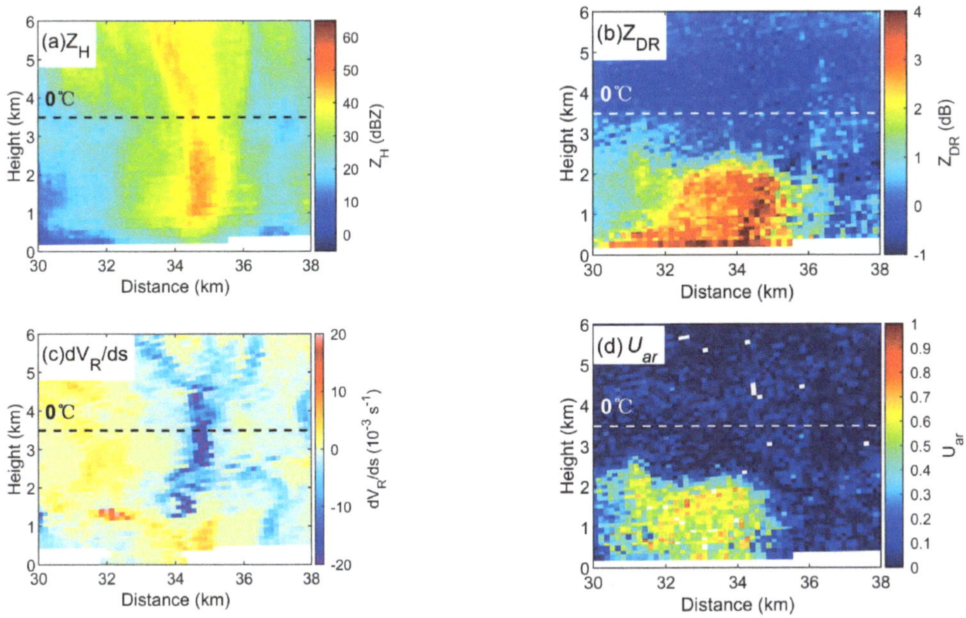

Figure 14. Variables after the Z_{DR} column disappeared. (**a**) Z_H, (**b**) Z_{DR}, (**c**) dV_R/ds, (**d**) U_{ar}.

3.2. Performance of U_{ar} on S-Band Volume Scans Radar Data

In this section, S-band operational dual polarization radar volume scan data are used to evaluate the application of U_{ar}. The involved case is a severe convective event that occurred in Shandong Province, China, during the evening on 17 May 2020. The radar is a dual polarization radar upgraded from CINRAD-SA type radar (station number: Z9532, 120.23°E, 35.99°N). The main characteristics of this radar are listed in Table 2. This radar performs a volume scan containing nine elevations from 0.5° to 19.5° (commonly called VCP-21 mode) in approximately 6 min. The vertical temperature profile used for the analysis was derived from the neighboring sounding station at 20:00 LST (station number: 54857, 120.33°E, 36.06°N).

Table 2. Characteristics of the S-band dual polarization radar used in this paper.

Attribute	Value	Attribute	Value
Antenna diameter	8.5 m	Pulse width	1.57/4.57 μs
Frequency	2.88 GHz	Linear dynamic range	>85 dB
Antenna gain	>45 dB	Beam width	0.93°
Peak power	650 kW	Radial resolution	250 m
Polarization	Horizontal/Vertical	Observation range	460 km

The mid-late stage data regarding the convective system development are selected, where there is a large range of stratiform cloud areas behind the convection line. PPI data with obvious ML features are shown in Figure 15. The BB signal in Z_H is not obvious (Figure 15a), while Z_{DR} and ρ_{hv} both have a ring area with extreme values and rough texture, indicating that the ML is between 0 and 11 °C (Figure 15b,c). For the high Z_H area, which is approximately 50 dBZ in the ML, it is noted that there is a negative Z_{DR} area on the east side (higher), which is consistent with the characteristics of snow riming in previous studies [27,28]. The U_{ar} shows an appearance similar to that of the X-band in Section 3.1, where a wide range of large values appears mostly just below the ML. This result indicates that U_{ar} also has the ability to identify raindrop areas in S-band volume scan data.

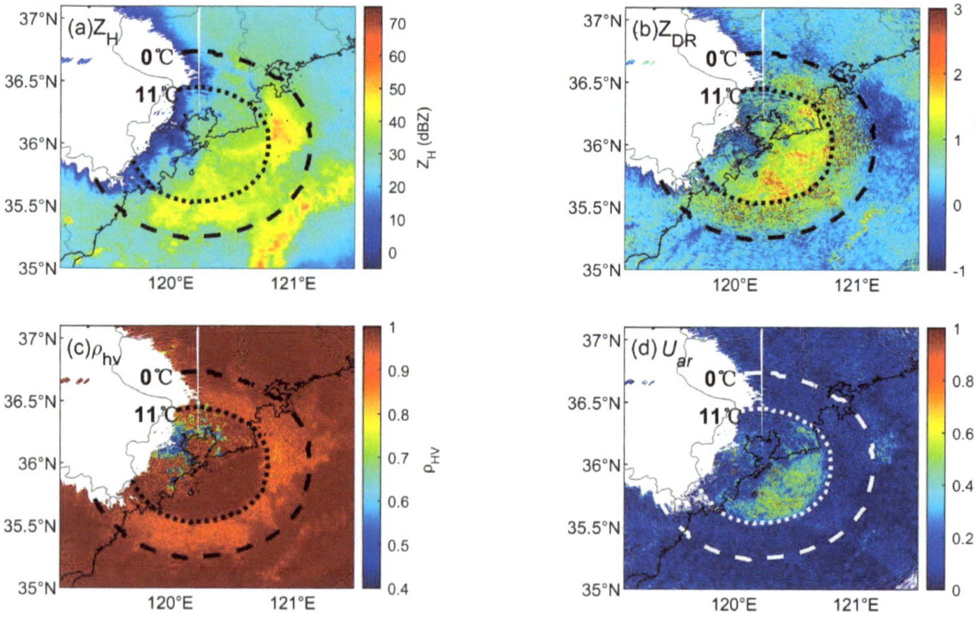

Figure 15. Sample PPI observations with BB from the S-band dual polarization weather radar. (The data time is 17 May 2020 23:51 LST. The elevation is 2.4°, which is the third level of a volume scan). (**a**) Z_H, (**b**) Z_{DR}, (**c**) ρ_{hv}, (**d**) U_{ar}.

To be further compared with the vertical structure in Section 3.1, a composite RHI is derived by interpolation, which covers both convective and stratiform areas (Figure 16). In the stratiform cloud area (distance from 0 to 60 km), the ML is visually estimated by the BB in Z_H, Z_{DR} and ρ_{hv}, whose bottom is at the height of 11 °C, and the high U_{ar} appears below the ML. The Z_{DR} column at a distance of 100 km also corresponds to a high value area of U_{ar}. An approximate U-shaped vertical structure of U_{ar}, whose two branches are at 70 and 100 km distances (Figure 16d), is more difficult to identify compared with that in X-band RHI data (Figure 12d). This may be due to the low elevation resolution of the volume scans. However, the general performances in the X-band and S-band are similar, regardless of the scanning mode.

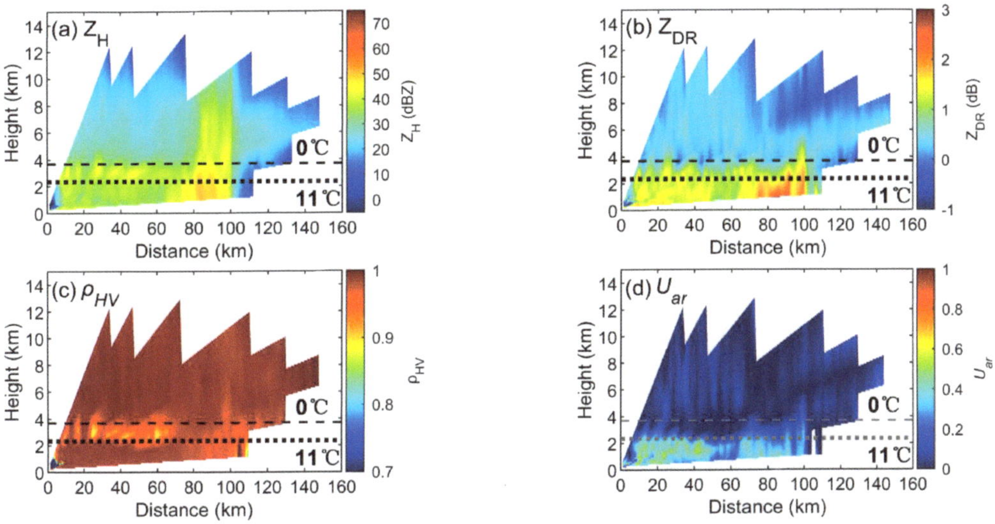

Figure 16. Composite RHI at 130° azimuth from the S-band dual polarization weather radar. (The data time is 17 May 2020 23:51 LST, the same as Figure 15. The radial data are smoothed by 10-point median filtering, and triple linear interpolation is used to derive this composite result). (**a**) Z_H, (**b**) Z_{DR}, (**c**) ρ_{hv}, (**d**) U_{ar}.

3.3. Identification Ratio of Raindrops in Stratiform Cloud Areas

One inevitable question is how accurate it is to use U_{ar} to identify raindrops. However, the accurate phase state of particles in clouds is not easy to obtain, especially in convective areas, which is also the key and difficult point in the study of weather radar remote sensing. After all, there is not always a cloud-penetrating detection by aircraft to make a space–time continuous observation. Therefore, a relatively reliable method is selected. Aiming at stratiform areas with ML, the bottom height of the visual ML boundary is selected as the dividing line. Parts below the ML bottom are divided into rain layers, while other parts are divided into nonrain layers (mixed and ice phases), so that the identification results can be examined quantitatively. The data in Figure 10 (Section 3.1) are selected as Case 1, with the height of 9 °C as the dividing line; the data in Figure 16 (Section 3.2) are selected as Case 2, with the height of 11 °C as the dividing line. Weak echoes with Z_H less than 20 dBZ are ignored in the statistics. The identification ratio of rain layers (S_{rain}), nonrain layers ($S_{nonrain}$) and overall ratio (S_{total}) are defined as follows:

$$S_{rain} = \frac{T_r}{N_r} \quad (26)$$

$$S_{nonrain} = \frac{T_i}{N_i} \quad (27)$$

$$S_{total} = \frac{T_r + T_i}{N_r + N_i} \quad (28)$$

where N_r is data count in rain layers; Ni is data count in nonrain layers; T_r is data count of the correct identification of the rain layer, where a data point has a U_{ar} larger than the given threshold and is located in rain layers; T_i is data count of correct identification of nonrain layer, where a data point has a U_{ar} equal to or less than the given threshold and is located in nonrain layers.

The results (Table 3) show that S_{rain} decreases as the U_{ar} threshold increases, probably because small raindrops have a smaller U_{ar} and cannot be identified, while the trend in $S_{nonrain}$ is the opposite. This results in a maximum value for S_{total}, where U_{ar} = 0.3 for Case 1 and U_{ar} = 0.2 for Case 2. In general, a more than 95% overall identification accuracy can be obtained when U_{ar} is set to 0.2~0.3.

Table 3. Identification ratio of rain and nonrain layers using different thresholds of U_{ar}.

Threshold of U_{ar}	Case 1			Case 2		
	S_{total}	S_{rain}	$S_{nonrain}$	S_{total}	S_{rain}	$S_{nonrain}$
0.1	0.88	1.00	0.79	0.90	0.98	0.88
0.2	0.95	1.00	0.91	0.95	0.93	0.96
0.3	0.96	0.97	0.95	0.93	0.73	0.99
0.4	0.95	0.91	0.98	0.86	0.38	1.00
0.5	0.88	0.74	0.99	0.78	0.03	1.00

3.4. Using U_{ar} as a Mask to Compute Composite Reflectivity

Composite reflectivity (CR) is a common data product of weather radar volume scan data and is a two-dimensional image derived from the Z_H maximum at the same horizontal position in each PPI for different elevations. By using the CR, the horizontal spatial distribution of the strong reflectivity population can be quickly observed, and a preliminary judgment of precipitation can be made, where it can avoid missing information caused by a partial occlusion in a single PPI or the uncertainty of the height of the reflectivity core. Using the same case in Section 3.2, Figure 17a shows a CR, where there is a severe convective line over 60 dBZ and a large-scale stratiform cloud area at the rear (northwest side). However, due to the high value and uneven horizontal distribution of Z_H in the ML, some spots in the ML may be misjudged as convective clouds, which would lead to misinterpretations of the precipitation situation. To avoid this, the CR must be recalculated after removing the effects of ML signals.

First, there is an example in which ML signals are not successfully removed. Suppose the lower bound temperature of the ML is 5 °C by assuming some experience that is not applicable to this case. After masking data with temperatures below 5 °C, a new CR is derived (Figure 17b), where the high Z_H spots still exist in the stratiform area since the ML signal is not completely removed from the original PPI. This problem can certainly be solved if the lower bound of the ML in this example is accurately obtained at 11 °C. However, inaccuracies in the temperature profile and ML boundary detection can make a difference.

Next, U_{ar} is taken as a mask template from Z_H in each PPI. The Z_H is masked where the U_{ar} is less than a certain value, and then, the CR is calculated, which is equivalent to calculating the CR of raindrop areas. A loose threshold is applied first by taking $U_{ar} < 0.2$ as the mask template to form the new CR (Figure 17c). The intensity of the CR in the stratiform cloud region to the northwest of the strong convective line is obviously smaller and more uniform and is basically not more than 45 dBZ, which is consistent with the common features of stratiform rainfall. If a stricter threshold is used, such as by taking $U_{ar} < 0.4$ as the mask template (Figure 17d), the entire convective system becomes fragmented, and only part of the stratiform cloud region remains. The parts masked in Figure 17d may contain both smaller raindrops and particles, such as graupel and hail, that are likely to exist in the convective line. However, due to the need for other ground observation instruments to verify the identification of graupel and hail, such content is not discussed in this paper. In general, taking a loose U_{ar} threshold as a mask (e.g., $U_{ar} < 0.2$) to calculate CR can preserve the horizontal spatial distribution characteristics of most precipitation systems while masking ML signals. Furthermore, this approach is more convenient and efficient than traditional methods that require temperature input and multithreshold management to first detect the ML.

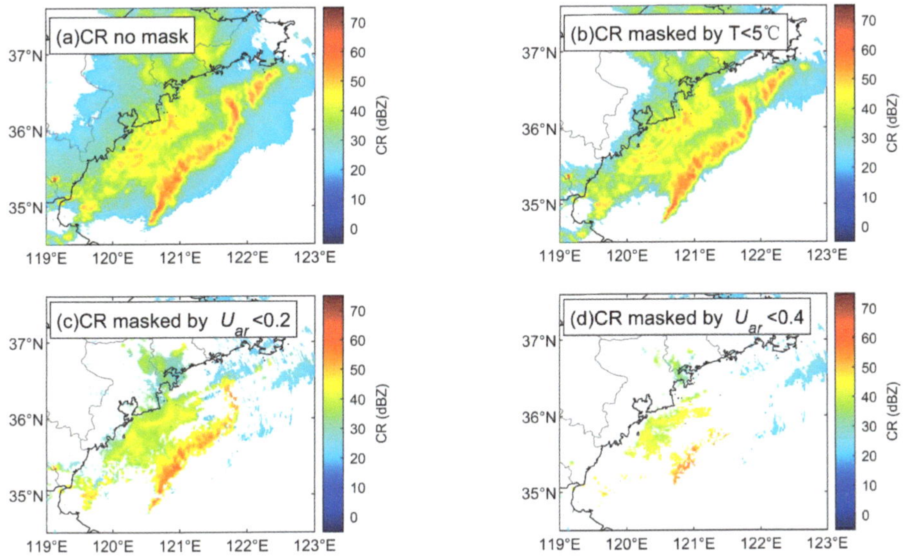

Figure 17. CR with different reflectivity masks. (The points are with Z_H > 20 dBZ. The time is the same as Figure 15. (**a**) CR no mask, (**b**) CR masked by T < 5 °C, (**c**) CR masked by U_{ar} < 0.2, (**d**) CR masked by U_{ar} < 0.4.

4. Discussion on Limitations of U_{ar}

There are also some limitations of using U_{ar} that were found in the course of this study. Here are three points to consider.

(1) If the Z_{DR} of the X-band does not undergo quality control (QC) and attenuation correction, there may be a large area of small Z_{DR} anomalies in the lower troposphere away from the radar side due to attenuation (Figure 18a). In addition, in the area where the signal-to-noise ratio is theoretically low at the end of the radar beam, there may be an abnormally large or small Z_{DR} value due to noise. These can result in a larger value and overestimate U_{ar} (Figure 18b), which inevitably affects the results of raindrop area identification. One solution is to set U_{ar} to 0 when Z_{DR} is less than 0 dB. Thus, a rough location of the raindrop areas can be obtained without waiting for a time-consuming quality control and attenuation correction process when collecting data.

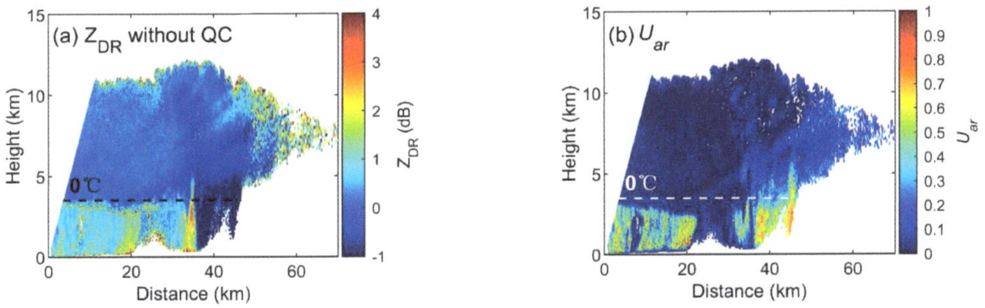

Figure 18. Sample of the impact on the U_{ar} calculation without data quality control and attenuation correction for Z_{DR} in the X-band. (**a**) Z_{DR} without QC, (**b**) U_{ar}.

(2) If there is a systematic bias in the Z_{DR} detected by the radar, it may cause U_{ar} to be unavailable. A test in which a 0.3 dB systematic bias is artificially added to Z_{DR} (Figure 19a) shows that U_{ar} in the stratiform area no longer exhibits mutation features as shown in Figure 10, but rather displays a valley band feature (Figure 19b) similar to ρ_{hv} in the ML and loses the capability to identify raindrop areas. This suggests that the premise of using U_{ar} is that the systematic deviation of Z_{DR} needs to be controlled within 0.3 dB. For these reasons, attention should be given to the calibration of dual polarization radars, especially for mobile X-band radars, when conducting field observation experiments.

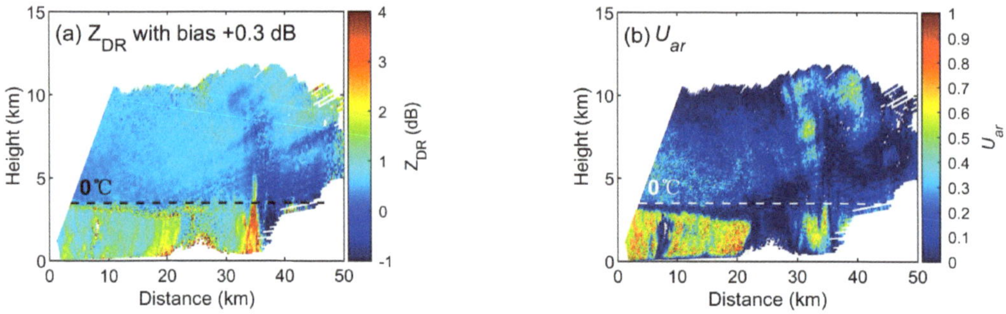

Figure 19. Sample of the impact on the U_{ar} calculation when there is a +0.3 dB systematic deviation in Z_{DR} at the X-band. (**a**) Z_{DR} with bias +0.3 dB, (**b**) U_{ar}.

(3) The premise of using U_{ar} is to eliminate or at least reduce the influence of the phase state and only to retrieve and utilize the axial ratio distribution characteristics of the particle group. However, there must be some extreme cases where the ice or snow particles exhibit a pronounced horizontal orientation under the action of the dominant wind. This could also lead to large values of U_{ar}, which could be confused with the raindrop area. However, there is no good example to illustrate this expected extreme situation, and this work needs to be carried out in depth in the future. In addition, the cases presented in this paper involve relatively low elevations, and observations at high elevations will result in a smaller Z_{DR} in the raindrop area and would need to be corrected. The impact of these factors on U_{ar} also needs to be explored in more cases.

5. Conclusions and Summary

A uniformity index for hydrometeor axis ratios (U_{ar}) derived from dual polarization weather radar data is proposed in this paper. Backscattering numerical simulations are used to find available relationships to derive U_{ar} and show its theoretical features for the identification of raindrops. Then, observation data from X-band and S-band radar are used to show and examine the performance of U_{ar} under real conditions and carry out initial applications. The main conclusions are as follows.

(1) U_{ar} is close to 0 for ice particles with varying shapes and orientations and is close to 1 for raindrops theoretically, which gives U_{ar} the ability to identify raindrop areas.

(2) In the real observations, U_{ar} is basically consistent with its theoretical feature above. A more than 95% overall identification ratio can be obtained in stratiform cloud areas when the threshold of U_{ar} is set to 0.2~0.3. Thus, the raindrop area can be more easily identified instead of identifying ML first by inputting a temperature profile and setting multiple thresholds.

(3) In the demonstration using X-band radar RHI data, high U_{ar} in the convective areas presents a U-shaped vertical structure, which indicates the process of ice particles melting into raindrops and then being transported upward by strong updraft and provides evidence for the formation mechanism of the Z_{DR} column.

(4) In the CR calculation demonstration using S-band radar, the impact of the ML signal on CR can be eliminated by setting $U_{ar} < 0.2$ as a mask template to avoid misjudging stratiform clouds at the rear of the convective line as convective clouds.

The application of U_{ar} still requires more in-depth research in the future. Due to the spatiotemporal limitations of RHI and volume scanning, the change process of the Z_{DR} column may not be fully captured. Thus, more studies are needed to better summarize the evolution of U_{ar} and other variables. Additionally, it is necessary to further evaluate and expand the application of U_{ar} methods in hydrometeor classification and in the quantitative retrieval of microphysical features.

Author Contributions: Conceptualization, Y.S.; methodology, Y.S.; software, L.F. and W.S.; validation, Y.S. and L.F.; formal analysis, Y.S.; investigation, Y.S.; resources, H.X., H.C. and H.Y. (Han Yao); data curation, H.Y. (Han Yao); writing—original draft preparation, Y.S.; writing—review and editing, H.X., H.C. and H.Y. (Huiling Yang); visualization, W.S.; supervision, H.X.; project administration, H.X.; funding acquisition, Y.S. and H.X. All authors have read and agreed to the published version of the manuscript.

Funding: This research was supported in part by the National Key Research and Development Plan of China (grant no. 2019YFC1510304), the National Natural Science Foundation of China (grant no. 42105127), and the Special Research Assistant Project of Chinese Academy of Sciences.

Data Availability Statement: Not applicable.

Conflicts of Interest: The authors declare no conflict of interest.

Appendix A

In Section 2.2 (Figure 5), R, MAE and MRE are used to evaluate the difference between $\rho_{hv}^{(Ideal)}$ and $\rho_{hv}^{(Aprrox)}$, which are defined as follows:

$$R = \frac{Cov(X, Y)}{\sqrt{Var(X) \cdot Var(Y)}} \tag{A1}$$

$$MAE = \frac{\sum_{i=1}^{n} |Y_i - X_i|}{n} \tag{A2}$$

$$MRE = 100\% \times \sum_{i=1}^{n} \left| \frac{|Y_i - X_i|}{Y_i} \right| \tag{A3}$$

where X is $\rho_{hv}^{(Aprrox)}$, Y is $\rho_{hv}^{(Aprrox)}$, $Cov(\ldots)$ denotes the covariance and $Var(\ldots)$ indicates the variance.

References

1. Bringi, V.N.; Chandrasekar, V. *Polarimetric Doppler Weather Radar: Principles and Applications*; Cambridge University Press: Cambridge, UK, 2001.
2. Zhang, G. *Weather Radar Polarimetry*; CRC Press, Inc.: Boca Raton, FL, USA, 2016.
3. Ryzhkov, A.V.; Zrnic, D.S. *Radar Polarimetry for Weather Observations*; Springer Nature Switzerland AG: Cham, Switzerland, 2019.
4. Seliga, T.A.; Bringi, V.N. Differential reflectivity and differential phase-shift-applications in radar meteorology. *Radio Sci.* **1978**, *13*, 271–275. [CrossRef]
5. Aydin, K.; Seliga, T.A.; Balaji, V. Remote-sensing of hail with a dual linear-polarization radar. *J. Appl. Meteorol. Climatol.* **1986**, *25*, 1475–1484. [CrossRef]
6. Vivekanandan, J.; Zrnic, D.S.; Ellis, S.M.; Oye, R.; Ryzhkov, A.V.; Straka, J. Cloud microphysics retrieval using S-band dual-polarization radar measurements. *Bull. Am. Meteorol. Soc.* **1999**, *80*, 381–388. [CrossRef]
7. Straka, J.M.; Zrnic, D.S.; Ryzhkov, A.V. Bulk hydrometeor classification and quantification using polarimetric radar data: Synthesis of relations. *J. Appl. Meteorol.* **2000**, *39*, 1341–1372. [CrossRef]
8. Liu, H.P.; Chandrasekar, V. Classification of hydrometeors based on polarimetric radar measurements: Development of fuzzy logic and neuro-fuzzy systems, and in situ verification. *J. Atmos. Oceanic Technol.* **2000**, *17*, 140–164. [CrossRef]
9. Zrnic, D.S.; Ryzhkov, A.; Straka, J.; Liu, Y.D.; Vivekanandan, J. Testing a procedure for automatic classification of hydrometeor types. *J. Atmos. Oceanic Technol.* **2001**, *18*, 892–913. [CrossRef]

10. Park, H.; Ryzhkov, A.V.; Zrnic, D.S.; Kim, K.E. The Hydrometeor Classification Algorithm for the Polarimetric WSR-88D: Description and Application to an MCS. *Weather Forecast.* **2009**, *24*, 730–748. [CrossRef]
11. Dolan, B.; Rutledge, S.A. A Theory-Based Hydrometeor Identification Algorithm for X-Band Polarimetric Radars. *J. Atmos. Oceanic Technol.* **2009**, *26*, 2071–2088. [CrossRef]
12. Snyder, J.C.; Bluestein, H.B.; Zhang, G.F.; Frasier, S.J. Attenuation Correction and Hydrometeor Classification of High-Resolution X-band, Dual-Polarized Mobile Radar Measurements in Severe Convective Storms. *J. Atmos. Oceanic Technol.* **2010**, *27*, 1979–2001. [CrossRef]
13. Ryzhkov, A.V.; Kumjian, M.R.; Ganson, S.M.; Khain, A.P. Polarimetric Radar Characteristics of Melting Hail. Part I: Theoretical Simulations Using Spectral Microphysical Modeling. *J. Appl. Meteorol. Climatol.* **2013**, *52*, 2849–2870. [CrossRef]
14. Ryzhkov, A.V.; Kumjian, M.R.; Ganson, S.M.; Zhang, P.F. Polarimetric Radar Characteristics of Melting Hail. Part II: Practical Implications. *J. Appl. Meteorol. Climatol.* **2013**, *52*, 2871–2886. [CrossRef]
15. Ortega, K.L.; Krause, J.M.; Ryzhkov, A.V. Polarimetric Radar Characteristics of Melting Hail. Part III: Validation of the Algorithm for Hail Size Discrimination. *J. Appl. Meteorol. Climatol.* **2016**, *55*, 829–848. [CrossRef]
16. Zhang, G.F.; Mahale, V.N.; Putnam, B.J.; Qi, Y.C.; Cao, Q.; Byrd, A.D.; Bukovcic, P.; Zrnic, D.S.; Gao, J.D.; Xue, M.; et al. Current Status and Future Challenges of Weather Radar Polarimetry: Bridging the Gap between Radar Meteorology/Hydrology/Engineering and Numerical Weather Prediction. *Adv. Atmos. Sci.* **2019**, *36*, 571–588. [CrossRef]
17. White, A.B.; Gottas, D.J.; Strem, E.T.; Ralph, F.M.; Neiman, P.J. An automated brightband height detection algorithm for use with Doppler radar spectral moments. *J. Atmos. Oceanic. Technol.* **2002**, *19*, 687–697. [CrossRef]
18. Sanchez-Diezma, R.; Zawadzki, I.; Sempere-Torres, D. Identification of the bright band through the analysis of volumetric radar data. *J. Geophys. Res. Atmos.* **2000**, *105*, 2225–2236. [CrossRef]
19. Zhang, J.; Langston, C.; Howard, K. Brightband Identification Based on Vertical Profiles of Reflectivity from the WSR-88D. *J. Atmos. Oceanic Technol.* **2008**, *25*, 1859–1872. [CrossRef]
20. Zhang, J.; Qi, Y.C. A Real-Time Algorithm for the Correction of Brightband Effects in Radar-Derived QPE. *J. Hydrometeorol.* **2010**, *11*, 1157–1171. [CrossRef]
21. Matrosov, S.Y.; Clark, K.A.; Kingsmill, D.E. A polarimetric radar approach to identify rain, melting-layer, and snow regions for applying corrections to vertical profiles of reflectivity. *J. Appl. Meteorol. Climatol.* **2007**, *46*, 154–166. [CrossRef]
22. Giangrande, S.E.; Krause, J.M.; Ryzhkov, A.V. Automatic designation of the melting layer with a polarimetric prototype of the WSR-88D radar. *J. Appl. Meteorol. Climatol.* **2008**, *47*, 1354–1364. [CrossRef]
23. Boodoo, S.; Hudak, D.; Donaldson, N.; Leduc, M. Application of Dual-Polarization Radar Melting-Layer Detection Algorithm. *J. Appl. Meteorol. Climatol.* **2010**, *49*, 1779–1793. [CrossRef]
24. Wolfensberger, D.; Scipion, D.; Berne, A. Detection and characterization of the melting layer based on polarimetric radar scans. *Q. J. R. Meteorol. Soc.* **2016**, *142*, 108–124. [CrossRef]
25. Allabakash, S.; Lim, S.H.; Jang, B.J. Melting Layer Detection and Characterization based on Range Height Indicator-Quasi Vertical Profiles. *Remote Sens.* **2019**, *11*, 2848. [CrossRef]
26. Brandes, E.A.; Ikeda, K. Freezing-level estimation with polarimetric radar. *J. Appl. Meteorol.* **2004**, *43*, 1541–1553. [CrossRef]
27. Kumjian, M.R.; Mishra, S.; Giangrande, S.E.; Toto, T.; Ryzhkov, A.V.; Bansemer, A. Polarimetric radar and aircraft observations of saggy bright bands during MC3E. *J. Geophys. Res. Atmos.* **2016**, *121*, 3584–3607. [CrossRef]
28. Ryzhkov, A.; Zhang, P.F.; Reeves, H.; Kumjian, M.; Tschallener, T.; Tromel, S.; Simmer, C. Quasi-Vertical Profiles-A New Way to Look at Polarimetric Radar Data. *J. Atmos. Oceanic Technol.* **2016**, *33*, 551–562. [CrossRef]
29. Gorgucci, E.; Chandrasekar, V.; Bringi, V.N.; Scarchilli, G. Estimation of raindrop size distribution parameters from polarimetric radar measurements. *J. Atmos. Sci.* **2002**, *59*, 2373–2384. [CrossRef]
30. Ray, P.S. Broad-band complex refractive indexes of ice and water. *Appl. Opt.* **1972**, *11*, 1836–1844. [CrossRef]
31. Bohren, C.F.; Battan, L.J. Radar backscattering by inhomogeneous precipitation particles. *J. Atmos. Sci.* **1980**, *37*, 1821–1827. [CrossRef]
32. Mishchenko, M.I. Calculation of the amplitude matrix for a nonspherical particle in a fixed orientation. *Appl. Opt.* **2000**, *39*, 1026–1031. [CrossRef]
33. Pruppach, H.; Beard, K.V. A wind tunnel investigation of internal circulation and shape of water drops falling at terminal velocity in air. *Q. J. R. Meteorol. Soc.* **1970**, *96*, 247–256. [CrossRef]
34. Brandes, E.A.; Zhang, G.F.; Vivekanandan, J. Experiments in rainfall estimation with a polarimetric radar in a subtropical environment. *J. Appl. Meteorol.* **2002**, *41*, 674–685. [CrossRef]
35. Brandes, E.A.; Zhang, G.F.; Vivekanandan, J. Experiments in rainfall estimation with a polarimetric radar in a subtropical environment (Corrigendum). *J. Appl. Meteorol.* **2005**, *44*, 186. [CrossRef]
36. Tang, Q.; Xiao, H.; Guo, C.W.; Feng, L. Characteristics of the raindrop size distributions and their retrieved polarimetric radar parameters in northern and southern China. *Atmos. Res.* **2014**, *135*, 59–75. [CrossRef]
37. Liu, L.; Bringi, V.N.; Chandrasekar, V.; Mueller, E.A.; Mudukutore, A. Analysis of the copolar correlation-coefficient between horizontal and vertical polarizations. *J. Atmos. Oceanic Technol.* **1994**, *11*, 950–963. [CrossRef]
38. Schuur, T.; Ryzhkov, A.V.; Heinselman, P.L.; Zrnic, D.S.; Scharfenberg, K.A. Observations and Classification of Echoes with the Polarimetric WSR-88D radar. In Proceedings of the Report of the National Severe Storms Laboratory, Norman, OK, USA, 1 October 2003.

39. Ulbrich, C.W. Natural variations in the analytical form of the raindrop size distribution. *J. Clim. Appl. Meteorol.* **1983**, *22*, 1764–1775. [CrossRef]
40. Brandes, E.A.; Zhang, G.F.; Vivekanandan, J. Comparison of polarimetric radar drop size distribution retrieval algorithms. *J. Atmos. Oceanic Technol.* **2004**, *21*, 584–598. [CrossRef]
41. Anagnostou, M.N.; Anagnostou, E.N.; Vivekanandan, J.; Ogden, F.L. Comparison of two raindrop size distribution retrieval algorithms for X-band dual polarization observations. *J. Hydrometeorol.* **2008**, *9*, 589–600. [CrossRef]
42. Feng, L.; Xiao, H.; Wen, G.; Li, Z.F.; Sun, Y.; Tang, Q.; Liu, Y.N. Rain Attenuation Correction of Reflectivity for X-Band Dual-Polarization Radar. *Atmosphere* **2016**, *7*, 164. [CrossRef]
43. Sun, Y.; Xiao, H.; Yang, H.L.; Feng, L.; Chen, H.N.; Luo, L. An Inverse Mapping Table Method for Raindrop Size Distribution Parameters Retrieval Using X-band Dual-Polarization Radar Observations. *IEEE Trans. Geosci. Remote Sens.* **2020**, *58*, 7611–7632. [CrossRef]
44. Hall, M.P.M.; Cherry, S.M.; Goddard, J.W.F.; Kennedy, G.R. Rain drop sizes and rainfall rate measured by dual-polarization radar. *Nature* **1980**, *285*, 195–198. [CrossRef]
45. Feng, L.; Xiao, H.; Sun, Y. A Study on Hydrometeor Classification and Application Based on X-band Dual-polarization Radar Measurements. *Clim. Environ. Res.* **2018**, *23*, 366–386. (In Chinese)
46. Conway, J.W.; Zrnic, D.S. A study of embryo production and hail growth using dual-doppler and multiparameter radars. *Mon. Weather Rev.* **1993**, *121*, 2511–2528. [CrossRef]
47. Brandes, E.A.; Vivekanandan, J.; Tuttle, J.D.; Kessinger, C.J. A study of thunderstorm microphysics with multiparameter radar and aircraft observations. *Mon. Weather Rev.* **1995**, *123*, 3129–3143. [CrossRef]
48. Bringi, V.N.; Knupp, K.; Detwiler, A.; Liu, L.; Caylor, I.J.; Black, R.A. Evolution of a Florida thunderstorm during the convection and precipitation/electrification experiment: The case of 9 August 1991. *Mon. Weather Rev.* **1997**, *125*, 2131–2160. [CrossRef]
49. Kumjian, M.R.; Ganson, S.M.; Ryzhkov, A.V. Freezing of Raindrops in Deep Convective Updrafts: A Microphysical and Polarimetric Model. *J. Atmos. Sci.* **2012**, *69*, 3471–3490. [CrossRef]
50. van Lier-Walqui, M.; Fridlind, A.M.; Ackerman, A.S.; Collis, S.; Helmus, J.; MacGorman, D.R.; North, K.; Kollias, P.; Posselt, D.J. On Polarimetric Radar Signatures of Deep Convection for Model Evaluation: Columns of Specific Differential Phase Observed during MC3E. *Mon. Weather Rev.* **2016**, *144*, 737–758. [CrossRef]

remote sensing MDPI

Article

Evaluating Simulated Microphysics of Stratiform and Convective Precipitation in a Squall Line Event Using Polarimetric Radar Observations

Yuting Sun [1,2], Zhimin Zhou [1,*], Qingjiu Gao [2], Hongli Li [1] and Minghuan Wang [1]

1 China Meteorological Administration Basin Heavy Rainfall Key Laboratory, Hubei Key Laboratory for Heavy Rain Monitoring and Warning Research, Institute of Heavy Rain, China Meteorological Administration, Wuhan 430205, China
2 Key Laboratory of Meteorological Disaster (KLME), Ministry of Education & Collaborative Innovation Center on Forecast and Evaluation of Meteorological Disasters (CIC-FEMD), Nanjing University of Information Science & Technology, Nanjing 210044, China
* Correspondence: zhouzm@whihr.com.cn

Abstract: Recent upgrades to China's radar network now allow for polarimetric measurements of convective systems in central China, providing an effective data set with which to evaluate the microphysics schemes employed in local squall line simulations. We compared polarimetric radar variables derived by Weather Research and Forecasting (WRF) and radar forward models and the corresponding hydrometeor species with radar observations and retrievals for a severe squall line observed over central China on 16 March 2022. Two microphysics schemes were tested and were able to accurately depict the contrast between convective and stratiform regions in terms of the drop size distribution (DSD) and reproduce the classical polarimetric signatures of the observed differential reflectivity (Z_{DR}) and specific differential phase (K_{DP}) columns. However, for the convective region, the simulated DSDs in both schemes exhibited lower proportions of large drops and lower liquid water content; by contrast, for the stratiform region, the proportion of large drops was found to be too high in the Morrison (MORR) scheme. The underprediction of ice-phase processes in the convective region, particularly the riming processes associated with graupel and hail, was likely responsible for the bias toward large raindrops at low levels. In the stratiform region, raindrop evaporation in the WRF Double-Moment 6-Class (WDM6) scheme, which partially offsets the overestimation of ice-phase processes, produced ground DSDs that more closely matched the observational data, and did not exhibit the overly strong warm-rain collisional growth processes of MORR.

Keywords: polarimetric radar signatures; microphysical schemes; model evaluation; polarimetric radar forward operator

1. Introduction

The modeling of microphysical processes remains a major obstacle to achieving reliable numerical weather predictions and climate simulations, in part due to gaps in our knowledge of cloud processes [1]. A vast amount of microphysical information can be extracted using observational data from polarimetric weather radars. The dual-polarization upgrades of the S-band Weather Surveillance Radar 1988 Doppler (WSR-88D) network in the USA and the C-band network in Germany and other European countries have been successively completed [2]. China has been enhancing its operational radar network to polarimetry in recent years [3], thus providing new opportunities for the extensive evaluation of, and potential improvements to, existing microphysical parameterization schemes.

Polarimetric radar is capable of determining disparities in horizontal and vertical backscatter, along with the phase shift of propagation, correlation, and depolarization [4]. The logarithm of the ratio of the power returned from horizontally and vertically polarized backscatter, known as differential reflectivity (Z_{DR}), depends mainly on the median shape and size of hydrometeors. The specific differential phase (K_{DP}), defined as the range derivative of the differential phase shift (Φ_{DP}) between the horizontal and vertical pulse phases, is dependent on the number concentration. The copolar correlation coefficient (ρ_{hv}) is a normalized measure of the correlation of horizontal and vertical backscattered power within a radar sample volume and can help distinguish meteorological from non-meteorological targets and ice particles from rain. The polarimetric variables are susceptible to alterations in hydrometeors' shape, spatial orientation, and phase composition, which can occur during a variety of microphysical processes. Hence, these multivariate polarimetric "fingerprints" can provide insight into microphysical processes that include size sorting, breakup, aggregation, riming, melting, and hail production [5–7]. For instance, a commonly observed microphysical feature of mesoscale convective systems (MCSs) is that in convective regions ice growth is dominated by riming whereas in stratiform regions deposition and aggregation are the primary mechanisms [8–10]. Warm-rain processes such as raindrop growth from collision-coalescence and the gathering of cloud droplets are highly efficient in MCS precipitation processes in a relatively moist stable environment [11,12]. Microphysical characteristics also vary with terrain elevation and the life cycle stage of the convective system involved [13–15]. Dual-polarization radar observations have also been used to evaluate the microphysics schemes used in various numerical models using forward operators, e.g., [16–18]. The polarimetric radar forward operators, using the T-matrix algorithm, generate synthetic polarimetric observations from the model output variables and compare them with the actual observations. Studies have leveraged these radar forward operators to validate raindrop size properties simulated by microphysics schemes and to reproduce meaningful polarimetric signatures in convective storms, including a Z_{DR} column in the convective updraft region and a depressed ρ_{hv} ring in the storm-inflow region [19–21]. It has also been found that the simulated polarimetric radar variables are highly affected using different microphysics schemes with varying moments. The single-moment microphysics used in the modeling studies of Jung et al. [21] was barely able to reproduce certain signatures, such as mid-level Z_{DR} and ρ_{hv} rings; however, these features were reproduced realistically in the same model using the corresponding three-moment scheme [22] by Snyder et al. [23] and Fan et al. [5]. Likewise, different polarimetric signatures have been exploited for quantifiable examinations and in-depth analyses of liquid- and ice-phase processes [10,24–27]. For example, polarimetric fingerprints in warm clouds associated with size sorting and collision–coalescence processes are reproduced to varying extents among simulations with different microphysics schemes [26,27]. By contrast, it has been noted that the simulated polarimetric variables above the melting layer lack explicit variability and are explained in part by the limited diversity of particles in the model, as well as the inability of the T-matrix approach to replicate the polarimetric fingerprints of ice processes involving snow and graupel [20,27]. However, using retrievals of ice-water content profiles and hydrometeor types, Chen et al. [24] quantitatively evaluated the ice processes of three widely used microphysics schemes. They found that the total ice-water content values for convective regions were overpredicted, where the overestimation of graupel and snow by riming was possibly responsible for extremely large raindrops below freezing levels. These quantitative retrievals of ice properties strongly depend on assumptions regarding the habit and properties of particles that govern polarimetric scattering, and thus some uncertainties exist. Morrison et al. [1] highlighted incorporating uncertainty concerning ice density, habit, and size distribution into these retrievals. In the present study, we use the same assumptions regarding ice density and size distribution from the microphysics schemes for the polarimetric radar forward operators and uniformly adopt the assumptions established by Ryzhkov et al. [17] regarding the shape and orientation angle of ice particles to constrain uncertainty in the evaluation.

Squall lines are linearly organized MCSs that serve as useful test cases for microphysics schemes since their convective strength and precipitation organization are sensitive to cloud microphysical parameterization [5,28,29]. Previous studies have found that the simulations vary depending on the varieties of ice species and certain ice process parameterizations used [30–33]. Compared with the use of a graupel category in the simulation, the use of faster-falling hail commonly narrows the convection region [33] but limits the development of bow echoes due to the weaker surface cold pools and rear inflow jets [32]. An approach that uses one category of ice with adjustable properties to cover all types of ice species can somewhat reduce the impact of the uncertainty associated with distinguishing graupel/hail particles in MCS simulations [34–36]. The flexible characterization of ice properties, according to Zhao et al. [29], has a substantial influence on whether a transition zone arises in a simulated squall line. The impact of microphysical–dynamical feedbacks on squall line simulations has also been considered in several studies [5,37,38]. For instance, latent heating/cooling can influence vertical upward motion by directly dominating cloud-scale buoyancy distributions, which can have large impacts on MCS structures and lifetimes [37]. In addition, it remains important to constrain particle size distributions in microphysics schemes such as the popular two-moment bulk schemes, which were found to have issues with correctly representing the size distribution of rain particles through the comparison of polarimetric radar variables [20,38,39].

Previous studies have focused mainly on MCSs in mid-latitude North America and Europe and East and South China, but microphysical characteristics within MCSs vary according to the climate regime and synoptic environment. The successively upgraded dual-polarization function of the operational radar network in China can provide observational data that are maximally effective for evaluating and constraining microphysics schemes in various background environments over the East Asian monsoon region and can help to address the gaps in our understanding of the microphysics involved. In this study, we are using the new data to gain insight into the microphysics of a spring squall line over central China and to validate the microphysical process parameterization used for squall line numerical simulations. To our knowledge, few studies have identified the main microphysical processes that cause variations in polarimetric signatures in squall line precipitation over central China. Our study aims to do this, using numerical models and a forward operator. This method may inspire more analysis of cases in central China using dual-polarization radar measurements.

On 16 March 2022, a heavy rainfall event of a squall line occurred in central China. This study focuses on convective and stratiform precipitation processes during the mature stage, specifically from 1200 UTC to 1500 UTC. Section 2 introduces the observational data set and model simulation setup used in this study. It also describes the polarimetric radar operator and the methodologies used for the classification of types of rainfall and the identification of hydrometers. Section 3 describes the synoptic background and mesoscale evolution of the squall line. Microphysical characteristics of squall lines from polarimetric radar observations and numerical simulations are presented and discussed in Section 4. A summary of the findings and future effort directions are given in Section 5.

2. Observational Data and Methods

2.1. Observational Data Sets

To examine the microphysical features, observational information from an S-band polarimetric radar (S-POL) at the Huaihua site (the location of which is indicated by an unfilled triangle in Figure 1) was used. S-band polarimetric radar is part of China's operational weather radar (hereafter referred to as CINRAD-SAD) network and operates in the volume coverage pattern 21 volume-scan strategy. The volume scan consists of nine elevation angles between 0.5° and 19.5° every 6 min, with an unambiguous range (velocity) of 150 km (26.5 m/s). The S-POL data used in this study were subject to both a quality control and calibration procedure to remove non-meteorological artifacts and any systematic bias. Non-meteorological echoes for which $\rho_{hv} < 0.85$ were excluded [40]. Although some beneficial data may have been discarded due to this ρ_{hv} threshold, the overall statistical findings should remain unaffected. While stringent internal calibration procedures similar to WSR-88D are applied for CINRAD-SAD radars, small fluctuations in the differential reflectivity are still known to occur. An external calibration procedure was performed for the S-POL radar used in the study to remove any unresolved systematic bias in Z_{DR}. Since the CINRAD-SAD radars, which have a fixed observation strategy, cannot point vertically, the bias was estimated using light rain echoes at lower elevation angles, beginning at 20 km from the radar to avoid ground clutter, based on the method described by Cunningham et al. [41]. After thresholding based on ρ_{hv} and calibration, a despeckling procedure was performed following Bell et al. [42], whereby valid data from at least four contiguous range gates along the radar beam were required to remove isolated radar echoes. The K_{DP} was estimated by calculating the range derivative of Φ_{DP} following Lang et al. [43]. The Φ_{DP} was filtered using a finite-impulse response filter (in the Colorado State University (CSU) RadarTools; [44]) before computing K_{DP}.

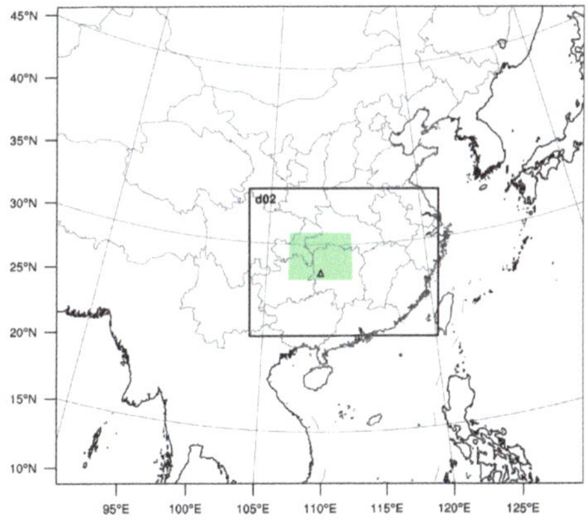

Figure 1. The model domains used. The outer region is the parent domain (9 km), and d02 is the inner domain (3 km). The innermost box, shaded green, denotes the area of the squall line system (approximately 27°–31°N, 107°–113°E), and the triangle at the bottom of the green area indicates the location of the Huaihua station.

Polarimetric variables were then interpolated from polar to Cartesian coordinates at 1 km horizontal and 500 m vertical resolution using a Barnes scheme in an open-source radar toolkit Py-ART [45]. The radius of influence of the interpolation grows with the distance from the S-POL radar, based on a minimum radius of influence of 1 km and a virtual beam width of 0.957° (equal to the S-POL beam width). The gridded data make it simple to compare the polarimetric signatures to their modeled counterparts.

Radiosonde data at the S-POL site can provide temperature profiles for hydrometeor classification and characterize environmental conditions for convective initiation. Additional observations used for evaluation were three-dimensional radar mosaic reflectivity, surface precipitation, and temperature. The former combined all available radars in the study region, with a grid spacing of 0.01° ($\sim$1 km) horizontally and varying vertical resolutions (0.5–19 km) updated every 6 min. Surface precipitation and temperature data were gathered from thousands of weather stations.

2.2. Model Setup

Simulations of the selected squall line event were conducted employing the Weather Research and Forecasting (WRF) Model version 4.2 [46] with a 9–3 km two-way nested domain configuration (see Figure 1 for domain coverage and Table 1 for a summary of configuration options). The ERA5 reanalysis data [47] with 3 h intervals and 0.25° grid resolution were used to generate initial and lateral boundary conditions. Numerical experiments were integrated from 0000 UTC to 1800 UTC on 16 March 2022 for 18 h, with the model outputs saved at 15 min intervals. Spectral nudging [48] was applied to the long-wave spectral regimes (wavelength > 1000 km) of temperature, geopotential height, and wind fields at every integral time step above the planetary boundary layer (PBL) over the WRF outer domain to keep the simulated state close to the ERA5 data at large scales [20].

In the present study, two widely used bulk microphysics schemes in WRF, double-moment six-class scheme WDM6 [49] and Morrison scheme MORR [28], are tested, and both simulations are compared with radar observational data. The WDM6 scheme is not truly double moment and only prognoses both the mass and number mixing ratios for cloud droplets and rain, and the mass mixing ratio of ice, snow, and graupel. As opposed to WDM6, the MORR scheme predicts mass and number mixing ratios for rain, cloud ice, snow, and graupel and the mass mixing ratio of cloud droplets. In WDM6, the size distributions in rain and cloud droplets are represented by a generalized four-parameter gamma function; ice microphysics are identical to those in the corresponding single-moment scheme following Hong et al. [50]. In MORR, the size distribution of cloud droplets is assumed to be in the form of a gamma function with a variable shape parameter [51], while the other hydrometeors conform to the inverse-exponential size distribution. Despite the variations in rain size distributions for the WDM6 and MORR schemes, the rain evaporation scheme is kept consistent between different schemes. The relationship between fall speed and size for graupel/hail varies between the two schemes, but the melting scheme for graupel/hail remains the same.

Table 1. Summary of WRF configuration options.

Configuration Options		
Domains	parent domain	d02
Grid points	541 × 493	583 × 457
Grid spacing	9 km	3 km
Vertical layers	51 layers	
Cumulus scheme	Kain–Fritsch [52]	Turned off
PBL scheme	Mellor–Yamada–Janjic [53]	
Longwave radiation	RRTMG [54]	
Shortwave radiation	Dudhia [55]	
Surface layer	Eta similarity [56]	
Land surface	Thermal diffusion scheme [57]	
microphysics	WDM6 scheme [49] MORR scheme [28]	

Other physics parameterizations are identical between two experiments when using either the WDM6 or MORR scheme (Table 1).

It should be noted that, as an ongoing effort to improve operational forecast models in central China, this study used the same horizontal grid spacings (3 km) for simulations. We acknowledge the limitation in the comparison with 1 km-mesh radar observations and thereby concentrate more on examining stratiform and convective scale features.

2.3. Dual-Polarization Forward Operator for WRF

The forward model of the Polarimetric Radar Retrieval and Instrument Simulator (POLARRIS-f)—a radar forward operator that was developed by scientists at Colorado State University and NASA (https://earth.gsfc.nasa.gov/meso/models/polarris (accessed on 27 March 2021))—was used to simulate the S-POL variables [16].

The Polarimetric Radar Retrieval and Instrument Simulator incorporates the WRF microphysics assumptions, such as particle sizes, densities, and liquid/ice phase from different microphysics schemes. Additional uncertainties in the polarimetric estimates arise due to characteristics of hydrometeors that are not usually included in the model, such as aspect ratio, orientation angle distributions, and dielectric constant. For this study, particle aspect ratio and canting angles uniformly followed the assumptions described by Ryzhkov et al. [17], and effective dielectric properties were estimated using the Garnett [58] mixing formula for air–ice and air–ice–liquid mixtures. However, given the intricate nature of ice shapes and spatial orientations, differences between modeled and observed data for the above-freezing-level Z_{DR} and K_{DP} are not examined.

2.4. Rainfall Type Categorization and Hydrometeor Identification

For objective model–observation comparisons, the convective/stratiform partitioning algorithm proposed by Powell et al. [59] was applied consistently to the 3 km-height gridded reflectivity data. It is an update to the method described by Steiner et al. [60], based on the horizontal radar reflectivity gradient (often referred to as a reflectivity texture scheme). The new algorithm allows for echo objects in the immediate vicinity of convective cores to be appropriately allocated to a new "mixed" category, while in the old algorithm these echoes are considered convective. Furthermore, the new algorithm more often correctly recognizes periods in which stratiform rain is absent. Hence, it is more appropriate for the study of squall lines.

In addition, a fuzzy logic hydrometeor identification (HID) algorithm in CSU Radar-Tools [43,61] was adopted to categorize hydrometeors present in the squall line. Apart from polarimetric variables, the temperature is also applied to the algorithm to separate liquid and ice regions. For model data, prognostic air temperatures were extracted from the WRF model field. For radar observations, the radiosonde data from the Huaihua site were interpolated to the S-POL analysis grids. Ten hydrometeor species were included in the study: drizzle, rain, ice crystals, aggregates, wet snow, vertically aligned ice, low-density graupel, high-density graupel, hail, and big drops.

3. Environmental Conditions and Case Description

Figure 2 shows the synoptic conditions before the squall line of interest formed. A deep trough at 500 hPa (Figure 2a) was above southwest China, along with a downstream weak ridge, which resulted in a strong southwesterly flow. In the low-level troposphere, the paralleled 850-hPa jet (20 m/s), was advecting the warm, moist air toward central China (Figure 2b). In the region in which the squall line occurred, total column water was as high as 40 kg/m^2, and the equivalent potential temperature was also high.

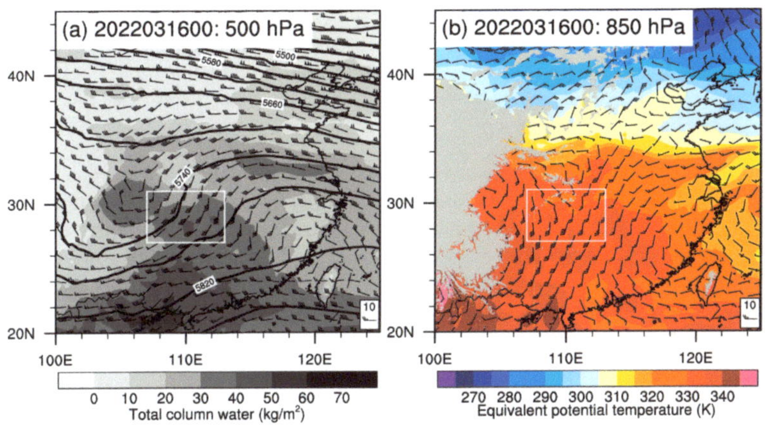

Figure 2. The ERA5 data for (**a**) 500 hPa geopotential height (solid black, contoured every 30 m) and wind barbs (m/s) superimposed on total column water (kg/m^2; grayscale shading) and (**b**) 850 hPa equivalent potential temperature (K; color shading) and wind barbs (m/s) at 0000 UTC on 16 March 2022. The gray shading in (**b**) denotes terrain height > 1.5 km. The white rectangles in (**a**,**b**) represent the area of the squall line system.

Upper-air soundings (Figure 3; the location of which is represented by an unfilled triangle in Figure 1) were also used to examine the environmental conditions associated with the storm. The total precipitable water (TWP) of 45.1 mm could be considered moderate considering the intense rainfall associated with squall line events. Compared with the corresponding values from plum rain or typhoon events, this value is fairly low and, in relative terms, is indicative of a drier environment ([11,62]. A convective available potential energy (CAPE) value as high as 1097 J/kg was recorded in Huaihua, with a lifting condensation level (Plcl) of 931.5 hPa ahead of the squall line, indicating suitable environmental conditions for convective initiation, with only a minor degree of lifting required.

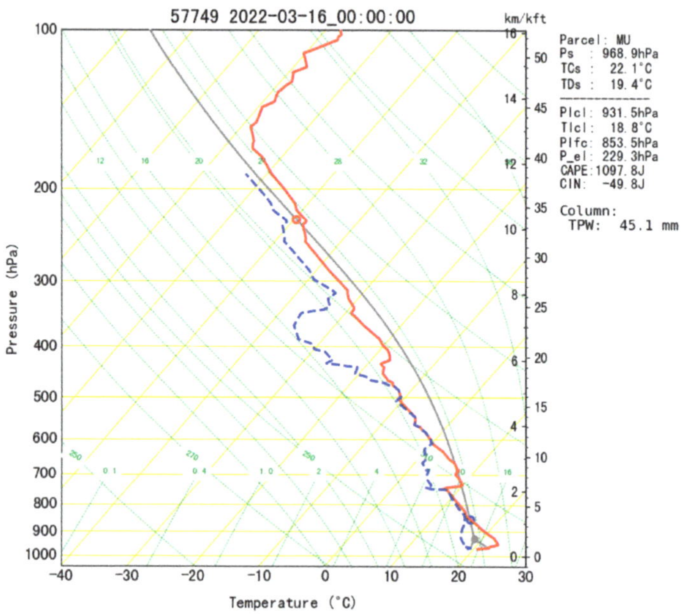

Figure 3. Skew T-logP diagram for Huaihua (triangle in Figure 1) at 0000 UTC on 16 March 2022. The red and blue lines show temperature and dewpoint temperature profiles, respectively. The gray curve represents the ascending path of the most unstable parcel. The yellow background line, sloped at a 45° angle, denotes temperature lines.

Horizontal distributions of composite reflectivity, from observations that were recorded every 6 min, distinctly illustrate the evolution of the MCSs (Figure 4). The convection line was initiated northwest of Huaihua (not shown), some distance from site, in the afternoon at ~0630 UTC (LST = UTC + 8); it developed quickly into a leading convective line accompanied by stratiform structures 2 h later (Figure 4a) and moved gradually east-southeast (Figure 4b–d). During the intensifying stage (0900–1200 UTC), the convective and stratiform areas expanded rapidly (compare Figure 4a,b). During the mature stage (1200–1500 UTC), the leading edge was marked by deep and intense convective cells, accompanied by heavy rainfall (Figure 5). The other typical features are a cold pool defined by a temperature drop of larger than 2 °C, and a transition zone with lower radar reflectivity, parallelly located right behind the leading convective line (Figure 6a–c). The linear MCS began to weaken at around 1530 UTC, and a gradual decrease in mesoscale organization was observed in the radar reflectivity data (Figure 4d). In this study, we focused on the mature stage of the squall line.

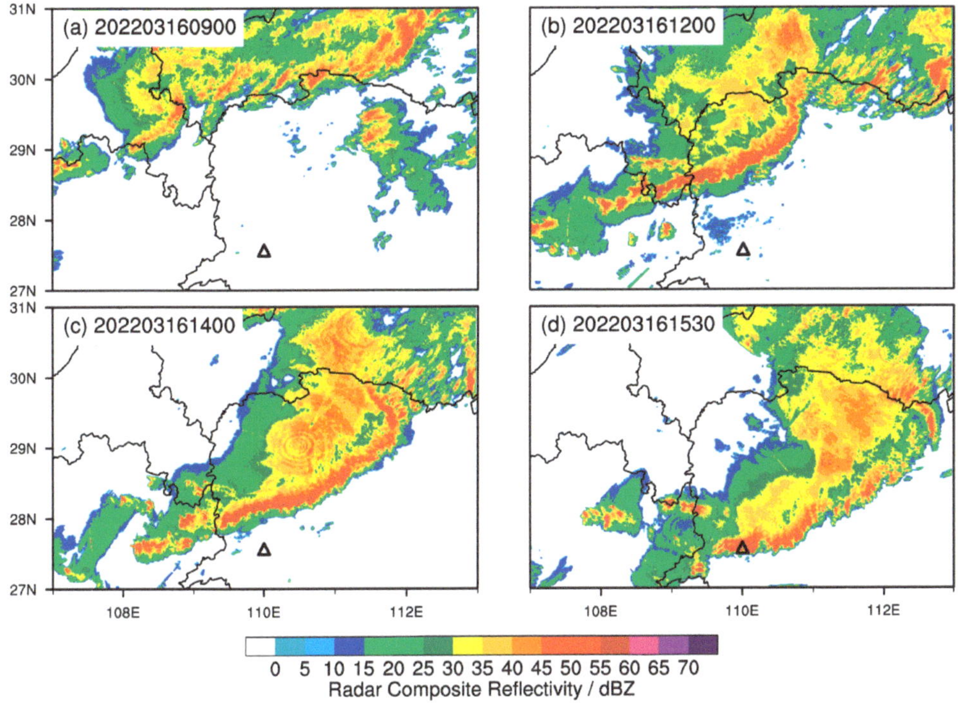

Figure 4. Radar mosaic showing radar reflectivity (dBZ) at (**a**) 0900 UTC, (**b**) 1200 UTC, (**c**) 1400 UTC, and (**d**) 1530 UTC on 16 March 2022. The triangle denotes the location of the S-POL radar.

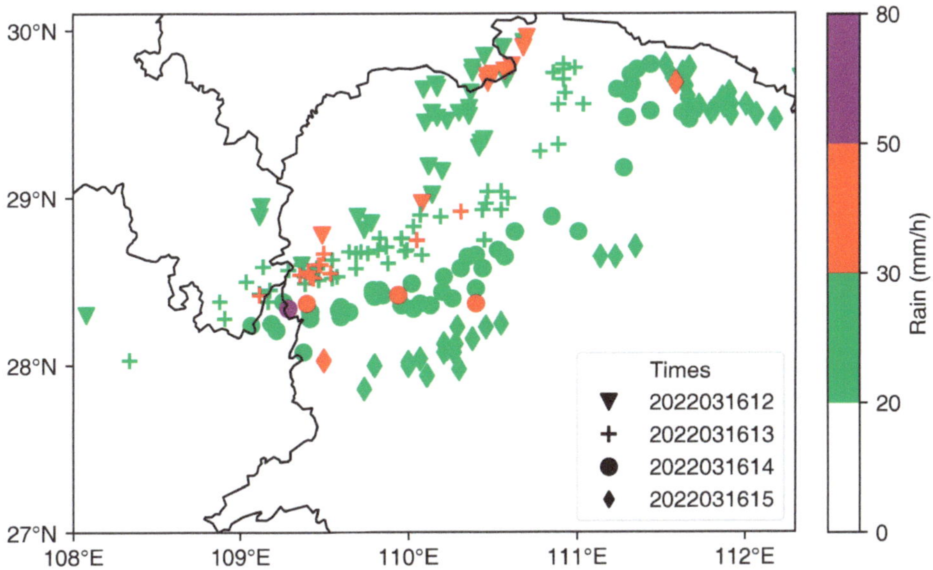

Figure 5. Spatial distribution of hourly rainfall above 20 mm/h during the mature stage of the squall line. The time values are given in the legend.

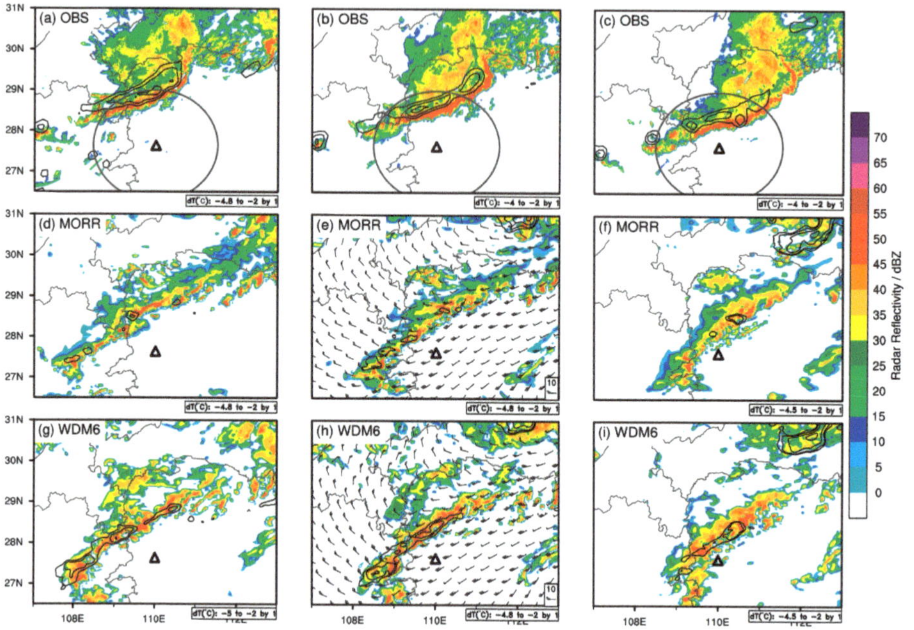

Figure 6. Temporal evolution of observed and simulated radar reflectivities (dBZ) for the 16 March 2022 squall line case at 1200 UTC (left column), 1300 UTC (middle column), and 1400 UTC (right column) at an elevation of 3 km. Black contours show the hourly temperature change, indicating the location of the surface cold pool. The triangle in each panel indicates where S-POL is situated. The dark gray circles in the first three panels indicate the 150 km range from the radar.

4. Results and Discussion

4.1. General Evaluation of the Simulated Squall Line

First, we examined the essential features of the model-simulated convective system using the 3 km composite reflectivity data and the hourly temperature variation data from 1200 to 1400 UTC. Overall, the orientation and movement of the squall line produced by both the MORR and WDM6 schemes resembled the radar observations, but modeled radar reflectivity values were generally underestimated compared with observations on the leading convective lines, especially for the MORR scheme (Figure 6d–i). Figure 6 also shows that the simulated systems advanced southeast earlier than the observed system. In terms of organizational morphology, both simulations produced a broken, less continuous, and narrower stratiform area compared with the observed uniform, continuous, and wide structure, and lacked a distinct low-reflectivity transition zone. The WDM6 scheme exhibited an hourly temperature drop of about 3 °C (Figure 6g−i), which was consistent with the cold pool intensity. However, the leading edge of the surface cold pool (the −2 °C isotherm) was not trapped behind the maximum radar reflectivity line, as seen in the observational data, but ahead of the leading convective line. The MORR scheme failed to generate a substantial cold pool; temperatures decreased by less than 2 °C (Figure 6d−f). This weak cold pool should be associated with inverse-exponential size distribution in the MORR scheme (as mentioned in Section 2.2), which dominates the rain evaporation rates.

4.2. Evaluation of Polarimetric Signatures

4.2.1. Horizontal Distributions

For comparison of synthetic polarimetric variables (horizontal reflectivity Z, Z_{DR}, and K_{DP}) with radar observations, we selected simulated samples from the Huaihua radar coverage area (dark gray circles in Figure 6a–c, 150 km radius). Figure 7 shows a snapshot of the observed and simulated Z, Z_{DR}, and K_{DP} at 1300 UTC at an elevation of 3.0 km. The area classified as a convective region is emphasized by black dots in Figure 7a,d,g. Overall, the modeled squall line showed a similar spatial distribution for all three polarimetric variables compared with the observed system, but with a narrower stratiform area and wider leading convective line. The transition zone, which is characterized by a thin gap of low Z and Z_{DR}, was not well reproduced in MORR or WDM6. The peak Z value, which exceeded 60 dBZ in the observations of the leading convective line, was underestimated in both MORR and WDM6 (~50dBZ), consistent with the characteristics illustrated in Figure 6. High Z_{DR} values (>3 dB) occurred in the leading convective line; this is a common polarimetric feature linked to size sorting [13,63], indicating the involvement of large raindrops. Low Z_{DR} (<1.5 dB) values were observed in the stratiform region (Figure 7b), in the absence of the intense updraft. The simulated Z_{DR} values for the WDM6 scheme were in general agreement with the observations, with Z_{DR} values being high near the leading convective line, and lower in areas away from it, but the high values were distributed more loosely (Figure 7b,h). In MORR, Z_{DR} showed no discernible pattern of the convective line (Figure 7e). There was considerable variation in the magnitude of the Z_{DR} values within the convective region: the simulated peak Z_{DR} value was as high as 6.0 dB for a few grid points along the leading edge, while the lowest value was only 0.2 dB for the surrounding points. Nonetheless, the difference in Z_{DR} that distinguishes the DSDs of the convective and stratiform regions was still evident in MORR.

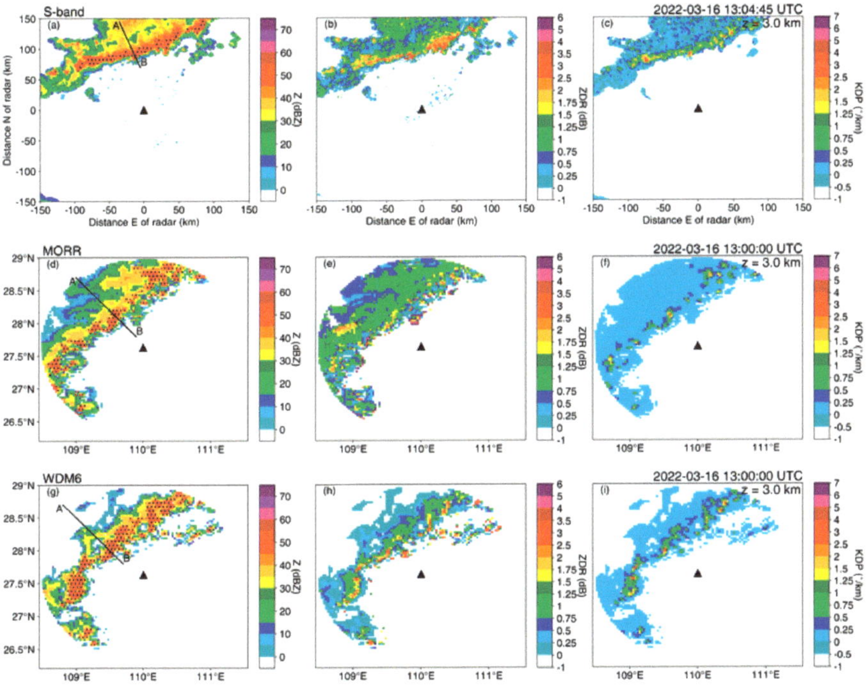

Figure 7. (**a**,**d**,**g**) Radar reflectivity (Z; dBZ), (**b**,**e**,**h**) differential reflectivity (Z_{DR}; dB), and (**c**,**f**,**i**) specific differential phase (K_{DP}; °/km) from (**a**–**c**) the S-POL radar observations and the POLARRIS-f simulations converted from the WRF output using (**d**–**f**) MORR and (**g**–**i**) WDM6 microphysics schemes, at 1300 UTC on 16 March 2022 at an elevation of 3 km. The convective region was divided based on the criteria revised by Powell et al. [59] and is indicated by black dots. The triangle in each panel indicates where S-POL is situated. Lines A-B in (a), (d), and (g) indicate cross-section lines of Figure 9.

The difference between observed and simulated values was less pronounced for K_{DP} than Z_{DR} (Figure 7c,f,i). The high K_{DP} values produced by MORR and WDM6 were roughly the same, and the convective line was reproduced, yet the simulated values were generally lower than the observed values (mostly < 1.5°/km; Figure 7c,f,i). A few more intense convective areas (Z > 55 dBZ) in the simulations had K_{DP} values higher than 2.5°/km (Figure 7f,i), which agreed more closely with the observations (Figure 7c). Throughout the stratiform region, both simulations exhibited low (but consistent) K_{DP} compared with the observations.

To statistically analyze and further evaluate the microphysics of the warm cloud layers, joint probability density functions (PDFs) in Z-Z_{DR} and Z-K_{DP} space were constructed and normalized by peaks in frequency. Each PDF encompassed 3 h of radar observations or POLARRIS-f simulations, covering the Huaihua radar coverage (150 km radius), and spanning from the surface to 3 km altitude to ensure only liquid drops. Peak frequencies of the observed Z and Z_{DR} are at 30 dBZ and 0.85 dB for stratiform samples (black contours in Figure 8a) and 50 dBZ and 2.5 dB for convective samples (colored shading in Figure 8a). For the radar observations of the convective samples, the modal distribution of Z and Z_{DR} values with frequencies higher than 30% extended from 43 dBZ to 55 dBZ, and from 1.5 to 3.2 dB, respectively, further supporting the conclusion drawn from Figure 7a,b that the clouds of the convective region are composed of larger raindrops. For convective samples in MORR, the simulated Z and Z_{DR} values with frequencies higher than 30% were confined to ~48 dBZ and ~2.2 dB (Figure 8b), indicating that the modal distributions occupy a smaller phase space compared to the observations. Despite effectively reproducing the peaks in frequency of moderate raindrops, MORR fails to capture the frequencies of over 30% of large raindrops in the top right side of the observational phase space. For observed stratiform precipitation, the distribution of Z and Z_{DR} values with frequencies over 30% extended from 15 to 38 dBZ, and from 0.2 to 1.2 dB, respectively. The MORR scheme exhibits high frequencies for large Z and Z_{DR} values that are not observed in reality; values with frequencies higher than 30% exceeded 40 dBZ and 2.0 dB, implying that the rain rate and drop sizes in the stratiform region are both exaggerated [64]. By contrast, the modal distributions of the WDM6 scheme match the radar observations more closely and exhibit a similarly broad range of Z values for both convective and stratiform regions, despite an apparent narrower distribution of Z_{DR} for a given reflectivity than the observations. The WDM6 simulations have a truncated Z_{DR}, which is in agreement with the findings of earlier studies [19], due to the use of a different shape parameter.

The specific differential phase is highly responsive to the amount of liquid water. Consistent with the K_{DP} below 0.5°/km in Figure 7c,f,i, the PDFs obtained from both observations and simulations for stratiform samples exhibit high frequencies of close to zero K_{DP} across a broad range of Z values (10 to 40 dBZ) (contours in Figure 8d–f). The low K_{DP} and similarly low Z_{DR} in both WDM6 and the observational data are indicative of a low concentration of small- to moderate-sized drops in the stratiform region. The MORR scheme showed similar K_{DP} but higher Z_{DR} values compared with the observations of the stratiform region, indicating DSDs with a small number of much larger raindrops collocated with low water content overall (Figure 8b,e). For convective samples, the modal distribution of K_{DP} values with a frequency greater than 30% ranged from 0.2 to 2.2°/km in the observations; in both simulations, these values were lower than 1.5°/km, consistent with the findings shown in Figure 7 and similar to those reported by Putnam et al. [26]. Compared with the observations, the MORR scheme exhibited narrower K_{DP} and Z_{DR} distributions for a given reflectivity and performed well for moderate convection, further demonstrating the absence of an adequate quantity of large raindrops in the convective region. Similarly, the low K_{DP} values in WDM6 indicate that there is an overall bias toward lower numbers of large raindrops and lower water contents in the two double-moment model microphysics schemes.

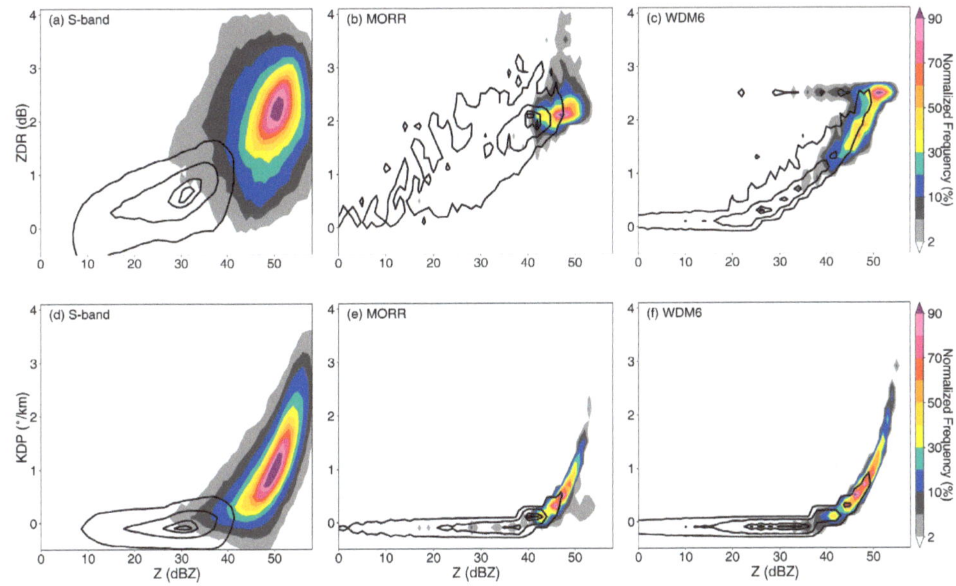

Figure 8. Joint radar reflectivity–differential reflectivity (Z-Z_{DR}; top) and reflectivity– specific differential phase (Z-K_{DP}; bottom) frequency distributions, normalized by maximum frequency, for 1200–1500 UTC (percent, color shadings from 2% to 100% for the convective region; black contours at 5, 30, 70, and 90% for the stratiform region). (**a**,**d**) The S-POL radar observations (below 3 km elevation) and POLARRIS-f simulations converted from the WRF output using the (**b**,**e**) MORR and (**c**,**f**) WDM6 microphysics schemes. The statistics from the simulations were limited to below approximately 2.75 km. Z was binned from 0 to 60 dBZ every 1 dBZ and both Z_{DR} and K_{DP} were binned from −0.5 to 4.5 dB (or °/km) every 0.05 dB (or °/km).

4.2.2. Vertical Cross Sections

Vertical cross sections oriented perpendicular to the squall line can be used to examine the structure of the system in more detail (Figure 9). The dashed black line in Figure 9 indicates the freezing level, which was obtained from a 1200 UTC sounding at the same site as S-POL and is located at ~4 km. In the cross sections of Z and Z_{DR} from S-POL (Figure 9a,b), the trailing stratiform region is split from the leading convective line by a weak echo transition zone with lower Z and Z_{DR} values. The melting layer in the stratiform part is marked by a well-pronounced abrupt drop in Z_{DR} with altitude (Figure 9b). Another notable polarimetric signature is the Z_{DR} column associated with the convective updraft, which is located at the front of the squall line and extends up to a height of 5 km (i.e., ~1 km higher than the freezing level). The maximum Z_{DR} (~3.5–4 dB) was concentrated between the surface and a height of 2 km (Figure 9b), corresponding to slightly lower K_{DP} (~1.5–2°/km) beneath an overhang structure in the K_{DP} profile (Figure 9c), which indicates a moderate concentration of large raindrops near the ground. This is supposed to be the result of size sorting, since large raindrops enlarge by acquiring cloud droplets when falling, whereas smaller drops are carried upwards by the rising air [65]. K_{DP} values in excess of 1°/km were observed at the upper part of the Z_{DR} column; this phenomenon, known as the KDP Column, is attributed to the occurrence of supercooled raindrops or water-coated hail adjacent to the updraft [66,67]. The overhang of K_{DP} below the freezing level is associated with high Z values (>60 dBZ), which is likely due to the melting of graupel and hail into raindrops, thus resulting in increased K_{DP}.

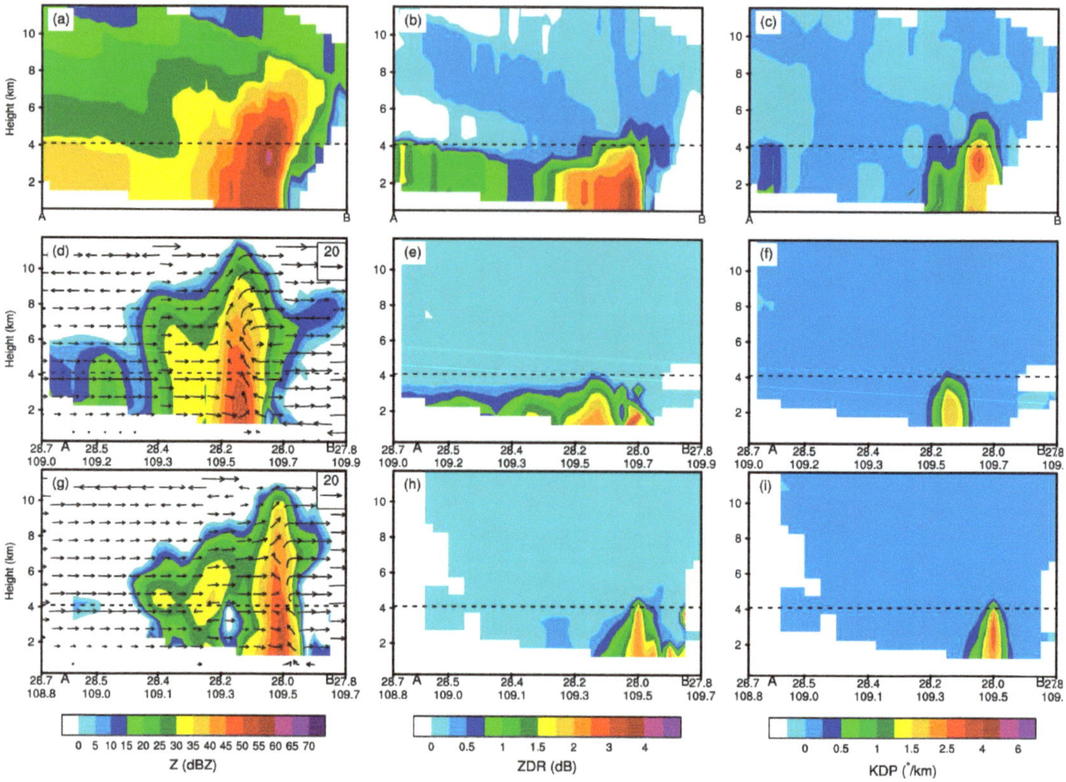

Figure 9. Cross section of (**a**,**d**,**g**) radar reflectivity (Z; dBZ), (**b**,**e**,**h**) differential reflectivity (Z_{DR}; dB), and (**c**,**f**,**i**) K_{DP} (°/km) from the (**a**–**c**) S-POL radar and (**d**–**f**) MORR and (**g**–**i**) WDM6 microphysics schemes. The dashed black line indicates the freezing level and the black arrows represent wind (m/s). The black lines in Figure 7 a,d,g indicate the locations of the cross-section plots.

The simulated Z distributions and the coupling of updrafts to high Z values (Figure 9d,g) are consistent with radar observations. Both schemes also reproduce a Z_{DR} column-like feature stretching up to the freezing level in the convective core (Figure 9e,h), though the simulated Z_{DR} columns are narrower and weaker (1.5–2.5 dB) than observed. The MORR scheme additionally exhibits a Z_{DR} peak (>3 dB; near 28.0°N, 109.7°E) near the surface ahead of the Z_{DR} column, along with a low Z value, suggesting the existence of large raindrops within a region of weak convection, where this would be unexpected. Previous studies [26,38] found similar Z_{DR} spikes in MORR and attributed them to the fact that the parameterization of the rain breakup rate in MORR depends on the value of the mixing ratio [28], so that the low efficiency of rain breakup in low-precipitation regions with a small mixing ratio could result in locally large drops. In the stratiform region, the melting layer that manifests as a decrease in Z_{DR} with height is better captured by MORR than WDM6, though its height is slightly lower than observed. The K_{DP} columns, which are co-located with the Z_{DR} columns, are located adjacent to the updraft region in MORR and WDM6 (Figure 9f,i), in agreement with expectations. The K_{DP} values for the K_{DP} column in both schemes were close to the radar observation values (0–3°/km) as with, e.g., [68,69], but neither scheme captured the distinct K_{DP} overhang structure. As with the Z_{DR} columns, the K_{DP} column in the simulations is clearly narrower than in the observations, especially in WDM6.

4.2.3. Composite Vertical Structure

The accurate temporal and spatial modeling of an observed squall line system is challenging; to further assess the effectiveness of the schemes under study, we produced a statistical comparison of the simulated polarimetric variables and the radar measurements for similar storm evolution phases (1200–1500 UTC) using contoured-frequency-with-altitude diagrams (CFADs) and composite profiles.

The CFADs of Z generated from S-POL radar observations and the MORR and WDM6 schemes for the convective and stratiform regions are shown in Figure 10. Close to the surface (i.e., below 3 km) in the convective region (Figure 10a–c), the maximum frequency contours of the MORR and WDM6 schemes (Figure 10b,c) tilt slightly to the right, with Z values increasing from ~42 dBZ at 3 km to 45–50 dBZ at the surface. Such a negative slope is not visible in the radar data (Figure 10a). Nevertheless, these simulated Z values (45–50 dBZ) near the surface are still slightly lower than the observed (50 dBZ). For the middle-upper levels (above 4 km) of the convective region, the observed and simulated Z values at the frequency maximum both decrease sharply with height. This fingerprint is associated with ice growth processes, suggesting precipitable particles increase in size by deposition, aggregation, or riming [24,70]. In the stratiform region (Figure 10d–f), the simulated Z exhibits a broader distribution throughout the altitude layer compared with observations, extending to around 45 dBZ (and >45 dBZ in WDM6) at a height of 6 km, with values between 30 and 45 dBZ occurring frequently below the freezing level, which is an overestimate of about 5 dBZ. Another distinct discrepancy between the model and radar stratiform Z CFADs occurs below 2 km, with a significantly negative slope in the frequency contours at reflectivity thresholds > 40 dBZ near the surface in simulations (Figure 10e,f).

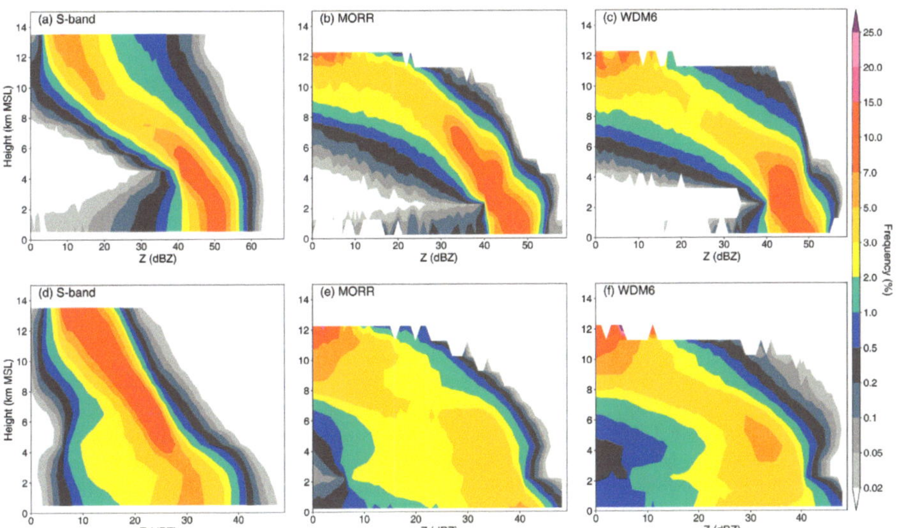

Figure 10. Contoured frequency by altitude diagrams (CFADs) for reflectivity over (top panels) convective and (bottom panels) stratiform regions, from (**a**,**d**) S-POL radar and the (**b**,**e**) MORR and (**c**,**f**) WDM6 microphysics schemes. The time period of the analysis is 1200–1500 UTC.

Figure 11a,b show the vertical profiles of median Z values and the interquartile range for each altitude layer. The median Z values for the convective region (Figure 11a) at 12 km (~10 dBZ) are nearly 10 dBZ lower than the radar value (~20 dBZ) for both MORR and WDM6, indicating the underprediction of convection by both schemes. For the stratiform region (Figure 11b), the underestimation of median Z values above the −20 °C levels is also evident, suggesting weaker deposition in MORR and WDM6 compared with observations, interpreted as a consequence of the weaker residual buoyancy [8] due to the underprediction of previous convection. However, the median Z values between the −20 °C level and the freezing level, and the increases in those values, were greater in both schemes than in the radar observations (Figure 11b). These extreme Z values within 3 km above the freezing level are consistent with the HID classification of snow aggregates and graupel in the simulations (see Section 4.3 for a detailed discussion) and indicate exaggerated aggregation and riming [8,20,68]. Also, below 2 km the median profile increased more rapidly in MORR and WDM6 than in the S-POL radar data—except for the final 1 km inversion in WDM6 (Figure 11b)—which is consistent with Figure 10e,f and could be associated with the overestimated collision–coalescence of raindrops [71].

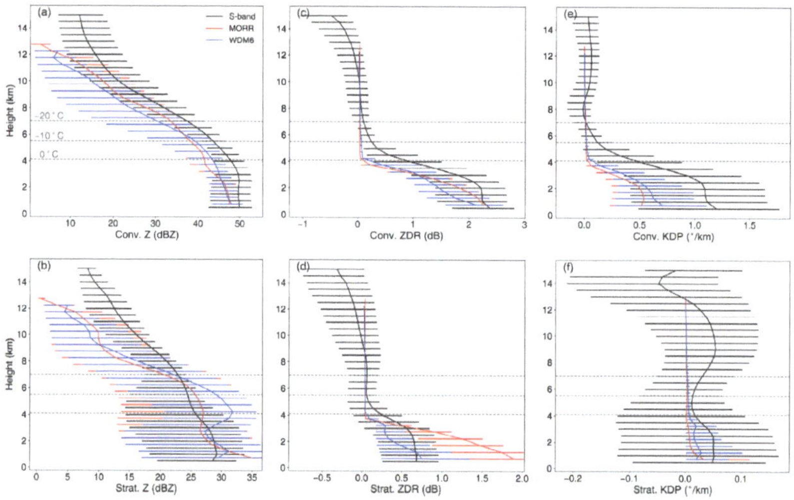

Figure 11. Comparison of (**a**,**b**) Z, (**c**,**d**) differential reflectivity (Z_{DR}), and (**e**,**f**) specific differential phase (K_{DP}) median profiles over (top) convective and (bottom) stratiform regions for S-POL radar and the MORR and WDM6 microphysics schemes in the sampling period 1200–1500 UTC. Error bars denote the interquartile range for each altitude layer. The gray dashed lines indicate the 0 °C, −10 °C, and −20 °C levels as recorded by soundings, with the lowest value at the bottom and the highest at the top.

The vertical profiles of Z_{DR} and K_{DP} median values and interquartile ranges for radar observations and the two schemes are compared in Figure 11c–f. The observed convective Z_{DR} median increases sharply from around 0.4 dB at 1 km above the freezing level to 2.2 dB at the 2 km level and then remains almost constant toward the surface (Figure 11c). The positive Z_{DR} values above the freezing level are primarily attributable to supercooled raindrops that are transported by the strong updrafts [10,27]. This is also reflected in the observed convective K_{DP} (the black line with bars in Figure 11e), which is partially related to the liquid water content, with an upper quartile value of approximately 0.4°/km at 1 km above the freezing level. Focusing on the warm cloud layers, MORR and WDM6 reproduce the observed Z_{DR} and K_{DP} profile shapes in the convective region reasonably well. It is clear that Z, Z_{DR}, and K_{DP} all markedly increase within the initial 1–2 km beneath the freezing level as a result of melting and subsequent collision–coalescence processes. Below 2 km, the increases diminish in the observations but are persistent for Z_{DR} in the two schemes and K_{DP} in WDM6. These findings suggest that the raindrop evaporation and/or breakup and coalescence processes are largely canceled out in the observational data, whereas the coalescence process is dominant in both schemes, where the conversion of cloud drops to raindrops is much more active in the WDM6 scheme, leading to an increase in K_{DP} [71,72]. However, the Z_{DR} and K_{DP} median values are consistently lower throughout the warm cloud layers in the convective region for the two schemes than in the radar observations, suggesting a significant underestimation of raindrop size and liquid water content in this region.

The most distinct features of Z_{DR} in the stratiform regions (Figure 11d) are that the Z_{DR} values and the rates of increase in median Z_{DR} (0.1–1.9 dB) within the warm cloud layers are highest in the MORR scheme, such that the Z_{DR} maximum is actually close to the typical value for convection regions (Figure 11c). As mentioned earlier (Figure 8b), this suggests that MORR exaggerates the size of raindrops in the stratiform region. Meanwhile, such a dramatic increase in Z_{DR} and a slight increase in K_{DP} below 2 km further support the conclusion evidenced by the plots shown in Figures 10d–f and 11b that the collision–coalescence of raindrops in this region is overestimated in the MORR scheme. The rate of Z_{DR} increase in WDM6 is also higher than that of the radar observations, indicating more efficient collision–coalescence processes; however, Z_{DR} values are underestimated by about 0.3 dB for the initial 2 km beneath the freezing level in the former. The interquartile ranges of K_{DP} over stratiform regions in WDM6 exhibit a narrow distribution, peaking at around 0.1°/km with a slight negative offset in the lowest 1 km, suggesting limited liquid water content and slightly dominant evaporation processes.

4.3. Statistical Comparison of Hydrometeors

To gain further insight into the differences in microphysical characteristics, we created stacked frequency by altitude diagrams (SFADs) of HID integrated over mature storm stages for the radar and model data (Figure 12). The probability-based analysis aids in mitigating the ambiguities of the HID algorithm and ascertains the predominant hydrometeor type in specific vertical layers [10,16].

The HID SFAD for the observed convective rain (Figure 12a) is marked by a large percentage of heavily rimed particles (hail, low- and high-density graupel) above the freezing level, with the highest proportion reaching almost 100% between the –20 °C level and the freezing level and still accounting for ~15–65% above the –20 °C level. Although both the MORR and WDM6 schemes tend to produce similar distributions of rimed particles, MORR predicts negligible amounts of hail. Both MORR and WDM6 significantly underestimated the fractions of high-density graupel at the mid-levels (28% and 39% vs. 63% on average) and low-density graupel at the upper levels (17% and 16% vs. 29% on average; Figure 12a,c,e). These discrepancies suggest much more active ice-phase processes in the observed convection, especially the riming processes of ice particles collecting liquid/supercooled drops between the freezing and −10 °C levels. The more efficient riming process and subsequent melting in the S-POL observations commonly result in larger quantities of graupel and stronger raindrop growth within the melting layer, which advances high proportions of big drops (40%) below the melting layer (Figure 12a). By contrast, less than 20% of hydrometeors in the convective samples were determined to be large drops in the two schemes (Figure 12c,e). According to Leinonen and von Lerber [73], the number and size of the melted raindrops vary with the degree of riming. Rime can trap a considerable quantity of meltwater due to its porous structure, whereas if the particles are unrimed or lightly rimed the meltwater is instantly visible when the melting begins, resulting in a lower melting rate.

For the stratiform samples, ice crystals and aggregates each account for almost 50% of the hydrometeors above the freezing level in the S-POL radar observations (Figure 12b), whereas aggregates dominate in the MORR and WDM6 schemes (Figure 12d,f) except at the upper levels (above 11 km) where high fractions of ice crystals are observed. Additionally, simulations from both MORR and WDM6 tend to overestimate the graupel fractions at the mid-levels compared with the observed HID SFADs (Figure 12b,d,f; 36% and 43% vs. 7% on average) as also reflected in the biases in Z at these altitudes (Figure 11b). In stratiform regions, these rimed aggregates may be left over from collapsing convection [8]. In principle, there are three ways to induce a high bias in reflectivity if the particles are overly dense, numerous, or oversize. Due to their large size, a large fraction of snow aggregates would be expected to dominate the radar signal in this layer, but their fractions are similar among the three stratiform regions. It should be noted that the snow density assumed by the MORR and WDM6 schemes (and correspondingly in the forward simulator) is 100 kg/m^3, which is much smaller than the assumed density of more rimed graupel (400 and 500 kg/m^3). It appears likely that the overestimation of Z values in simulations is due to either the larger assumed graupel density or a larger graupel fraction. As expected, a rain layer of drizzle and moderate-sized drops (indicated by RN in Figure 12) is also observed in the stratiform region, as well as a melting layer largely composed of wet snow rather than graupel (Figure 12b). However, wet snow, an indicator of a melting layer [8], only accounts for around 10% in MORR and less than 5% in WDM6. Below the melting layer in the MORR simulation, an appreciable increase in moderate-sized drop fraction occurs at the lowest 2 km level, indicating enhancing warm rain growth processes in this layer, which may explain the distinct increases in the polarimetric radar variables at this altitude (Figure 11b,d,f).

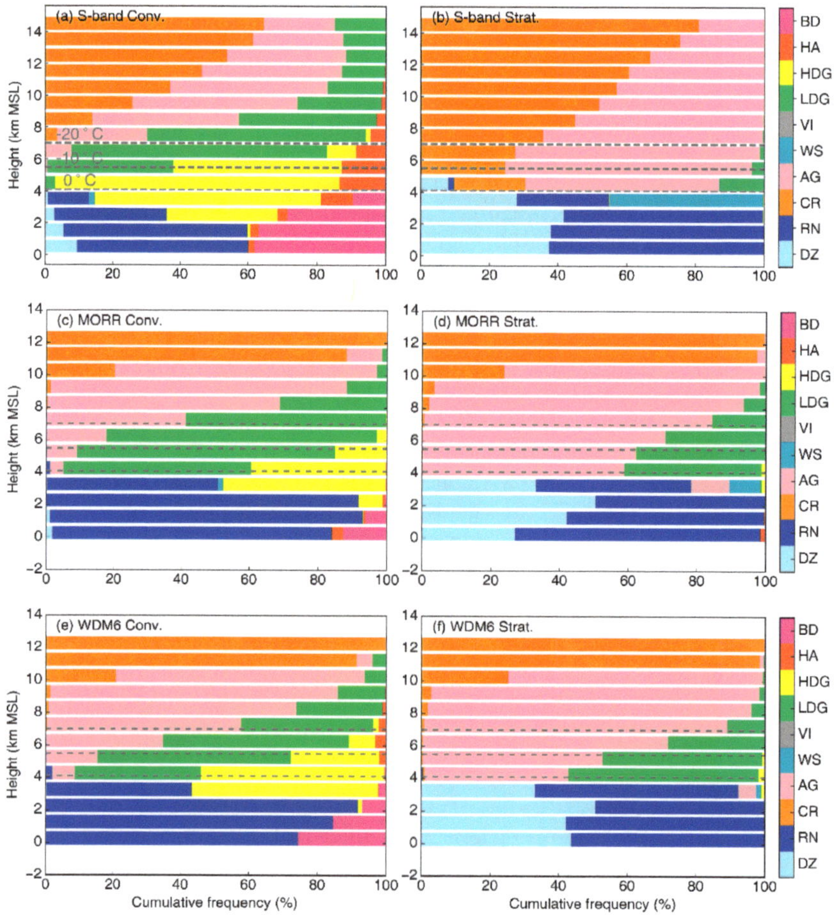

Figure 12. Hydrometeor identification frequency by height for convective (left) and stratiform (right) regions from (**a**,**b**) S-POL radar and the (**c**,**d**) MORR and (**e**,**f**) WDM6 microphysics schemes during 1200–1500 UTC. DZ = drizzle, RN = rain, CR = ice crystals, AG = aggregates, WS = wet snow, VI = vertical ice, LDG = low-density graupel, HDG = high-density graupel, HA = hail, BD = big drops/melting hail. The 0, −10, and −20 °C levels are indicated by gray dashed lines, as in Figure 11.

Note that the depth and microphysical processes of warm cloud layers of the squall line vary by region [5,38] and season [11] and can strongly influence the potential for intense rainfall. The balanced warm-rain processes for the squall line from central China, as outlined earlier, are consistent with the findings in East China [11]. However, the coalescence process is weaker because the depth of the warm cloud layer in spring is shallower than in summer (~4 km vs. 5 km). Many problems still exist in the simulation, such as the weaker intensity of the cold pool in MORR compared to WDM6. This result is different from the findings of Qian et al. [31] for a squall line in southeast China under a weak-forcing environment but is consistent with those of Zhou et al. [38] for a merger-formation bow echo under the same conditions. The source of these scheme biases requires further investigation. There are some uncertainties that could have an impact on the findings of the present study. For instance, the radar forward operator, which converts model output data into simulated polarimetric variables, contains uncertainties. Assumptions must be made regarding the aspect ratio, orientation, and shape of the particles in order to calculate scattering properties. This information is not usually included in the model and often limits the simulation of polarization variables above the melting layer.

5. Conclusions

In this study, we compared polarimetric radar variables derived by the WRF and POLARRIS-f models and the corresponding hydrometeor species with radar observations and retrievals for a severe squall line observed over central China on 16 March 2022. Two double-moment microphysics schemes, MORR and WDM6, were tested, and both simulations were able to capture the major structure of the squall line, including the leading convective line and the trailing stratiform structure and a cool pool, despite the absence of the transition zone with low Z and low Z_{DR}. However, the dominant polarimetric features and microphysical variability simulated in both schemes differed considerably from the polarimetric radar observations.

The model-simulated drop size distributions (DSDs) present realistic differences in patterns between convective and stratiform regions, but for convective regions they produce lower frequencies of large drops and lower liquid water content than observed. Conversely, the high Z and Z_{DR} values in stratiform regions indicate that drop size was overestimated in the DSDs for stratiform rain within the MORR scheme.

Both simulations reproduce the classical polarimetric signatures of the observed Z_{DR} and K_{DP} columns, which are colocated with convective updrafts, though the corresponding widths are narrower than those in the radar observations due to underpredicted convection. Weaker convective updrafts are incapable of lofting enough supercooled raindrops up to the levels of subfreezing temperatures to produce the wider Z_{DR} and K_{DP} columns as observed.

Ice-phase processes were underestimated in the convective region, particularly riming processes. The highest median Z values were observed in the radar data at all vertical levels, positive values of Z_{DR} and K_{DP} were evident above the freezing level, and the highest fractions of graupel and hail were observed between the −20 °C level and the freezing level, indicating active riming processes of ice particles collecting liquid/supercooled drops. By contrast, Z values were lowest in MORR between the −20 °C and 0 °C levels, followed by WDM6; this underestimation of Z is consistent with the results obtained from the comparison of the degree of riming, where MORR predicts negligible amounts of hail and underestimates the occurrence of high-density graupel at the mid-levels similar to WDM6. The limited degree of riming can lead to reduced fall speeds and melting rates and, consequently, sequentially lower frequencies of large drops and lower liquid water content in WDM6 and MORR than the radar observations, even though warm-rain collision–coalescence processes are overestimated in both schemes.

As for stratiform rain, both schemes underpredict deposition above the −20 °C levels, producing a lower ice crystal fraction and lower Z values, and are unable to reproduce a melting layer consisting mainly of wet snow as observed. We believe that the overprediction of Z values associated with graupel between the −20 °C level and the freezing level is a consequence of higher assumed graupel densities in MORR and WDM6 or larger graupel fractions. This is also an important determinant in the overprediction of raindrop size by MORR below the melting layer. In addition, collisional growth in warm-rain processes is slightly dominant in MORR, while the raindrop evaporation behavior in WDM6 partially offsets the overestimation of ice-phase processes and produces ground DSDs that more closely resemble the radar observations.

Author Contributions: Conceptualization, Y.S.; methodology, Y.S.; Software, Y.S.; validation, Y.S. and Z.Z.; formal analysis, Y.S.; investigation, Y.S. and M.W.; resources, H.L.; writing—original draft preparation, Y.S.; writing—review and editing, Q.G. and Z.Z.; visualization, Y.S.; supervision, Z.Z. and Q.G. All authors have read and agreed to the published version of the manuscript.

Funding: This research was funded by the National Natural Science Foundation of China (grant number 42230612), the Joint Open Project of KLME and CIC-FEMD, NUIST (grant number KLME202211) and the Special Program for Innovation and Development of China Meteorological Administration (CXFZ2022J010).

Data Availability Statement: The polarimetric radar data from the operational S-band radar in Huaihua are available for research from the China Meteorological Administration (CMA) upon request. The radar mosaic reflectivity, radiosonde, and surface data presented in this paper are available from the authors upon request. The data used for the model runs are available from ERA5 web interface (https://cds.climate.copernicus.eu/cdsapp#!/search?text=ERA5 (accessed on 1 January 2021)).

Acknowledgments: The authors thank the support of the open-source radar library Py-ART, CSU RadarTools (https://github.com/CSU-Radarmet/CSU_RadarTools.git (accessed on 27 March 2021)) and POLARRIS-f (https://earth.gsfc.nasa.gov/meso/models/polarris (accessed on 27 March 2021)) regarding the processing of radar data and the simulation of radar variables, respectively. We acknowledge Anwei Lai and Muyun Du (Institute of Heavy Rain, CMA) for their valuable suggestions.

Conflicts of Interest: The authors declare no conflict of interest.

References

1. Morrison, H.; Lier-Walqui, M.; Fridlind, A.M.; Grabowski, W.W.; Harrington, J.Y.; Hoose, C.; Korolev, A.; Kumjian, M.R. Milbrandt, J.A.; Pawlowska, H.; et al. Confronting the Challenge of Modeling Cloud and Precipitation Microphysics. *J. Adv Model. Earth Syst.* **2020**, *12*, e2019MS001689. [CrossRef] [PubMed]
2. Trömel, S.; Simmer, C.; Blahak, U.; Blanke, A.; Doktorowski, S.; Ewald, F.; Frech, M.; Gergely, M.; Hagen, M.; Janjic, T.; et al. Overview: Fusion of radar polarimetry and numerical atmospheric modelling towards an improved understanding of cloud and precipitation processes. *Atmos. Chem. Phys.* **2021**, *21*, 17291–17314. [CrossRef]
3. Zhao, K.; Huang, H.; Wang, M.; Lee, W.C.; Chen, G.; Wen, L.; Wen, J.; Zhang, G.; Xue, M.; Yang, Z.; et al. Recent Progress in Dual-Polarization Radar Research and Applications in China. *Adv. Atmos. Sci.* **2019**, *36*, 961–974. [CrossRef]
4. Ryzhkov, A.V.; Zrnic, D.S. *Radar Polarimetry for Weather Observations*; Springer: Cham, Switzerland, 2019; p. 486.
5. Fan, J.; Han, B.; Varble, A.; Morrison, H.; North, K.; Kollias, P.; Chen, B.; Dong, X.; Giangrande, S.E.; Khain, A.; et al. Cloud resolving model intercomparison of an MC3E squall line case: Part I—Convective updrafts. *J. Geophys. Res. Atmos.* **2017**, *122* 9351–9378. [CrossRef]
6. Li, H.; Moisseev, D.; von Lerber, A. How Does Riming Affect Dual-Polarization Radar Observations and Snowflake Shape *J. Geophys. Res. Atmos.* **2018**, *123*, 6070–6081. [CrossRef]
7. Li, H.; Tiira, J.; Von Lerber, A.; Moisseev, D. Towards the connection between snow microphysics and melting layer: Insights from multifrequency and dual-polarization radar observations during BAECC. *Atmos. Chem. Phys.* **2020**, *20*, 9547–9562. [CrossRef]
8. Barnes, H.C.; Houze, R.A. Precipitation hydrometeor type relative to the mesoscale airflow in mature oceanic deep convection of the Madden-Julian Oscillation. *J. Geophys. Res. Atmos.* **2014**, *119*, 13990–914014. [CrossRef]
9. Huang, H.; Zhao, K.; Chan, J.C.L.; Hu, D. Microphysical Characteristics of Extreme-Rainfall Convection over the Pearl River Delta Region, South China from Polarimetric Radar Data during the Pre-summer Rainy Season. *Adv. Atmos. Sci.* **2022**. [CrossRef]
10. Chen, G.; Zhao, K.; Lu, Y.; Zheng, Y.; Xue, M.; Zhang, S.; Fan, X. Variability of microphysical characteristics in the " 21 · 7 " Henan extremely heavy rainfall event. *Sci. China Earth Sci.* **2022**, *65*, 1861–1871. [CrossRef]
11. Chen, G.; Zhao, K.; Wen, L.; Wang, M.; Huang, H.; Wang, M.; Yang, Z.; Zhang, G.; Zhang, P.; Lee, W.C. Microphysical characteristics of three convective events with intense rainfall observed by polarimetric radar and disdrometer in Eastern China *Remote Sens.* **2019**, *11*, 2004. [CrossRef]

112. Chang, W.Y.; Lee, W.C.; Liou, Y.C. The kinematic and microphysical characteristics and associated precipitation efficiency of subtropical convection during SoWMEX/TiMREX. *Mon. Weather Rev.* **2015**, *143*, 317–340. [CrossRef]
113. Wen, J.; Zhao, K.; Huang, H.; Zhou, B.; Yang, Z.; Chen, G.; Wang, M.; Wen, L.; Dai, H.; Xu, L.; et al. Evolution of microphysical structure of a subtropical squall line observed by a polarimetric radar and a disdrometer during OPACC in Eastern China. *J. Geophys. Res.* **2017**, *122*, 8033–8050. [CrossRef]
114. Friedrich, K.; Kalina, E.A.; Aikins, J.; Gochis, D.; Rasmussen, R. Precipitation and cloud structures of intense rain during the 2013 great Colorado flood. *J. Hydrometeorol.* **2016**, *17*, 27–52. [CrossRef]
115. Houze, R.A. Orographic effects on precipitating clouds. *Rev. Geophys.* **2012**, *50*, 1–47. [CrossRef]
116. Matsui, T.; Dolan, B.; Rutledge, S.A.; Tao, W.K.; Iguchi, T.; Barnum, J.; Lang, S.E. POLARRIS: A POLArimetric Radar Retrieval and Instrument Simulator. *J. Geophys. Res. Atmos.* **2019**, *124*, 4634–4657. [CrossRef]
117. Ryzhkov, A.; Pinsky, M.; Pokrovsky, A.; Khain, A. Polarimetric radar observation operator for a cloud model with spectral microphysics. *J. Appl. Meteorol. Climatol.* **2011**, *50*, 873–894. [CrossRef]
118. Xie, X.; Shrestha, P.; Mendrok, J.; Carlin, J.; Trömel, S.; Blahak, U.; Bonn Polarimetric Radar forward Operator (B-PRO). CRC/TR32 Database (TR32DB). 2021. Available online: https://www.tr32db.uni-koeln.de/search/view.php?doiID=115 (accessed on 7 July 2021).
119. Brown, B.R.; Bell, M.M.; Frambach, A.J. Validation of simulated hurricane drop size distributions using polarimetric radar. *Geophys. Res. Lett.* **2016**, *43*, 910–917. [CrossRef]
120. Köcher, G.; Zinner, T.; Knote, C.; Tetoni, E.; Ewald, F.; Hagen, M. Evaluation of convective cloud microphysics in numerical weather prediction models with dual-wavelength polarimetric radar observations: Methods and examples. *Atmos. Meas. Technol.* **2022**, *15*, 1033–1054. [CrossRef]
121. Jung, Y.; Xue, M.; Zhang, G. Simulations of polarimetric radar signatures of a supercell storm using a two-moment bulk microphysics scheme. *J. Appl. Meteorol. Climatol.* **2010**, *49*, 146–163. [CrossRef]
122. Milbrandt, J.A.; Yau, M.K. A multimoment bulk microphysics parameterization. Part II: A proposed three-moment closure and scheme description. *J. Atmos. Sci.* **2005**, *62*, 3065–3081. [CrossRef]
123. Snyder, J.C.; Bluestein, H.B.; Dawson, D.T.; Jung, Y. Simulations of polarimetric, X-band radar signatures in supercells. Part II: ZDR columns and rings and KDP columns. *J. Appl. Meteorol. Climatol.* **2017**, *56*, 2001–2026. [CrossRef]
124. Chen, G.; Zhao, K.; Huang, H.; Yang, Z.; Lu, Y.; Yang, J. Evaluating Simulated Raindrop Size Distributions and Ice Microphysical Processes with Polarimetric Radar Observations in a Meiyu Front Event Over Eastern China. *J. Geophys. Res. Atmos.* **2021**, *126*, e2020JD034511. [CrossRef]
125. You, C.R.; Chung, K.S.; Tsai, C.C. Evaluating the performance of a convection-permitting model by using dual-polarimetric radar parameters: Case study of SoWMEX IOP8. *Remote Sens.* **2020**, *12*, 3004. [CrossRef]
126. Putnam, B.J.; Xue, M.; Jung, Y.; Zhang, G.; Kong, F. Simulation of polarimetric radar variables from 2013 CAPS spring experiment storm-scale ensemble forecasts and evaluation of microphysics schemes. *Mon. Weather Rev.* **2017**, *145*, 49–73. [CrossRef]
127. Shrestha, P.; Trömel, S.; Evaristo, R.; Simmer, C. Evaluation of modelled summertime convective storms using polarimetric radar observations. *Atmos. Chem. Phys.* **2022**, *22*, 7593–7618. [CrossRef]
128. Morrison, H.; Thompson, G.; Tatarskii, V. Impact of cloud microphysics on the development of trailing stratiform precipitation in a simulated squall line: Comparison of one- and two-moment schemes. *Mon. Weather Rev.* **2009**, *137*, 991–1007. [CrossRef]
129. Zhao, X.; Lin, Y.; Luo, Y.; Qian, Q.; Liu, X.; Liu, X.; Colle, B.A. A Double-Moment SBU-YLIN Cloud Microphysics Scheme and Its Impact on a Squall Line Simulation. *J. Adv. Model. Earth Syst.* **2021**, *13*, e2021MS002545. [CrossRef]
130. Wu, D.; Dong, X.; Xi, B.; Feng, Z.; Kennedy, A.; Mullendore, G.; Gilmore, M.; Tao, W.K. Impacts of microphysical scheme on convective and stratiform characteristics in two high precipitation squall line events. *J. Geophys. Res. Atmos.* **2013**, *118*, 11119–111135. [CrossRef]
131. Qian, Q.; Lin, Y.; Luo, Y.; Zhao, X.; Zhao, Z.; Luo, Y.; Liu, X. Sensitivity of a Simulated Squall Line During Southern China Monsoon Rainfall Experiment to Parameterization of Microphysics. *J. Geophys. Res. Atmos.* **2018**, *123*, 4197–4220. [CrossRef]
132. Adams-Selin, R.D.; Van Den Heever, S.C.; Johnson, R.H. Impact of graupel parameterization schemes on idealized bow echo simulations. *Mon. Weather Rev.* **2013**, *141*, 1241–1262. [CrossRef]
133. Bryan, G.H.; Morrison, H. Sensitivity of a simulated squall line to horizontal resolution and parameterization of microphysics. *Mon. Weather Rev.* **2012**, *140*, 202–225. [CrossRef]
134. Han, B.; Fan, J.; Varble, A.; Morrison, H.; Williams, C.R.; Chen, B.; Dong, X.; Giangrande, S.E.; Khain, A.; Mansell, E.; et al. Cloud-Resolving Model Intercomparison of an MC3E Squall Line Case: Part II. Stratiform Precipitation Properties. *J. Geophys. Res. Atmos.* **2019**, *124*, 1090–1117. [CrossRef]
135. Morrison, H.; Milbrandt, J.A. Parameterization of cloud microphysics based on the prediction of bulk ice particle properties. Part I: Scheme description and idealized tests. *J. Atmos. Sci.* **2015**, *72*, 287–311. [CrossRef]
136. Naeger, A.R.; Colle, B.A.; Zhou, N.; Molthan, A. Evaluating warm and cold rain processes in cloud microphysical schemes using Olympex field measurements. *Mon. Weather Rev.* **2020**, *148*, 2163–2190. [CrossRef]
137. Feng, Z.; Leung, L.R.; Houze Jr, R.A.; Hagos, S.; Hardin, J.; Yang, Q.; Han, B.; Fan, J. Structure and Evolution of Mesoscale Convective Systems: Sensitivity to Cloud Microphysics in Convection-Permitting Simulations Over the United States. *J. Adv. Model. Earth Syst.* **2018**, *10*, 1470–1494. [CrossRef]

38. Zhou, A.; Zhao, K.; Lee, W.C.; Ding, Z.; Lu, Y.; Huang, H. Evaluation and Modification of Microphysics Schemes on the Cold Poo Evolution for a Simulated Bow Echo in Southeast China. *J. Geophys. Res. Atmos.* **2022**, *127*, e2021JD035262. [CrossRef]
39. Khain, A.P.; Beheng, K.D.; Heymsfield, A.; Korolev, A.; Krichak, S.O.; Levin, Z.; Pinsky, M.; Phillips, V.; Prabhakaran, T.; Telle A.; et al. Representation of Microphysical Processes in Cloud-Resolving Models: Spectral (bin) Microphysics Versus Bul Parameterization. *Rev. Geophys.* **2015**, *53*, 247–322. [CrossRef]
40. Chen, H.; Chandrasekar, V.; Bechini, R. An Improved Dual-Polarization Radar Rainfall Algorithm (DROPS2.0): Application i NASA IFloodS Field Campaign. *J. Hydrometeorol.* **2017**, *18*, 917–937. [CrossRef]
41. Cunningham, J.G.; Zittel, W.D.; Lee, R.R.; Ice, L.; Hoban, N.P. Methods for Identifying Systematic Differential Reflectivit (Zdr) Biases on the Operational WSR-88D Network. In Proceedings of the 36th Conference on Radar Meteorology; America Meteorological Society: Brekenridge, CO, USA, 2013; Volume 9, pp. 1–24.
42. Bell, M.M.; Lee, W.C.; Wolff, C.A.; Cai, H. A solo-based automated quality control algorithm for airborne tail Doppler radar dat *J. Appl. Meteorol. Climatol.* **2013**, *52*, 2509–2528. [CrossRef]
43. Lang, T.J.; Ahijevych, D.A.; Nesbitt, S.W.; Carbone, R.E.; Rutledge, S.A.; Cifelli, R. Radar-Observed Characteristics of Precipitatin Systems during NAME 2004. *J. Clim.* **2007**, *20*, 1713–1733. [CrossRef]
44. Lang, T.; Dolan, B.; Guy, N.; Gerlach, C.A.M.; Hardin, J. *CSU-Radarmet/CSU_RadarTools: CSU_RadarTools, v1.3*; Zenodo: Genève Switzerland, 2019. [CrossRef]
45. Heistermann, M.; Collis, S.; Dixon, M.J.; Giangrande, S.; Helmus, J.J.; Kelley, B.; Koistinen, J.; Michelson, D.B.; Peura, M.; Pfaff, T et al. The Emergence of Open-Source Software for the Weather Radar Community. *Bull. Am. Meteorol. Soc.* **2015**, *96*, 117–12 [CrossRef]
46. Skamarock, C.; Klemp, B.; Dudhia, J.; Gill, O.; Liu, Z.; Berner, J.; Wang, W.; Powers, G.; Duda, G.; Barker, D.M.; et al. *A Descriptio of the Advanced Research WRF Model Version 4*; National Center for Atmospheric Research: Boulder, CO, USA, 2019. [CrossRef]
47. Hersbach, H.; Bell, B.; Berrisford, P.; Hirahara, S.; Horányi, A.; Muñoz-Sabater, J.; Nicolas, J.; Peubey, C.; Radu, R.; Schepers, D et al. The ERA5 global reanalysis. *Q. J. R. Meteorol. Soc.* **2020**, *146*, 1999–2049. [CrossRef]
48. Cha, D.-H.; Jin, C.-S.; Lee, D.-K.; Kuo, Y.-H. Impact of intermittent spectral nudging on regional climate simulation using Weathe Research and Forecasting model. *J. Geophys. Res.* **2011**, *116*, 1–11. [CrossRef]
49. Lim, K.-S.S.; Hong, S.-Y. Development of an Effective Double-Moment Cloud Microphysics Scheme with Prognostic Clou Condensation Nuclei (CCN) for Weather and Climate Models. *Mon. Weather Rev.* **2010**, *138*, 1587–1612. [CrossRef]
50. Hong, S.Y.; Dudhia, J.; Chen, S.H. A revised approach to ice microphysical processes for the bulk parameterization of clouds an precipitation. *Mon. Weather Rev.* **2004**, *132*, 103–120. [CrossRef]
51. Martin, G.M.; Johnson, D.W.; Spice, A. The Measurement and Parameterization of Effective Radius of Droplets in Warn Stratocumulus Clouds. *J. Atmos. Sci.* **1994**, *51*, 1823–1842. [CrossRef]
52. Kain, J.S. The Kain–Fritsch Convective Parameterization: An Update. *J. Appl. Meteorol.* **2004**, *43*, 170–181. [CrossRef]
53. Janjić, Z.I. The Step-Mountain Eta Coordinate Model: Further Developments of the Convection, Viscous Sublayer, and Turbulenc Closure Schemes. *Mon. Weather Rev.* **1994**, *122*, 927–945. [CrossRef]
54. Iacono, M.J.; Delamere, J.S.; Mlawer, E.J.; Shephard, M.W.; Clough, S.A.; Collins, W.D. Radiative forcing by long-lived greenhous gases: Calculations with the AER radiative transfer models. *J. Geophys. Res. Atmos.* **2008**, *113*, D13103. [CrossRef]
55. Dudhia, J. Numerical Study of Convection Observed during the Winter Monsoon Experiment Using a Mesoscale Two-Dimensiona Model. *J. Atmos. Sci.* **1989**, *46*, 3077–3107. [CrossRef]
56. Janjić, Z.I. The Step-Mountain Coordinate: Physical Package. *Mon. Weather Rev.* **1990**, *118*, 1429–1443. [CrossRef]
57. Dudhia, J. A Multi-layer Soil Temperature Model for MM5. Proceedings of Paper Presented at 6th Annual MM5 Users Worksho Boulder, CO, USA, 27–30 June 1996.
58. Garnett, J.C.M. Colours in Metal Glasses and in Metallic Films. *Philos. Trans. R. Soc. London. Ser. A Contain. Pap. A Math. Or Phy Character* **1904**, *203*, 385–420. [CrossRef]
59. Powell, S.W.; Houze, R.A.; Brodzik, S.R. Rainfall-type categorization of radar echoes using polar coordinate reflectivity dat *J. Atmos. Ocean. Technol.* **2016**, *33*, 523–538. [CrossRef]
60. Steiner, M.; Houze, R.A.; Yuter, S.E. Climatological characterization of three-dimensional storm structure from operational rada and rain gauge data. *J. Appl. Meteorol.* **1995**, *34*, 1978–2007. [CrossRef]
61. Dolan, B.; Rutledge, S.A.; Lim, S.; Chandrasekar, V.; Thurai, M. A robust C-band hydrometeor identification algorithm an application to a long-term polarimetric radar dataset. *J. Appl. Meteorol. Climatol.* **2013**, *52*, 2162–2186. [CrossRef]
62. He, Z.; Zhang, Q.; Zhao, K.; Hu, H. Initiation and Evolution of Elevated Convection in a Nocturnal Squall Line Along the Meiy Front. *J. Geophys. Res. Atmos.* **2018**, *123*, 7292–7310. [CrossRef]
63. Park, H.S.; Ryzhkov, A.V.; Zrnić, D.S.; Kim, K.-E. The Hydrometeor Classification Algorithm for the Polarimetric WSR-88 Description and Application to an MCS. *Weather Forecast.* **2009**, *24*, 730–748. [CrossRef]
64. Chen, G.; Zhao, K.; Zhang, G.; Huang, H.; Liu, S.; Wen, L.; Yang, Z.; Yang, Z.; Xu, L.; Zhu, W. Improving Polarimetric C-Ban Radar Rainfall Estimation with Two-Dimensional Video Disdrometer Observations in Eastern China. *J. Hydrometeorol.* **201** 1375–1391. [CrossRef]
65. Bringi, V.N.; Burrows, D.A.; Menon, S.M. Multiparameter Radar and Aircraft Study of Raindrop Spectral Evolution in Warm-base Clouds. *J. Appl. Meteorol. Climatol.* **1991**, *30*, 853–880. [CrossRef]

66. Hubbert, J.; Bringi, V.N.; Carey, L.D.; Bolen, S. CSU-CHILL Polarimetric Radar Measurements from a Severe Hail Storm in Eastern Colorado. *J. Appl. Meteorol.* **1998**, *37*, 749–775. [CrossRef]
67. Loney, M.L.; Zrnić, D.S.; Straka, J.M.; Ryzhkov, A.V. Enhanced Polarimetric Radar Signatures above the Melting Level in a Supercell Storm. *J. Appl. Meteorol. (1988–2005)* **2002**, *41*, 1179–1194. [CrossRef]
68. Straka, J.M.; Zrnić, D.S.; Ryzhkov, A.V. Bulk Hydrometeor Classification and Quantification Using Polarimetric Radar Data: Synthesis of Relations. *J. Appl. Meteorol.* **2000**, *39*, 1341–1372. [CrossRef]
69. Ryzhkov, A.V.; Schuur, T.J.; Burgess, D.W.; Zrnic, D.S. Polarimetric Tornado Detection. *J. Appl. Meteorol.* **2005**, *44*, 557–570. [CrossRef]
70. Sun, Y.; Dong, X.; Cui, W.; Zhou, Z.; Fu, Z.; Zhou, L.; Deng, Y.; Cui, C. Vertical Structures of Typical Meiyu Precipitation Events Retrieved from GPM-DPR. *J. Geophys. Res. Atmos.* **2020**, *125*, e2019JD031466. [CrossRef]
71. Houze, R.A., Jr. *Cloud Dynamics*, 2nd ed.; Elsevier/Academic Press: Oxford, UK, 2014; pp. 141–165.
72. Kumjian, M.R.; Prat, O.P. The impact of raindrop collisional processes on the polarimetric radar variables. *J. Atmos. Sci.* **2014**, *71*, 3052–3067. [CrossRef]
73. Leinonen, J.; von Lerber, A. Snowflake Melting Simulation Using Smoothed Particle Hydrodynamics. *J. Geophys. Res. Atmos.* **2018**, *123*, 1811–1825. [CrossRef]

remote sensing

MDPI

Article

Improving the Completion of Weather Radar Missing Data with Deep Learning

Aofan Gong [1], Haonan Chen [2] and Guangheng Ni [1,*]

[1] State Key Laboratory of Hydroscience and Engineering, Department of Hydraulic Engineering, Tsinghua University, Beijing 100084, China; gaf20@mails.tsinghua.edu.cn

[2] Electrical and Computer Engineering, Colorado State University, Fort Collins, CO 80523, USA; haonan.chen@colostate.edu

* Correspondence: ghni@tsinghua.edu.cn

Abstract: Weather radars commonly suffer from the data-missing problem that limits their data quality and applications. Traditional methods for the completion of weather radar missing data, which are based on radar physics and statistics, have shown defects in various aspects. Several deep learning (DL) models have been designed and applied to weather radar completion tasks but have been limited by low accuracy. This study proposes a dilated and self-attentional UNet (DSA-UNet) model to improve the completion of weather radar missing data. The model is trained and evaluated on a radar dataset built with random sector masking from the Yizhuang radar observations during the warm seasons from 2017 to 2019, which is further analyzed with two cases from the dataset. The performance of the DSA-UNet model is compared to two traditional statistical methods and a DL model. The evaluation methods consist of three quantitative metrics and three diagrams. The results show that the DL models can produce less biased and more accurate radar reflectivity values for data-missing areas than traditional statistical methods. Compared to the other DL model, the DSA-UNet model can not only produce a completion closer to the observation, especially for extreme values, but also improve the detection and reconstruction of local-scale radar echo patterns. Our study provides an effective solution for improving the completion of weather radar missing data, which is indispensable in radar quantitative applications.

Keywords: weather radar; missing data; data completion; deep learning

1. Introduction

Modern weather radars are powerful tools in today's real-time weather monitoring. Thanks to their high spatial resolution and short scanning interval, radars can usually obtain more comprehensive and finer-grained observations in regions than rain gauges and satellites. Despite the advantages of radars, they suffer from the data-missing problem that limits their data quality. A significant cause of radar missing data is beam blockage, which occurs when radar beams are obstructed by terrain objects like mountains and buildings, resulting in wedge-shaped blind zones behind the objects. Beam blockage is more likely to arise at low elevations, which are most useful for precipitation estimation because lower-elevation radar observations are nearer to the ground [1]. Therefore, this problem directly limits the application of radars in regions with large elevation variations, such as mountainous regions.

Plenty of methods have been explored to solve the data-missing problem mainly caused by beam blockage. A direct solution is to install the radar on the mountaintop and use negative elevation angles. This approach was proven to gain much higher detection of precipitation systems at all ranges [2,3] but was limited by the expensive cost of transportation, installation, and maintenance [4]. Another solution is to refer to a nearby unobstructed radar. However, the density of radar networks in mountainous areas is often not enough. Take the China Next Generation Weather Radar (CINRAD) network as an example. It was reviewed that for low elevations (1 or 2 km), the effective detection ranges of the CINRAD network can hardly cover each other in areas with complex terrains, such as Qinghai–Tibet Plateau and Yunnan–Guizhou Plateau [5].

Other studies shift their perspectives to unobstructed vertical observations from the same instrument. Researchers have established models to describe the change in radar reflectivity with altitude, known as the vertical profile of radar reflectivity (VPR) [6–9]. The VPR can be deduced and calibrated from various-elevation radar reflectivity and other atmospheric observations and then applied to the extrapolation of low elevations or even ground level. A limitation of VPR methods is the locality and temporal variability in VPR, which increases the difficulty of their generalization in diverse combinations of regions and seasons, especially for isolated convective storms [6,10].

The mechanism of radar beam propagation has also attracted researchers' interest. Radar beam propagation is highly influenced by the local topography, which can be described by the high-resolution digital elevation model (DEM). The obstructed reflectivity can be estimated by calculating the power loss because of beam shielding under certain atmospheric conditions based on the geometrical relationship between radar beams and topography [11–14]. DEMs can also be applied to radar beam blockage identification and, therefore, can serve as a preprocessing of VPR methods [15,16]. The limitations of DEM-based methods include the impact of anomalous propagation (such as super-refraction and sub-refraction) and microscale terrain features (such as buildings and vegetation).

Besides the above methods based on weather radar physics, researchers have also applied statistical methods to radar data interpolation and correction. Yoo et al. [17] applied the multivariate linear regression method to correct the mean-field bias of radar rain rate data. Kvasovet al. [18] proposed a bilinear interpolation method to enrich radar imaging details for better real-time radar data visualization. Foehnet al. [19] compared and evaluated several spatial interpolation methods in geostatistics for radar-based precipitation field interpolation, including inverse distance weighting, regression inverse distance weighting, regression kriging, and regression co-kriging. Although these studies have promoted the progress of radar interpolation and correction, statistical methods that can effectively improve radar data completion are still insufficient.

In recent years, deep learning (DL) has made rapid progress and has been successfully applied in various fields. Deep learning methods have shown prominent advantages over traditional methods in hydrological and meteorological applications, including runoff forecasting [20], precipitation nowcasting [21–23], quantitative precipitation estimation [24,25], cloud-type classification [26], tropical cyclone tracking [27], etc. Several researchers have made attempts to apply deep learning models in radar missing data completion. Yinet al. [10] split an occlusion area into several sections and filled the radar echoes in different sections using a multi-layer neural network with different parameters. However, the method in this study is unsuitable for large-range data-missing situations. Geiss and Hardin [28] proposed a deep generative model for solving the data-missing problem caused by beam blockage, low-level blind zone, and instrument failure. However, since the method in this study focuses more on image fidelity, it has a larger bias than traditional data completion methods in terms of data completion accuracy.

This study proposes a dilated and self-attentional UNet (DSA-UNet) model to improve the quality of radar missing data completion. The model is built based on the popular U-Net [29] and adopts dilation convolution and self-attentional modules to improve performance. Our DSA-UNet model is compared with several effective methods in relative studies, including the multivariate linear regression (MLG) method, the bilinear interpolation (BI) method, and the UNet++ GAN model [28], based on several widely used evaluation metrics. The rest of this article is organized as follows: Section 2 introduces the data and the study area. Section 3 illustrates the architecture of the DSA-UNet model, the baseline methods, the evaluation metrics, and the experimental settings. Section 4 displays and analyzes the results of the experiments. Section 5 discusses the findings from the results. The conclusions of this study are summarized in Section 6.

2. Data and Study Area

The radar observations in this study are collected from one of the S-band CINRAD radars located in Yizhuang, Beijing, China. Figure 1 illustrates the details of the Yizhaung radar. It has a maximum detection range of 460 km, a radial resolution of 1 km, and an azimuthal resolution of 1°, covering the entire area of Beijing, Tianjin, and Hebei, as well as parts of Liaoning, Shandong, Shanxi, Henan, and Inner Mongolia. It works with the VP21 volume scan mode, with a scanning interval of 6 min and a total of nine elevation angles (0.5°, 1.5°, 2.5°, 3.5°, 4.5°, 6.0°, 10.0°, 15.0°, and 19.5°). It was observed that the beam blockage mainly occurs in the northwest direction of the detection area because of the surrounding topography, from the azimuth of 255° to 5° in the elevation of 0.5°. The radar observations on rainy days during warm seasons (from May to September) from 2017 to 2019 were selected, containing 86 rainy days and 17,666 plan position indicator (PPI) scans in total.

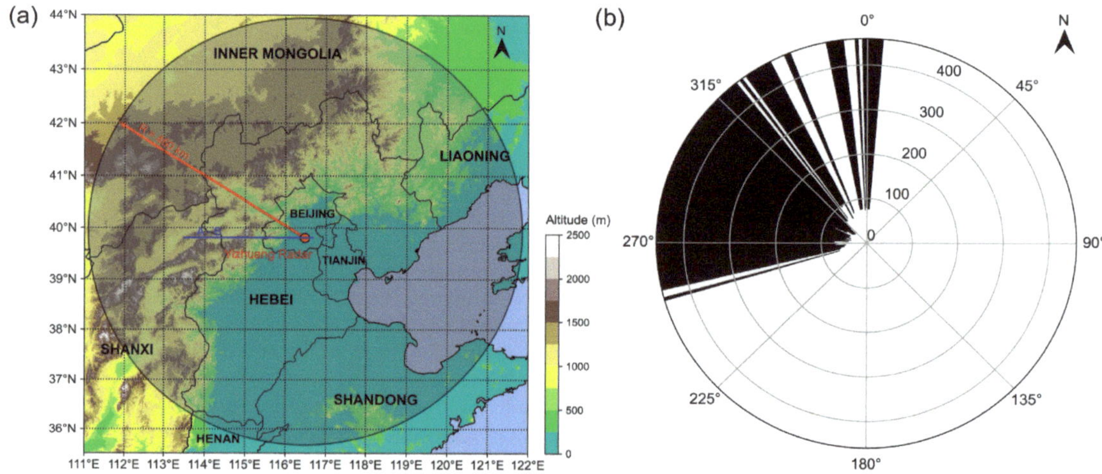

Figure 1. Details of the Yizhuang S-band radar. (**a**) The location and the detection range of the Yizhuang radar. (**b**) Beam blockage range at 0.5° of the Yizhaung radar. The radar observations at the elevation angle of 0.5° are totally blocked in the black areas from the azimuth of 255° to 5°, while in the white areas, the 0.5° observations are available.

To help understand the data-missing problem caused by beam blockage more intuitively, we plotted a rough zonal terrain profile along the latitude line passing through the Yizhuang radar (the blue line in Figure 1a, marked with "A,B"), which depicts the topography along the profile and the blockage of radar beams (Figure 2). For radar beams of higher elevation angles, such as 1.5° and 2.5°, the radar beams will never be blocked by the terrain, so the observations at these elevation angles are visible. However, radar beams of 0.5° will be blocked by a mountain with the local maximum altitude point near the radar site, which leads to a large blocked area behind the mountains and a short visible radial distance in front of the mountains.

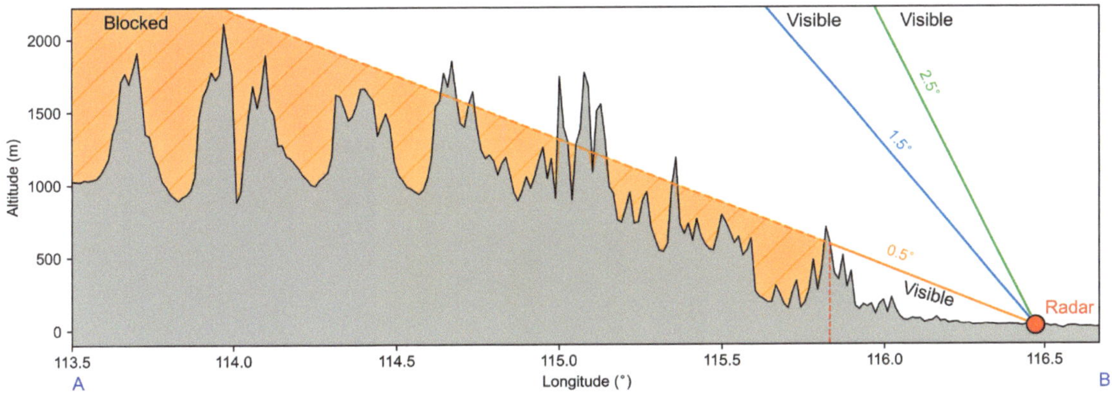

Figure 2. A rough zonal terrain profile along the latitude line passing through the Yizhuang radar. The location of the Yizhuang radar is marked with a red circle. The terrain along the profile is drawn in black lines and filled in grey. The radar beams are represented by orange, blue, and green lines, corresponding to the elevation angles of 0.5°, 1.5°, and 2.5°, respectively. The orange dashed line and the orange-filled region illustrate that radar beams of 0.5° are blocked by the mountain peak near the radar site, leading to a large data-missing area.

3. Methods

3.1. Model Architecture

The architecture of DSA-UNet is illustrated in Figure 3. The model follows the U-Net structure, which can be divided into an encoder and a decoder. The encoder accepts the multi-elevation reflectivity and the corresponding mask as its input (orange rectangle), which are compressed into a tensor with the shape of $C \times H \times W$ (channel, height, and width). The input tensor is processed with a feature recombination block (1×1 convolution, orange arrows) before being fed into four sequential downscaling blocks (red arrows). The feature maps are halved in dimensions H and W and doubled in dimension C through every downscaling block. The downscaling blocks are followed by four dilated convolution blocks (violet arrows) to capture the multi-scale information aggregation of the encoding feature maps [30]. The decoder consists of four upscaling blocks (green arrows) and a 1×1 convolution block (orange arrow), similar to the encoder. One difference is that the decoding feature maps of the first two upscaling blocks are processed with two additional self-attentional blocks (black arrows) to learn patterns using cues from all feature locations for image generation or completion [31]. Each decoding feature map is concatenated by the encoding feature map of an equal scale copied from the encoder through skip connections (grey arrows). The sizes of encoding and decoding feature maps are also marked out next to the rectangles, which will be explained in the following subsection. The DSA-UNet model has about 18.17 million trainable parameters in total.

The structures of the downscaling blocks and upscaling blocks are shown in Figure 4. The downscaling blocks consist of double convolutional blocks, each including a 3×3 convolution layer, a batch normalization (BN) layer, and a rectified linear unit (ReLU) layer. The first 3×3 convolution layer serves as the downscaling operator, which reshapes the input tensor from $C \times H \times W$ to $2C \times \frac{H}{2} \times \frac{W}{2}$. The feature maps keep the same size in the following layers. The upscaling operation in the upscaling blocks is implemented by a 3×3 transpose convolution layer instead of the 3×3 convolution layer in downscaling blocks.

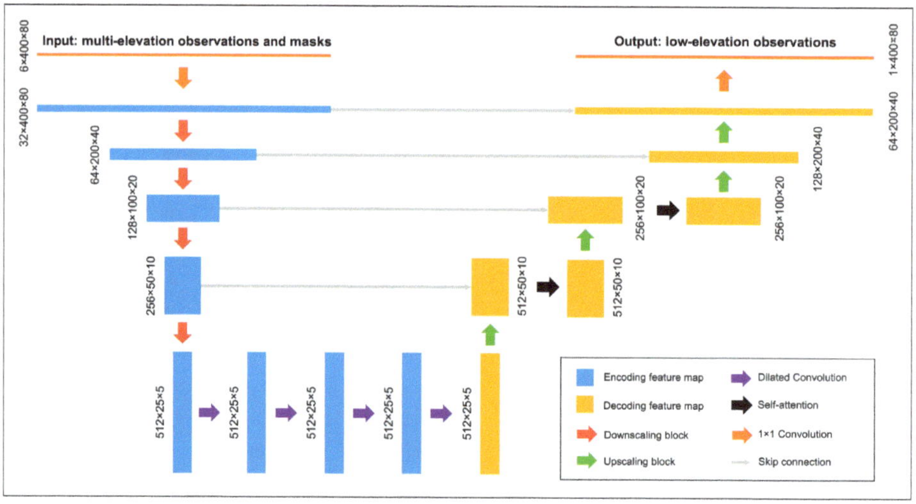

Figure 3. The architecture of DSA-UNet. The intermediate feature maps, scaling blocks, and other modules are represented with rectangles and arrows in corresponding colors. The legend can be found in the lower right corner.

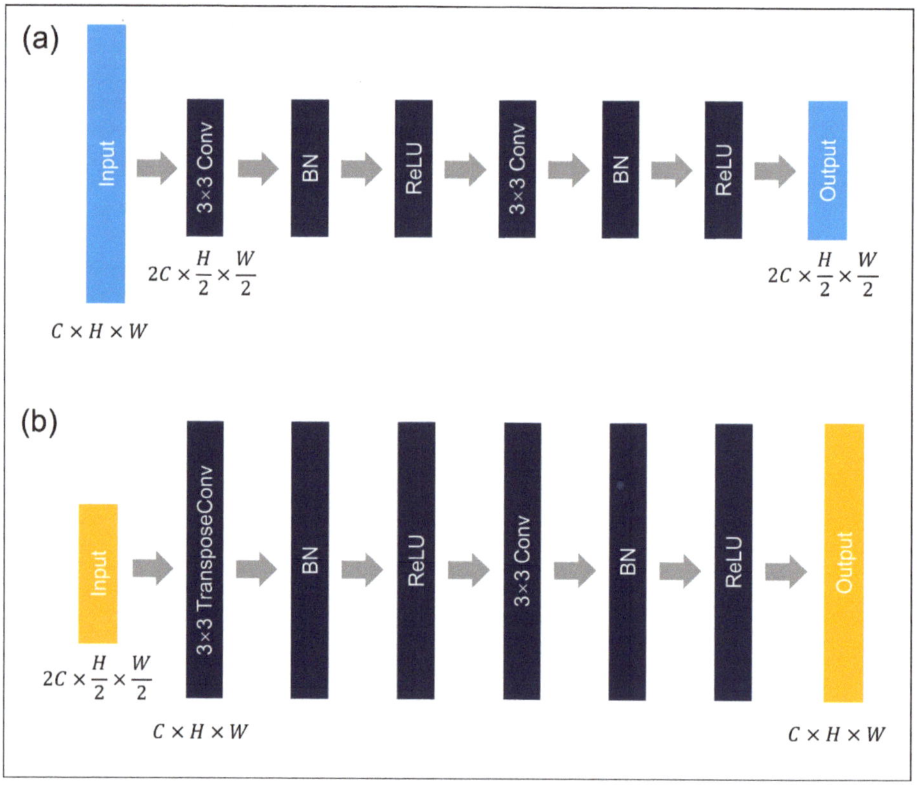

Figure 4. The structures of the downscaling blocks (**a**) and upscaling blocks (**b**). The light-blue and yellow rectangles denote the input and output feature maps, and the dark-blue rectangles denote the neural network layers.

The structure of the self-attentional blocks is shown in Figure 5. The input tensor is flattened and fed into three independent 1 × 1 convolution layers. The feature maps of the first two convolution layers are combined through a dot product and a softmax layer to generate an attention map that records a full relationship between all pixels in the spatial dimensions. It is then multiplied by the feature map from the last 1 × 1 convolution and transferred to the original size.

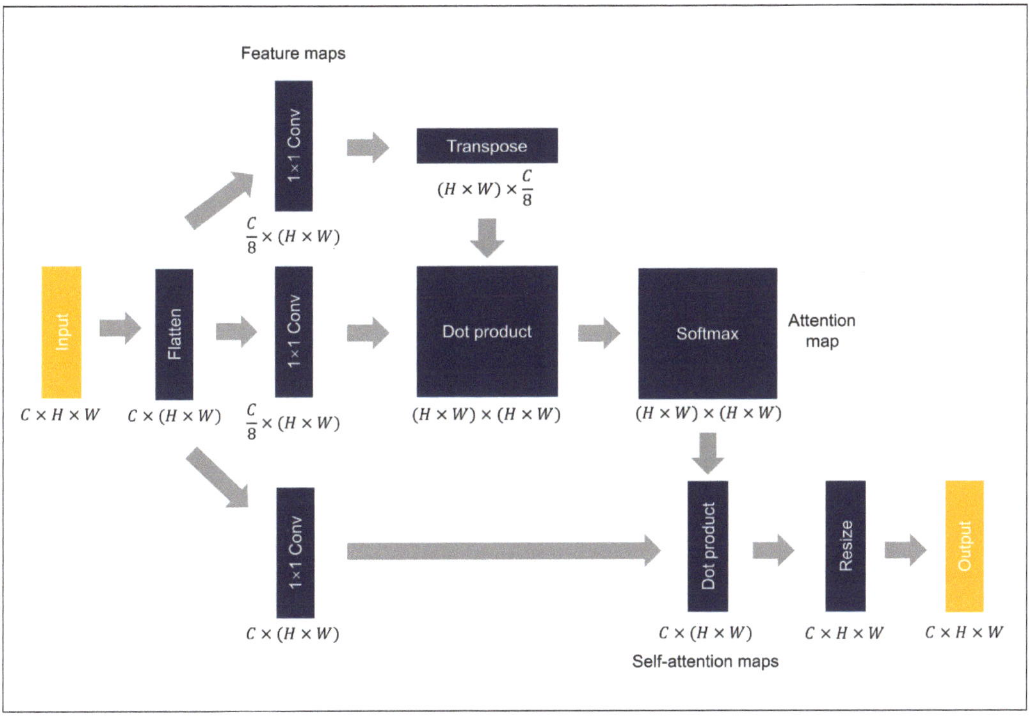

Figure 5. The structure of the self-attentional blocks. The yellow rectangles denote the input and output feature maps, and the dark-blue rectangles denote the neural network layers.

3.2. Data Processing and Dataset Construction

The radar raw data were first processed with a quality control process, which is always implemented in weather monitoring applications to eliminate systematic observation errors of radars. The processed radar data had a minimum value of −33 dBZ, which indicates clear air, and a maximum value of 69 dBZ, which indicates an extreme amount of precipitation particles. The processed radar reflectivity values were clipped between 0 and 70 dBZ and scaled to the range of 0–1 with the linear min–max normalization. The lower bound of the min–max normalization was 0 dBZ because, according to several Z-R relations, when the radar reflectivity is lower than 0 dBZ, the estimated precipitation is lower than 0.1 mm, which can be regarded as none precipitation [32]. This clipping operation has also been applied in some related studies [28,33]. As we introduced in Section 2, the original radar data have a size of 360° × 460 km with a resolution of 1° × 1 km in the polar coordinate system centered at the Yizhuang radar site. Considering that the radar beams broaden with distance and altitude and the phase of precipitation particles changes above the melting layer [10], the radar data are restricted to a maximum distance of 80 km and a maximum elevation angle of 3.5°.

The dataset for training and evaluation is built based on radar data of 1.5°, 2.5°, and 3.5° because the data quality of 0.5° observations is limited by considerable beam blockage.

Observations at these three elevation angles are inputted into the model, and the 1.5° observations are set as the training target. To simulate varying degrees of data-missing situations, a series of random sector masks are generated for the 1.5° observations, where the values of each mask are filled with zero in data-missing regions and one elsewhere. The location of the masked area is randomly determined for each 1.5° observation. The azimuthal distance of the masked area is also randomly sampled from 10° to 40°. The radial distance of the mask is fixed to 80 km. The random sector mask is also regarded as part of the input to provide necessary information about the locations of missing pixels for the neural network. The model's output is split by the mask and merged with the input, which can be described by

$$\hat{\mathbf{X}} = \mathbf{M} \odot \mathbf{X} + (1 - \mathbf{M}) \odot \mathcal{F}(\mathbf{X}), \tag{1}$$

where $\mathbf{X}$ and $\mathbf{M}$ represent the input radar image tensor and the mask; $\mathcal{F}$ represents the neural network; $\mathcal{F}(\mathbf{X})$ and $\hat{\mathbf{X}}$ represent the raw output and the final output of the neural network; and the symbol $\odot$ represents the element-wise product operation. Besides the mask of the 1.5° observation, two additional masks filled with one were also fed into the neural network to let it understand that the 2.5° and 3.5° radar observations serve as auxiliary information but not the training target. The channel sizes of the input and output tensors are 6 and 1, respectively.

In deep learning tasks, structured data are usually converted into a tensor format in a Cartesian coordinate system for convolutional neural networks. This is reasonable for convolution operations in most image completion tasks because of the translation invariance in convolution operators, while for the polar coordinate system, where the elements are in sequential order along the azimuth direction, the head (0°) and the tail (359°) elements are geographically adjacent but are far apart from each other in the tensor. Therefore, an additional padding operation to both the head and the tail of the tensor is necessary. For a tensor at a certain time step, the elements within 0°–20° are attached to the tail of the tensor; meanwhile, those within 339°–359° are padded before the head, which indicates that the azimuthal size of the tensor is expanded to 400. The radial size of the tensor is 80 since its radial distance is 80 km and its radial resolution is 1 km. Therefore, the full sizes of the input tensor and the output tensor are $6 \times 400 \times 80$ and $1 \times 400 \times 80$, respectively, as shown in Figure 3.

The random sector masks were also used to augment the dataset for better training. Each 1.5° observation is linked to four random sector masks with different locations and ranges. In this way, the size of the dataset was expanded to 4 times the original size. The augmented dataset with a total of 70,664 samples was sorted in chronological order and was then split into the training set (45,225 samples), validation set (11,306 samples), and test set (14,133 samples).

3.3. Baseline Methods

The multivariate linear regression (MLG) method, the bilinear interpolation (BI) method, and the UNet++ GAN model were selected as the baseline methods. The MLG method is one of the most basic statistical methods for multivariate correlation analysis. The BI method is a non-parametric statistical interpolation method for multi-dimensional variables, which is widely used in image processing. The UNet++ GAN model directly aims at the radar data completion task, which was designed based on the conditional generative adversarial network [34] and UNet++ [35].

For the MLG method, observations from all the elevation angles that are above 1.5° (including 2.5°, 3.5°, 4.5°, 6.0°, 10.0°, 15.0°, and 19.5°) were used to fit the 1.5° observations. For the BI method, the unmasked values in the 1.5° observations served as the source for interpolating masked values. The UNet++ GAN model shares the same dataset and input–output settings as the DSA-UNet model. Its architectures and details can be found in [28].

3.4. Training

The original L1 loss function is a commonly used training goal for deep learning models, which is defined as the average of the absolute difference between the prediction and the observation. It usually works when the data follow a balanced or regular distribution, for example, a Gaussian distribution. However, the dataset in this study was not in this condition. According to the principle of Doppler weather radars, observations with higher radar reflectivity are usually instructive to heavier precipitation or more fierce storm events. Although the data were collected during rainy days in warm seasons, the number of high-intensity values was significantly fewer than that of low-intensity values, causing the dataset to be highly unbalanced. Figure 6 describes the distribution of the raw radar reflectivity data. It can be observed that the density of radar reflectivity values over 40 dBZ was significantly less than that of values below 10 dBZ. Since the original L1 loss function was unsuitable for the highly unbalanced dataset, we designed a weighted L1 loss function for training the DSA-UNet model to overcome this problem, as shown in Equation (2). The weighted L1 loss of the prediction ($\mathbf{X}$) and the observation ($\hat{\mathbf{X}}$) is scaled by the weight ($\mathbf{W}$) determined by the observation's value. Larger weights are set for intervals with higher intensity but lower distribution density based on the characteristics of the data distribution. The correspondence between the radar reflectivity values and the weights is exhibited in Figure 6.

$$\mathcal{L}(\hat{\mathbf{X}}, \mathbf{X}; \mathbf{W}) = \underset{\rho,\theta}{\text{mean}}\, \mathbf{W} \odot \left|\hat{\mathbf{X}} - \mathbf{X}\right| \tag{2}$$

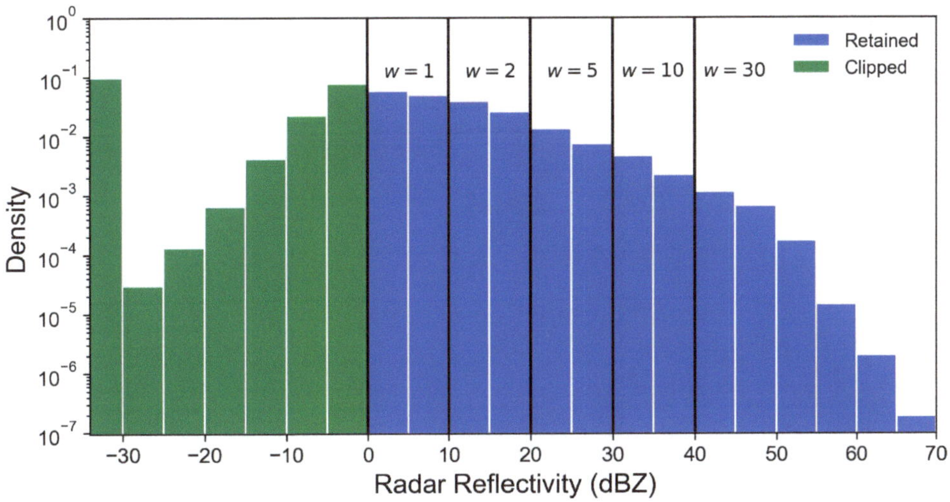

Figure 6. The distribution histogram of the raw radar reflectivity data. The green bar indicates that values below 0 dBZ were clipped, while the blue bar indicates that values between 0 and 70 dBZ were retained. The weights of the L1 loss function for different intervals were placed above the bars.

The learnable parameters of each layer in the DSA-UNet model were initialized with the default initialization scheme and then trained by an Adam optimizer [36] with a learning rate of 0.0001 and a max iteration of 100,000. The parameters of the DSA-UNet were updated via backpropagation after each iteration with the mini-batch training skill. The dataset was segmented into mini-batches with a size of 32. The validation set with 11,306 samples was employed for optimal parameter selection. The early stopping training skill and the L2 regularization were adopted to avoid the overfitting of the neural network. The generators and discriminators of the UNet++ GAN model shared the optimizer, the learning rate, and the mini-batch size. It was trained using a combination of the original GAN loss and the weighted L1 loss. The optimal parameters of the MLG method were also approximated with an Adam optimizer because the dataset was too large for directly computing the theoretical solution. All of the experiments in this study were implemented on an Nvidia Tesla A100 graphic card based on the open-source machine learning framework PyTorch. Other detailed settings are listed in Table 1.

Table 1. Training settings of the DSA-UNet model, the UNet++ GAN model, and the MLG method. The non-parametric BI method is not included.

Model	Learning Rate	Applying Sector Masks	Applying Azimuthal Padding	Loss Function	Regularization Weight Decay
DSA-UNet	0.0001	Yes	Yes	Weighted L1	0.0001
UNet++ GAN	0.0001	Yes	Yes	BCE + weighted L1	0.01
MLG	0.001	No	No	L2	0

3.5. Evaluation Metrics

Since the aim of this study is to better complete the missing regions of the radar observations, the evaluation metrics should be capable of measuring the differences between the predicted values and the observed values in the masked areas. The mean bias error (MBE), the mean absolute error (MAE), and the root mean squared error (RMSE) are popular evaluation metrics for estimating the differences or bias. However, since the area of the annular sector grows rapidly as the radial distance increases, the evaluation metrics must take the variability of pixel area into consideration. The above metrics were weighted by a tensor $\mathbf{W}$ determined by the areas of the annular sectors at different radial distances, as defined in Equation (3). The weighted metrics (WMBE, WMAE, and WRMSE) were selected as the evaluation metrics, which are defined in Equations (4)–(6).

$$\mathbf{W}_{\rho,\theta} = \frac{(\rho+1)^2 - \rho^2}{R^2\Theta} \quad (\rho = 0, 1, \cdots, R) \tag{3}$$

$$\mathrm{WMBE}(\hat{\mathbf{X}}, \mathbf{X}; \mathbf{M}, \mathbf{W}) = \sum_{\rho,\theta} \mathbf{W} \odot (\mathbf{M} \odot \hat{\mathbf{X}} - \mathbf{M} \odot \mathbf{X}) \tag{4}$$

$$\mathrm{WMAE}(\hat{\mathbf{X}}, \mathbf{X}; \mathbf{M}, \mathbf{W}) = \sum_{\rho,\theta} \mathbf{W} \odot |\mathbf{M} \odot \hat{\mathbf{X}} - \mathbf{M} \odot \mathbf{X}| \tag{5}$$

$$\mathrm{WRMSE}(\hat{\mathbf{X}}, \mathbf{X}; \mathbf{M}, \mathbf{W}) = \sqrt{\sum_{\rho,\theta} \mathbf{W} \odot (\mathbf{M} \odot \hat{\mathbf{X}} - \mathbf{M} \odot \mathbf{X})^2} \tag{6}$$

In the above equations, $\mathbf{X}$ represents the observation and $\hat{\mathbf{X}}$ represents the prediction; $\mathbf{M}$ represents the sector mask; $\odot$ represents the Hadamard product; the operator $\sum_{\rho,\theta}$ represents the sum operation along the radial (ρ) and the azimuthal (θ) axis; and R and Θ are the sizes of radial range (80) and azimuthal range (400 for DL models and 360 for traditional methods). The weight tensor $\mathbf{W}$ meets the condition of $\sum_{\rho,\theta} \mathbf{W} = 1$. Considering the unbalance of the dataset, we selected 10, 20, 30, 40, and 50 dBZ as the thresholds for the above metrics to better evaluate the bias in different radar reflectivity intervals.

In addition to the above metrics, we selected the PPI plots, the contrast scatter (CS) plots, and the power spectral density (PSD) plots as evaluation methods for further analysis. The PPI plot is a radar display that gives a conical section of radar observations at a certain elevation angle. The CS plot describes the correlation between the predicted values and the observed values. The PSD plot presents the relationship between the power and frequency of a signal, which was used in radar nowcasting studies to evaluate the performance of representing diverse-scale weather patterns [22,23]. The above metrics and plots were also used for model evaluation in related studies [10,28,33].

4. Results

4.1. Performance on the Test Set

The overall data completion performance of the DSA-UNet and its baseline models was evaluated on the test set with a total of 14,133 samples. The WMAE, WRMSE, and WMBE of the models in different radar reflectivity intervals were calculated and are illustrated in Figure 7. It can be observed that the MLG method produced the highest overall WMAE and WRMSE, as well as the lowest overall WMBE that was far below zero, implying a severe systematic underestimation problem. The BI method's systematic error indicated by the WMBE was lower than the MLG method, and so did its absolute error indicated by the WMAE and the WRMSE. DL models (UNet++ GAN and DSA-UNet) performed better than the MLG method and the BI method when comparing the overall metrics. Among all of the models, the DSA-UNet reached the lowest overall WMAE, the lowest overall WRMSE, and the nearest overall WMBE to zero, which indicated that the DSA-UNet could generate a lower error between its predictions and the observations than the other baseline models.

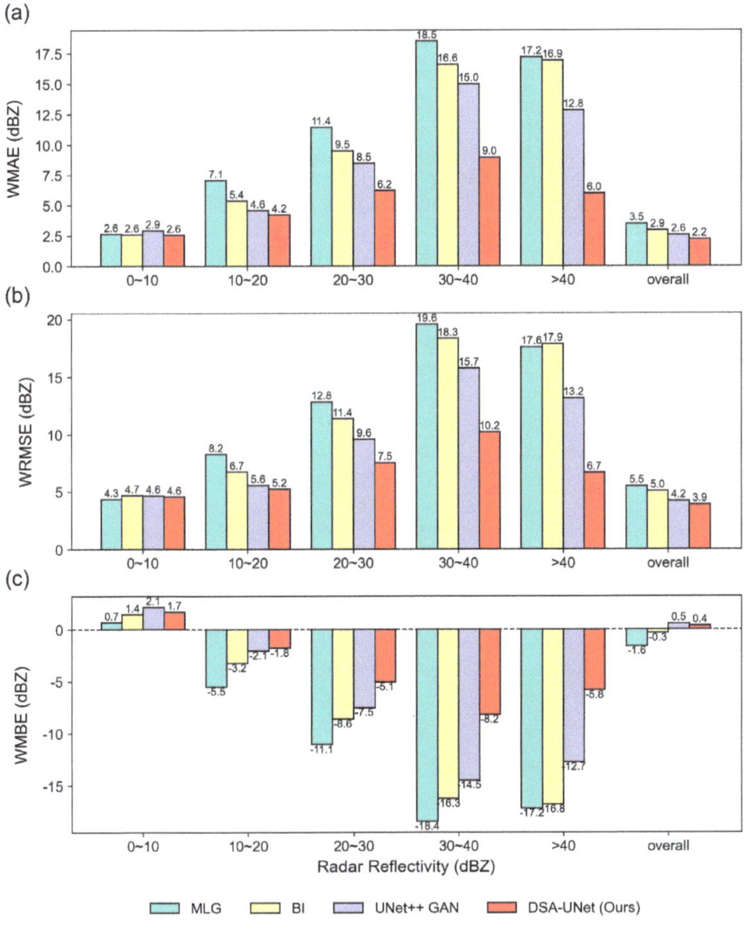

Figure 7. The evaluation metrics of the models on the test set. The horizontal axis represents different radar reflectivity intervals and the overall performance. The vertical axis represents the evaluation metrics, including WMAE (**a**), WRMSE (**b**), and WMBE (**c**).

The evaluation metrics for different radar reflectivity intervals were also illustrated in Figure 7. Generally, the results show that the DL models and the traditional methods performed the best for radar reflectivity in 0–10 dBZ, but the worst in 30–40 dBZ, where all of them met a notable systematic underestimation. Their errors were close to each other when the radar reflectivity was low (0–20 dBZ). Specifically, the MLG method and the BI method had even lower errors than DL models for radar reflectivity in 0–10 dBZ. However, when the radar reflectivity exceeded 20 dBZ, the errors of the baseline methods grew rapidly as the radar reflectivity increased. Their performance became considerably worse when the radar reflectivity was higher than 30 dBZ. Compared to the baseline methods, the DSA-UNet model achieved significantly lower WMAE and WRMSE as well as the absolute value of WMBE, especially for high radar reflectivity values, which suggests that it had a lower systematic error and a lower absolute bias than its baseline methods.

4.2. Case Study

The data completion performance of the DSA-UNet and its baseline models was also evaluated and further analyzed with two cases that were selected from the test set. Case 1 was selected from a moderate-intensity widespread rainfall event at 13:00 UTC on 22 July 2019. Case 2 was selected from a high-intensity squall line event at 15:00 UTC on 6 August 2019. The evaluation was implemented only for the masked areas.

4.2.1. Case 1

The PPI images of the 1.5° radar reflectivity observation and the predictions generated by the DSA-UNet and its baseline models for Case 1 are shown in Figure 8. The azimuthal boundaries of the sector mask for Case 1 were set as 315° and 355°, which were marked with dashed lines. The PPI image of the observation shows that the observed radar reflectivity of the masked area was mostly between 20 and 40 dBZ. The azimuthal boundaries of the sector mask for Case 1 were set as 270° and 310°, which were marked with dashed lines. The PPI image of the observation (Figure 8a) shows that the observed radar reflectivity of the masked area was mostly between 20 and 40 dBZ. It can be found that the radar reflectivity values predicted by the MLG method (Figure 8b) were significantly lower than the observed values, implying a severe systematic underestimation. The prediction of the BI method (Figure 8c) in the masked area showed an abnormal striped feature that did not exist in the observation. This was because when the beams in the radial direction were completely blocked, the values were interpolated merely along the azimuthal direction based on the visible values outside the mask boundaries. The predictions of the DL models were closer to the observation compared to the MLG method and the BI method. However, the predictions of the UNet++ GAN model (Figure 8d) had fewer local-scale patterns than the observation, or to be more intuitive, its prediction was smoother and blurrier. The prediction of the DSA-UNet model (Figure 8e) was the closest to the observation. It not only had the most similar reflectivity shape to the observation but also had the richest local-scale details. The merged PPI image of the DSA-UNet model's prediction also appeared more seamless, which indicates that the predicted values in the masked area and the observed values in the unmasked area were intensely close.

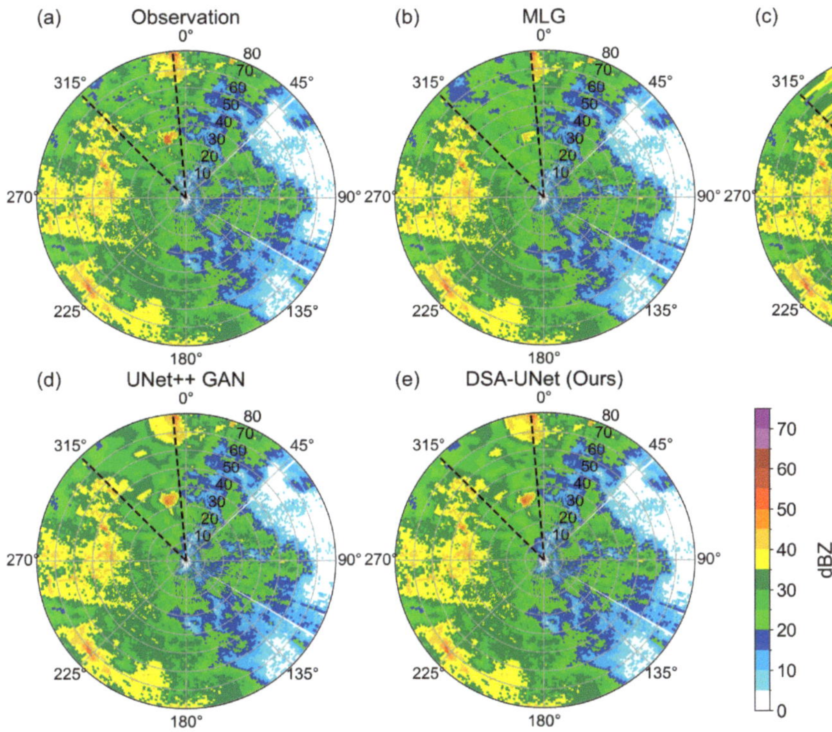

Figure 8. The PPI images of the 1.5° radar reflectivity observation and the predictions for Case 1. (**a**) Observation; (**b**) prediction of MLG; (**c**) Prediction of BI; (**d**) prediction of UNet++ GAN; (**e**) prediction of DSA-UNet. The boundaries of the sector mask are marked with dashed lines.

Figure 9 shows the evaluation metrics of the models for Case 1. The results show that the BI method had the highest overall systematic error and absolute error, especially for radar reflectivity over 30 dBZ. The MLG method produced the second-highest systematic error absolute error. Its performance was limited by a systematic underestimation of high radar reflectivity values that were over 30 dBZ. The overall error of the UNet++ GAN model was lower than the traditional methods but was higher than the DSA-UNet model. The UNet++ GAN model had the worst performance for radar reflectivity below 10 dBZ, but it significantly exceeded the traditional methods for high radar reflectivity. The DSA-UNet model achieved the best performance for most radar reflectivity intervals except 0–10 dBZ where it was a bit inferior to the BI method.

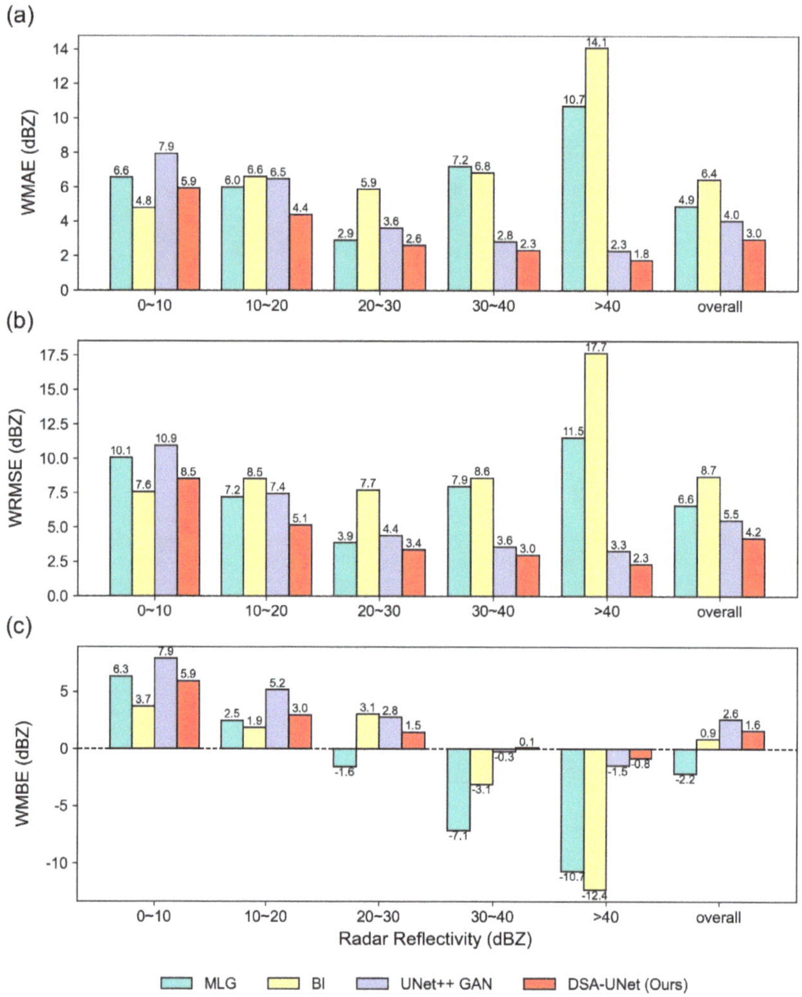

Figure 9. The evaluation metrics of the models for Case 1. The horizontal axis represents different radar reflectivity intervals and the total performance. The vertical axis represents the evaluation metrics, including WMAE (**a**), WRMSE (**b**), and WMBE (**c**).

Figure 10 shows the CS plot of the observed and predicted values in the masked areas for Case 1. In a CS plot, the horizontal axis and the vertical axis represent the observation and the prediction, respectively. The points on the 45° line imply perfect predictions, yet the points below or beyond this line signify underestimations or overestimations. The colors of the points are determined with the probability density provided by the Gaussian kernel density estimation. The results show that most of the observation values lay in the range of 20–40 dBZ, which is consistent with Figure 8. The MLG method (Figure 10a) and the BI method (Figure 10b) were limited by a significant underestimation and a high prediction variance, respectively. The scatters of the UNet++ GAN model (Figure 10c) and the DSA-UNet model (Figure 10d) were distributed closer to the 45° line, indicating their better performance than traditional methods. The scatters of the DSA-UNet model were more tightly concentrated on the 45° line than the UNet++ GAN model, which indicates that the DSA-UNet model could achieve a lower prediction systematic error and variance for Case 1. It can also be found that the DL models tended to overestimate low radar reflectivity values in 0–10 dBZ, which is consistent with Figure 9.

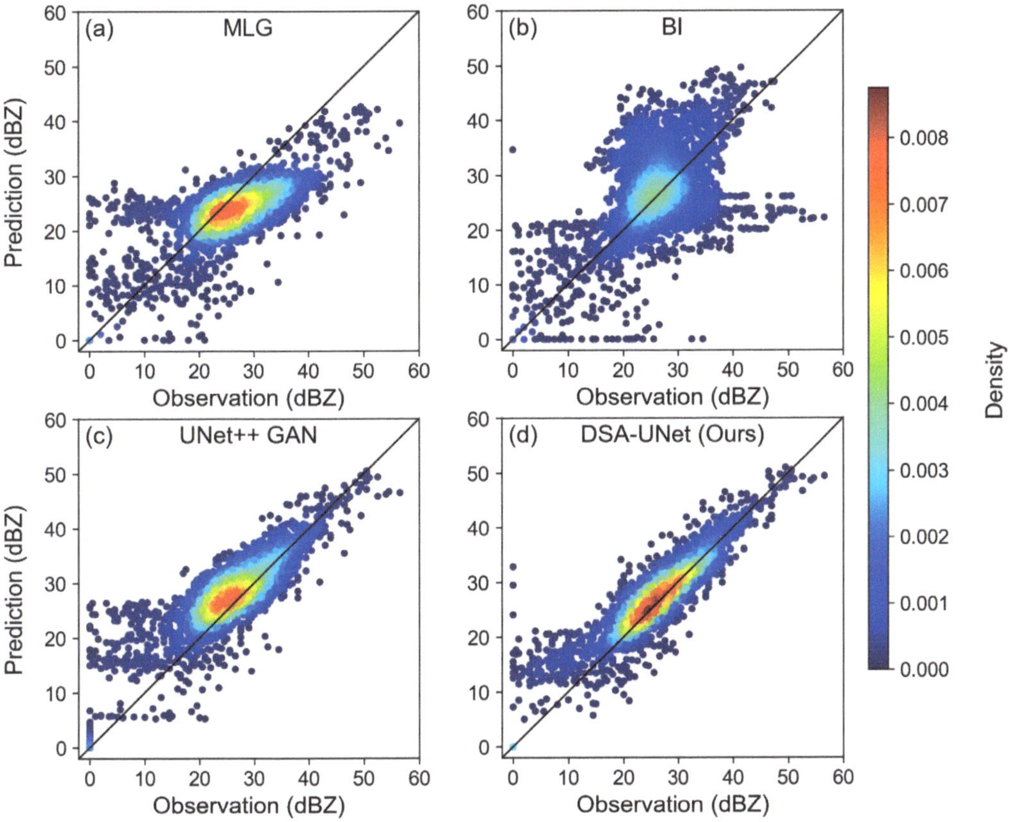

Figure 10. The CS plot of the observed and predicted values in the masked areas for Case 1. (**a**) MLG; (**b**) BI; (**c**) UNet++ GAN; (**d**) DSA-UNet.

To further explain the above results, we plotted the PSDs of the observation and predictions along the radial and azimuthal axes in the masked areas for Case 1, as shown in Figure 11. In this study, the horizontal axis of the PSD plot represents the spatial wavelength, which is positively correlated with the spatial scale. The vertical axis represents the PSD at certain spatial wavelengths. A higher PSD generally indicates richer local details when the spatial wavelength is short. Generally, both radial and azimuthal PSDs decrease as the spatial wavelength reduces. When the spatial wavelength is long (over 16 km or 16°), the radial and azimuthal PSDs of the DL models' predictions were closer to the PSD of the observation than those of the MLG method and the BI method. When the spatial wavelength is shorter than 4 km or 4°, the PSDs of the MLG method exceeded the PSDs of the DL models. The radial PSD of the BI method is significantly dissimilar to the azimuthal PSD due to a lack of radial information. The DSA-UNet model reached a higher PSD than the UNet++ GAN model, especially at short spatial wavelengths, indicating that the DSA-UNet model had a better capability of reconstructing local-scale spatial patterns, which is consistent with the results in the PPI plot and the CS plot.

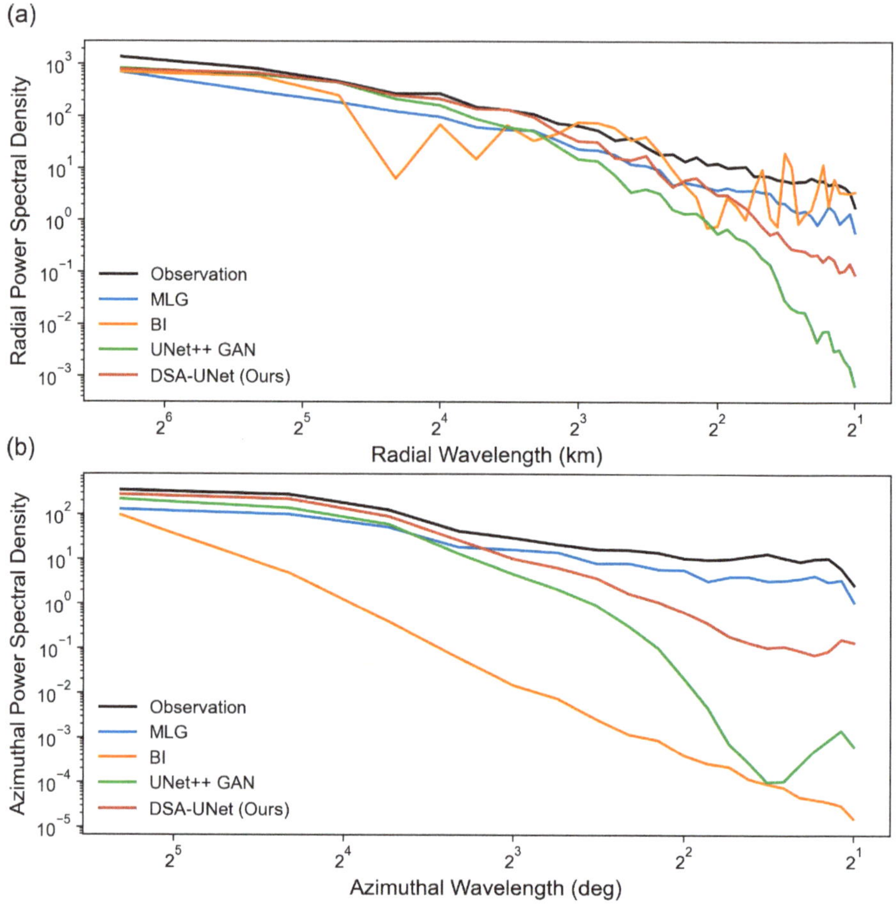

Figure 11. The PSD plot of the observation and predictions in the masked areas for Case 1. (**a**) Radial PSD; (**b**) azimuthal PSD.

4.2.2. Case 2

The PPI images of the observation and the predictions for Case 2 are shown in Figure 12. The sector mask covered an azimuthal range of 40°–80°. Figure 12a shows that a local high-intensity radar echo region was located in the northeast of the masked area, with a maximum reflectivity of over 50 dBZ, while in the remaining part of the masked area, the radar reflectivity was much lower (below 10 dBZ). The MLG method (Figure 12b) and the BI method (Figure 12c) met the systematic underestimation and striped pattern problems, respectively, which were also found in Case 1. The DL models made better predictions than the above two methods in the aspect of both general echo intensity and shape. The DL models made better predictions than the above two methods in the aspect of both general echo intensity and shape. The UNet++ GAN model (Figure 12d) fell behind the DSA-UNet model (Figure 12e) in predicting the location of the peak value.

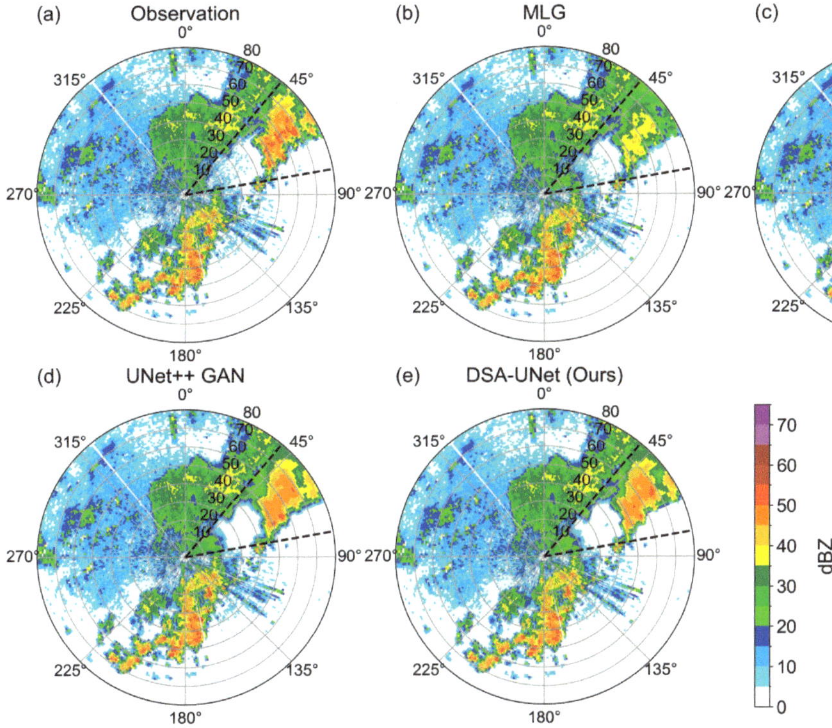

Figure 12. The PPI images of the 1.5° radar reflectivity observation and the predictions for Case 2. (**a**) Observation; (**b**) prediction of MLG; (**c**) prediction of BI; (**d**) prediction of UNet++ GAN; (**e**) prediction of DSA-UNet. The boundaries of the sector mask are marked with dashed lines.

Figure 13 shows the evaluation metrics of the predictions of the models for Case 2. Similar to Case 1, the BI method had the highest overall prediction error in all of the metrics (WMAE, WRMSE, and WMBE), especially when the radar reflectivity was over 40 dBZ. The performance of the MLG method was also unacceptable due to its high bias and underestimation of high radar reflectivity. In contrast to the traditional methods, both DL models could produce better overall performance. However, the performance of the UNet++ GAN model was inferior to the MLG method when the radar reflectivity was below 20 dBZ. Instead, the DSA-UNet model achieved a lower systematic and absolute error for both high and low radar reflectivity than all of the other baseline models.

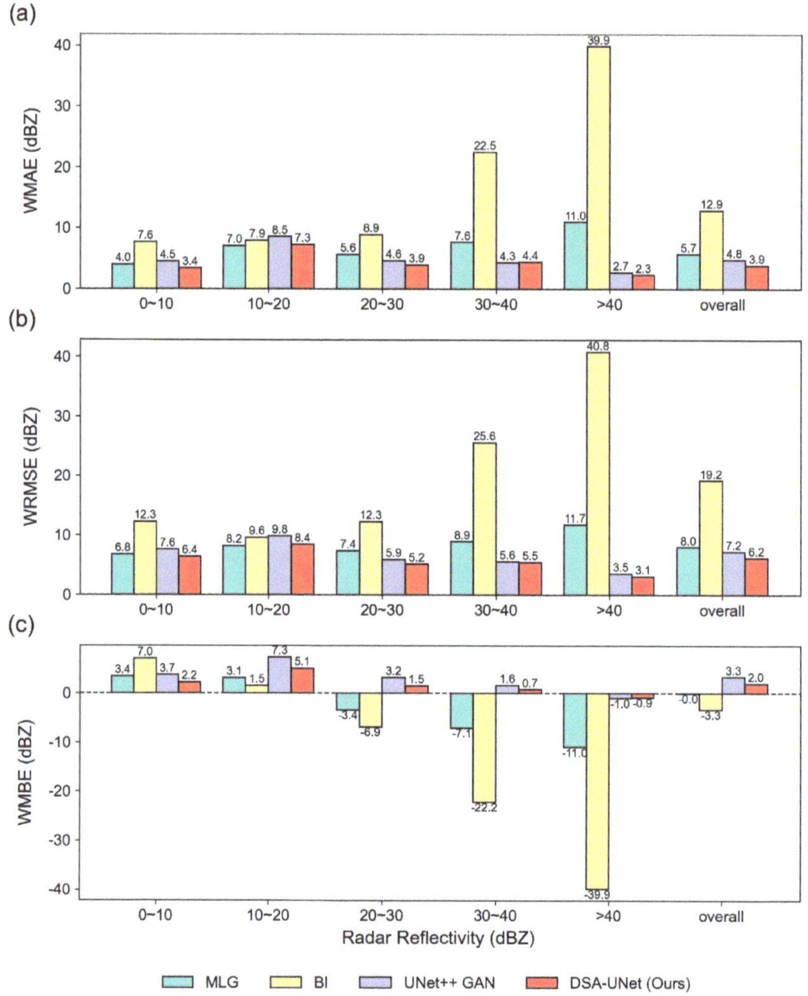

Figure 13. The evaluation metrics of the models for Case 2. The horizontal axis represents different radar reflectivity intervals and the total performance. The vertical axis represents the evaluation metrics, including WMAE (**a**), WRMSE (**b**), and WMBE (**c**).

Figure 14 displays the CS plot of the observed and predicted values in the masked areas for Case 2. Different from Case 1, the radar reflectivity values of the observation in Case 2 were mainly clustered in the 0–10 dBZ. Meanwhile, the amount of observed values over 50 dBZ in Case 2 was larger. The scatters of the MLG method (Figure 14a) deviated from the 45° line, especially for observed values that were over 30 dBZ, indicating the aforementioned underestimation problem for high-intensity radar data completion. The scatters of the BI method (Figure 14b) were dispersed over the entire radar reflectivity value ranges, which confirms its poor performance, as shown in Figures 12 and 13. The scatters of the UNet++ GAN model (Figure 14c) were more centered around the 45° line, but this model tended to overestimate the values. The DSA-UNet model (Figure 14d) produced a better prediction than the other baseline models. The scatters of the DSA-UNet model were the most concentrated in the 45° line and had the highest cluster density (dark-red points). This model also had a systematic overestimation for low radar reflectivity values (below 10 dBZ), but this problem was remarkably alleviated for moderate and high radar reflectivity (over 20 dBZ).

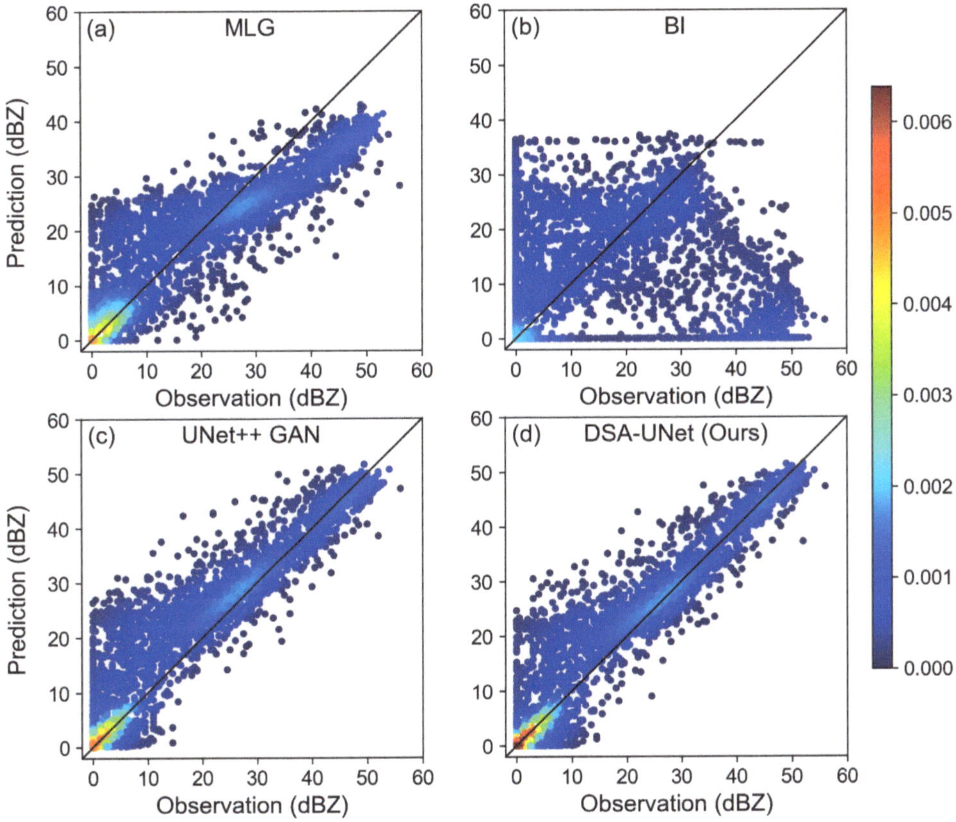

Figure 14. The CS plot of the observed and predicted values in the masked areas for Case 2. (**a**) MLG; (**b**) BI; (**c**) UNet++ GAN; (**d**) DSA-UNet.

The radial and azimuthal PSDs of the predictions and the observation in the masked area for Case 2 were calculated and exhibited in Figure 15. Similar to Case 1, the MLG method achieved the highest radial and azimuthal PSD among all of the methods. The azimuthal PSD of the BI method is significantly lower than the other models. Meanwhile, the radial PSD of the BI method oscillates as the radial wavelength reduces due to a lack of radial information. The UNet++ GAN model performed the worst, particularly for short spatial wavelengths that imply local-scale patterns. The DSA-UNet model achieved higher radial and azimuthal PSDs than the UNet++ GAN model, which indicates that this model was more qualified for capturing and completing diverse-scale patterns in data-missing areas. However, there is still a gap between the PSDs of the DSA-UNet model and the MLG method for very short spatial wavelengths.

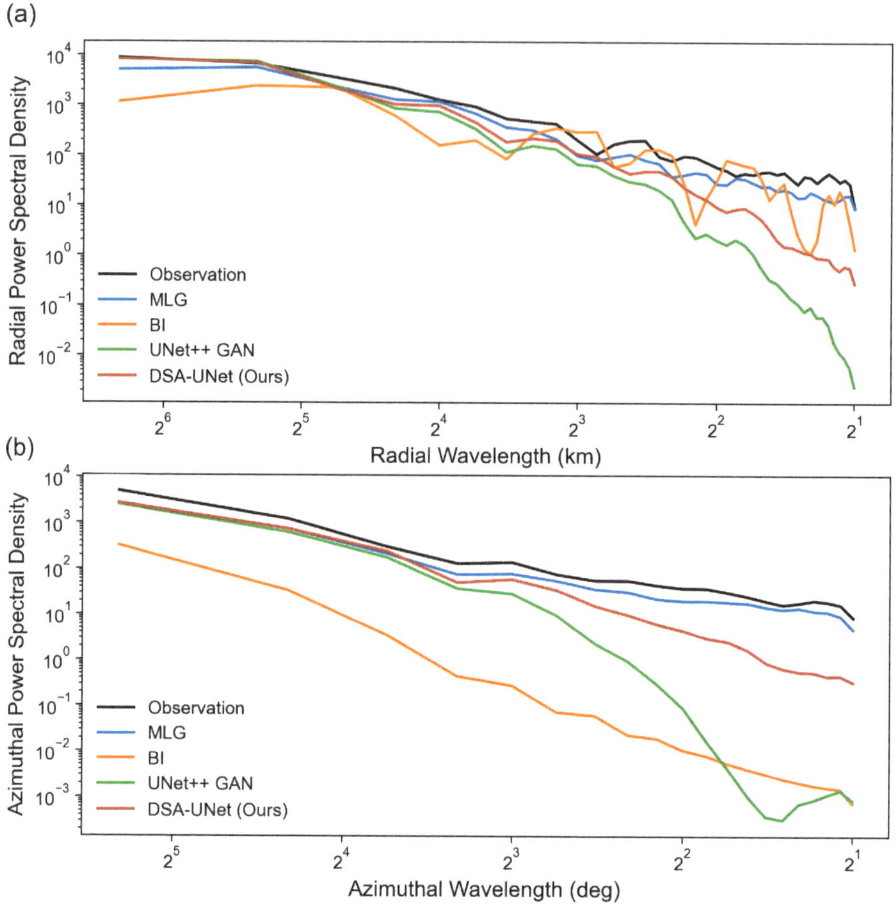

Figure 15. The PSD plot of the observation and predictions in the masked areas for Case 2. (**a**) Radial PSD; (**b**) azimuthal PSD.

5. Discussion

In Section 4, the data completion performance of our proposed DSA-UNet model was evaluated on the entire test set and further analyzed with two cases, where the model was also compared with the MLG method, the BI method, and the UNet++ GAN model. The evaluation methods consisted of three quantitative metrics (WMBE, WMAE, and WRMSE) and three diagrams (PPI plot, CS plot, and PSD plot) for further analysis of the case study.

Generally, the quantitative evaluation results revealed that the DL models performed better than the traditional methods. The MLG method benefited from its simple principle and strong interpretability, which was based on the assumption that the radar reflectivity values at different elevation angles were linearly correlated. However, the MLG method ran into a systematic underestimation problem, especially for areas with high radar reflectivity values, which made it unsuitable for completing radar observations in heavy rainfall scenarios. The non-parametric BI method was found to suffer from a striped pattern problem that caused the method to fail in high-reflectivity data completion. Compared to the traditional methods, the general errors between the DL models' predictions and the observations were significantly lower; the predicted intensity and position of high-value areas were also more accurate.

It was also revealed that the DSA-UNet model could bring improvements to radar missing data completion over the current DL model proposed in previous research. Compared to the UNet++ GAN model, the DSA-UNet model achieved better performance on both the entire test and the case study. It provided the completion closer to the real observation in almost all radar reflectivity intervals, especially for extreme values, which play an important role in signifying severe storms. The PPI and PSD diagrams in the case study illustrated that the DSA-UNet model could improve the modeling and reconstruction of local-scale radar echo patterns over the UNet++ GAN model. These improvements were possibly facilitated by the dilated convolutional modules and self-attentional modules in the DSA-UNet model, which could strengthen the model's ability to learn local and subtle information from the data. The improvements of the DSA-UNet model over the baseline models are beneficial for finer weather monitoring.

Although the DSA-UNet model has shown a surpassing performance over the baseline models in weather radar data completion, it still meets several limitations. The quantitative results and the CS diagrams revealed that the DSA-UNet model tended to slightly overestimate low radar reflectivity values, especially for values below 10 dBZ, which was similar to the UNet++ GAN model. Under this overestimation, the model has the potential to involve abnormal clutters when completing missing data. Furthermore, the PSD diagrams indicated that although the DSA-UNet was able to narrow the gap between the predicted and observed radar reflectivity values in local-scale radar echo patterns, it still lagged behind traditional statistical methods, suggesting that there is still room for improvement. The above drawbacks were also noticed and summarized as the blurry effect of DL models in studies on precipitation nowcasting [22,23,37]. The improving methods proposed in these studies might be constructive for eliminating the drawbacks of the DSA-UNet model. Moreover, the data in this study were collected only from warm seasons, which were mainly composed of convective precipitation samples instead of stratiform precipitation samples. Meanwhile, the lowest elevation angle of the radar data for the experiments was selected as 1.5° instead of 0.5°, which was unavailable because of severe beam blockage and noises. The generalization of the DSA-UNet model to different precipitation types and lower elevation angles still needs further assessment.

6. Conclusions

The data-missing problem is one of the major factors that limit the quality of weather radar data and subsequent applications. Traditional solutions based on radar physics and statistics in previous studies have shown obvious defects in various aspects. Researchers have applied deep learning (DL) techniques to the completion of weather radar missing data, but their methods were limited by low accuracy. In this study, we proposed a dilated and self-attentional UNet (DSA-UNet) model to improve the completion quality of weather radar missing data. The model was trained and evaluated on a radar dataset built from the Yizhuang radar observations during the warm seasons from 2017 to 2019. It was further analyzed with two cases based on three quantitative metrics (WMBE, WMAE, and WRMSE) and three diagrams (PPI plot, CS plot, and PSD plot). Several baseline methods and models were selected and compared with our proposed DSA-UNet model, including the MLG method, the BI method, and the UNet++ GAN model. The major findings of this study are as follows:

- The DL models can outperform traditional statistical methods by reducing the general errors between their predictions and the observation and by predicting the intensity and position of high radar reflectivity values more accurately.
- Compared to the UNet++ GAN model, the DSA-UNet model can produce a better completion that is closer to the real observation in almost all radar reflectivity intervals, especially for extreme values.
- The DSA-UNet model can better capture and reconstruct local-scale radar echo patterns over the UNet++ GAN model.

- The limitations of the DSA-UNet model include the slight underestimation of low values and the local-scale details.

This study provides an effective solution for improving the completion of weather radar missing data and also reveals the great potential of deep learning in weather radar applications. Future studies will improve the network architecture to eliminate the current drawbacks and incorporate radar data that contain various precipitation types and lower elevation angles to assess the generalization of our method.

Author Contributions: Conceptualization, A.G. and G.N.; methodology, A.G.; software, A.G.; validation, A.G.; formal analysis, A.G.; investigation, A.G.; resources, A.G.; data curation, A.G.; writing—original draft preparation, A.G.; writing—review and editing, H.C. and G.N.; visualization, A.G.; supervision, G.N.; project administration, G.N.; funding acquisition, G.N. All authors have read and agreed to the published version of the manuscript.

Funding: This work was supported by the National Key Research and Development Program of China (2022YFC3090604) and the Fund Program of State Key Laboratory of Hydroscience and Engineering (61010101221).

Data Availability Statement: The data in this study is unavailable due to privacy restrictions. The codes for the experiments are available at https://github.com/THUGAF/Radar-Completion (accessed on 14 September 2023).

Acknowledgments: The authors gratefully acknowledge the anonymous reviewers for providing careful reviews and comments on this article.

Conflicts of Interest: The authors declare no conflict of interest.

Abbreviations

The following abbreviations are used in this manuscript:

CINRAD	China Next Generation Weather Radar
VPR	Vertical profile of radar reflectivity
DEM	Digital elevation model
DL	Deep learning
DSA	Dilated and self-attentional
MLG	Multivariate linear regression
BI	Bilinear interpolation
GAN	Generative adversarial network
MBE	Mean bias error
MAE	Mean absolute error
RMSE	Root mean squared error
WMBE	Weighted mean bias error
WMAE	Weighted mean absolute error
WRMSE	Weighted root mean squared error
PPI	Plan position indicator
CS	Contrast scatter
PSD	Power spectral density

References

1. Agrawal, S.; Barrington, L.; Bromberg, C.; Burge, J.; Gazen, C.; Hickey, J. Machine Learning for Precipitation Nowcasting from Radar Images. *arXiv* **2019**, arXiv:1912.12132.
2. Brown, R.A.; Wood, V.T.; Barker, T.W. Improved detection using negative elevation angles for mountaintop WSR-88Ds: Simulation of KMSX near Missoula, Montana. *Weather Forecast.* **2002**, *17*, 223–237. [CrossRef]
3. Wood, V.T.; Brown, R.A.; Vasiloff, S.V. Improved detection using negative elevation angles for mountaintop WSR-88Ds. Part II: Simulations of the three radars covering Utah. *Weather Forecast.* **2003**, *18*, 393–403. [CrossRef]
4. Germann, U.; Boscacci, M.; Clementi, L.; Gabella, M.; Hering, A.; Sartori, M.; Sideris, I.V.; Calpini, B. Weather radar in complex orography. *Remote Sens.* **2022**, *14*, 503. [CrossRef]
5. Min, C.; Chen, S.; Gourley, J.J.; Chen, H.; Zhang, A.; Huang, Y.; Huang, C. Coverage of China new generation weather radar network. *Adv. Meteorol.* **2019**, *2019*, 5789358. [CrossRef]

6. Vignal, B.; Galli, G.; Joss, J.; Germann, U. Three methods to determine profiles of reflectivity from volumetric radar data to correct precipitation estimates. *J. Appl. Meteorol. Climatol.* **2000**, *39*, 1715–1726. [CrossRef]
7. Andrieu, H.; Creutin, J.D. Identification of vertical profiles of radar reflectivity for hydrological applications using an inverse method. Part I: Formulation. *J. Appl. Meteorol. Climatol.* **1995**, *34*, 225–239. [CrossRef]
8. Vignal, B.; Andrieu, H.; Creutin, J.D. Identification of vertical profiles of reflectivity from volume scan radar data. *J. Appl. Meteorol. Climatol.* **1999**, *38*, 1214–1228. [CrossRef]
9. Joss, J.; Pittini, A. Real-time estimation of the vertical profile of radar reflectivity to improve the measurement of precipitation in an Alpine region. *Meteorol. Atmos. Phys.* **1991**, *47*, 61–72. [CrossRef]
10. Yin, X.; Hu, Z.; Zheng, J.; Li, B.; Zuo, Y. Study on Radar Echo-Filling in an Occlusion Area by a Deep Learning Algorithm. *Remote Sens.* **2021**, *13*, 1779. [CrossRef]
11. Bech, J.; Codina, B.; Lorente, J.; Bebbington, D. The sensitivity of single polarization weather radar beam blockage correction to variability in the vertical refractivity gradient. *J. Atmos. Ocean. Technol.* **2003**, *20*, 845–855. [CrossRef]
12. Lang, T.J.; Nesbitt, S.W.; Carey, L.D. On the correction of partial beam blockage in polarimetric radar data. *J. Atmos. Ocean. Technol.* **2009**, *26*, 943–957. [CrossRef]
13. Bech, J.; Gjertsen, U.; Haase, G. Modelling weather radar beam propagation and topographical blockage at northern high latitudes. *Q. J. R. Meteorol. Soc. J. Atmos. Sci. Appl. Meteorol. Phys. Oceanogr.* **2007**, *133*, 1191–1204. [CrossRef]
14. Shakti, P.; Maki, M.; Shimizu, S.; Maesaka, T.; Kim, D.S.; Lee, D.I.; Iida, H. Correction of reflectivity in the presence of partial beam blockage over a mountainous region using X-band dual polarization radar. *J. Hydrometeorol.* **2013**, *14*, 744–764.
15. Andrieu, H.; Creutin, J.; Delrieu, G.; Faure, D. Use of a weather radar for the hydrology of a mountainous area. Part I: Radar measurement interpretation. *J. Hydrol.* **1997**, *193*, 1–25. [CrossRef]
16. Creutin, J.; Andrieu, H.; Faure, D. Use of a weather radar for the hydrology of a mountainous area. Part II: Radar measurement validation. *J. Hydrol.* **1997**, *193*, 26–44. [CrossRef]
17. Yoo, C.; Park, C.; Yoon, J.; Kim, J. Interpretation of mean-field bias correction of radar rain rate using the concept of linear regression. *Hydrol. Process.* **2014**, *28*, 5081–5092. [CrossRef]
18. Kvasov, R.; Cruz-Pol, S.; Colom-Ustáriz, J.; Colón, L.L.; Rees, P. Weather radar data visualization using first-order interpolation. In Proceedings of the 2013 IEEE International Geoscience and Remote Sensing Symposium-IGARSS, Melbourne, Australia, 21–26 July 2013; pp. 3574–3577.
19. Foehn, A.; Hernández, J.G.; Schaefli, B.; De Cesare, G. Spatial interpolation of precipitation from multiple rain gauge networks and weather radar data for operational applications in Alpine catchments. *J. Hydrol.* **2018**, *563*, 1092–1110. [CrossRef]
20. Li, B.; Li, R.; Sun, T.; Gong, A.; Tian, F.; Khan, M.Y.A.; Ni, G. Improving LSTM hydrological modeling with spatiotemporal deep learning and multi-task learning: A case study of three mountainous areas on the Tibetan Plateau. *J. Hydrol.* **2023**, *620*, 129401. [CrossRef]
21. Shi, X.; Chen, Z.; Wang, H.; Yeung, D.Y.; Wong, W.K.; Woo, W.C. Convolutional LSTM network: A machine learning approach for precipitation nowcasting. In Proceedings of the Advances in Neural Information Processing Systems, Montreal, ON, Canada, 7–12 December 2015; pp. 802–810.
22. Ravuri, S.; Lenc, K.; Willson, M.; Kangin, D.; Lam, R.; Mirowski, P.; Fitzsimons, M.; Athanassiadou, M.; Kashem, S.; Madge, S.; et al. Skilful precipitation nowcasting using deep generative models of radar. *Nature* **2021**, *597*, 672–677. [CrossRef]
23. Ayzel, G.; Scheffer, T.; Heistermann, M. RainNet v1.0: A convolutional neural network for radar-based precipitation nowcasting. *Geosci. Model Dev.* **2020**, *13*, 2631–2644. [CrossRef]
24. Sadeghi, M.; Asanjan, A.A.; Faridzad, M.; Nguyen, P.; Hsu, K.; Sorooshian, S.; Braithwaite, D. PERSIANN-CNN: Precipitation estimation from remotely sensed information using artificial neural networks–convolutional neural networks. *J. Hydrometeorol.* **2019**, *20*, 2273–2289. [CrossRef]
25. Pan, B.; Hsu, K.; AghaKouchak, A.; Sorooshian, S. Improving precipitation estimation using convolutional neural network. *Water Resour. Res.* **2019**, *55*, 2301–2321. [CrossRef]
26. Afzali Gorooh, V.; Kalia, S.; Nguyen, P.; Hsu, K.l.; Sorooshian, S.; Ganguly, S.; Nemani, R.R. Deep neural network cloud-type classification (DeepCTC) model and its application in evaluating PERSIANN-CCS. *Remote Sens.* **2020**, *12*, 316. [CrossRef]
27. Song, T.; Li, Y.; Meng, F.; Xie, P.; Xu, D. A novel deep learning model by Bigru with attention mechanism for tropical cyclone track prediction in the Northwest Pacific. *J. Appl. Meteorol. Climatol.* **2022**, *61*, 3–12.
28. Geiss, A.; Hardin, J.C. Inpainting radar missing data regions with deep learning. *Atmos. Meas. Tech.* **2021**, *14*, 7729–7747. [CrossRef]
29. Ronneberger, O.; Fischer, P.; Brox, T. U-net: Convolutional networks for biomedical image segmentation. In Proceedings of the International Conference on Medical Image Computing and Computer-Assisted Intervention, Munich, Germany, 5–9 October 2015; Springer: Berlin/Heidelberg, Germany, 2015; pp. 234–241.
30. Yu, F.; Koltun, V. Multi-scale context aggregation by dilated convolutions. *arXiv* **2015**, arXiv:1511.07122.
31. Zhang, H.; Goodfellow, I.; Metaxas, D.; Odena, A. Self-attention generative adversarial networks. In Proceedings of the International Conference on Machine Learning, Long Beach, CA, USA, 9–15 June 2019; PMLR: Cambridge, MA, USA, 2019; pp. 7354–7363.
32. Ma, Y.; Ni, G.; Chandra, C.V.; Tian, F.; Chen, H. Statistical characteristics of raindrop size distribution during rainy seasons in the Beijing urban area and implications for radar rainfall estimation. *Hydrol. Earth Syst. Sci.* **2019**, *23*, 4153–4170. [CrossRef]

33. Tan, S.; Chen, H. A Conditional Generative Adversarial Network for Weather Radar Beam Blockage Correction. *IEEE Trans. Geosci. Remote Sens.* **2023**, *61*, 4103014. [CrossRef]
34. Mirza, M.; Osindero, S. Conditional generative adversarial nets. *arXiv* **2014**, arXiv:1411.1784.
35. Zhou, Z.; Rahman Siddiquee, M.M.; Tajbakhsh, N.; Liang, J. Unet++: A nested u-net architecture for medical image segmentation. In Proceedings of the Deep Learning in Medical Image Analysis and Multimodal Learning for Clinical Decision Support: 4th International Workshop, DLMIA 2018, and 8th International Workshop, ML-CDS 2018, Held in Conjunction with MICCAI 2018, Granada, Spain, 20 September 2018; Springer: Berlin/Heidelberg, Germany, 2018; pp. 3–11.
36. Kingma, D.P.; Ba, J. Adam: A method for stochastic optimization. *arXiv* **2014**, arXiv:1412.6980.
37. Gong, A.; Li, R.; Pan, B.; Chen, H.; Ni, G.; Chen, M. Enhancing Spatial Variability Representation of Radar Nowcasting with Generative Adversarial Networks. *Remote Sens.* **2023**, *15*, 3306. [CrossRef]

remote sensing

MDPI

Article

A Novel Optimization Strategy of Sidelobe Suppression for Pulse Compression Weather Radar

Jiaqi Hu [1,2], Xichao Dong [1,2,*], Weiming Tian [1,2], Cheng Hu [1,2], Kai Feng [3] and Jun Lu [4,5]

1 School of Information and Electronics, Beijing Institute of Technology, Beijing 100081, China; 3120185414@bit.edu.cn (J.H.); tianweiming@bit.edu.cn (W.T.); cchchb@bit.edu.cn (C.H.)
2 The Key Laboratory of Electronic and Information Technology in Satellite Navigation, Ministry of Education, Beijing Institute of Technology, Beijing 100081, China
3 The Aerospace Newsky Technology Co., Ltd., Wuxi 214000, China; feng.kai@js1959.com
4 Beijing Institute of Technology Chongqing Innovation Center, Chongqing 401120, China; lujun@bitcq-radarlab.com
5 Chongqing Key Laboratory of Novel Civilian Radar, Chongqing 401120, China
* Correspondence: xcdong@bit.edu.cn

Abstract: The solid-state transmitters are widely adopted for weather radars, where pulse compression is operated to provide the required sensitivity and range resolution. Therefore, effective sidelobe suppression strategies must be employed, especially for weather observation. Currently, many methods can suppress the sidelobe to a very low level in the case of point targets or uniformly distributed targets. However, in strong convection weather process, the weather echo amplitude lies in a wide dynamic range and the main lobe of weak target is prone to being contaminated by the sidelobe of strong target, causing the degradation of weather fundamental data estimation, even generating artifacts and affecting the quantitative precipitation evaluation. In this paper, we propose a novel strategy which is the further processing of a general pulse compression radar to mitigate the effects of sidelobes. The proposed method is called the predominant component extraction (PCE), in which the re-weighting processing is operated after pulse compression, and then the echo of each bin is optimized and its energy will approach the real targets in each bin. It can improve the estimation of weak signals or even eliminate the artifact at the edge of the scene. Numerical simulation experiments and real-data verifications are implemented to validate the feasibility and superiority. It is noted that the proposed method has no requirement on the transmitted waveform and can be realized only by adding a step after pulse compression in the actual system.

Keywords: weather radar; sidelobe suppression; pulse compression

1. Introduction

Pulse compression technology can meet the requirements of high-range resolution and long-range detection at the same time, which has been widely used in weather radar systems with the application of solid-state transmitter.

The commonly used pulse compression waveforms include linear frequency modulation (LFM), nonlinear frequency modulation (NLFM), phase coded signal [1–3], etc. The LFM waveform which is the most widely used has high-range sidelobe (-13.26 dB) after matched filtering. A window function is often utilized to suppress range sidelobes, resulting in the expansion of the main lobe and the loss of the signal to noise ratio (SNR) [4,5]. However, the reflectivity of precipitations ranges from about -10 to 75 dBZ [6,7] and the radar sensitivity is one of the most critical factors for distributed precipitations in weather observations. Therefore, in order to reduce the SNR loss of LFM signal after windowing, the NLFM waveform was developed decades ago. The concept of NLFM waveform was put forward by Fowle and Brandon in 1959 [8,9]. The energy spectrum (i.e., the square of the spectrum) of NLFM can be designed as a window function, namely, the effect of windowing can be transferred to signal modulation, and then the matching filter can be performed directly to achieve low sidelobe performance, in order to avoid the SMR loss caused by windowing processing [7]. The peak sidelobe ratio (PSLR) can be reduced to below −40 dB after matched filtering, which can achieve the same results of window function weighting processing [10]. At the same time, NLFM has better detection rate characteristics and more accurate range detection performance than LFM [11].

However, some researches have shown that the adjacent reflectivity in extreme weather targets varies dramatically and the gradient often ranges from 30 to 40 dBZ within 1 km, sometimes even reaching 55 dBZ/km [12–14]. In order to realize the detection of weather targets with high gradient-reflectivity phenomenon, a very effective sidelobe suppression strategy must be adopted to avoid the artifact caused by range sidelobes [15], while the sidelobe performances of NLFM waveform are not enough to meet the accuracy of the quantitative detection of distributed scatterers, and the sidelobe needs to be further suppressed.

To reach the ultra-low range sidelobes, a mismatch filter can be used at the cost of losing SNR [5,16,17]. Argenti et al. [12] designed transmit waveforms and receive filters using the quadratic nonlinear optimization method by minimizing both PSLR and integrated sidelobe ratio (ISLR) of the waveform at the receiver output, and the PSLR and ISLR can reach −80 dB and −70 dB, respectively with the loss of resolution degradation. Beauchamp et al. [18] discussed the optimal design of pulse compression waveform/filter pairs for use with near-nadir spaceborne radar in low Earth orbit for the observation of clouds and precipitation. It was demonstrated that the LFM waveforms provide superior performance over NLFM waveforms for the application subject to unmitigated Doppler shifts and the PSLR and ISLR could reach −56 dB and −34 dB, respectively using the minimum integrated sidelobe (ISL) mismatch filter. Kurdzo et al. [15] designed NLFM waveform using a genetic algorithm that took into account individual system characteristics and performance measures in order to design a low SNR loss (high power efficiency) waveform for use with weather radar utilizing pulse compression. In addition, the waveform was implemented in the X-band transportable solid-state dual-polarized weather radar system (PX-1000), and the PSLR and ISLR in the actual system can reach −52 dB and −37 dB, respectively. For weather radar, the received signal of one radar resolution volume (RRV) is the sum of scattered signals from the ensemble of particles in this RRV. Therefore, the velocities of the particles in RVV continuously distribute in an interval and the Doppler spectrum or power spectral density (PSD) generally has Gaussian shape [19]. The spectral moments, i.e., reflectivity, mean radial velocity, and spectrum width (ZVW) are called fundamental weather parameters and are defined as the top three order moments of PSD. Bharadwaj et al. [20] used frequency diversity waveform and minimum ISL mismatch filter for pulse compression. The simulation results show that the estimation errors of fundamental weather parameters, i.e., reflectivity, mean radial velocity, and spectrum width (ZVW) for weak targets will increase when the reflectivity changes greatly, which had been verified using CASA (Center for Collaborative Adaptive Sensing of the Atmosphere) X-band dual-polarized radar.

The sidelobe of the point target from the above-mentioned pulse compression filters can be compressed to a very low level, while it can only be achieved for point or uniformly distributed targets, such as layered precipitation. In the process of strong convection weather phenomenon, the reflectivity will appear as a large gradient, and the sidelobe energy in the strong target range bins will be superimposed into the weak target range bins. When the reflectivity gradient reaches a certain range (such as 40 dBZ/km), the sidelobe energy of the strong target is equal to the main lobe energy of the weak target which will be contaminated, causing the unacceptable estimation errors of ZVW, even generating false targets and affecting the estimation of precipitation [15,21].

Let us assume that the probing of the meteorological object is conducted with frequency modulated waveform (such as LFM or NLFM) at a fixed antenna orientation. After pulse compression of the return signal by the optimal filter (OF), such as matched filter or mismatched filter, some realization of a random echo-signal is received. This signal is a mixture of the amplitudes of the main and sidelobes of the response of the OF, resulting in the increase in estimation error of ZVW. To isolate the reaction of the OF only by the main lobe, in this paper, an artificial realization of the echo-signal is initially modeled. This realization consists only of rectangular pulses with amplitudes obtained by smoothing the amplitudes of the real realization of the echo-signal, which is called basic waveform transfer function (B-WTF) to describe the main lobe signals. Then, considering the parameters of the waveform and OF, the frequency-modulated pulse with amplitudes obtained by smoothing the amplitudes of the real realization of the echo-signal is modeled to describe the sidelobe signals, which is called pulse-compression waveform transfer function (PC-WTF) and is supplemented by the main lobe signals. Thereafter, the cost function of each range bin can be constructed by B-WTF and PC-WTF for the calculations of optimal extraction matrix (OEM) for re-weighting of original real realization. The optimized echoes are used to calculate the relevant data quality index according to some criterions. If the criterions are not satisfied, the iteration procedure is repeated with slightly modified initial conditions until the matching criterion is met. Through the above-mentioned procedure, the predominant signal in each range bin can be extracted from the mixed signal and the sidelobes are cleared.

This method is called the predominant component extraction (PCE), which is operated after pulse compression for a normal weather radar with LFM or NLFM waveform to improve sidelobe suppression. After PCE processing, the energy of optimized echoes will be close to the main lobe energy of the actual targets, achieving the sidelobe suppression of the weather targets with large gradient reflectivity.

When the reflectivity gradient is large, the proposed method can obtain the weak target signal and reduce the estimation errors of ZVW. This optimization effect is particularly clear for the large gradient reflectivity scene in the typhoon or other strong convective weather phenomenon. In addition, this method can significantly improve the data quality and eliminate the artifact at the edge of the scene, which contributes to obtaining the accurate estimation of precipitation and other parameters. It is worth mentioning that the proposed method is operated after pulse compression, and it has no requirement on the radar transmitting waveform, which can be realized by adding a step after pulse compression in the actual radar system.

This article is organized as follows. The optimization model is presented in Section 2. Section 3 describes the calculation of the OEM and the iterative optimization process of the PCE. Some results of simulations and verifications based on the real measured data are shown in Section 4. The discussion is drawn in Section 5. Finally, Section 6 provides the conclusion and research perspectives.

2. Problem Statement and Modeling

Considering one range-direction signal, the normalized echo matrix of unit amplitude transmitted modulation waveform, such as LFM or NLFM after pulse compression is defined as the pulse-compression waveform transfer function (PC-WTF) $\mathbf{s}_{MF}$, and the normalized echo matrix with unit amplitude rectangular-pulse transmitted waveform is defined as the basic waveform transfer function (B-WTF) $\mathbf{s}_{SP}$, which are expressed as follows:

$$\mathbf{s}_{MF} = \begin{bmatrix} \mathbf{s}_{MF,1} \\ \mathbf{s}_{MF,2} \\ \vdots \\ \mathbf{s}_{MF,k} \\ \vdots \\ \mathbf{s}_{MF,K} \end{bmatrix} = \begin{bmatrix} s_{MF,1}(1), s_{MF,1}(2), \cdots, s_{MF,1}(p), \cdots, s_{MF,1}(P) \\ s_{MF,2}(1), s_{MF,2}(2), \cdots, s_{MF,2}(p), \cdots, s_{MF,2}(P) \\ \vdots \\ s_{MF,k}(1), s_{MF,k}(2), \cdots, s_{MF,k}(p), \cdots, s_{MF,k}(P) \\ \vdots \\ s_{MF,K}(1), s_{MF,K}(2), \cdots, s_{MF,K}(p), \cdots, s_{MF,K}(P) \end{bmatrix} \tag{1}$$

$$\mathbf{s}_{SP} = \begin{bmatrix} \mathbf{s}_{SP,1} \\ \mathbf{s}_{SP,2} \\ \vdots \\ \mathbf{s}_{SP,k} \\ \vdots \\ \mathbf{s}_{SP,K} \end{bmatrix} = \begin{bmatrix} s_{SP,1}(1), s_{SP,1}(2), \cdots, s_{SP,1}(p), \cdots, s_{SP,1}(P) \\ s_{SP,2}(1), s_{SP,2}(2), \cdots, s_{SP,2}(p), \cdots, s_{SP,2}(P) \\ \vdots \\ s_{SP,k}(1), s_{SP,k}(2), \cdots, s_{SP,k}(p), \cdots, s_{SP,k}(P) \\ \vdots \\ s_{SP,K}(1), s_{SP,K}(2), \cdots, s_{SP,K}(p), \cdots, s_{SP,K}(P) \end{bmatrix} \quad (2)$$

where each row of (1) and (2) represents the range direction sample sequence and P is the number of range bins of echoes. Each column represents the normalized amplitude of each target applied to the corresponding range bin, and K is the number of targets. In contrast to point targets, the entire radar beam is usually filled with weather targets; therefore, the number of range bins can be treated as equivalent to the number of targets, i.e., $P = K$. In addition, $\mathbf{s}_{MF,k}$ is the PC-WTF of the kth target and $\mathbf{s}_{SP,k}$ is the B-WTF of the kth target as follows:

$$\mathbf{s}_{MF,k} = [s_{MF,k}(1), s_{MF,k}(2), \cdots, s_{MF,k}(p), \cdots, s_{MF,k}(P)] \quad (3)$$

$$\mathbf{s}_{SP,k} = [s_{SP,k}(1), s_{SP,k}(2), \cdots, s_{SP,k}(p), \cdots, s_{SP,k}(P)] \quad (4)$$

Due to the absence of sidelobes of rectangular-pulse waveform and the presence of sidelobes of pulse compression waveform, $s_{MF,k}(p)$ and $s_{SP,k}(p)$ which are the members of (3) and (4), respectively, can be expressed as follows:

$$s_{MF,k}(p) = \begin{cases} a_0 & p = k \\ b(p) & p \neq k \end{cases} \quad (5)$$

$$s_{SP,k}(p) = \begin{cases} a_0 & p = k \\ 0 & p \neq k \end{cases} \quad (6)$$

where a_0 is the main lobe value after pulse compression of kth target's echo, k is the range bin position where the main lobe is located, and $b(p)$ is the sidelobe value at other range bin locations of kth target's echo.

Therefore, the values of a_0 and $b(p)$ are only related to the waveform parameters, the radar system parameters, and the method of pulse compression, which are independent of the target characteristics. In particular, as long as the waveform parameters, the radar system parameters, and the method of pulse compression are determined, $\mathbf{s}_{MF}$ and $\mathbf{s}_{SP}$ will be determined. Considering the PC-WTF as an example, its schematic diagram is shown in Figure 1:

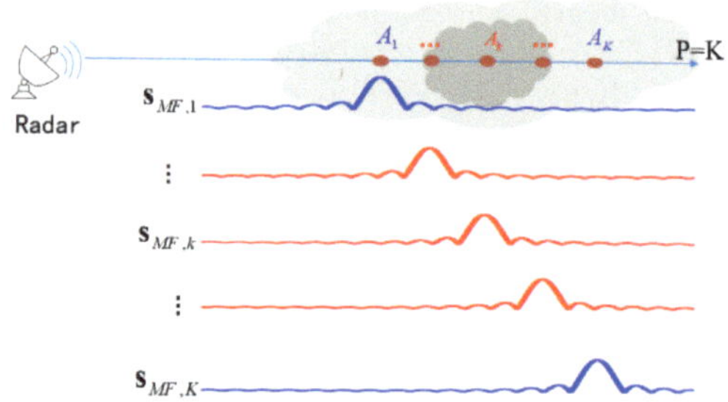

Figure 1. The schematic diagram of PC-WTF.

Let us assume that $\mathbf{A} = \begin{bmatrix} A_1 & \cdots & A_k & \cdots & A_K \end{bmatrix}$ is the backscattering coefficient matrix, and A_k is the complex backscattering coefficient of the kth target, where the amplitude is related to the scattering intensity of the target and the phase includes the information of doppler velocity of the target. The received echo is the superposition from all targets as shown in Figure 2, which can be expressed as (7):

$$\begin{aligned}\mathbf{P}_0 &= \mathbf{A}\mathbf{s}_{MF} \\ &= \left[\sum_{k=1}^{K} A_k s_{MF,k}(1), \sum_{k=1}^{K} A_k s_{MF,k}(2), \cdots, \sum_{k=1}^{K} A_k s_{MF,k}(P)\right]\end{aligned} \tag{7}$$

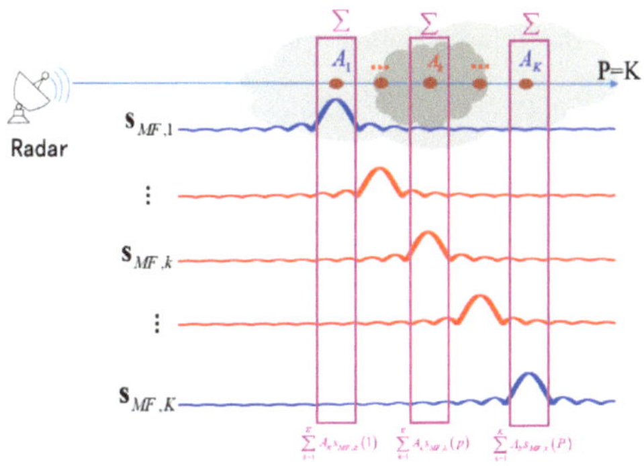

Figure 2. The schematic diagram of received echoes.

It can be seen from (7) that the energy of each range bin is the superposition of the main lobe of this bin and the sidelobes from other bins. When the energy of the extra added sidelobes is comparable to that of the original main lobe, it will affect the estimations of ZVW. Therefore, it is considered to re-optimize the signal of each range bin and its energy will approach that of the actual targets in each bin; therefore, we call it the predominant component extraction (PCE) method.

Considering the operation of the weighting processing of the signals in each range bin, we can obtain the weighted signal as follows:

$$\mathbf{P}_{est} = \mathbf{A}\left(\mathbf{W}_{opt} \odot \mathbf{s}_{MF}\right) \tag{8}$$

where $\mathbf{W}_{opt}$ is the OEM with $K \times P$ dimension as follows:

$$\mathbf{W}_{opt} = \begin{bmatrix} w_{opt,1}(1) & w_{opt,2}(1) & \cdots & w_{opt,p}(1) & \cdots & w_{opt,P}(1) \\ w_{opt,1}(2) & w_{opt,2}(2) & \cdots & w_{opt,p}(2) & \cdots & w_{opt,P}(2) \\ \vdots & \vdots & \cdots & \vdots & \cdots & \vdots \\ w_{opt,1}(k) & w_{opt,2}(k) & \cdots & w_{opt,p}(k) & \cdots & w_{opt,P}(k) \\ \vdots & \vdots & \vdots & \vdots & \vdots & \vdots \\ w_{opt,1}(K) & w_{opt,2}(K) & \cdots & w_{opt,p}(K) & \cdots & w_{opt,P}(K) \end{bmatrix} \tag{9}$$

where $\mathbf{w}_{opt,p} = \left[w_{opt,p}(1) \quad \cdots \quad w_{opt,p}(K)\right]^{\mathrm{T}}$ is the optimization coefficient of pth range bin and $\odot$ is Hadamard product. The signal is weighed using $\mathbf{w}_{opt,p}$ in each range bin in order that the energy will approach that of the actual targets. Next, an optimization problem is established by B-WTF to obtain $\mathbf{W}_{opt}$.

Since the echoes of rectangular pulse signal only have the main lobe and there is no sidelobe to be superimposed in the other range bins, the calculated signal energies are optimal in theory. Therefore, we can use rectangular waveforms to model the expected output echo:

$$\begin{aligned}\mathbf{P}_{\exp} &= \mathbf{A}\mathbf{s}_{SP} \\ &= \left[\sum_{k=1}^{K} A_k s_{SP,k}(1), \sum_{k=1}^{K} A_k s_{SP,k}(2), \cdots, \sum_{k=1}^{K} A_k s_{SP,k}(P)\right] \\ &= \left[A_1 s_{SP,1}(1), A_2 s_{SP,2}(2), \cdots, A_P s_{SP,P}(P)\right]\end{aligned} \tag{10}$$

In order to obtain $\mathbf{P}_{\exp}$, $\mathbf{W}_{opt}$ needs to meet the following condition:

$$\begin{aligned}\min_{\mathbf{W}_{opt}} \quad & p(\mathbf{W}_{opt}) \\ s.t. \quad & p(\mathbf{W}_{opt}) = \|\mathbf{P}_{est} - \mathbf{P}_{\exp}\|_2^2\end{aligned} \tag{11}$$

To date, the optimization model has been established.

It should be noted that, in this paper, we assume that the echoes are entirely from meteorological targets and the patterns of targets are determined by the backscattering coefficient matrix and waveform transfer function. However, the PCE method can be operated regardless of how the target is distributed. We do not need to know the value of backscattering coefficient matrix in advance when modeling PC-WTF and B-WTF. As long as the waveform and radar system parameters are determined, PC-WTF and B-WTF can be modeled, which can be realized in the actual processing process. After modeling PC-WTF and B-WTF, what we need to solve is how to optimize (7) to make it closer to (10), which needs to be realized by the PCE method proposed in this paper. The block diagram of the PCE algorithm is presented in Figure 3.

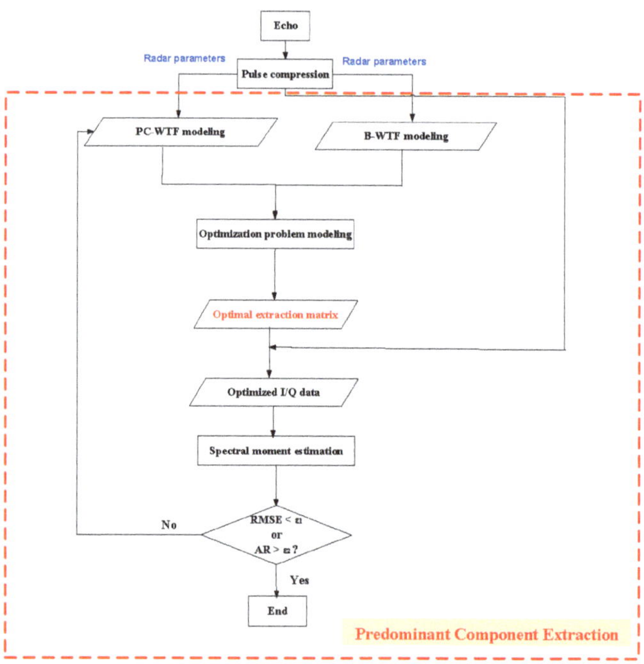

Figure 3. The block diagram of the PCE algorithm.

3. Optimization Process

3.1. The Solution of OEM

After obtaining the above-mentioned optimization model, $\mathbf{W}_{opt}$ can be calculated by the derivative of $p(\mathbf{W}_{opt})$ and setting the derivative to zero. Next, we will introduce the calculation process in detail.

Expanding (11), we can achieve:

$$\begin{aligned} \min_{\mathbf{W}_{opt}} \quad & p(\mathbf{W}_{opt}) \\ s.t. \quad & p(\mathbf{W}_{opt}) = \left\|\mathbf{A}\left(\mathbf{W}_{opt} \odot \mathbf{s}_{MF}\right) - \mathbf{A}\mathbf{s}_{SP}\right\|_2^2 \end{aligned} \tag{12}$$

For distributed meteorological targets, P = K can be assumed and A is a matrix that does not change in one coherent processing interval (CPI). At this time, a unit matrix **I** can be introduced to facilitate the derivative calculation of Hadamard product. Then, the above-mentioned optimization problem can be written as:

$$\begin{aligned} \min_{\mathbf{W}_{opt}} \quad & p(\mathbf{W}_{opt}) \\ s.t. \quad & p(\mathbf{W}_{opt}) = \left\|\left(\mathbf{W}_{opt}\mathbf{I} \odot \mathbf{s}_{MF}\right) - \mathbf{s}_{SP}\right\|_2^2 \end{aligned} \tag{13}$$

Additionally, $\mathbf{W}_{opt}$ can be calculated from $\frac{d[p(\mathbf{W}_{opt})]}{d\mathbf{W}_{opt}} = 0$, as shown in (14), where the detailed calculation process is in Appendix A.

$$\mathbf{W}_{opt} = \left(\mathbf{I} \odot \mathbf{s}_{MF} \odot \mathbf{s}_{MF}\mathbf{I}^{\mathrm{T}}\right)^{-1}\left(\mathbf{I} \odot \mathbf{s}_{MF}\mathbf{s}_{SP}\right) \tag{14}$$

3.2. Iteration Process

After obtaining $\mathbf{W}_{opt}$, the echoes can be optimized through formula (8). However, the initial matrix **A** needs to be known. Thereafter, the data quality can be improved through iteration. If the actual received echo is $\mathbf{S}_r$, we can express it as:

$$\mathbf{S}_r = \mathbf{A}\mathbf{s}_{MF} \tag{15}$$

A can be calculated as:

$$\mathbf{A} = \left[\left(\mathbf{s}_{MF}^{\mathrm{T}}\right)^{-1}\mathbf{S}_r^{\mathrm{T}}\right]^{\mathrm{T}} \tag{16}$$

Then, the estimated echo by the PCE method can be expressed as:

$$\mathbf{S}_{est} = \mathbf{A}\left(\mathbf{W}_{opt} \odot \mathbf{s}_{MF}\right) \tag{17}$$

Thereafter, the OEM of (17) is recalculated and optimized iteratively until the stop criterion is met. Next, we set the appropriate stop criterion to make the data quality meet the requirements.

Since the proposed method in this paper is not aimed at the point target sidelobe, the performance of PSLR and ISLR cannot be compared. Therefore, other indicators that can reflect the data quality need to be set as the stop criterion.

- Data error iteration condition:

The root mean square error (RMSE) reflects the degree of data deviation from the ground truth. In the quantitative analysis of simulation, the ground truth is given; therefore, the stop criterions can be set according to the RMSE. The RMSE of one range direction can be calculated as:

$$RMSE_E = \sqrt{\frac{1}{P}\sum_{p=1}^{P}\left|E_{est}(p) - E_{gt}(p)\right|^2} \tag{18}$$

where E_{est} represents the ZVW (i.e., reflectivity Z_{est}, radial velocity V_{est}, and spectral width W_{est}) calculated from the optimized echoes and E_{gt} is the ground truth of ZVW.

Since the energy optimization effect of each range bin is our concern in this paper, we can set the RMSE of the reflectivity less than a certain threshold ε_1 as the stop criterion, as follows:

$$RMSE_{Z_{est}} < \varepsilon_1 \tag{19}$$

- Data fluctuation iteration condition:

In real-data verification, the ground truth cannot be obtained, while the moving average can be used as the ground truth to calculate SD. Therefore, the SD can reflect the fluctuation of radar data and we call it data fluctuation, which can be used to set the stop criterion. The standard deviation (SD) of ZVW in the pth range bin can be calculated by the data on n range bins before and n range bins after, and the first n range bins and the last n range bins do not participate in the calculation of SD [22]:

$$\begin{aligned} SD_p &= \sqrt{\frac{1}{2n+1}\sum_{j=0}^{2n+1}\left(E_{p+j-n} - \overline{E}_p\right)^2}, 0 < n < \tfrac{P}{2}, n < p < P-n \\ \overline{E}_p &= \frac{1}{2n+1}\sum_{j=0}^{2n+1} E_{p+j-n}, 0 < n < \tfrac{P}{2}, n < p < P-n \end{aligned} \tag{20}$$

where E represents the ZVW (i.e., reflectivity $\mathbf{Z}_{est}$, radial velocity $\mathbf{V}_{est}$, and spectral width $\mathbf{W}_{est}$) calculated from the optimized echoes. $\overline{E}_p$ is the moving average calculated from $2n+1$ range bins.

Setting the percentage of data with the SD less than 1 in the total data greater than ε_2 as the stop criterion, we can make the data fluctuation meet the requirements. This percentage is defined as the data fluctuation qualification rate AR, as follows:

$$AR = \frac{\text{sum}(SD_p < 1)}{P} \times 100\% > \varepsilon_2, 0 < \varepsilon_2 < 100\% \tag{21}$$

In summary, Algorithm 1 can be summarized as follows.

Algorithm 1: The PCE algorithm

◆ *STEP 1*: Model the PC-WTF and B-WTF according to the radar system and waveform parameters.
◆ *STEP 2*: Calculate $\mathbf{W}_{opt}$ according to (14).
◆ *STEP 3*: Calculate the initial matrix $\mathbf{A}$ according to (16).
◆ *STEP 4*: Calculate the first optimization echo $\mathbf{S}_{est}$ according to (17) and estimate the spectral moments, i.e., $\mathbf{Z}_{est}$, $\mathbf{V}_{est}$, and $\mathbf{W}_{est}$.
◆ *STEP 5*: In the quantitative analysis of simulation stage, calculate the RMSE and AR of $\mathbf{Z}_{est}$ and set the iteration thresholds ε_1 and ε_2 Judge whether the stop criterion of (19) and (21) are met. In the real-data verification stage, calculate the AR of $\mathbf{Z}_{est}$ and set the iteration threshold ε_2 Judge whether the stop criterion of (21) is met.
◆ *STEP 6*: Repeat steps 3 to 5 until the stop criterion is satisfied.

---

4. Results

Before the distributed target simulation, first, the point target scene is simulated and the performances of PSLR and ISLR of four different pulse compression methods are compared, i.e., LFM waveform with windowing, LFM waveform with mismatch filtering, NLFM waveform with windowing, and NLFM waveform with mismatch filtering, wherein the NLFM signal is constructed according to [23] and the mismatch filter is constructed according to the method in [24]. Then, the input ZVW is used to simulate one-dimensional echoes of distributed targets in the range direction. The pulse compression method with the best performance is used for processing. The pulse pair processing method (PPP) is used for spectral moment estimation [21]. This processing set can explain the influence of sidelobe superposition of distributed meteorological targets. Thereafter, the PCE method is used and the RMSE and AR are calculated through (18) and (21) to quantitatively analyze the advantages of the proposed method. Finally, the real in-phase and quadrature-phase (I/Q) data of an actual supercell from a ground-based weather radar and global precipitation measurement (GPM) precipitation data are obtained, which are applied to the proposed method for verification.

4.1. Verifications Based on Numerical Simulation Experiments

4.1.1. Sidelobe Superposition Effects

First, the echo of point target is simulated, and the parameters are shown in Table 1. The processing results using different pulse compression methods are in Figure 4, where the window function is Hamming window function.

Table 1. The simulation parameters of point target.

Parameters	Values
Frequency (GHz)	13.6
Band width (MHz)	4
Pulse width (μs)	64
Sample frequency (MHz)	16
Target distance (km)	15

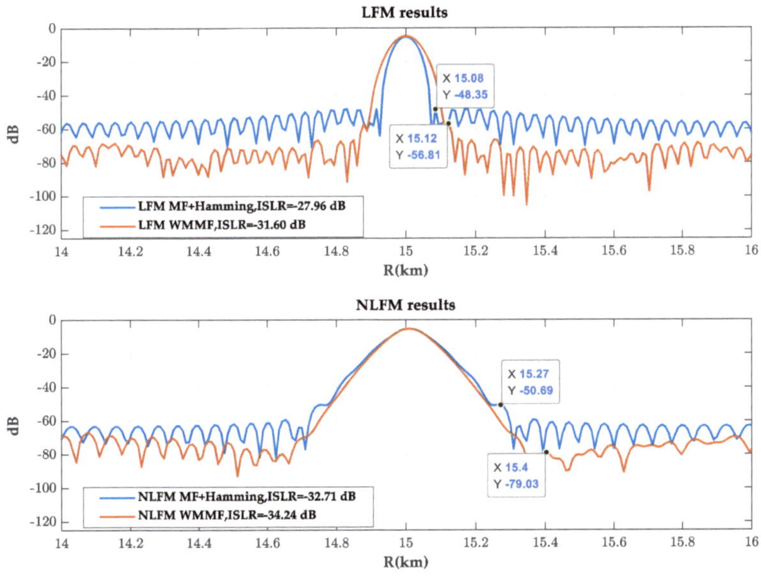

Figure 4. The pulse compression results of point target.

The performances of the four pulse compression methods are shown in Table 2, where "W" refers to the windowing process. It can be indicated that the windowing process will cause serious SNR loss and the loss of SNR can be mitigated by using NLFM waveform. In addition, the performances of the combination of NLFM waveform and mismatch filter are best, which can suppress PSLR and ISLR below −60 dB and −35 dB, respectively.

Table 2. The performances of the four pulse compression methods.

	LFM (MF + W)	LFM (WMMF)	NLFM (MF + W)	NLFM (WMMF)
PSLR (dB)	−53.34	−55.31	−60.97	−67.89
ISLR (dB)	−25.66	−27.84	−28.94	−37.23
SNR loss (dB)	6.12	1.84	3.16	0.14

Next, the one-dimensional echoes of distributed targets with different reflectivity gradients are simulated. The simulation parameters are shown in Table 3, where the SNR of the input signal is 60 dB. The target number is equal to 350, which is equivalent to the total range bin number. The range revolution is 37.5 m and the detection distance is 13.125 km. In addition, the weak targets account for 7.5 km and the strong targets account for 3.6 km. The transmit waveform is NLFM in [23] and the mismatch filter in [24] is used for pulse compression. The input Z with different gradients is shown in Figure 5. In addition, the reflectivity, mean velocity, and spectral width can be obtained by the PPP method in Figures 6–8, respectively.

Table 3. The simulation parameters of one-dimensional echoes of distributed targets with different reflectivity gradients.

Parameters	Values	Parameters	Values
Frequency (GHz)	13.6	Pulse width (μs)	64
Band width (MHz)	4	Sample frequency (MHz)	16
Accumulation pulse number	64	PRF (Hz)	1860/1395
Target number	350	Gradient of Z (dBZ/km)	15/25/40
Target spectral width (m/s)	Randi [1, 5]	Target velocity (m/s)	Int [−7, 7]

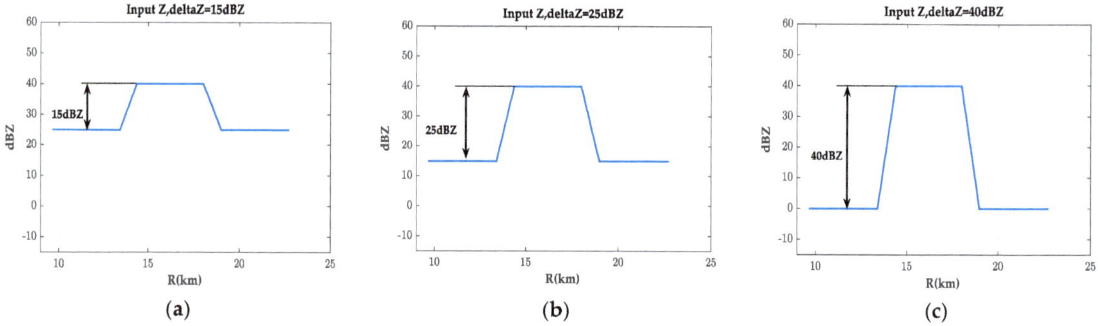

Figure 5. The input Z with different gradients. (**a**) Gradient of Z is 15 dBZ/km; (**b**) gradient of Z is 25 dBZ/km; (**c**) gradient of Z is 40 dBZ/km.

Although the performance of the pulse compression method is good enough, when the reflectivity gradient is large (such as 40 dBZ/km), the energy of the sidelobe of the strong target is equal to or even greater than the energy of the weak target, resulting in the estimation errors of the ZVW in the weak target area.

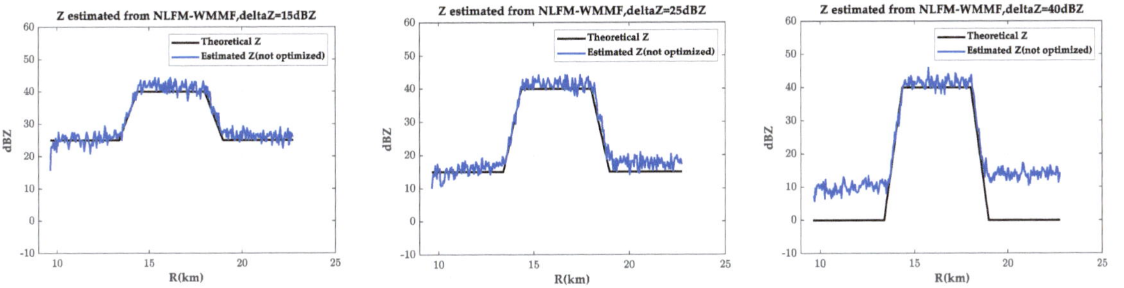

(a) (b) (c)

Figure 6. The reflectivity calculated from unoptimized echoes. (**a**) Gradient of Z is 15 dBZ/km; (**b**) gradient of Z is 25 dBZ/km; (**c**) gradient of Z is 40 dBZ/km.

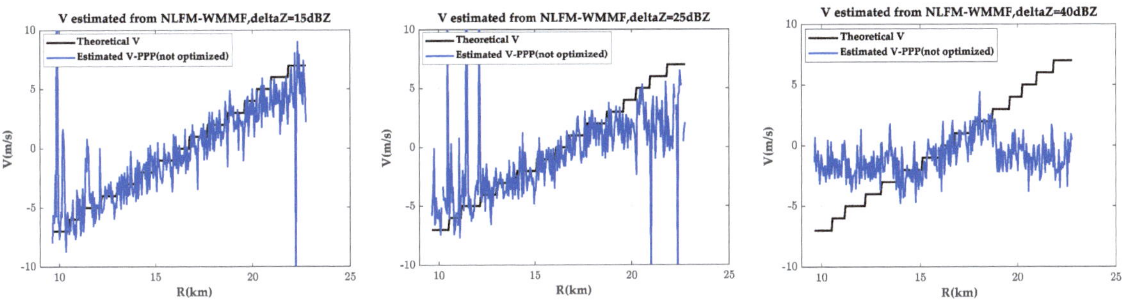

(a) (b) (c)

Figure 7. The mean velocity calculated from unoptimized echoes. (**a**) Gradient of Z is 15 dBZ/km; (**b**) gradient of Z is 25 dBZ/km; (**c**) gradient of Z is 40 dBZ/km.

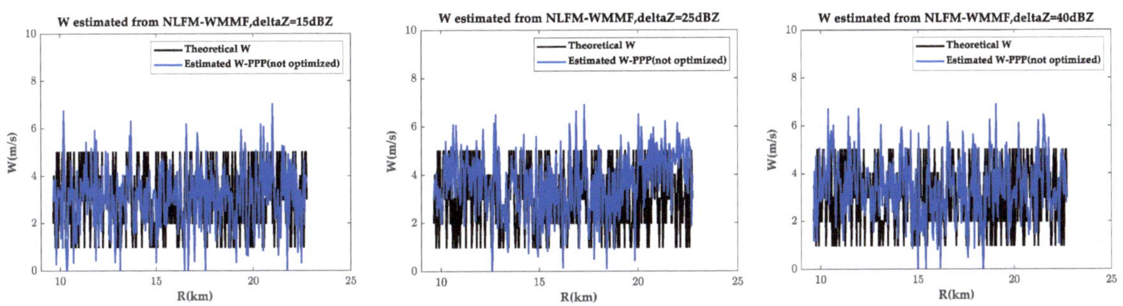

(a) (b) (c)

Figure 8. The spectral width calculated from unoptimized echoes. (**a**) Gradient of Z is 15 dBZ/km; (**b**) gradient of Z is 25 dBZ/km; (**c**) gradient of Z is 40 dBZ/km.

4.1.2. Optimization Process

In order to solve the above-mentioned problems, the PCE algorithm is used for each range bin. The comparison results are shown in Figures 9–11. The RMSE and AR of ZVW before and after optimization can be calculated through (18) and (21), respectively. The results are shown in Table 4.

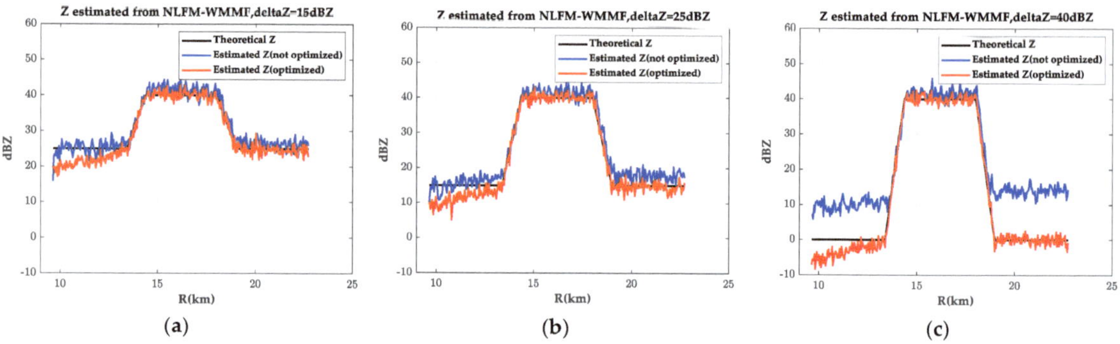

Figure 9. The comparison results of reflectivity. (**a**) Gradient of Z is 15 dBZ/km; (**b**) gradient of Z is 25 dBZ/km; (**c**) gradient of Z is 40 dBZ/km.

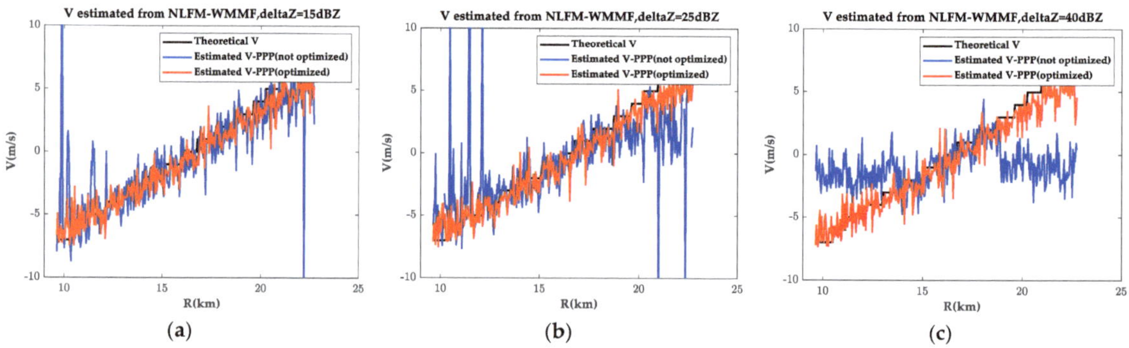

Figure 10. The comparison results of mean velocity. (**a**) Gradient of Z is 15 dBZ/km; (**b**) gradient of Z is 25 dBZ/km; (**c**) gradient of Z is 40 dBZ/km.

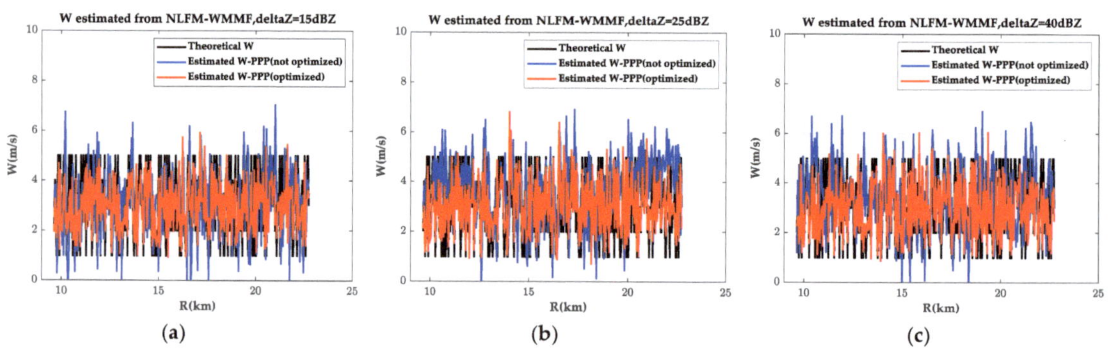

Figure 11. The comparison results of spectral width. (**a**) Gradient of Z is 15 dBZ/km; (**b**) gradient of Z is 25 dBZ/km; (**c**) gradient of Z is 40 dBZ/km.

Table 4. The RMSE and AR of ZVW (not optimized/optimized).

Gradient of Z (ΔZ, dBZ/km)		15	25	40
Z (dBZ)	RMSE (dBZ)	2.65/2.32	3.30/2.34	10.42/4.23
	AR (%)	24.57/50.01	32.37/45.66	28.90/28.61
V (m/s)	RMSE (m/s)	2.02/1.14	2.93/1.20	4.13/1.61
	AR (%)	39.02/85.84	28.61/82.95	36.99/49.42
W (m/s)	RMSE (m/s)	1.92/0.72	1.99/0.69	2.01/1.56
	AR (%)	36.13/50.87	45.66/48.84	39.60/59.54

It can be indicated that the RMSE and AR can be improved by the proposed method. Moreover, we calculate the RMSE and AR changing with the gradient of reflectivity in the range of 10~40 dBZ/km. The results are shown in Figures 12 and 13. When the reflectivity gradient is large (≥30 dBZ/km), the Z and V estimations are seriously deviated from the ground truth due to the fact that the RMSEs are large. At this time, the data are invalid and the calculated AR is not referential. The proposed method can greatly reduce the RMSE, in order that the invalid data can become valid. When the reflectivity gradient is not large (<30 dBZ/km), the RMSE is acceptable and the data are valid. At this time, we focus on the results of AR. The ARs of optimized data are more acceptable and the data quality is effectively improved.

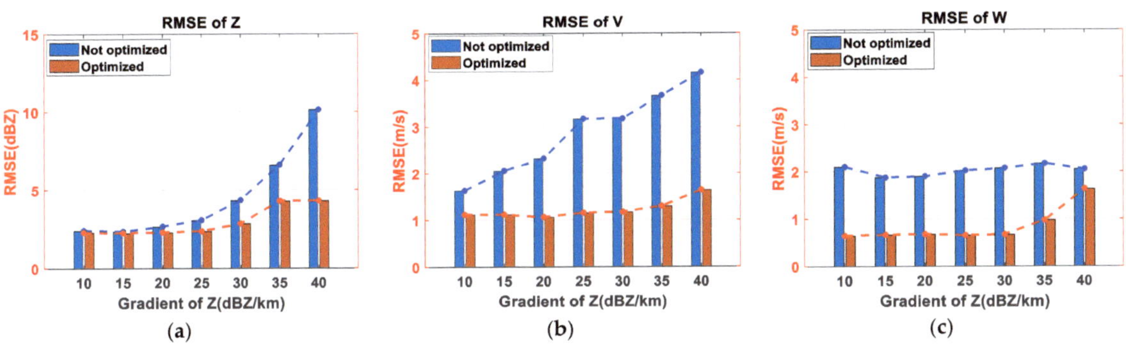

Figure 12. The RMSE of ZVW changing with the gradient of reflectivity. (**a**) Reflectivity; (**b**) mean velocity; (**c**) spectral width.

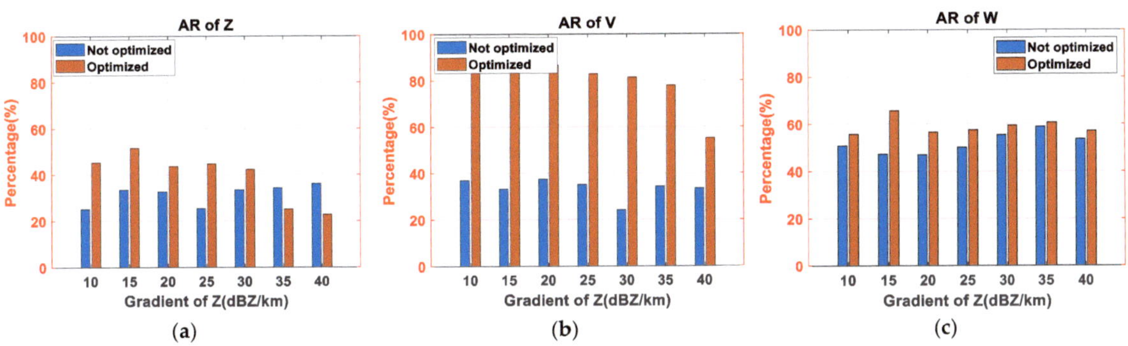

Figure 13. The AR of ZVW changing with the gradient of reflectivity. (**a**) Reflectivity; (**b**) mean velocity; (**c**) spectral width.

In addition, we compared the SNR loss caused by the proposed method. In contrast to point targets, the distributed targets cannot directly compare the energy of the main lobe. Therefore, the SNR loss is defined as the mean difference SNR before and after optimization:

$$SNRloss = \frac{1}{P}\sum_{p=1}^{P}\left[SNR_{before}(p) - SNR_{after}(p)\right] \tag{22}$$

The SNR loss changing with reflectivity gradient is shown in Figure 14. It can be indicated that with the increase in reflectivity gradient, SNR loss will become serious. Therefore, the proposed method is to improve the data quality at the cost of loss of SNR.

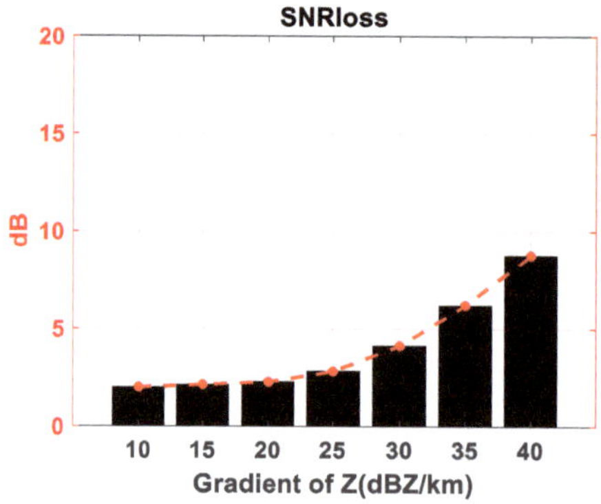

Figure 14. The SNR loss changing with reflectivity gradient.

4.2. Verifications Based on Real Data

4.2.1. Verifications by Ground-Based Weather Radar Data

The radar is located at WRCP station (32.75°N, 119.35°E) with an altitude of 32.3 m and works in the conical scanning mode with the azimuth angle of 0 to 360 degrees. The operating parameters are listed in Table 5.

Table 5. The operating parameters of ground-based weather radar.

Parameters	Value	Parameters	Value
Frequency (GHz)	5.5	Peak Power (W)	1710
Antenna gain (dB, T/R)	34/38	Beam width (°, AZ/EL)	1/3
Noise figure (dB)	3	Pulse number	32
Pulse width (μs)	100	Bandwidth (MHz)	1
Dual PRF (Hz)	900/1200	Elevation angle (°)	2

Supercells, as one of the important mesoscale weather systems, can form severe convective weather, such as heavy precipitation, thunderstorm wind, hail, and tornado. In the radar map, the supercell appears as a tightly organized image of high reflectivity, which may have a hook echo. The in-phase and quadrature(I/Q) data originated from a supercell appearing in 22:52 UTC on 17 July 2020. The echo data contained 902 CPI, 32 pulses per CPI, and 652 range gates per pulse. The transmit waveform is LFM and the matched filtering and Hamming window are used.

The ZVWs of the supercell are shown in Figure 15 and the results after optimization by PCE are shown in Figure 16. Due to the large scale of the supercell, the contrast result is not clear in the complete image. Since the proposed method improved the reflectivity (i.e., echo power) significantly, we enlarged the local reflectivity for comparison as shown in Figure 17. We can see that the artifacts at the scene edge are basically removed after optimization and the target edge is clearer due to the fact that the sidelobe is suppressed.

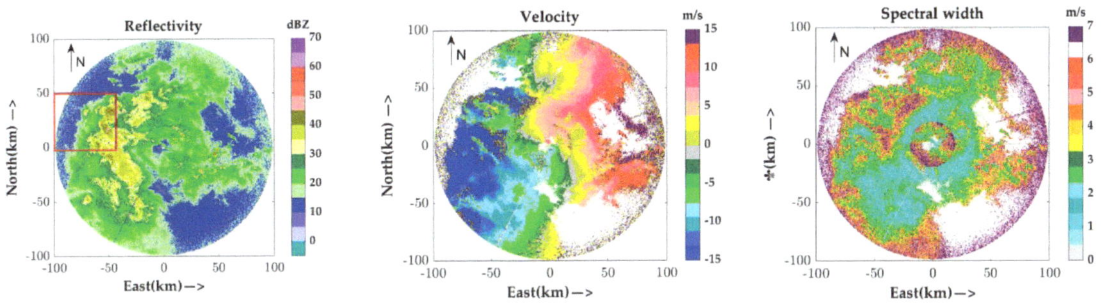

(a) (b) (c)

Figure 15. The ZVW of the supercell. ((**a**) Reflectivity; (**b**) velocity; (**c**) spectral width).

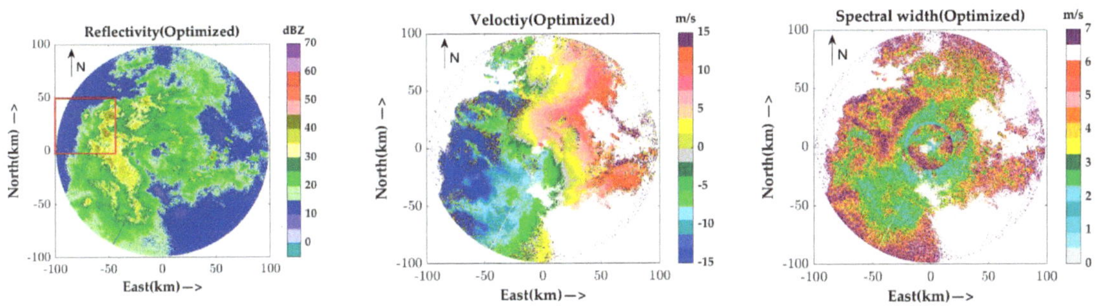

(a) (b) (c)

Figure 16. The results after optimization by PCE. ((**a**) Reflectivity; (**b**) velocity; (**c**) spectral width).

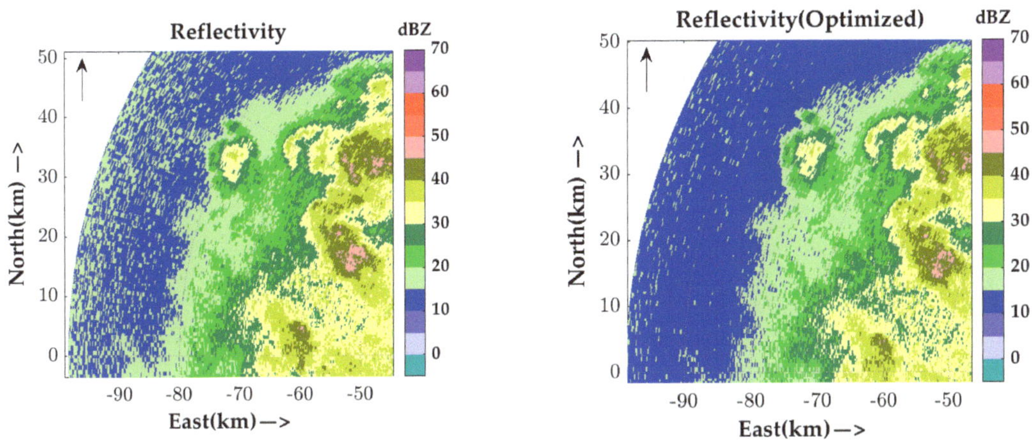

(a) (b)

Figure 17. Local reflectivity before and after optimization. ((**a**) Not optimized; (**b**) optimized).

The ARs of the ZVW are calculated in Table 6. It can be indicated that the percentages of ZVW with the SD less than 1 in total data are increased after optimization, which indicated the effective improvement of the data quality.

Table 6. The AR of the ZVW.

	Z	V	W
The AR before optimization (%)	62.68	64.90	76.39
The AR after optimization (%)	70.66	70.20	84.54

4.2.2. Verifications by Quantitative Precipitation Estimation (QPE) Data

In order to verify the application value of the proposed method, we use the Z–I relationship to estimate the precipitation in positions A and B, in which Z is the reflectivity calculated by the unoptimized data and optimized data, respectively, and I is the precipitation intensity in mm/h. The same Z–I relationship is used for both, which is $Z = 70I^{1.38}$ fitted from a supercell in [25].

The ground truth of precipitation is the global precipitation measurement (GPM) from Goddard Earth Sciences Data and Information Services Center (GES) DISK) [26] with a time resolution of 0.5 h and a spatial resolution of $0.1° \times 0.1°$.

In positions A and B, we select 3×3 pixels with an adjacent distance of 10 km. For each pixel, the reflectivity of the range bins within the four surrounding spatial resolutions $(0.1° \times 0.1°)$ is extracted to calculate the precipitations which are averaged successively, and the precipitation estimates at the pixel can be obtained. The selected region is shown in Figure 18. The precipitations at nine pixel points in the two scenes are calculated using the optimized and unoptimized reflectivity and the results are compared as shown in Figure 19. Then, the differences between the estimation and ground truth for optimized and unoptimized data are calculated to obtain the estimation errors as shown in Figure 20.

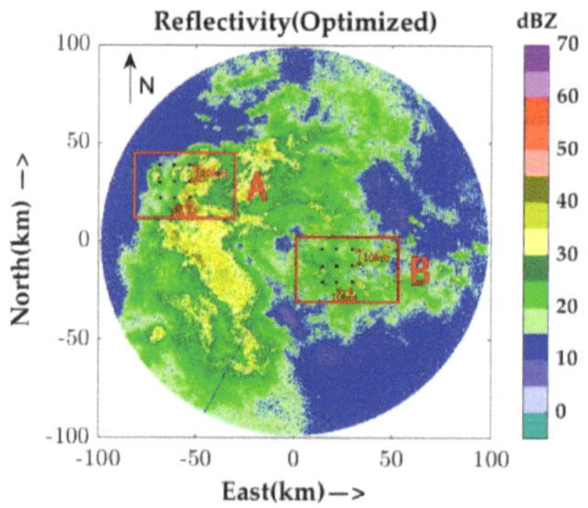

Figure 18. Selected region for QPE.

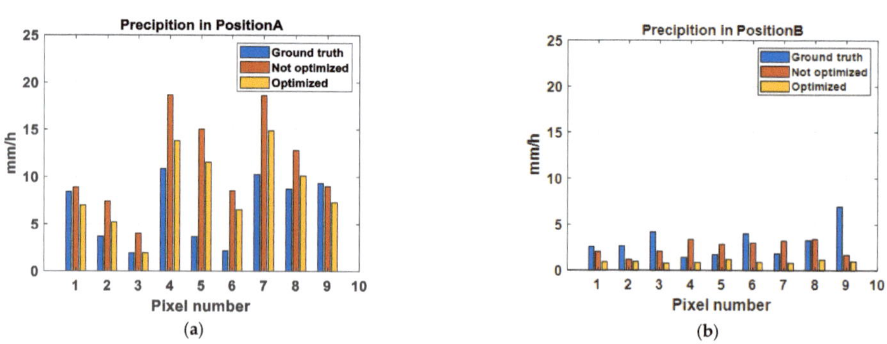

Figure 19. The comparison results of precipitations. (**a**) Position A; (**b**) position B.

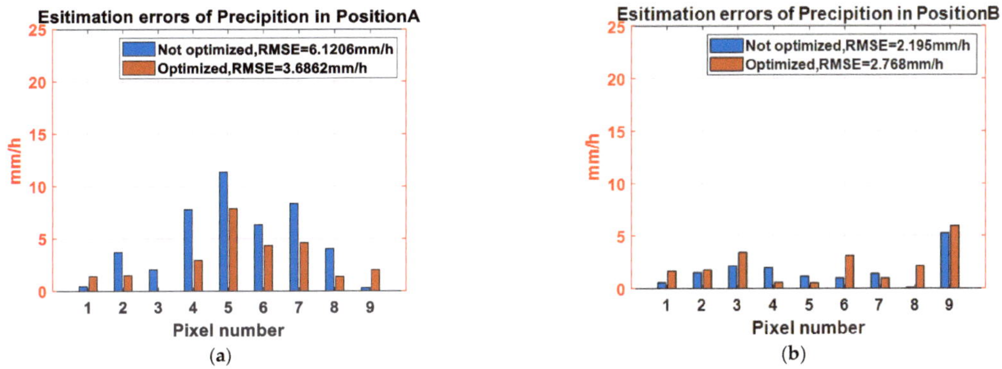

Figure 20. The estimation errors of precipitations. (**a**) Position A; (**b**) position B.

It is indicated that the precipitations calculated by the optimized reflectivity are more accurate and the RMSE is effectively decreased for the convective region (Position A) that has large gradient reflectivity, which indicates that the proposed method can improve the quality of precipitation estimation. For the stratified precipitation with low rainfall (Position B), the improvement effect is not clear. Therefore, the proposed method is more suitable for strong convection weather targets.

To date, the application value of the proposed method in this paper is illustrated by the real data of ground-based weather radar and QPE data. This method can eliminate artifacts at the edge of the scene and improve the data quality of ZVW and precipitation.

5. Discussion

In the numerical simulation experiments, the performances of the combination of NLFM waveform and mismatch filter are best, which can suppress PSLR and ISLR below −60 dB and −35 dB, respectively. Then, this pulse compression method is used in the distributed targets with different reflectivity gradients, while the energy of the sidelobe of the strong target is equal to or even greater than the energy of the weak target when the reflectivity gradient is large, resulting in the estimation errors of the ZVW in the weak target area, as shown in Figures 6–8. It is indicated that the point target performance (such as PSLR and ISLR) of the pulse compression method is good enough, while it cannot meet the requirement of the high-gradient reflectivity distributed targets. At this time, if we operate the PCE algorithm after pulse compression, the weak targets can be reconstructed and the ZVW can be calculated more accurately, as shown in Figures 12 and 13. In addition, the RMSE and AR results in Figures 12 and 13 can verify the better performance of the PCE algorithm.

Furthermore, in real-data verification, the ground-based weather radar data and QPE data are used to validate the feasibility and superiority of the proposed method. First, the PCE algorithm can eliminate the artifact in the scene and the contour edge of reflectivity is smoother, as shown in Figure 16. In addition, the percentages of ZVW with the SD less than 1 in total data are increased after optimization, as shown in Table 6. These results indicated the effective improvement of the data quality after PCE algorithm. Finally, the precipitations calculated by the optimized reflectivity are more accurate, especially for the convective region as shown in Figures 19 and 20, indicating the advancement of the proposed method in the precipitation estimation.

6. Conclusions

In this paper, a novel sidelobe suppression strategy based on the extraction and iteration of weather radar called PCE is proposed. The cost function is constructed by modeling the transfer function of each range bin for the calculations of OPE. Through PCE processing, the energy of optimized echoes will be close to the main lobe energy of the actual targets in this range bin, achieving the sidelobe suppression of the weather targets with large gradient reflectivity. The proposed method is operated after pulse compression and it has no requirement on the radar transmitting waveform, which can be realized by adding a step after pulse compression in the actual radar system.

It is indicated from numerical simulation experiments that when the reflectivity gradient is large, the proposed method can greatly reduce the estimation errors for weak target, which is at the cost of loss of SNR. When the reflectivity gradient is not large, the data qualities are also effectively improved. The real-data verifications indicate that the proposed method can eliminate artifacts at the edge of the scene and improve the data quality of ZVW and precipitation.

Furthermore, although we only used the measured radar data of supercells to verify the proposed method in this paper, the method can also be applied to other strong convective weather phenomenon such as eyewall of a hurricane and a convective cell. These will be studied in the follow-up work.

Author Contributions: Conceptualization, J.H. and X.D.; methodology, J.H. and X.D.; software, J.H.; validation, J.H. and X.D.; formal analysis, J.H. and X.D.; investigation, J.H. and W.T.; resources, J.H. and C.H.; data curation, J.H. and K.F.; writing—original draft preparation, J.H.; writing—review and editing, J.H. and X.D.; visualization, J.H. and J.L.; supervision, X.D. and C.H.; project administration, W.T.; funding acquisition, X.D. and C.H. All authors have read and agreed to the published version of the manuscript.

Funding: This research was funded by "the Special Fund for Research on National Major Research Instruments (NSFC Grant Nos. 61827901)" and "the National Science Fund for Distinguished Young Scholars (Grant Nos. 62225104)" and "the National Natural Science Foundation of China (Grant Nos. 61971039, 61960206009, 61971037)" and "the Natural Science Foundation of Chongqing (Grant No. cstc2020jcyj-msxmX0621)" and "Distinguished Young Scholars of Chongqing (Grant No. cstc2020jcyj-jqX0008)" and "National Ten-thousand Talents Program 'Young top talent' (Grant No. W03070007)".

Data Availability Statement: The data are unavailable due to privacy.

Acknowledgments: The authors would like to thank Aerospace Newsky Technology Co., Ltd. for providing real ground-based weather radar data.

Conflicts of Interest: The authors declare no conflict of interest.

Appendix A

The optimal extraction matrix can be calculated as follows.

According to (13) $p(\mathbf{W}_{opt})$ can be expanded as:

$$\begin{aligned} p(\mathbf{W}_{opt}) &= \left\| \left(\mathbf{W}_{opt}\mathbf{I} \odot \mathbf{s}_{MF} \right) - \mathbf{s}_{SP} \right\|_2^2 \\ &= \left[\left(\mathbf{W}_{opt}\mathbf{I} \odot \mathbf{s}_{MF} \right) - \mathbf{s}_{SP} \right] \left[\left(\mathbf{W}_{opt}\mathbf{I} \odot \mathbf{s}_{MF} \right) - \mathbf{s}_{SP} \right]^T \\ &= \left\| \left(\mathbf{W}_{opt}\mathbf{I} \odot \mathbf{s}_{MF} \right) \right\|_2^2 - 2\left(\mathbf{W}_{opt}\mathbf{I} \odot \mathbf{s}_{MF} \right)\mathbf{s}_{SP} - \|\mathbf{s}_{SP}\|_2^2 \end{aligned} \tag{A1}$$

Considering the derivative and setting it to zero, the optimal $\mathbf{W}_{opt}$ can be obtained. Next, we consider the derivative of each term in:

$$\begin{aligned} &\mathbf{S} = \mathbf{W}_{opt}\mathbf{I} \odot \mathbf{s}_{MF} \\ &\mathbf{f} = \|\mathbf{W}_{opt}\mathbf{I} \odot \mathbf{s}_{MF}\|_2^2 = \mathbf{S} : \mathbf{S} \\ &\mathbf{g} = 2\left(\mathbf{W}_{opt}\mathbf{I} \odot \mathbf{s}_{MF} \right)\mathbf{s}_{SP} = 2\mathbf{S}\mathbf{s}_{SP} \end{aligned} \tag{A2}$$

where ":" represents the matrix inner product.

- Calculation of $d\mathbf{S}$:

According to the property of the derivative of Hadamard product, $d\mathbf{S}$ can be calculated as:

$$d\mathbf{S} = \left(d\mathbf{W}_{opt}\mathbf{I}\right) \odot \mathbf{s}_{MF} + \mathbf{W}_{opt}\mathbf{I} \odot d\mathbf{s}_{MF} \tag{A3}$$

Since $\mathbf{s}_{MF}$ is independent of $\mathbf{W}_{opt}$, we can achieve:

$$d\mathbf{S} = \left(d\mathbf{W}_{opt}\right)\mathbf{I} \odot \mathbf{s}_{MF} \tag{A4}$$

- Calculation of $\frac{d\mathbf{f}}{d\mathbf{W}_{opt}}$:

  According to the properties of matrix inner product and matrix differential [27]:

$$\mathbf{A} : \mathbf{B} = \mathbf{tr}\left(\mathbf{A}^{\mathrm{T}}\mathbf{B}\right) \tag{A5}$$

$$d\mathbf{f} = \mathbf{tr}\left[\left(\frac{d\mathbf{f}}{d\mathbf{X}}\right)^{\mathrm{T}} d\mathbf{X}\right] \tag{A6}$$

  Then, $d\mathbf{f}$ can be calculated from (A4) and (A5):

$$\begin{aligned} d\mathbf{f} &= d\mathbf{S} : \mathbf{S} + \mathbf{S} : d\mathbf{S} \\ &= 2\left(\mathbf{W}_{opt}\mathbf{I} \odot \mathbf{s}_{MF}\right) : \left[\left(d\mathbf{W}_{opt}\right)\mathbf{I} \odot \mathbf{s}_{MF}\right] \\ &= 2\left(\mathbf{W}_{opt}\mathbf{I} \odot \mathbf{s}_{MF} \odot \mathbf{s}_{MF}\mathbf{I}^{\mathrm{T}}\right) : \left(d\mathbf{W}_{opt}\right) \\ &= \mathbf{tr}\left[2\left(\mathbf{W}_{opt}\mathbf{I} \odot \mathbf{s}_{MF} \odot \mathbf{s}_{MF}\mathbf{I}^{\mathrm{T}}\right)^{\mathrm{T}}\left(d\mathbf{W}_{opt}\right)\right] \end{aligned} \tag{A7}$$

  Comparing (A7) with (A6), we can achieve:

$$\frac{d\mathbf{f}}{d\mathbf{W}_{opt}} = 2\mathbf{W}_{opt}\mathbf{I} \odot \mathbf{s}_{MF} \odot \mathbf{s}_{MF}\mathbf{I}^{\mathrm{T}} \tag{A8}$$

- Calculation of $\frac{d\mathbf{g}}{d\mathbf{W}_{opt}}$:

  Similarly, $\frac{d\mathbf{g}}{d\mathbf{W}_{opt}}$ can be calculated as:

$$\frac{d\mathbf{g}}{d\mathbf{W}_{opt}} = 2\left(\frac{d\mathbf{S}}{d\mathbf{W}_{opt}}\right)\mathbf{s}_{SP} + 2\mathbf{S}\frac{d\mathbf{s}_{SP}}{d\mathbf{W}_{opt}} \tag{A9}$$

  Since $\mathbf{s}_{SP}$ is independent of $\mathbf{W}_{opt}$, we can achieve:

$$\begin{aligned} \frac{d\mathbf{g}}{d\mathbf{W}_{opt}} &= 2\left(\frac{d\left(\mathbf{W}_{opt}\mathbf{I} \odot \mathbf{s}_{MF}\right)}{d\mathbf{W}_{opt}}\right)\mathbf{s}_{SP} \\ &= 2\left(\frac{d\left(\mathbf{W}_{opt}\mathbf{I}\right)}{d\mathbf{W}_{opt}} \odot \mathbf{s}_{MF} + \mathbf{W}_{opt}\mathbf{I} \odot \frac{d\mathbf{s}_{MF}}{d\mathbf{W}_{opt}}\right)\mathbf{s}_{SP} \\ &= 2\mathbf{I} \odot \mathbf{s}_{MF}\mathbf{s}_{SP} \end{aligned} \tag{A10}$$

In addition, since $\mathbf{s}_{SP}$ is independent of $\mathbf{W}_{opt}$, the derivative of $\|\mathbf{s}_{SP}\|_2^2$ is zero. Therefore, in combination with (A8) and (A10), the derivative of (A1) can be calculated as:

$$\begin{aligned} \frac{d\left[p\left(\mathbf{W}_{opt}\right)\right]}{d\mathbf{W}_{opt}} &= \frac{d\mathbf{f}}{d\mathbf{W}_{opt}} - \frac{d\mathbf{g}}{d\mathbf{W}_{opt}} \\ &= 2\mathbf{W}_{opt}\mathbf{I} \odot \mathbf{s}_{MF} \odot \mathbf{s}_{MF}\mathbf{I} - 2\mathbf{I} \odot \mathbf{s}_{MF}\mathbf{s}_{SP} \end{aligned} \tag{A11}$$

Additionally, $\mathbf{W}_{opt}$ can be calculated from $\frac{d[p(\mathbf{W}_{opt})]}{d\mathbf{W}_{opt}} = 0$ and can be expressed as:

$$\mathbf{W}_{opt} = \left(\mathbf{I} \odot \mathbf{s}_{MF} \odot \mathbf{s}_{MF}\mathbf{I}^{\mathrm{T}}\right)^{-1} (\mathbf{I} \odot \mathbf{s}_{MF}\mathbf{s}_{SP}) \tag{A12}$$

References

1. Skolnik, M.I. *Radar Handbook*, 3rd ed.; McGraw-Hill: New York, NY, USA, 2008.
2. Richards, M.A. *Fundamentals of Radar Signal Processing*; McGraw-Hill Processional: New York, NY, USA, 2005.
3. Cook, E.; Bernfeld, M. *Radar Signals—An Introduction to Theory and Applications*; Artech House: Norwood, MA, USA, 1993.
4. Griffiths, H.D.; Vinagre, L. Design of Low-Sidelobe Pulse Compression Waveforms. *Electron. Lett.* **1994**, *30*, 1004–1005. [CrossRef]
5. Cilliers, J.E.; Smit, J.C. Pulse Compression Sidelobe Reduction by Minimization of Lp-Norms. *IEEE Trans. Aerosp. Electron. Syst.* **2007**, *43*, 1238–1247. [CrossRef]
6. Bringi, V.N.; Chandrasekar, V. *Polarimetric Doppler Weather Radar: Principles and Applications*; Cambridge University Press: Cambridge, UK, 2001.
7. Pang, C.; Hoogeboom, P.; Le Chevalier, F.; Russchenberg, H.W.; Dong, J.; Wang, T.; Wang, X. A Pulse Compression Waveform for Weather Radars with Solid-State Transmitters. *IEEE Geosci. Remote Sens. Lett.* **2015**, *12*, 2026–2030. [CrossRef]
8. Brandon, P.S. The Design of a Nonlinear Pulse Compression System to Give a Low Loss High Resolution Radar Performance. *Marconi Rev.* **1973**, *36*, 1–45.
9. Fowle, E. The Design of FM Pulse Compression Signals. *IEEE Trans. Inform. Theory* **1964**, *10*, 61–67. [CrossRef]
10. Patton, L.K. On the Satisfaction of Modulus and Ambiguity Function Constraints in Radar Waveform Optimization for Detection. Ph.D. Thesis, Wright State University, Dayton, OH, USA, 2009.
11. Luszczyk, M.; Labudzinski, A. Side Lobe Level Reduction for Complex Radar Signals with Small Base. In Proceedings of the 2012 13th International Radar Symposium, Warsaw, Poland, 23–25 May 2012; pp. 146–149.
12. Argenti, F.; Facheris, L. Radar Pulse Compression Methods Based on Nonlinear and Quadratic Optimization. *IEEE Trans. Geosci. Remote Sens.* **2021**, *59*, 3904–3916. [CrossRef]
13. Hu, J.; Dong, X.; Hu, C. Research on Nonlinear Wind Field Retrieval Method Based on Modified VAD Analysis. *J. Signal Process.* **2021**, *37*, 284–291. [CrossRef]
14. Wang, X.; Zhai, W.; Greco, M.; Gini, F. Cognitive Sparse Beamformer Design in Dynamic Environment via Regularized Switching Network. *IEEE Trans. Aerosp. Electron. Syst.* **2022**, *59*, 1816–1833. [CrossRef]
15. Kurdzo, J.M.; Cheong, B.L.; Palmer, R.D.; Zhang, G.; Meier, J.B. A Pulse Compression Waveform for Improved-Sensitivity Weather Radar Observations. *J. Atmos. Ocean. Technol.* **2014**, *31*, 2713–2731. [CrossRef]
16. Ackroyd, M.H.; Ghani, F. Optimum Mismatched Filters for Sidelobe Suppression. *IEEE Trans. Aerosp. Electron. Syst.* **1973**, *AES-9*, 214–218. [CrossRef]
17. Cohen, M.N.; Baden, J.M. *Pulse Compression Coding Study*; Georgia Tech Research Institute: Atlanta, GA, USA, 1983.
18. Beauchamp, R.M.; Tanelli, S.; Peral, E.; Chandrasekar, V. Pulse Compression Waveform and Filter Optimization for Spaceborne Cloud and Precipitation Radar. *IEEE Trans. Geosci. Remote Sens.* **2017**, *55*, 915–931. [CrossRef]
19. Cavallaro, S. Statistical Properties of Polarimetric Weather Radar Returns for Nonuniformly Filled Beams. *IEEE Geosci. Remote Sens. Lett.* **2017**, *14*, 1584–1588. [CrossRef]
20. Bharadwaj, N.; Chandrasekar, V. Wideband Waveform Design Principles for Solid-State Weather Radars. *J. Atmos. Ocean. Technol.* **2012**, *29*, 14–31. [CrossRef]
21. Dong, X.; Hu, J.; Hu, C.; Chen, Z.; Li, Y. Rapid Identification and Spectral Moment Estimation of Non-Gaussian Weather Radar Signal. *IEEE Trans. Geosci. Remote Sens.* **2022**, *60*, 1–12. [CrossRef]
22. Li, S.; Yang, M.; Li, L. Evaluation on Data Quality of X-Band Dual Polarization Weather Radar Based on Standard Deviation Analysis. *J. Arid Meteorol.* **2019**, *37*, 467–476.
23. Levanon, N.; Mozeson, E. *Radar Signals*; Wiley: Hoboken, NJ, USA, 2004.
24. Sun, Y.; Liu, Q.; Cai, J.; Long, T. A Novel Weighted Mismatched Filter for Reducing Range Sidelobes. *IEEE Trans. Aerosp. Electron. Syst.* **2019**, *55*, 1450–1460. [CrossRef]
25. He, K.; Fan, Q.; Li, K. Z-R Relation with Its Application to Typhoon Precipitation in Zhoushan. *J. Appl. Meteorol.* **2007**, *18*, 573–576.
26. Huffman, G.J.; Stocker, E.F.; Bolvin, D.T.; Nelkin, E.J.; Tan, J. *GPM IMERG Final Precipitation L3 Half Hourly 0.1 Degree x 0.1 Degree V06 2019*; Goddard Earth Sciences Data and Information Services Center (GES DISC): Greenbelt, MD, USA, 2019.
27. Zhang, X. *Matrix Analysis and Applications*; Tsinghua University Press: Beijing, China, 2013.

remote sensing

MDPI

Article

Correction of Dual-PRF Velocity for Operational S-Band Doppler Weather Radar

So-Yeon Park, Sung-Hwa Jung * and Kwang-Ho Kim

Weather Radar Center, Korea Meteorological Administration, Seoul 07062, Republic of Korea; soyeoni1990@korea.kr (S.-Y.P.); khkim777@korea.kr (K.-H.K.)
* Correspondence: shjung95@korea.kr; Tel.: +82-2-2181-0862

Abstract: The Weather Radar Center (WRC) of the Korea Meteorological Administration (KMA) has been providing three-dimensional radar wind fields based on the "WInd Synthesis System using Doppler Measurements (WISSDOM)" in real time since February 2019. Its accuracy is significantly affected by the quality of the Doppler velocity, such as velocity aliasing. For the de-aliasing of Doppler velocity, the dual-pulse repetition frequency (dual-PRF) technique is commonly utilized for commercial Doppler weather radar. The Doppler weather radars of the KMA have extended their Nyquist velocity up to 132 m s^{-1} using a dual PRF of 5:4. However, the dual-PRF technique produces significant noise and loss of radial velocity. Therefore, we developed a technique for noise cancelation and recovery of radial velocity to improve the quality of WISSDOM wind fields. The proposed approach identifies and removes speckles of abnormal radial velocity by comparing the sign of the median radial velocity with the surrounding radar bins. We then recovered the eliminated radial velocity using median interpolation. To recover the losses of radial velocity over a wide area using the dual-PRF technique, we used the Velocity Azimuth Display curve-fitting technique. These techniques are straightforward, preserve spatial gradients, and suppress local extrema. We tested this technique, verified its performance, and applied it to the operational radar quality control system of the WRC from August 2021. We concluded that the process helps improve the quality of the radial velocity and the accuracy of the WISSDOM wind fields.

Keywords: quality control; Doppler weather radar; dual-PRF; radial velocity; noise elimination

1. Introduction

Three-dimensional precipitation and wind information with high spatiotemporal resolution based on nationwide weather radar networks is essential for real-time monitoring and dynamic analysis of mesoscale severe weather. For a more detailed analysis and routine weather forecasts, the Weather Radar Center (WRC) of the Korea Meteorological Administration (KMA) has provided three-dimensional wind fields, including wind vectors, convergence/divergence, vorticity, and storm relative helicity, at a horizontal resolution of 1 km over the entire Korean Peninsula since February 2019. The 3D wind fields are produced every 5 min by real-time implementation of the "WInd Synthesis System using Doppler Measurements (WISSDOM)," which is developed by National Central University (NCU) in Taiwan, on a supercomputer system [1,2]. In producing radar wind fields, radial velocity data are an important factor affecting the quality of wind data as input data for wind calculations.

Aliasing (or velocity folding) is one of the quality problems with radar radial velocity for the use of radial winds from Doppler weather radars. Previous studies have corrected aliased Doppler radar velocity data [3–10]. For the de-aliasing of Doppler velocity, the dual-pulse repetition frequency (dual-PRF) technique is commonly utilized for commercial Doppler Weather radar [11,12]. The Doppler weather radars operated by KMA also use a dual PRF of 5:4, which were all upgraded to dual-polarization radar from 2014 to 2019. In addition, the observational variables (such as reflectivity and radial velocity) are generated by applying signal processing algorithms using raw signal data before passing through the signal processor, thresholds, and signal quality filters optimized by the observation strategy. Thus, the Nyquist velocity has been extended up to 132 m s^{-1}, which provides stable information without velocity folding in the event of severe weather phenomena, such as typhoons. However, the dual-PRF technique produces significant noise and widely spaced gates that have no radial velocity data but still contain valid radar reflectivity data, owing to the limited number of pulse samples used. Therefore, the Doppler weather radars of the KMA require a quality control process of radial velocity to improve the accuracy of radar wind.

Holleman and Beekhuis [13] developed a post-processing algorithm for dual-PRF velocity data, which included comparing each velocity data point of the azimuth scan with the median velocity calculated from the data point itself and the surrounding data point to eliminate noise. Joe and May [14] proposed the Laplacian technique to handle dual-PRF errors and compared it with the median filtering technique. Median filtering is straightforward and preserves gradients, whereas the Laplacian technique is a detection-correction scheme that attempts to detect and correct errors. The Laplacian filter is preferred because it preserves local peaks, whereas the median filter may have problems when the velocity data straddle the extended Nyquist velocity. Both techniques are robust to missing and sparse data. Hengstebeck et al. [15] used a Laplacian filter based on Joe and May [14] within the quality assurance module for the mesocyclone detection algorithm of the German weather radar network. Park and Lee [16] compared the radial velocity of a given gate with the mean value of the radial velocity of a 15 × 7 area of neighboring gates to reduce speckles due to dual PRF velocity errors as pre-processing for radar wind retrieval. Altube et al. [17] proposed a novel method that uses circular statistics in the phase space rather than the velocity space to identify and correct dual-PRF errors. This method improves the past dual-PRF correction method that has been used to improve the accuracy of radial velocity, and it is able to accurately detect and correct dual-RPF errors even in the presence of severe aliasing. It relies on the continuity of the local field but uses circular statistics and does not require prior global dealiasing. However, this technique is limited to dealing with outliers caused by aliasing. It is not suitable for other causes of outliers, such as errors in the range or velocity measurements. Additionally, the technique is limited to dual-frequency pulse-Doppler radar and is not suitable for other types of radar, such as single-frequency pulse-Doppler or chirp radar.

In this study, we propose techniques for noise correction of the observed radial velocity field and restoration of the lost radial velocity to improve the quality of WISSDOM wind fields. These techniques are straightforward, preserve spatial gradients, and suppress local extrema. The remainder of the paper is structured as follows: Section 2 describes the KMA operational radar network; Section 3 introduces the radial velocity quality control technique; Section 4 describes the results of the application of this technique are described and validated the quality-controlled radial velocity using wind profiler data; and finally, Section 5 presents the summary and conclusions.

2. Data

We utilized KMA's ten S-band Doppler weather radar data. Figure 1 shows the S-band dual-polarization weather radar network and domain for WISSDOM. WISSDOM was used to compare the wind fields calculated with radial velocity before and after quality control. KMA established a homogeneous S-band dual-polarization network with the same specifications. Table 1 presents a list of radar observation characteristics. All strategies consisted of nine elevation angles for volume scanning and were performed every 5 min. The radar transmits a long pulse width of 2 μs at low elevation angles (less than 3°) to enhance the detectability of weak low-level echoes, such as winter snowstorms, whereas a short pulse of 1 μs pulse width is used at higher elevation angles (>3°) to increase the Nyquist velocity using a high PRF. In addition, the WRC developed CLEANER [18] for the quality control algorithm of radar observation data, which classifies precipitation and non-precipitation echoes using dual polarization parameters, and removes non-precipitation (such as ground clutter, Anomalous Propagation, second trip, and chaff echoes). CLEANER also removes non-precipitation areas from the radial velocity field.

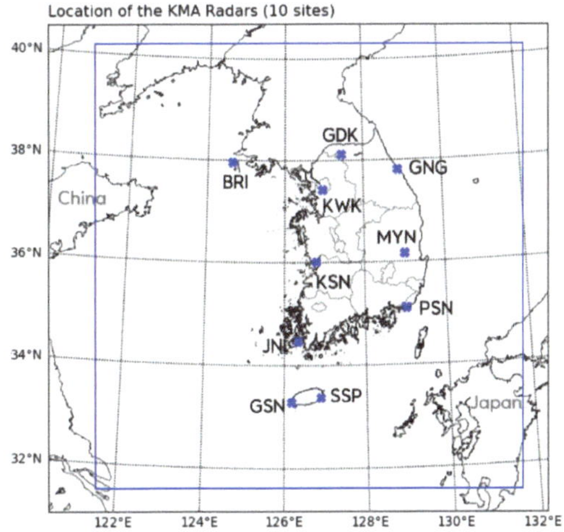

Figure 1. Operational S-band Doppler weather radar network of the KMA. The blue marks and gray circles represent the locations of individual radars and their observational ranges, respectively. The blue box stands for the domain of WISSDOM.

Table 1. Characteristics of the KMA's S-band radar.

Parameters	Values								
Dual polarization	Simultaneous H and V								
Frequency (MHz)	2718								
Wave length (cm)	11.02								
Range resolution (m)	250								
Beam width (°)	0.95								
Elevation angles (°)	Nine elevation angles (minimum: −0.8°, maximum: 15.0°)								
Elevation number	1	2	3	4	5	6	7	8	9
Pulse width (μs)	2.0	2.0	2.0	2.0	2.0	2.0	1.0	1.0	1.0
Observation range (km)	240	240	240	240	240	240	180	120	60
High PRF (Hz)	620	620	620	620	620	620	830	1200	1200
Dual-PRF ratio	5:4	5:4	5:4	5:4	5:4	5:4	5:4	5:4	5:4
Nyquist Velocity (m s^{-1})	68.4	68.4	68.4	68.4	68.4	68.4	91.5	132.4	132.4
Sampling Number	45	45	45	45	45	45	45	45	45

All elevation angles used the 5:4 dual-PRF mode to increase the Nyquist velocity to mitigate velocity aliasing. However, this technique not only generates noise in the process of receiving data but also loses radial velocity observations compared to reflectivity observations. Figure 2 shows the ratio of radial velocity observations to reflectivity observations for one month (August 2020) on 10 weather radars operated by the KMA. Here, it was possible to determine the lost radial velocity for each radar site. The radial velocity relative to the reflectivity was observed to be at least 50–80%, on average, at all elevation angles, indicating that the radial velocity observations relative to the reflectivity observations were lost from at least 20% to up to 50%. In particular, even though strong reflectivity values were observed owing to the influx of a strong precipitation system, the wind calculated from the radial velocity could be underestimated or a discontinuous wind distribution could appear in the area where the observed radial velocity is lost.

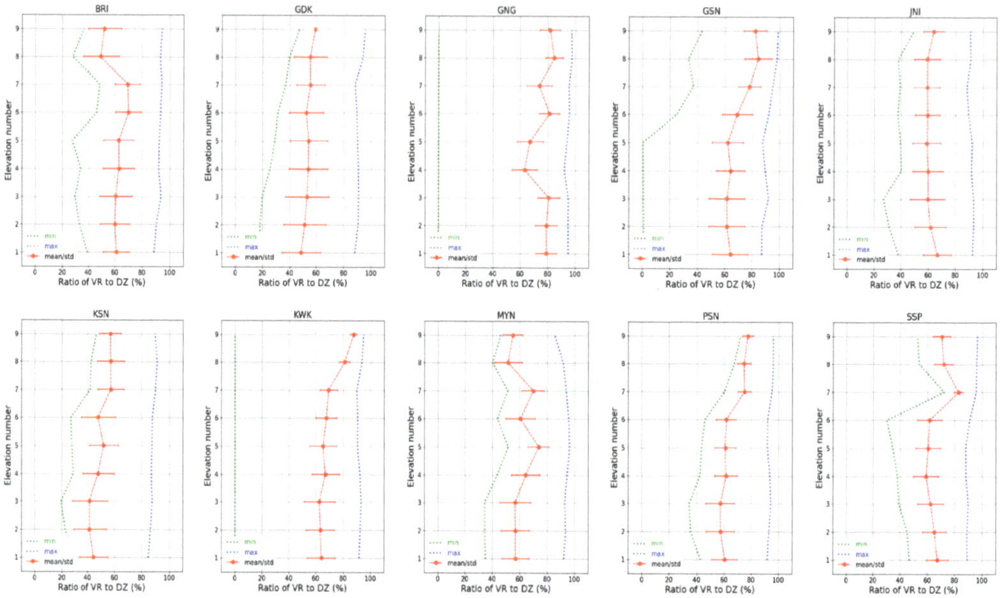

Figure 2. Ratio of radar bins with radial velocity (VR) observation to radar reflectivity (DZ) observation as a function of elevation angles for August 2020 from KMA's ten weather radars (BRI, GDK, GNG, GSN, JNI, KSN, KWK, MYN, PSN, and SSP). The red dots and bars represent the average ratio and standard deviations, respectively. The blue and green dotted lines indicate the maximum and minimum ratios at each elevation angle, respectively.

3. Quality Control of Radial Velocity

The entire process for quality control of the radar radial velocity is shown in Figure 3. The proposed approach identifies and removes speckles of abnormal radial velocity by comparing the sign of the median radial velocity with the surrounding radar bins. We then recovered the eliminated radial velocity using the median interpolation technique. To recover the losses of radial velocity over a wide area using the dual-PRF technique, we used the Velocity Azimuth Display (VAD) [19] curve-fitting technique for radial velocity.

3.1. Noise Elimination and Correction

We developed a noise filter that utilizes the sign and median values of radial velocity observations to detect and correct radial velocity noise. For an easier understanding, an example of the noise detection conditions and restoration process concepts is presented in Figure 4, along with a flowchart. An example of this concept assumes that the window size is three gates × three rays. Preferentially, in step 1, it is examined regardless of whether the radial velocity at the center of the window is valid (a value other than NaN). If it is NaN, it moves to the following azimuth scan. Otherwise, the noise is determined according to the following three steps: step 2) if the ratio of the valid radial velocity in the window is less than the threshold, it would be replaced by the NaN. Conversely, when the ratio is larger than the threshold, the median value is calculated using the radial velocity in the window, excluding the value at the centroid of the window. If the sign of the median value is not identical to the sign of the center radial velocity in the window. In that case, the center radial velocity is identified as noise, and the median value replaces it in step 3. If the sign of the center radial velocity is equal to the median sign, the noise is identified by the difference between the radial velocity and the median value in step 4. The median value replaces the radial velocity if its difference is larger than the threshold (e.g., 20 m s^{-1}).

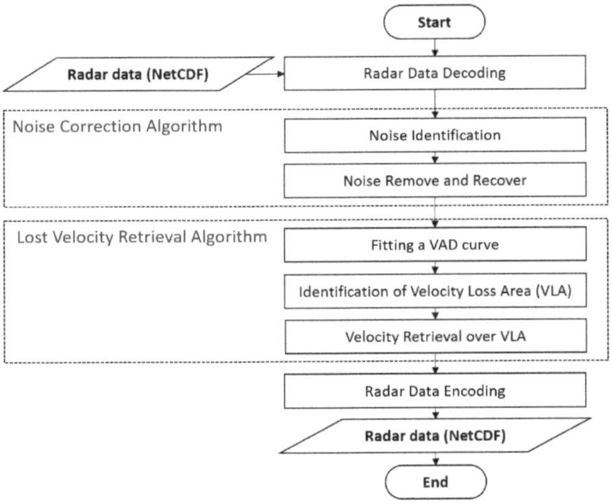

Figure 3. Flowchart for a quality control algorithm for radial velocity.

We conducted a sensitivity experiment to optimize the thresholds for each step. The thresholds for each condition were determined empirically based on noise cancelation performance and computational times for real-time implementation. The calculation time was mainly proportional to the window size. As the window size increased, the misidentification of normal echoes increased; however, the noise was not eliminated for a small window. The window size was set to seven gates × seven rays in consideration of the calculation time and noise correction rate. The ratio of the valid radial velocity in the window was set to 20% or less. The larger the threshold of the ratio, the higher the probability that a reasonable radial velocity, not a noise, will be removed. The absolute value of the difference between the median value and the center radial velocity in the window was set to 20 m s^{-1}. We found that the number of eliminated noises was reduced dramatically if the threshold of the absolute value was set to a value larger than 20 m s^{-1}.

3.2. Restoration Lost Radial Velocity

In this study, to restore the lost radial velocity, we utilized the VAD curve fitting function, which is the mean Doppler velocity of precipitation particles observed along the radar antenna rotated through a 360° azimuth scan as a function of the azimuth angle at a constant elevation angle of the Doppler radar. A flowchart of the restoration of the lost radial velocity is shown in Figure 5. First, we combined several adjacent azimuth scans to obtain a more reasonably fitted VAD curve and calculated the average radial velocity of radar bins within a given range along the radial direction across the azimuth scans. Based on the empirical experiments, ten azimuth scans were selected around the corresponding azimuth scan. Thus, the average for twenty-one azimuth scans, including the corresponding azimuth, was calculated. The number of azimuth scans used may be set differently depending on radar scan strategies, weather phenomena, etc.

The VAD restoration function of the averaged values was then calculated based on a Fourier series. Next, to determine the lost radial velocity area (VLA), it was compared with the radar reflectivity observation field. If the radial velocity observation value did not exist in the area where the reflectivity observation value exists, it was replaced by the value of the restoration function. Then, among the observed radial velocities, a value with a large difference from the restoration function was determined as an outlier. Outliers were discriminated using the multiples ($Error$) of the standard deviation and the average ($Mean_{diff}$) of the difference between the radial velocity value and the restoration function value. The outlier value was converted into the value of the restoration function. The threshold values of each condition for restoring the loss region of the radial velocity were empirically optimized through sensitivity experiments.

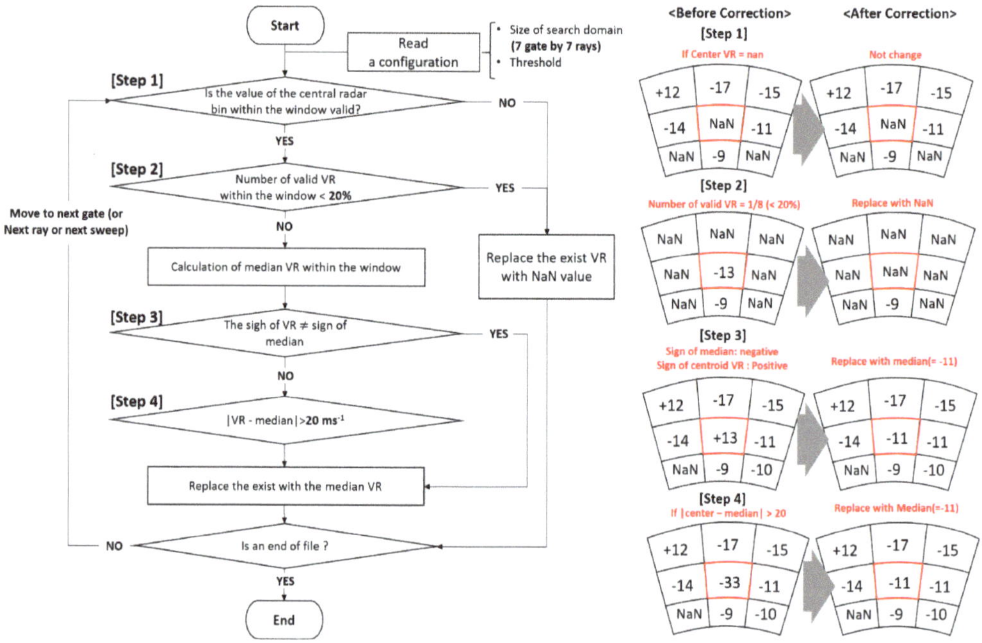

Figure 4. Flowchart for noise correction and conceptual diagram for noise correction based on the sign of median radial velocity.

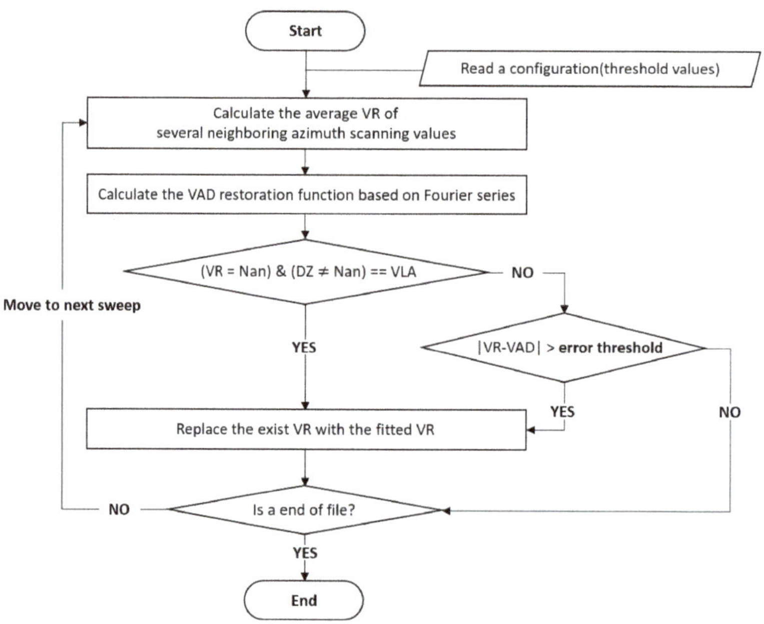

Figure 5. Flowchart of a retrieval algorithm for lost radial velocity.

Figure 6 shows why the average value of the radial velocity of several adjacent azimuth scans is used when calculating the VAD fitting function. Figure 6a shows an example of the lost radial velocity region of KWK at 0400 LST on 24 July 2020. In particular, when restoring by calculating each VAD fitting function from all azimuth scans, the texture of the entire radial velocity field appeared to be unnaturally restored, as shown in Figure 6b. To prevent discontinuous restoration of the radial velocity in this manner, the VAD fitting function calculated from the average value of the peripheral radial velocities was used to restore the radial velocity with continuity, as shown in Figure 6c.

Figure 6. Example of restoring the radial velocity of the KWK radar site at 0400 local standard time (LST) on 24 July 2020. (**a**) Before restoring radial velocity. (**b**) Restoration using each VAD fitting function calculated from all azimuth scans. (**c**) Restoration using a VAD fitting function calculated from the mean of several neighboring azimuth scans. The two black lines indicate range lines at the radius of 150 km and 200 km from the radar, respectively.

To express the radial velocity as a Fourier series, the radial velocity V_r in the spherical coordinate system is expressed by Equation (1) [19]. Here, θ represents the azimuth angle; ϕ represents the elevation angle; and u, v, and w represent the east-west, north-south, and vertical components of the particle, respectively. In addition, if the radial velocity at the center of the radar at a given altitude is expressed as a component of the linear wind field, Equation (2) can be obtained, where r is the distance from the center and u_0 and v_0 are the velocity components in the east-west and north-south directions from the center, respectively.

$$V_r = \mathrm{u} \sin\theta \cos\varnothing + v \cos\theta \sin\varnothing + w \sin\varnothing \tag{1}$$

$$\begin{aligned} V_r = {} & \tfrac{1}{2} r \cos^2\varnothing \left(\tfrac{\partial u}{\partial x} + \tfrac{\partial v}{\partial y}\right) + w \sin\varnothing + u_0 \cos\theta \cos\varnothing + v_0 \sin\theta \cos\varnothing \\ & + \tfrac{1}{2} r \cos^2\varnothing \left(\tfrac{\partial u}{\partial x} - \tfrac{\partial v}{\partial y}\right) \cos 2\theta + \tfrac{1}{2} r \cos^2\varnothing \left(\tfrac{\partial u}{\partial x} + \tfrac{\partial v}{\partial y}\right) \sin 2\theta \end{aligned} \tag{2}$$

The radial velocity, expressed as a linear wind field component, can be expressed as a quadratic Fourier series, as shown in Equation (3). Compared to the right side of Equation (2) and the right side of Equation (3), each Fourier coefficient is equal to Equations (4)–(8).

$$V_r = \sum_{n=0}^{2} (a_n \cos n\theta + b_n \sin n\theta) \tag{3}$$

$$a_0 = \frac{1}{2} r \cos^2\varnothing \left(\frac{\partial u}{\partial x} + \frac{\partial v}{\partial y}\right) + w \sin\varnothing \tag{4}$$

$$\mathrm{a}_1 = u_0 \cos\varnothing \tag{5}$$

$$\mathrm{b}_1 = v_0 \cos\varnothing \tag{6}$$

$$a_2 = \frac{1}{2} r \cos^2\varnothing \left(\frac{\partial u}{\partial x} - \frac{\partial v}{\partial y}\right) \tag{7}$$

$$b_0 = \frac{1}{2} r \cos^2\varnothing \left(\frac{\partial u}{\partial y} + \frac{\partial v}{\partial x}\right) \tag{8}$$

Equation (2) can be re-expressed as Equation (9) according to the coefficients, and the coefficients a_0 to b_2 are calculated by solving the Fourier series using numerical analysis. According to Equation (9), the radial velocity obtained by fitting each azimuth (VAD fitting) can be calculated.

$$V_r = a_0 + a_1 \cos\theta + b_1 \sin\theta + a_2 \cos 2\theta + b_2 \sin 2\theta \quad (9)$$

It cannot be considered an outlier in all cases where a positive(negative) area is observed in the negative (positive) area of the radial velocity. Thus, we considered the effect on the observed value of the radial velocity due to small-scale meteorological phenomena, such as downbursts and tornadoes. As shown in Equation (10), if the observed radial velocity ($+VR$) is within the range of the error threshold ($Mean_{diff} + Error * \sigma$) from the VAD fitting curve, VR is not an outlier value. Otherwise, the negative value of the observed radial velocity ($-VR$) is checked within the range of the error threshold from the VAD fitting curve. $Mean_{diff}$ represents the average difference between the radial velocity and the restoration function value in an azimuth scan, σ denotes the standard deviation for each azimuth scan, and the $Error$ value is 3, which is a value determined through optimization. Figure 7 shows an example of determining outliers in the azimuth scans. The observed radial velocity ($+VR$) shown by the blue circle in Figure 7a remains the same because the negative values ($-VR$) are within the range of determining the outliers. Figure 7b shows that the observed and negative values were determined as outliers because they were all outside the range.

$$\text{VAD} - \left(Mean_{diff} + Error * \sigma\right) < \pm VR < \text{VAD} + \left(Mean_{diff} + Error * \sigma\right), \quad Error = 3 \quad (10)$$

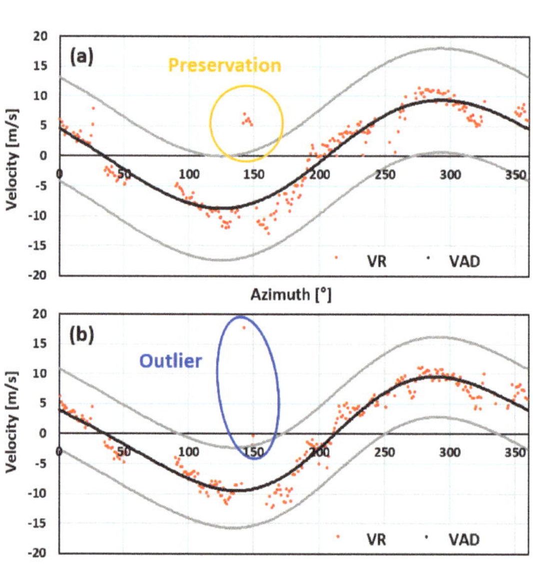

Figure 7. Examples of identifying outliers of radial velocity using VAD fits. (**a**) A case for not an outlier (in the orange circle). (**b**) A case for an outlier (in the blue circle). The red dots indicate individual radial velocities. The black and gray lines represent the VAD and error curves, respectively.

4. Results and Verification

4.1. Quality Control Results

4.1.1. Noise Elimination and Correction

Figures 8 and 9 show the results of the noise-removed and corrected radial velocity fields using the noise filter, with the sign and median of the observed radial velocity values developed in this study. Figure 8 shows the radial velocity fields observed by the KWK radar at 1200 LST on 28 February 2020. The noise correction technique was applied to all elevation angles of the radar observations, and the results are shown for elevation angles of $-0.2°$, $1.5°$, and $4.4°$. In this case, a precipitation echo was observed in most of the radar observation areas, and noise was spread in all directions. As a result of applying the noise filter, the noise was removed, and the radial velocity was corrected to the median value. Figure 9 shows the radial velocity fields observed using the MYN radar at 1610 LST on 6 June 2020. In this case, it can be confirmed that the radial velocity was corrected after removing only the noise of the radial velocity while maintaining the precipitation echo, even in the radial velocity of the small cell area. After removing and correcting the noise, spatially continuous radial velocity fields are obtained.

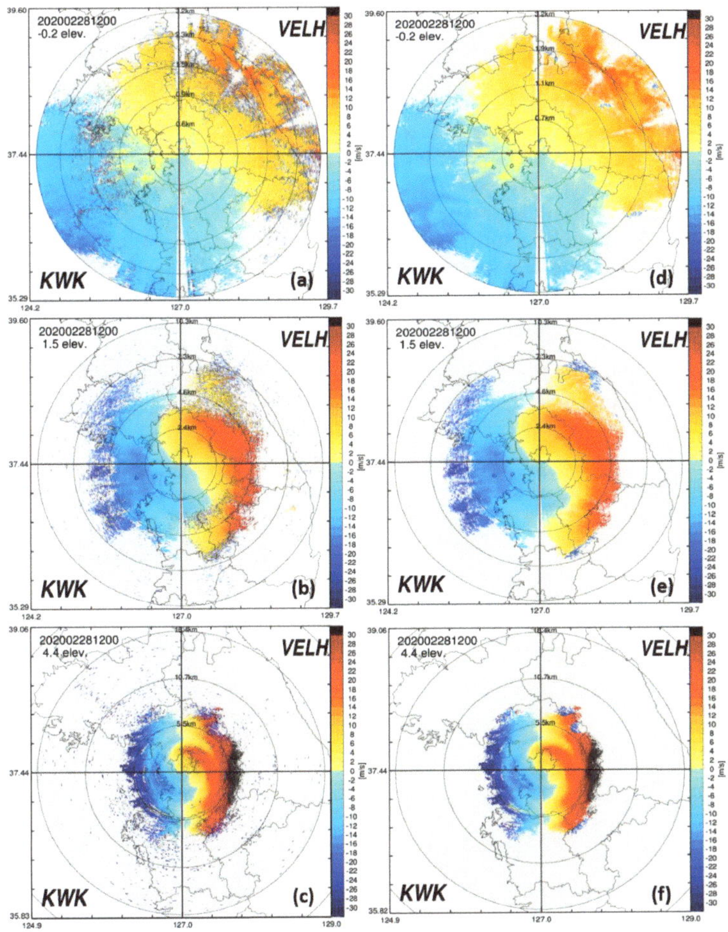

Figure 8. PPIs of radial velocity (**a–c**) before and (**d–f**) after quality control at elevation angles of $-0.2°$, $1.5°$, and $4.4°$ for KWK at 1200 LST on 28 February 2020.

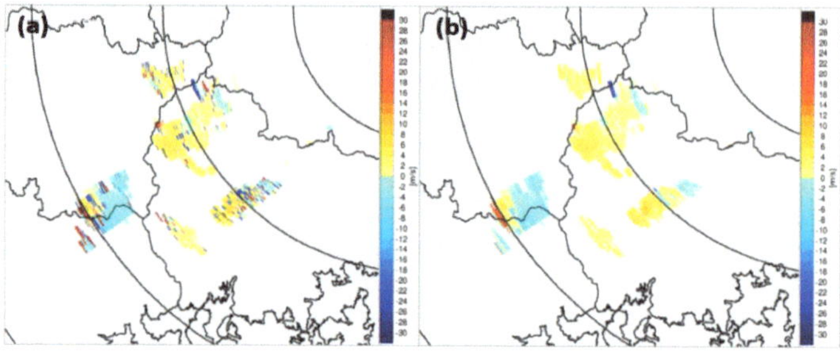

Figure 9. PPIs of radial velocity (**a**) before and (**b**) after noise correction at an elevation angle of $-0.8°$ for MYN at 1610 LST on 6 June 2020.

4.1.2. Restoration of Lost Radial Velocity

In this study, the VAD restoration function based on the Fourier series was calculated to restore the lost radial velocity compared to reflectivity. Figure 10 shows the results of applying the quality control technique step-by-step in the case of the radial velocity observed by the KWK radar at 0400 LST on 24 July 2020. Figure 10a shows the reflectivity field, showing strong reflectivity in the western region of the radar. Figure 10b, which presents the radial velocity field before quality control, shows that the radial velocity is lost over a wide area, including the strong reflectivity area. Figure 10c shows the result of applying the noise filter, and Figure 10d shows the result of restoring the lost radial velocity compared with the reflectivity observation using the VAD restoration function after applying the noise filter. By applying the restoration method based on the VAD, it was possible to obtain a spatially continuous radial velocity because it was completely restored even in areas where the radial velocity was widely lost.

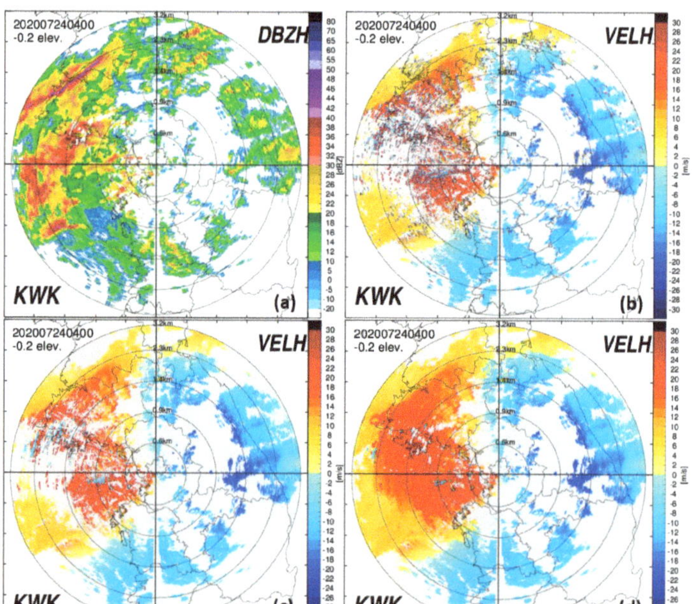

Figure 10. PPIs at the elevation angle of $-0.2°$ of KWK at 0400 LST on 24 July 2020; (**a**) radar reflectivity, (**b**) raw radial velocity, (**c**) radial velocity field after noise correction, and (**d**) radial velocity after noise correction and restoration of lost radial velocity.

Figure 11 shows a case in which a positive area was observed in the negative area of the radial velocity of the KWK radar site at 0400 LST on 24 July 2020. Figure 11a shows that a wide positive area of the radial velocity was observed, with outliers in the negative area of the radial velocity. Then, the positive radial velocity area shown in Figure 11b was preserved, as is, and only outliers were determined and corrected to the value of the VAD fitting function.

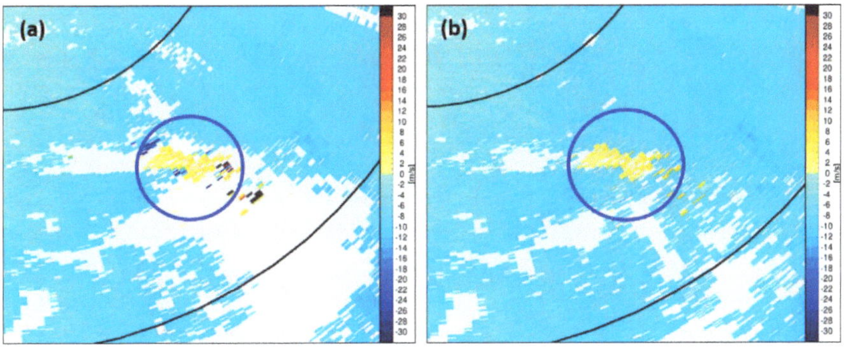

Figure 11. Examples of (**a**) before and (**b**) after outlier correction and retrieval of radial velocity using VAD fits at an elevation angle of $-0.2°$ for KWK at 0400 LST on 24 July 2020.

4.1.3. Utilization of Quality-Controlled Radial Velocity

To confirm the effect of the quality control technique on the radar radial velocity developed in this study, we compared the results of the WISSDOM wind calculated using the radial velocity before quality control and the radial velocity after quality control as input data. Figure 12 shows the reflectivity fields, raw radial velocity fields, and quality-controlled radial velocity fields on the BRI and GDK radar at 2100 LST on 3 July 2021. In the region where the echo of the reflectivity fields was observed, the loss of radial velocity in the radial velocity fields was seen, as marked by the blue dotted line. As for the quality-controlled radial velocity, the noise was clearly removed, and the lost area was restored, as shown in Figure 12c,f. When the BRI and GDK radars were synthesized to calculate the WISSDOM wind, the area where the radial velocity was lost in the BRI and GDK radars overlapped, as shown in Figure 12g. Figure 13 shows the wind speed and wind direction calculated using the BRI radar data and the GDK radar data as input data for WISSDOM, depending on whether the radial velocity quality is managed. Figure 13a,c show the wind speed and direction of WISSDOM calculated using the radial velocity before quality control as input data, and Figure 13b,d show the wind speed and direction of WISSDOM calculated using the radial velocity after quality control as input data. The wind field before quality control, as shown in Figure 13a,c, showed a spatially discontinuous distribution owing to the influence of noise and irregularly lost radial velocity. After quality control, the wind field showed a spatially continuous distribution, as shown in Figure 13b,d, and the wind speed in the lost radial velocity area was stronger in areas with strong reflectivity. The WISSDOM wind fields mainly changed depending on the presence or absence of restoration in the radial velocity loss area, and more spatially continuous WISSDOM wind fields could be calculated through the quality control technique of the radar radial velocity developed in this study.

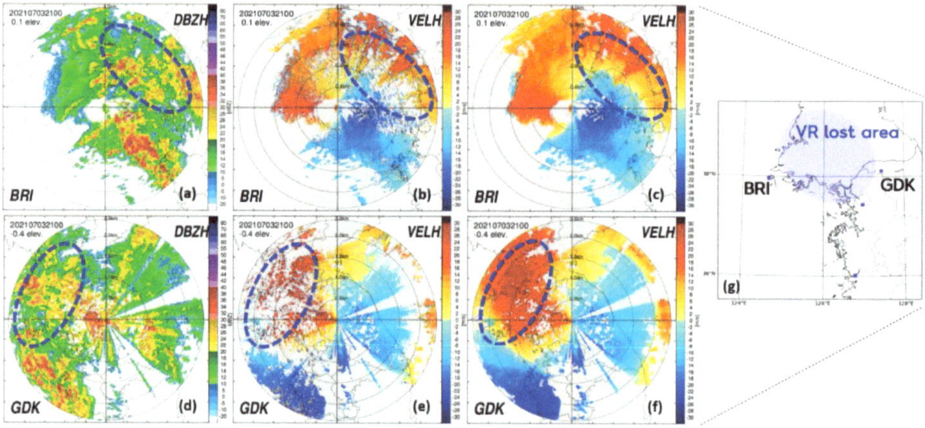

Figure 12. PPIs at the lowest elevation angles of (top) BRI and (bottom) GDK at 2100 LST on July 2021; (**a**,**d**) radar reflectivity, (**b**,**e**) raw radial velocity, (**c**,**f**) radial velocity after quality control, and (**g**) the simultaneously lost area of the radial velocities at both BRI and GDK.

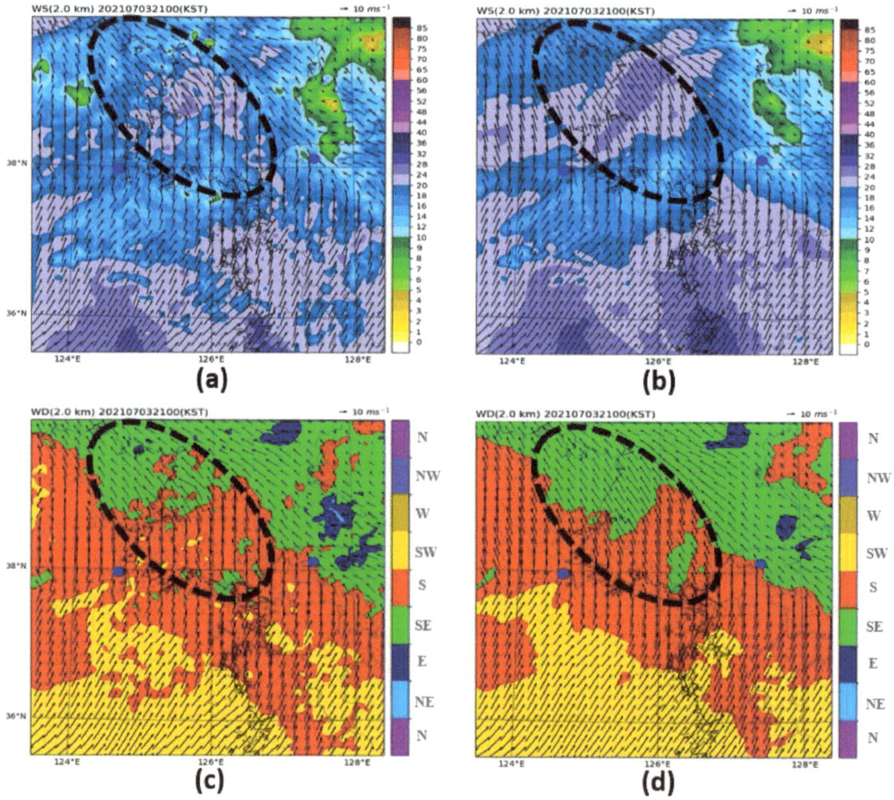

Figure 13. Comparison of (top) wind speed and (bottom) wind direction at an altitude of 2 km derived from WISSDOM at 2100 LST on 3 July 2021; (**a**,**c**) using raw radial velocity and (**b**,**d**) using quality-controlled radial velocity.

4.2. Verification

4.2.1. Data and Method

The quality-controlled radial velocity was quantitatively verified using observational data from a wind profiler. To compare the wind elements (u, v, w) observed by the wind profiler with the radar radial velocity, they were converted to the radial velocity in the direction observed by the radar [20]. The left side of Figure 14 shows the method for comparing the radial velocity of the radar and wind profiler data, where d represents the distance between the radar and wind profiler, r is the range from the radar to the point of interest [11], ϕ is the elevation angle of the radar beam measured upward from the horizontal plane through the radar, and θ is the azimuth angle of the radar beam measured clockwise from north. Five radars at KWK, GDK, KSN, MYN, and PSN and seven wind profilers at 47099, 47095, 47114, 47229, 47135, 47130, and 47155 points were used for verification, and their locations are shown on the right side of Figure 14. Considering the distance between the radar and the wind profilers and the radar observation radius, a comparative verification was performed with ten pairs of radars and wind profilers, as shown in Table 2.

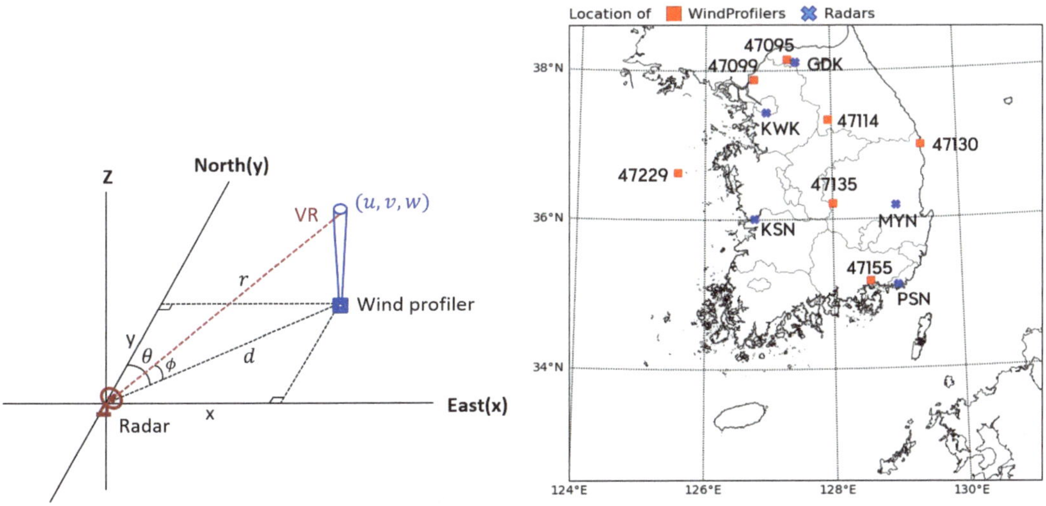

Figure 14. (**left**) Conceptual diagram of geometrical matching for radial velocity comparison between radar and wind profilers and (**right**) deployment of weather radars and wind profilers in Korea.

Table 2. Distance and azimuth of the wind profilers from the radars used for verification.

No.	Radar	Wind Profiler	Distance (km)	Azimuth (°)
1	GDK	47099	63.85	246
2	GDK	47114	97.83	152
3	KWK	47099	52.08	341
4	KWK	47114	87.65	97
5	KWK	47095	83.77	21
6	KSN	47229	129.12	302
7	KSN	47135	111.18	78
8	MYN	47130	97.63	22
9	MYN	47135	90.12	273
10	PSN	47155	39.25	278

To investigate the validity of the quality-controlled radar radial velocity, the radar radial velocity at each stage of quality control was compared with the observed value of the wind profilers. The CLEANER QC radial velocity, which is a step before radial velocity quality control (RAW), radial velocity after noise removal and correction (NOS), and radial velocity restored in the disappearance area (ALL), were calculated from the radial velocity and compared. During the verification period, only the corrected or restored radial velocity (QCD) was extracted through quality control, excluding the actual observation data, and compared with the radial velocity calculated by the wind profilers. The verification period was five months, from 0000 LST on 1 June 2021 to 2350 LST on 31 October 2021. The data were compared every 10 min, considering the time resolution of the wind profiler. The Root Mean Square Error (RMSE) and Correlation Coefficient (CC) were used as verification indicators to compare the radar radial velocity with the observed value of the wind profiler. The RMSE and CC were calculated using Equations (11) and (12), where R is the radial velocity of the radar at the point of the wind profiler and W is the radial velocity data calculated from the wind profiler.

$$\mathrm{RMSE} = \sqrt{\frac{1}{N}\sum_{i=1}^{n}(R_i - W_i)^2} \tag{11}$$

$$\mathrm{CC} = \frac{\sum_{i=1}^{n}(R_i - \overline{R_i})(W_i - \overline{W_i})}{\sqrt{\sum_{i=1}^{n}(R_i - \overline{R_i})^2 \sum_{i=1}^{n}(W_i - \overline{W_i})^2}} \tag{12}$$

4.2.2. Verification Results

Figure 15 shows the correlation distribution between the radial velocities of all radars and all wind profilers during the verification period at each quality control stage. The verification results of the RAW, NOS, ALL, and QCD stages are shown in Figure 15a–d. The raw radial velocity values (RAW), excluding non-precipitation echo, were distributed with a lot of noise, as shown in Figure 15a, and the RMSE was 7.99 m s^{-1} and CC was 0.75. Through the NOS step, it can be seen in Figure 15b that a significant amount of noise was eliminated and corrected, and the error was significantly reduced compared to the previous step. The RMSE was 4.78 m s^{-1}, which was less than that of the previous step of quality control, and the CC was 0.89, which was highly correlated with the radial velocity of the wind profilers. Even after restoring the lost radial velocity region, the RMSE was 4.71 m s^{-1}, which was less than the error in the previous step, even after restoring the lost radial velocity region. The CC was 0.89, which shows a high correlation with the radial velocity of the wind profilers, as shown in Figure 15c. In addition, only the quality-controlled radial velocity was extracted, excluding the actual observed value. The correlation distribution compared with the radial velocity of the wind profilers is shown in Figure 15d. The RMSE was 5.34 m s^{-1} and the CC was 0.85, which slightly increased the error compared to the values, including the actual observation data of the ALL stage. Although the correlation decreased, the results of the quality control technique were significant when compared to the radial velocity of the wind profilers.

Figure 16 shows the quantitative verification results for each stage of the quality control techniques for the 10 pairs of radar and wind profilers used in the verification. Comparing the verification values of RAW and NOS, the RMSE decreased in all pairs, and CC increased in all pairs. In ALL and QCD, the values of RMSE and CC were similar to those in the NOS step. For most of the verification pairs, the trend in the verification values is shown as KSN-47229 in Figure 17a–d. In the case of GDK-47114, the RMSE value was reduced during the quality control stage. Although the CC value after quality control increased, the CC value in RAW was 0.42, which means that the correlation was low. This is presumed to be due to the mountainous terrain between the GDK radar and the wind profiler at 47114 points and errors in the wind profiler. Nevertheless, the QCD distribution in Figure 17h and the distribution of the actual observation data in Figure 17e are similar, so it is considered effective to use the quality-controlled radial velocity. In the case of PSN-47155, the RMSE value in QCD was 7.60 m s^{-1}, which was larger than that of RAW, and the CC value in QCD was 0.68, which was lower than that of RAW. This is because the amount of data was small, and a few error values influenced the calculation of the verification values. Most of the QCD values in PNS-47155 were valid, as shown in Figure 17. Table 3 lists the values of the RMSE and CC for each stage of the quality control technique for the 10 pairs of radar and wind profilers used for verification.

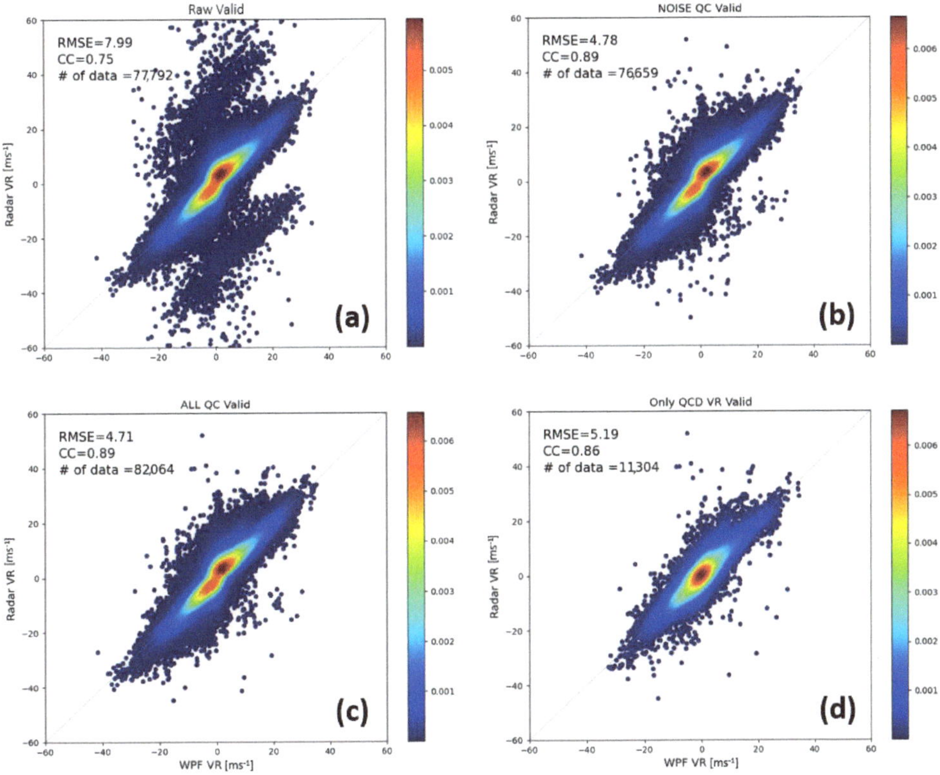

Figure 15. Two-dimensional normalized frequency distributions (NFDs) of radial velocity between the radars and the wind profilers (**a**) before noise correction (RAW), (**b**) after noise correction (NOS), (**c**) after quality control (ALL), and (**d**) only quality controlled radial velocity (QCD).

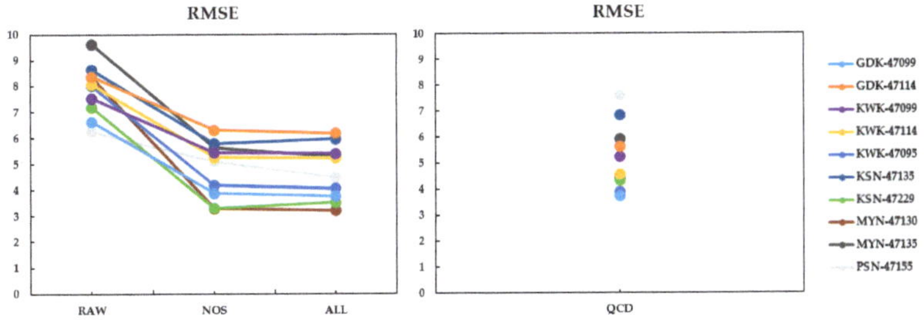

Figure 16. *Cont.*

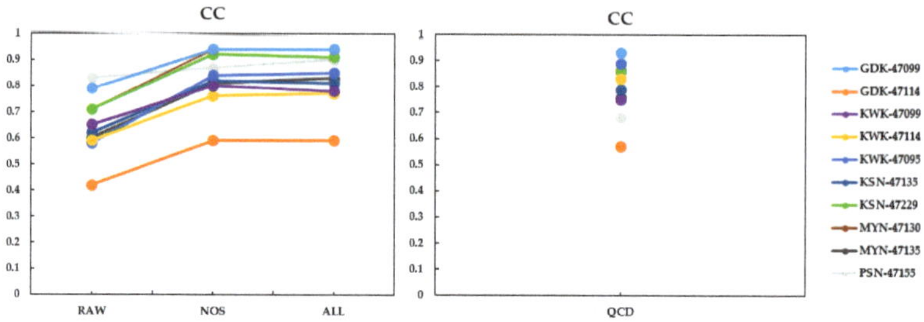

Figure 16. (**top**) RMSE and (**bottom**) CC for each QC stage and QCD of ten pairs. RAW is actual radial velocity, excluding non-precipitation echoes from raw data. NOS is noise-corrected radial velocity in RAW. ALL is the radial velocity after quality control. QCD is only quality-controlled radial velocity excluding actual observations.

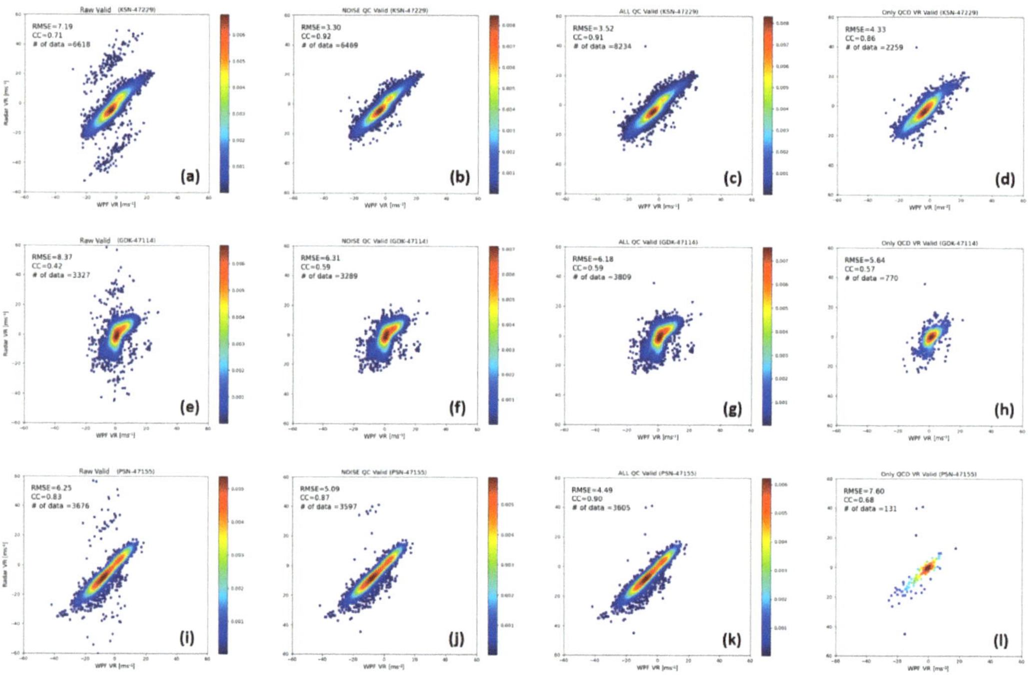

Figure 17. Same as Figure 15. (**a–d**) KSN-47229, (**e–h**) GDK-47114, (**i–l**) PSN-47155.

Table 3. Quantitative verification results for each stage of the radial velocity quality control technique.

RDR-WPF	RMSE				CC			
	RAW	NOS	ALL	QCD	RAW	NOS	ALL	QCD
GDK-47099	6.63	3.87	3.77	3.73	0.79	0.94	0.94	0.93
GDK-47114	8.37	6.31	6.18	5.64	0.42	0.59	0.59	0.57
KWK-47099	7.54	5.44	5.40	5.24	0.65	0.80	0.78	0.75
KWK-47114	8.07	5.26	5.23	4.57	0.59	0.76	0.77	0.83
KWK-47095	8.01	4.18	4.06	3.90	0.58	0.84	0.85	0.89
KSN-47229	7.19	3.30	3.52	4.33	0.71	0.92	0.91	0.86
KSN-47135	8.64	5.79	5.96	6.84	0.62	0.82	0.81	0.79
MYN-47130	8.35	3.29	3.21	4.42	0.71	0.94	0.94	0.86
MYN-47135	9.63	5.62	5.27	5.91	0.60	0.81	0.83	0.76
PSN-47155	6.25	5.09	4.49	7.60	0.83	0.87	0.90	0.68

5. Conclusions

A technique for quality control of radar radial velocity was developed based on WISSDOM to improve the accuracy of wind information, such as a three-dimensional radar wind field. This technique removes only the noise of radial velocity using a noise filter developed using the sign and median value of radial velocity observation, and then re-stores it to the median value while maintaining the precipitation echo. By spatiotemporal comparison of radial velocity between radar and adjacent wind profiler, the original RMSE of 6~10 m s^{-1} was dramatically reduced to 3~7 m s^{-1} after noise correction. The radial velocity was restored using the VAD restoration function to fill the blank (loss) area of the radial velocity. In particular, the distribution of radial velocity showed a more realistic and continuous spatial structure after the restoration procedure.

The quality-controlled radial velocity was used as input data to produce wind based on WISSDOM. After applying quality-controlled radial velocity to WISSDOM, it was confirmed that a more spatially continuous and accurate wind could be secured. This technique reflects the surrounding observations to some extent in the loss region because the VAD restoration function is calculated using several azimuth curves. However, wind analysis can be simplified by linearly restoring the lost radial velocity region due to the limitation of the VAD method, but it can be easier to predict and analyze meso-scale precipitation systems. This technique not only eliminates outliers by setting out an outlier range on the VAD curve but also considers that small-scale meteorological phenomena are not blurred in the area where the radial velocity is observed. In addition, each condition and threshold for the quality control algorithm can be flexibly changed in consideration of the user environment, such as radar type and observation strategy.

We operated on this technique as a test, verified its performance, and applied it to the operational radar quality control system of the WRC from August 2021. In the future, it is expected to be used for more accurate analysis and forecast support for various meteorological phenomena, such as precipitation development factor analysis, typhoon tracking, and local vortices.

6. Patents

Korean patent (Patent Number: 10-2212524) and United States patent (Patent Number: US 11,474,237 B2) "Method for retrieval of lost radial velocity in weather radar, recording medium and device for performing the method".

Author Contributions: This work was possible through significant contributions from all authors. Conceptualization, S.-Y.P., S.-H.J. and K.-H.K.; methodology, S.-Y.P., S.-H.J. and K.-H.K.; software, S.-Y.P., S.-H.J. and K.-H.K.; formal analysis, S.-Y.P.; investigation, S.-Y.P.; writing—original draft preparation, S.-Y.P.; writing—review and editing, S.-Y.P., S.-H.J. and K.-H.K.; visualization, S.-Y.P.; supervision, S.-H.J. and K.-H.K.; funding acquisition, S.-H.J. and K.-H.K.; All authors have read and agreed to the published version of the manuscript.

Funding: This research was supported by the "Development of radar based severe weather monitoring technology (KMA2021-03121)" of "Development of integrated application technology for Korea weather radar" project funded by the Weather Radar Center, Korea Meteorological Administration.

Institutional Review Board Statement: Not applicable.

Informed Consent Statement: Not applicable.

Data Availability Statement: Not applicable.

Conflicts of Interest: The authors declare no conflict of interest.

References

1. Liou, Y.C.; Chang, Y.J. A variational multiple-Doppler radar three-dimensional wind synthesis method and its impacts on thermodynamic retrieval. *Mon. Wea. Rev.* **2009**, *137*, 3992–4010. [CrossRef]
2. Liou, Y.C.; Chang, S.F.; Sun, J. An application of the immersed boundary method for recovering the three-dimensional wind fields over complex terrain using multiple-Doppler radar data. *Mon. Wea. Rev.* **2012**, *140*, 1603–1619. [CrossRef]

3. Gong, J.; Wang, L.L.; Xu, Q. A three-step dealiasing method for Doppler velocity data quality control. *J. Atmos. Oceanic Technol.* **2003**, *20*, 1738–1748. [CrossRef]
4. Gao, J.; Droegemeier, K.K. A Variational Technique for Dealiasing Doppler Radial Velocity Data. *J. Appl. Meteor.* **2003**, *43*, 934–940. [CrossRef]
5. Ming, W.; Liu, Y.; Jiawen, P.; Nan, L. New Velocity Dealiasing of Doppler Radar with the Interactive Method. In Proceedings of the WRI World Congress on Computer Science and Information Engineering, Los Angeles, CA, USA, 31 March–2 April 2009. [CrossRef]
6. Li, G.; He, G.X.; Zou, X.L.; Ray, P.S. A velocity dealiasing scheme for C-band weather radar systems. *Adv. Atmos. Sci.* **2014**, *31*, 17–26. [CrossRef]
7. Chu, A.; Wu, D.; Hu, H.; Xu, D.; Chen, A. A novel velocity dealiasing algorithm for S-Band weather radars. *Atmosfera* **2018**, *32*, 1–13. [CrossRef]
8. He, G.; Sun, J.; Ying, Z.; Zhang, L. A Radar Radial Velocity Dealiasing Algorithm for Radar Data Assimilation and its Evaluation with Observations from Multiple Radar Networks. *Remote Sens.* **2019**, *11*, 2457. [CrossRef]
9. Louf, A.; Protat, A.; Jackson, R.C.; Collis, S.M.; Helmus, J. UNRAVEL: A Robust Modular Velocity Dealiasing Technique for Doppler Radar. *J. Atmos. Oceanic Technol.* **2020**, *37*, 741–758. [CrossRef]
10. Zrnic, D.S.; Mahapatra, P. Two methods of ambiguity resolution in pulse Doppler weather radars. *IEEE Trans. Aerosp. Electron. Syst.* **1985**, *AES-21*, 470–483. [CrossRef]
11. Doviak, R.J.; Zrnic, D.S. *Doppler Radar and Weather Observations*, 2nd ed.; Academic: San Diego, CA, USA, 1993.
12. Tahanout, M.; Adane, A.E.H.; Chatelet, J.P.D. An improved m-PRT technique for spectral analysis of weather radar observations *IEEE Trans. Geosci. Remote Sens.* **2015**, *53*, 5572–5582. [CrossRef]
13. Holleman, I.; Beekhuis, H. Analysis and correction of dual PRF velocity data. *J. Atmos. Oceanic Technol.* **2003**, *20*, 443–453. [CrossRef]
14. Joe, P.; May, P.T. Correction of Dual PRF Velocity Errors for Operational Doppler Weather Radars. *J. Atmos. Oceanic Technol.* **2003**, *20*, 429–442. [CrossRef]
15. Hengstebeck, T.; Wapler, K.; Heizenreder, D. Radar Network-Based Detection of Mesocyclones at the German Weather Service. *J. Atmos. Oceanic Technol.* **2018**, *35*, 299–321. [CrossRef]
16. Park, S.-G.; Lee, D.-K. Retrieval of High-Resolution Wind Fields over the Southern Korean Peninsula Using the Doppler Weather Radar Network. *Weather. Forecast.* **2009**, *24*, 87–103. [CrossRef]
17. Altube, P.; Bech, J.; Argemi, O.; Rigo, T.; Pineda, N.; Collis, S.; Helmus, J. Correction of Dual-PRF Doppler Velocity Outliers in the Presence of Aliasing. *J. Atmos. Oceanic Technol.* **2017**, *34*, 1529–1543. [CrossRef]
18. Oh, Y.A.; Kim, H.L.; Suk, M.K. Clutter Elimination Algorithm for Non-Precipitation Echo of Radar Data Considering Meteorological and Observational Properties in Polarimetric Measurements. *Remote Sens.* **2020**, *12*, 3790. [CrossRef]
19. Browning, K.A.; Wexler, R. The determination of kinematic properties of a wind field using Doppler radar. *J. Appl. Meteor.* **1968**, *7*, 105–113. [CrossRef]
20. Miller, L.J.; Anderson, W. *Multiple Doppler Radar Wind Synthesis in CEDRIC*; CEDRIC Manual, Appendix F; National Center for Atmospheric Research: Boulder, CO, USA, 1991.

remote sensing

Article

Enhancing Spatial Variability Representation of Radar Nowcasting with Generative Adversarial Networks

Aofan Gong [1], Ruidong Li [1], Baoxiang Pan [2], Haonan Chen [3], Guangheng Ni [1,*] and Mingxuan Chen [4]

1 Department of Hydraulic Engineering, Tsinghua University, Beijing 100084, China; gaf20@mails.tsinghua.edu.cn (A.G.); lrd19@mails.tsinghua.edu.cn (R.L.)
2 Institute of Atmospheric Physics, Chinese Academy of Sciences, Beijing 100029, China; panbaoxiang@lasg.iap.ac.cn
3 Electrical and Computer Engineering, Colorado State University, Fort Collins, CO 80523, USA; haonan.chen@colostate.edu
4 Institute of Urban Meteorology, China Meteorological Administration, Beijing 100089, China; mxchen@ium.cn
* Correspondence: ghni@tsinghua.edu.cn

Abstract: Weather radar plays an important role in accurate weather monitoring and modern weather forecasting, as it can provide timely and refined weather forecasts for the public and for decision makers. Deep learning has been applied in radar nowcasting tasks and has exhibited a better performance than traditional radar echo extrapolation methods. However, current deep learning-based radar nowcasting models are found to suffer from a spatial "blurry" effect that can be attributed to a deficiency in spatial variability representation. This study proposes a Spatial Variability Representation Enhancement (SVRE) loss function and an effective nowcasting model, named the Attentional Generative Adversarial Network (AGAN), to alleviate this blurry effect by enhancing the spatial variability representation of radar nowcasting. An ablation experiment and a comparison experiment were implemented to assess the effect of the generative adversarial (GA) training strategy and the SVRE loss, as well as to compare the performance of the AGAN and SVRE loss function with the current advanced radar nowcasting models. The performances of the models were validated on the whole test set and inspected in two storm cases. The results showed that both the GA strategy and SVRE loss function could alleviate the blurry effect by enhancing the spatial variability representation, which helps the AGAN to achieve better nowcasting performance than the other competitor models. Our study provides a feasible solution for high-precision radar nowcasting applications.

Keywords: nowcasting; radar; generative adversarial network; spatial variability

1. Introduction

Detailed weather forecasting over a very short period that lasts from the present to the next few hours, which is also known as nowcasting, has significant benefits related to weather-related human activities, including public traffic, flood alarms, disaster warnings, emergency management, and risk prevention [1]. Based on current meteorological observations, accurate nowcasting can provide timely (up to the minute level) and refined (mesoscale or even microscale) weather forecasts for the public and decision makers [2].

Thanks to the rapid progress in meteorological observation technology, Doppler weather radars have become one of the most valuable tools for observing clouds, precipitation, and wind [3,4]. Since radars can detect larger areas than rain gauges and scan at a higher resolution in shorter intervals than satellites, they can better reflect the spatial and temporal variability of the above meteorological elements than rain gauges and satellites, which are also powerful tools for weather forecasting [5–8]. Traditional radar echo extrapolation methods are widely utilized as the basis of nowcasting systems, such as the following: Thunderstorm Identification, Tracking, And Nowcasting (TITAN) [9]; Storm Cell Identification and Tracking (SCIT) [10]; Tracking Reflectivity Echoes by Correlation

(TREC) [11]; the McGill Algorithm for Precipitation Nowcasting by Lagrangian Extrapolation (MAPLE) [12]; Dynamic and Adaptive Radar Tracking of Storms (DARTS) [13]; optical flow-based methods [14,15]; and methods for nowcasting the growth and decay of storms [16]. These traditional methods have exposed shortcomings due to the limits of their underlying assumptions and constraints, including the motion–field constancy of TREC-based methods and the spatial smoothness constraints of optical flow-based methods [17–19].

In recent years, deep learning (DL), which has seen remarkable advancements in diverse domains, such as computer vision [20], natural language processing [21], and geoscience [22], has also been applied in radar nowcasting by meteorological researchers [23]. In these studies, radar nowcasting was formulated as a spatiotemporal sequence extrapolation problem. Compared to conventional extrapolation models, DL models usually perform better because of their strong non-linear modeling capacity driven by large-scale historical radar echo datasets [24]. Generally, DL nowcasting models consist of two main types: the convolutional neural network (CNN) and the recurrent neural network (RNN). CNNs are widely adopted in image processing because of their translation invariance property, while RNNs feature in time series analysis because of their recurrent structure. Present CNN-based nowcasting models focus more on the spatial correlation of meteorological fields, while RNNs pay more attention to the sequential correlation [25]. For CNNs, several researchers have made progress in developing three-dimensional CNNs and their variants [26–28]. Klein, et al. [29] proposed a dynamic convolutional layer, but revealed the limits of predicting one echo frame in one step. Ayzel, et al. [30] designed an All Convolutional Neural Network and introduced a more effective model called "RainNet" with the U-Net [31] structure in their later work [32]. Trebing, et al. [33] designed the SmaAt-UNet with attentional modules and depthwise separable convolutions, which produced higher performance with fewer parameters than the original U-Net. For RNNs, the authors in [34] proposed the Convolutional Long Short-Term Memory (ConvLSTM), which replaced the full connection in the gates of the vanilla LSTM [35] with a convolutional operator. This work is regarded as the pioneer study of DL-based precipitation nowcasting. An encoding–forecasting network structure was built, based on the same authors' newly proposed Trajectory Gated Recurrent Unit (TrajGRU) in their following study [36]. Wang, et al. [37] proposed the PredRNN by expanding the original ConvLSTM with spatiotemporal memory flow and developed an enhanced model, "PredRNN++", in their following study [38]. Wu, et al. [39] proposed the MotionRNN, which significantly improves the ability to predict changeable motions and avoid motion vanishing for stacked multiple-layer nowcasting models.

Although DL models have shown advantages in radar nowcasting tasks, several researchers have noted a systematic deviation from current DL models and summarized them as "blurry" effects [32,36,40,41]. DL models were found to neglect high-intensity features and small-scale patterns of the weather system, causing the generated images to lose spatial variability and look blurry. This effect was attributed to the impact of convolutional operators contained in DL models, in that their inductive bias of the translation invariance would lead to a loss of spatiotemporal features for precipitation [42,43]. Under this deviation, although DL models can outperform traditional models in regard to most precision scores, they are weak when it comes to learning the spatial variability of the radar echo sequences. Since radar parameters have a strong relationship with precipitation, the deviation in the spatial variability representation of radar echoes causes an expanded error in the downstream applications, such as in urban flood simulation [44,45].

Researchers have adopted generative adversarial networks (GANs) [46] to alleviate the blurry effect. Jing, et al. [40] developed an Adversarial Extrapolation Neural Network (AENN), based on the ConvLSTM and CNN, for nowcasting at an interval of 30 min to generate accurate and realistic extrapolation echoes. Tian, et al. [41] proposed a Generative Adversarial Convolutional Gated Recurrent Unit (GA-ConvGRU) model that outperformed the original ConvGRU, but their model suffered due to training instability.

Xie, et al. [47] developed a more robust Energy-Based Generative Adversarial Forecaster and demonstrated the stability and accuracy of their model. Ravuri, et al. [48] compared their novel Deep Generative Model of Rainfall (DGMR) with ensemble optical flow and evaluated its effectiveness with both quantitative verification measures and qualitative cognitive assessments. However, few studies have paid attention to enhancing the local spatial variability representation ability of DL nowcasting models.

This study aimed to solve the above drawbacks of DL models for radar nowcasting. A Spatial Variability Representation Enhancement (SVRE) loss function and an Attentional Generative Adversarial Network (AGAN) are proposed to enhance the spatial variability representation of radar nowcasting. The SVRE loss implements regularization in the adversarial training process with representative spatial variability features. The SVRE loss function and the AGAN model are evaluated on a three-year radar observation dataset through an ablation experiment and a comparison experiment. Several state-of-the-art DL nowcasting models are selected as the baseline models, including the MotionRNN [49], the SmaAt-UNet [33], and a traditional ensemble nowcasting model, PySTEPS [50]. The rest of this article is organized as follows. Section 2 illustrates the principle of the SVRE loss function and the architecture of the AGAN. Details of the experiments are explained in Section 3, and the results of these experiments are presented and discussed in Section 4. The last section concludes this study and points out the direction of future work.

2. Methods

2.1. Problem Statement

As previous studies [34] summarized, the radar nowcasting problem can be abstracted as a spatiotemporal sequence extrapolation problem which aims to predict the next length-n sequence given a previous length-m observation. Let tensor $\mathcal{X} \in \mathbb{R}^{N\times H\times W}$ denote radar observations over a previous period (with length N, height H, and width W) and θ denote the parameters of a DL-based nowcasting model; then, the problem can be described by

$$\hat{\mathcal{X}}_{1:n} = \underset{\mathcal{X}_{1:n}}{\operatorname{argmax}}\, p(\mathcal{X}_{1:n}|\mathcal{X}_{1-m:0};\theta), \tag{1}$$

where the subscript index of $\mathcal{X}$ denotes the tensor's slice at the corresponding time step (0 represents the current time).

The perspective of generative models in machine learning can change to a probabilistic problem instead of a deterministic problem, which means that the prediction is no longer estimated by the maximum likelihood estimation of the conditional probabilistic distribution, but is, rather, sampled from the conditional probabilistic distribution given the prior distribution of a latent code $\mathcal{Z}$, used to describe the latent states of the system, and described as

$$\begin{aligned}\hat{\mathcal{X}}_{1:n} &\sim p(\mathcal{X}_{1:n}|\mathcal{X}_{1-m:0};\theta)\\ &= \int_{\mathcal{Z}} p(\mathcal{X}_{1:n},\mathcal{Z}|\mathcal{X}_{1-m:0};\theta)\mathrm{d}\mathcal{Z}\\ &= \int_{\mathcal{Z}} p(\mathcal{X}_{1:n}|\mathcal{X}_{1-m:0},\mathcal{Z};\theta)p(\mathcal{Z}|\mathcal{X}_{1-m:0})\mathrm{d}\mathcal{Z}\\ &= \mathbb{E}_{\mathcal{Z}}[p(\mathcal{X}_{1:n}|\mathcal{X}_{1-m:0},\mathcal{Z};\theta)].\end{aligned} \tag{2}$$

2.2. The Principle of GAN

Machine learning models generally consist of two paradigms: discriminative models and generative models. Fundamentally, discriminative models aim to draw the decision boundaries from the data space, while generative models learn a joint probability pattern based on Bayesian rules, thus learning and applying the mapping of the low-dimensional manifold to the high-dimensional data space [51,52]. Since generative models calculate the joint distribution of the input and the target variables before the derivation of the posterior distribution, they can learn more information and, thus, describe more indicative features of the data. A generative model can extract more information about the relationship between the input and the target variables than a discriminative model can, especially when latent variables exist. Classic generative models, including Gaussian mixture models, hidden Markov models, Boltzmann machines, and variational autoencoders, calculate the joint distribution by explicitly specifying the probabilistic density function and optimizing it with suitable optimization algorithms, such as gradient descent or variational inference [51].

The GAN was one of the generative models proposed by Goodfellow, et al. [53], which was designed to identify the joint distribution based on the adversarial learning theory. A GAN is composed of a generator that captures the data distribution and a discriminator that estimates the probability of where a sample came from. During adversarial learning, the generator is optimized with the guidance of the discriminator. The optimization goal of a GAN is to achieve a Nash equilibrium between the generator and the discriminator. The optimization of a GAN is equivalent to a min–max two-player game between the generator and the discriminator, expressed by

$$\min_{G}\max_{D} V(D,G) = \mathbb{E}_{\mathbf{x}\sim p_{data}(\mathbf{x})}[\log D(\mathbf{x})] + \mathbb{E}_{\mathbf{z}\sim p_{\mathbf{z}}(\mathbf{z})}[\log(1 - D(G(\mathbf{z})))], \quad (3)$$

where D and G represent discriminator and generator, respectively. $\mathbf{x}$ represents the sample and $\mathbf{z}$ denotes the latent random vector of the generator. The authors proved that the generator and the discriminator reach Nash equilibrium if, and only if, the estimated distribution equals the data distribution by simultaneously training the generator and the discriminator, given there are enough data (in practice, they have to be trained alternately instead). As a result of the outstanding performance of the GAN, this generative model has been applied as one of the mainstream tools for generative learning.

2.3. The SVRE Loss Function

As its name suggests, the SVRE loss function focuses on the direct enhancement of a model's spatial variability representation. The optimization of the original GAN proved to be equivalent to a sigmoid cross-entropy function [53], while Mao, et al. [54] pointed out that this function had a gradient vanishing problem, leading to quality loss in the generated images. Therefore, we adopted the least-squares loss (proposed by the above authors) as the basic adversarial loss. Beyond generative adversarial training, we added two additional regularization terms to the adversarial loss to improve the sharpness and spatial variability of the generated images. The first regularization term is L1-normalization, which has been used for encouraging less blurring in image translation tasks [55]. The second term is the L1-norm distance between the coefficient of variation (C_v) of the prediction and the index of the target sequence, which has never been used for regularization in previous related studies. In statistics, C_v is defined as the quotient of the standard deviation divided by the mean value of a group of samples. This metric has also been used in hydrological research to describe the precipitation variability and to compare the variability of different precipitation fields [56]. In our study, the standard deviation and the mean value were calculated along the spatial axis (H and W), instead of the temporal axis (N), to represent spatial variability, as is shown in Equation (4). This L1-norm term of C_v gives a quantification of the gap between the prediction's spatial variability and the observation's spatial variability. Both regularization terms were scaled by a corresponding hyperparameter λ. The loss functions for the adversarial training of the discriminator and the generator are illustrated in Equations (5) and (6).

$$C_v(\mathcal{X}_t) = \frac{\sigma_{h,w}(\mathcal{X}_t)}{\text{mean}_{h,w}(\mathcal{X}_t)} \quad (4)$$

$$\mathcal{L}_D(\phi) = \mathbb{E}_{\mathcal{X}}[D(\mathcal{X}_{1:n}|\mathcal{X}_{1-m:0};\phi) - 1]^2 + \mathbb{E}_{\mathcal{X},\mathcal{Z}}\left[D(G(\mathcal{Z}|\mathcal{X}_{1-m:0};\theta)|\mathcal{X}_{1-m:0};\phi)^2\right] \quad (5)$$

$$\mathcal{L}_G(\theta) = \mathbb{E}_{\mathcal{X},\mathcal{Z}}[D(G(\mathcal{Z}|\mathcal{X}_{1-m:0};\theta)|\mathcal{X}_{1-m:0};\phi) - 1]^2 + \lambda_r \mathbb{E}_{\mathcal{X},\mathcal{Z}}||G(\mathcal{Z}|\mathcal{X}_{1-m:0};\theta) - \mathcal{X}_{1:n}||_1 + \lambda_v \mathbb{E}_{\mathcal{X},\mathcal{Z}}||C_v(G(\mathcal{Z}|\mathcal{X}_{1-m:0};\theta)) - C_v(\mathcal{X}_{1:n})||_1 \quad (6)$$

Here, D and G represent discriminator and generator, respectively. The operator mean denotes the mean value across axis h and w of an image at time t and σ denotes its standard deviation. The discriminator loss is meant to minimize the sum of the averaged squared distance between a sample and its corresponding label (0 for fake and 1 for real) to distinguish the predicted sequence from the target. Meanwhile, the generator loss is meant to minimize the averaged squared distance between the fake sample and the real label, since it ultimately needs to confuse the discriminator.

2.4. The Architecture of the AGAN

We now present the details of the AGAN (Figure 1). Like other GANs, it consists of a generator and a discriminator. The generator (Figure 1a) follows a U-Net structure, which has been effective in related nowcasting applications. It accepts a 1-h (10 steps with a 6 min interval) historical radar observation sequence (orange rectangles) as the input and predicts the radar sequence for the next hour. The encoding part of the U-Net contains four downscaling blocks (red arrows), which are preceded by a 1 × 1 convolution (orange arrows) for temporal feature combination. The decoding part applies the same structure, containing four upscaling blocks and an additional 1 × 1 convolution. The multi-scale encoding feature maps (blue rectangles) are copied through skip connections (gray arrows) and concatenated with the corresponding decoding feature maps (yellow rectangles) in upscaling blocks (green arrows). The discriminator (Figure 1b) is a fully convolutional network composed of two convolutional blocks and six downscaling blocks, which accepts the concatenated sequences from historical and future data (orange rectangles) as input and returns a probability between 0 and 1 as output to guide the optimization of the generator through adversarial training.

The downscaling and the upscaling blocks of the AGAN are displayed in Figure 2a,b, respectively. For the downscaling blocks, the input feature maps are first reshaped by a 2 × 2 max-pooling layer (the red rectangle) and then fed into two 3 × 3 convolutional layers, each followed by a batch normalization layer (BN) and a rectified linear unit (ReLU) layer. For the upscaling blocks, the input features are reshaped by a 2 × 2 bilinear interpolation layer (the green rectangle) and concatenated with the corresponding encoding feature map copied from the encoder. A convolutional block attention module (CBAM) [57], which was developed for self-adaptive feature combination, is integrated at the end of the scaling block (the orange rectangle) to strengthen the model's attention-based feature refinement ability. This differentiates our model from the original U-Net, and that is why it is called "attentional".

The number of convolutional kernels doubles after downscaling and is reduced by half after upscaling in the generator (32, 64, 128, and 256 kernels in the corresponding blocks), which references the original U-Net. The AGAN contains, in total, about 6.40 million trainable parameters, among which 3.66 million are for the generator and 2.74 million are for the discriminator.

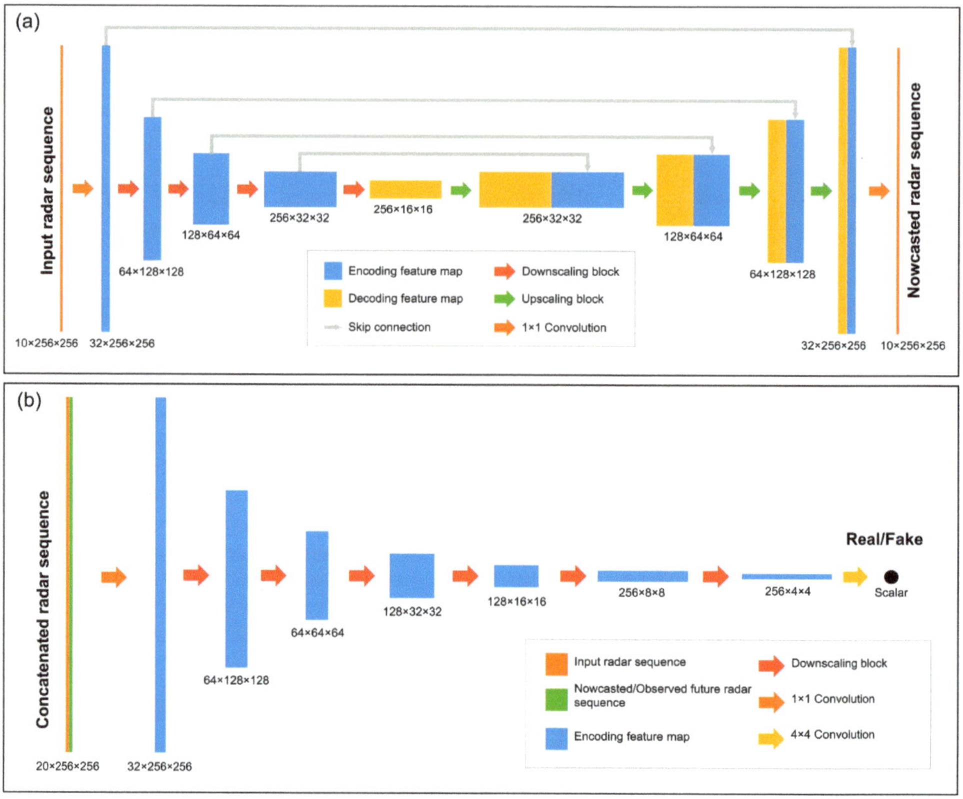

Figure 1. Architecture of AGAN. (**a**) AGAN's generator. (**b**) AGAN's discriminator.

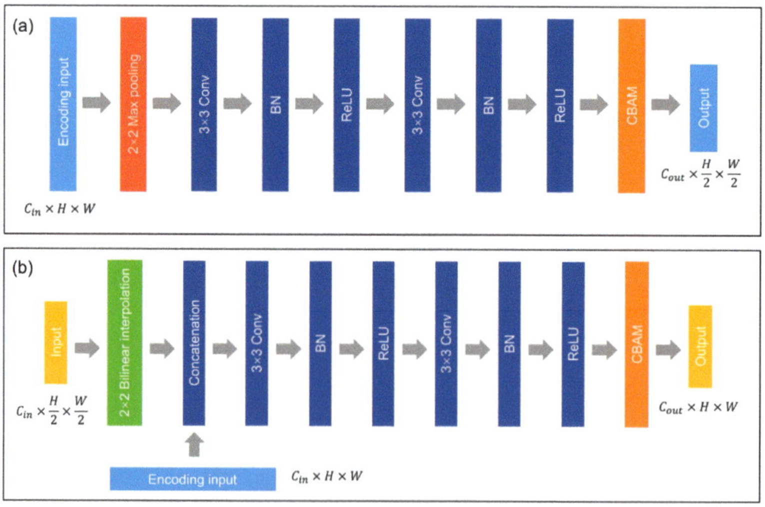

Figure 2. Scaling blocks in AGAN. (**a**) Downscaling blocks. (**b**) Upscaling blocks.

3. Experiments

3.1. Data and Study Area

To validate the effectiveness of the AGAN, we selected a radar composite reflectivity dataset from the Beijing Auto-Nowcast (BJ-ANC) System that was developed by the Institute of Urban Meteorology, China Meteorological Administration, Beijing [58]. The system collects observations from S-band and C-band Doppler weather radars that are part of the China Next Generation Weather Radar (CINRAD) network and produces radar mosaic images with quality control and mosaic generation algorithms. The radar composite reflectivity mosaic product covers a total area of 800 × 800 km^2, with a spatial resolution of 1 km and a temporal resolution of 6 min. Since we were concerned about the weather in Beijing and its surroundings, we defined the study area as a square region of 256 × 256 km^2 that had the Yizhuang Radar (one of the CINRAD radars) located at the exact center (39.81°N, 116.47°E) of the square (Yizhuang is near the center of Beijing City), as shown in Figure 3.

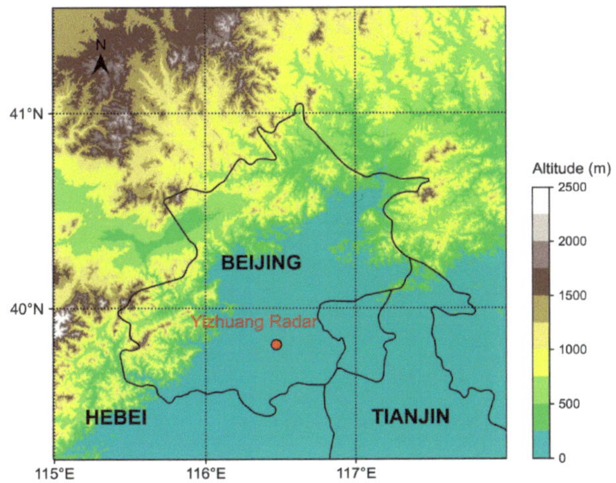

Figure 3. The study area (the whole square) and the location of Yizhuang Radar (red circle). The figure is plotted using UTM 50N coordination.

For dataset construction and processing, observations from three S-band radars on 86 rainy days that occurred during the warm seasons (from June to September) from 2017 to 2019 were selected, according to ground precipitation observations. The mosaic images were clipped between 0 and 70 dBZ and scaled with a min–max normalization which converted the values to a range from 0 to 1. According to the input and output settings, we packed observations that covered 20 consecutive time steps into 1 sequence and extracted a total of 18,892 sequences from the dataset. The sequences were sorted in chronological order and temporal overlapping sequences were dropped. Then, they were split into the training set, validation set, and test set with a ratio of approximately 7:1:2. To avoid data leakage, the sequences in the training and validation sets were only selected from the warm seasons in 2017 and 2018, yet all the sequences in the test set came from the warm season in 2019, as shown in Table 1.

Table 1. Split settings of the dataset.

Items	Training	Validation	Test	Removed	Total
Sequences	13,200	1888	3760	44	18,892
Rainy days	60	9	17	/	86
Proportion	0.699	0.1	0.199	0.003	1
Year	2017, 2018	2018	2019	/	2017–2019

3.2. Evaluation Metrics

The nowcasting performances of different models were evaluated from two aspects. The first one was forecasting accuracy, which has been focused on by most previous studies. The commonly used weather forecast metrics were adopted, including the probability of detection (POD), the false alarm ratio (FAR), and the critical success index (CSI) [59]. These metrics were calculated with a contingency table to count the frequency of hits (h), false alarms (f), and misses (m), as defined in Equations (7)–(9). For a certain threshold, a hit occurs when both the prediction and the target value exceed the threshold within the same grid. A false alarm occurs when the threshold is beyond the target value but under the prediction value, and a miss occurs in the opposite case. In this study, the threshold of these contingency metrics was set to 30 dBZ because this has been commonly used to distinguish heavy rainfall from light rainfall in related studies [60,61]. The mean bias error (MBE), the mean absolute error (MAE), and the root mean squared error (RMSE) were used to roughly estimate the forecasting bias, as defined in Equations (10)–(12).

$$\mathrm{POD} = \frac{h}{h+m} \tag{7}$$

$$\mathrm{FAR} = \frac{f}{h+f} \tag{8}$$

$$\mathrm{CSI} = \frac{h}{h+m+f} \tag{9}$$

$$\mathrm{MBE}(\hat{\mathcal{X}}_t, \mathcal{X}_t) = \mathrm{mean}_{h,w}(\hat{\mathcal{X}}_t - \mathcal{X}_t) \tag{10}$$

$$\mathrm{MAE}(\hat{\mathcal{X}}_t, \mathcal{X}_t) = \mathrm{mean}_{h,w}|\hat{\mathcal{X}}_t - \mathcal{X}_t| \tag{11}$$

$$\mathrm{RMSE}(\hat{\mathcal{X}}_t, \mathcal{X}_t) = \sqrt{\mathrm{mean}_{h,w}(\hat{\mathcal{X}}_t - \mathcal{X}_t)^2} \tag{12}$$

The second aspect was spatial variability representation. Three metrics were adopted to evaluate the spatial variability similarity between the observed precipitation field and the prediction of a certain model: (1) The Jensen–Shannon divergence (JSD), which measures the statistical difference between one probability distribution p and a second reference probability distribution $\hat{p}$. It can be proved that the JSD is symmetric to p and $\hat{p}$, and ranges from 0 to 1; (2) The structural similarity index, which measures (SSIM) [62] and is used to measure the similarity between two images from the three aspects of luminance, contrast, and structure; (3) The power spectral density (PSD), which presents the relationship between the power and frequency of a signal. It has been used in radar nowcasting tasks to evaluate a model's ability to represent diverse-scale weather patterns [32,48]. In this study, the PSD was calculated both with the height axis and the width axis of the radar images, defined in Equations (13) and (14). The other two metrics are defined in Equations (15) and (16).

$$\mathrm{PSD}_h(\mathcal{X}_t) = 10\log_{10}\mathrm{mean}_w|\mathcal{F}_h(\mathcal{X}_t)|^2 \tag{13}$$

$$\mathrm{PSD}_w(\mathcal{X}_t) = 10\log_{10}\mathrm{mean}_h|\mathcal{F}_w(\mathcal{X}_t)|^2 \tag{14}$$

$$\mathrm{JSD}(p_t\|\hat{p}_t) = \frac{1}{2}\sum_{h,w} p_t \log\frac{2p_t}{p_t+\hat{p}_t} + \frac{1}{2}\sum_{h,w}\hat{p}_t \log\frac{2\hat{p}_t}{p_t+\hat{p}_t} \tag{15}$$

$$\mathrm{SSIM}(\hat{\mathcal{X}}_t, \mathcal{X}_t) = \frac{\left[2\mathrm{mean}_{h,w}(\hat{\mathcal{X}}_t) \cdot \mathrm{mean}_{h,w}(\mathcal{X}_t) + C_1\right]\left[2\mathrm{cov}_{h,w}(\hat{\mathcal{X}}_t, \mathcal{X}_t) + C_2\right]}{\left[\mathrm{mean}_{h,w}(\hat{\mathcal{X}}_t)^2 + \mathrm{mean}_{h,w}(\mathcal{X}_t)^2 + C_1\right]\left[\mathrm{std}_{h,w}(\hat{\mathcal{X}}_t)^2 + \mathrm{std}_{h,w}(\mathcal{X}_t)^2 + C_2\right]} \quad (16)$$

In the above equations, $\mathcal{X}_t$ represents the observation radar map at time t and $\hat{\mathcal{X}}_t$ represents the prediction, p_t and $\hat{p}_t$ are the probability distributions of $\mathcal{X}_t$ and $\hat{\mathcal{X}}_t$. std represents the standard deviation operator. The value cov is the covariance between the observation and the prediction. The constants C_1 and C_2 were set to 1×10^{-4} and 9×10^{-4} in this research, which was the same as in [62]). Then, $\mathcal{F}_h(\cdot)$ and $\mathcal{F}_w(\cdot)$ denote the Fourier transform operation along the height and width axis, respectively.

3.3. Experiment Settings

The performances of the AGAN and the SVRE loss function were evaluated through an ablation experiment and a comparison experiment. In the ablation experiment, the two components, the GA strategy and SVRE, were tested by training the AGAN or the AGAN's generator with or without SVRE loss, respectively. The generator of the AGAN trained with the purely L1-norm loss function served as the control group, denoted by AGAN(g). The names of the models and their meanings are explained in Table 2.

Table 2. The names of the models and their meanings in the ablation experiment. The ✓represents "with" and the × represents "without".

Model	Trained with GA Strategy	Trained with SVRE Loss
AGAN(g)	×	×
AGAN(g) + SVRE	×	✓
AGAN	✓	×
AGAN + SVRE	✓	✓

For the AGAN, the generator and the discriminator were trained alternately with an Adam optimizer regularized by a decoupled weight decay of 0.01, where $\beta_1 = 0.9$ and $\beta_2 = 0.999$. The maximum training step was set to 100,000. The Two Time-Scale Update Rule (TTUR) [63] was adopted as one of the adversarial training strategies for better convergence. However, in a way that differed from the original TTUR skill, the learning rate was set to 1×10^{-4} for the generator and 5×10^{-5} for the discriminator to avoid the early convergence of the discriminator, based on preliminary experiments. The coefficients of the two regularization terms in the loss function were set to 10 and 1 (10 for the reconstruction term and 1 for the SVRE term) according to the performance on the validation set. The early stopping strategy was applied to prevent the model from overfitting. The integration over latent variables in Equation (2) was approximated with six latent random vectors at one training step, which was the same as in another related study [48]. Other settings included a batch size of 8 and the early stopping patience of 10 epochs. The AGAN(g) shared the same settings as the AGAN.

In the comparison experiment, the AGAN with SVRE were compared with three baseline models: the RNN-based MotionRNN [39], the CNN-based SmaAt-UNet [33], and the optical flow-based ensemble forecast system PySTEPS [50]. We kept the architecures of the deep learning baseline models (SmaAt-UNet and MotionRNN) unchanged and implemented the same training strategies on them as on the AGAN. For PySTEPS, the Lucas–Kanade motion tracking method and the semi-Lagrangian extrapolation method were selected.

All the experiments in this study were implemented on a computing platform with an Intel Xeon Gold 6226R CPU and an Nvidia Tesla A100 GPU, based on the open-source machine learning framework PyTorch (https://pytorch.org/ (accessed on 6 August 2022)).

4. Results and Discussion

4.1. Overall Performance

The overall performances of the models in the ablation experiment and the comparison experiment were evaluated on the whole test set. For each sample in the test set, we first extracted the last frames of the forecast sequence and the observed sequence. Since the time step of the last frame was 60 min ahead of the reference time, we determined the last frame to have a lead time of 60 min. The evaluation metrics of the observation and the prediction at the lead time of 60 min (+60 min observation and prediction) were calculated to reflect the models' performances for the sample. These metrics were averaged over all samples in the test set. The results of the experiments are listed in Table 3.

Table 3. Overall, +60 min nowcasting performances on the test set. The up and down arrows in the heading indicate whether the highest or the lowest was the best for different metrics. A bold number indicates that the model in its row had the best performance, evaluated with the metric in its column.

Model	POD ↑	FAR ↓	CSI ↑	MBE ↓	MAE ↓	RMSE ↓	SSIM ↑	JSD ↓
PySTEPS	0.299	**0.470**	0.204	**0.0**	**6.1**	**9.6**	0.292	0.601
SmaAt-UNet	0.631	0.521	0.351	13.8	14.6	18.8	0.365	0.516
MotionRNN	0.572	0.477	0.310	14.0	15.0	19.2	0.337	0.521
AGAN(g)	**0.749**	0.568	0.373	13.7	14.6	18.8	0.360	0.437
AGAN(g) + SVRE	0.643	0.484	0.377	13.2	14.4	18.6	0.377	0.477
AGAN	0.721	0.565	0.374	13.4	14.4	18.5	0.376	0.427
AGAN + SVRE	0.745	0.578	**0.380**	13.4	14.5	18.6	**0.387**	**0.421**

In Table 3, three baseline models (PySTEPS, SmaAt-UNet, and MotionRNN), the AGAN's generator trained with the ordinary adversarial loss function (AGAN(g)), the AGAN's generator trained with the SVRE loss function (AGAN(g) + SVRE), the AGAN trained with the ordinary adversarial loss function (AGAN), and the AGAN trained with the SVRE loss function (AGAN + SVRE) are included. For forecasting accuracy, it was found that, although the AGAN, AGAN(g)+SVRE, and AGAN + SVRE did not reach the highest POD or the lowest FAR, when considering hits and false alarms together, both the GA strategy and the SVRE loss function could increase the CSI. Compared with the AGAN(g), the AGAN + SVRE increased the CSI from 0.373 to 0.380, indicating that the combination of the GA strategy and SVRE could improve the general forecasting accuracy. It was also observed that SVRE slightly narrowed the MBE, MAE, and RMSE between the prediction and observation. For spatial variability representation, SVRE improved the performances of both the AGAN and AGAN(g) by concurrently increasing the image similarity (increasing the SSIM by 0.073 and 0.050) and reducing the distribution difference (reducing the JSD by 0.085 and 0.100) between the observation and the prediction. The effect of the GA strategy was less significant than for SVRE. Among all the models, the AGAN + SVRE reached the highest CSI, the highest SSIM, and the lowest JSD, indicating that the combination of the GA strategy and the SVRE loss function could help improve both forecast accuracy and spatial variability representation. The traditional optical flow method, PySTEPS, showed the least FAR and forecasting bias, while in other metrics it fell far behind our proposed model. The DL-based models (SmaAt-UNet and MotionRNN) performed worse in most metrics than the AGAN + SVRE.

4.2. Case Study

In this subsection, we selected two cases from the test set and further analyzed the nowcasting performances of different models for these cases.

4.2.1. Case 1

The first case was selected from a mesoscale squall line from the midsummer of 2019 on 6 August UTC. A 2-h (20-step) sequence from 15:00 to 16:54 was selected. The first half (from 15:00 to 15:54) of the sequence was fed into the model as the input, while the second half (from 16:00 to 16:54) served as the ground truth for evaluation. The +60 min radar images forecast by different models were visualized together with the observation in Figure 4. The top left subfigure indicates that the weather system evolved into a squall line trending from southwest to northwest at the lead time of 60 min. It can be easily observed that all DL models suffered from the blurry effect. The prediction of PySTEPS was closer to the observation in terms of peak reflectivity intensity, but it had significant errors in peak positions and echo shapes. The predictions of the SmaAt-UNet and MotionRNN had opposite systematic deviations compared to those of PySTEPS. The AGAN(g) performed better than the baseline models, but it exaggerated the extent of the storm center, particularly in areas where the reflectivity was over 35 dBZ. The AGAN + SVRE provided the closest prediction to the observation in both peak intensity and center location. Compared to the AGAN(g), the blurry effect could be alleviated when the GA strategy and the SVRE loss function were simultaneously implemented.

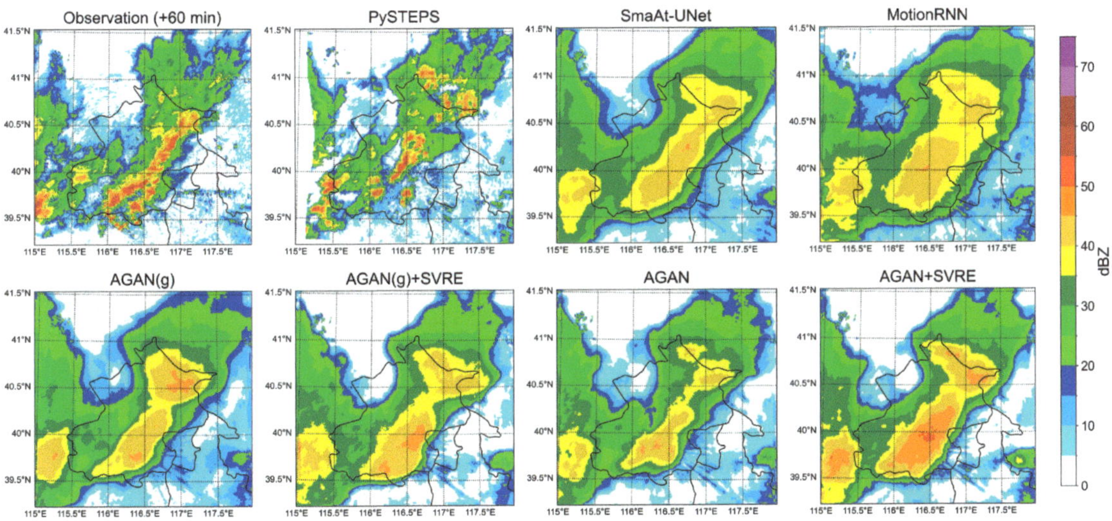

Figure 4. Visualization of the +60 min prediction and the observation in case 1.

A contrast scatter plot is presented in Figure 5 to evaluate the pixelwise similarity between the +60 min prediction and observation in case 1 for each model. The horizontal axis of each subfigure represents the observed value, and the vertical axis represents the predicted value. Points on the 45° line are regarded as perfect predictions, while points below or beyond this line correspond to underestimations or overestimations, respectively. The points are colored with their probability density provided by the Gaussian kernel density estimation. The results show that DL models tended to overestimate low-intensity pixels but underestimated high-intensity pixels, which coincided with the blurry effect in Figure 4. The positions of the peak probability density of the SmaAt-UNet and MotionRNN were beyond the 45° line, reflecting their systematic overestimations of mid-intensity pixels (between 20 and 30 dBZ). The AGAN(g), AGAN(g) + SVRE, AGAN, and AGAN + SVRE had lower forecast biases than the DL models. Their positions of peak probability density were more concentrated around the 45° line. PySTEPS had the most balanced prediction with the least general forecast bias, but its performance was significantly limited in pixels over 35 dBZ, which might cause a failure in extreme precipitation scenarios.

Table 4 enumerates the models' nowcasting performances for case 1. It was found that both the AGAN and AGAN(g) offered better comprehensive forecasting accuracy (higher CSI) than the baseline models. They also achieved a smaller forecasting bias (lower

MBE, MAE, and RMSE) and a closer spatial variability to the observation (higher SSIM and lower JSD). When trained with the SVRE loss function, the AGAN and AGAN's generator obtained a higher SSIM and lower JSD than the SmaAt-UNet and MotionRNN, as well as a slightly reduced CSI and increased FAR. The AGAN + SVRE reached the highest SSIM (0.358) and the lowest JSD (0.616) among all the models. Although PySTEPS reached the lowest bias for the MBE, MAE, and RMSE, it was limited in regard to the POD, CSI, SSIM, and JSD because of the misplacement of the peak intensity location, mentioned above.

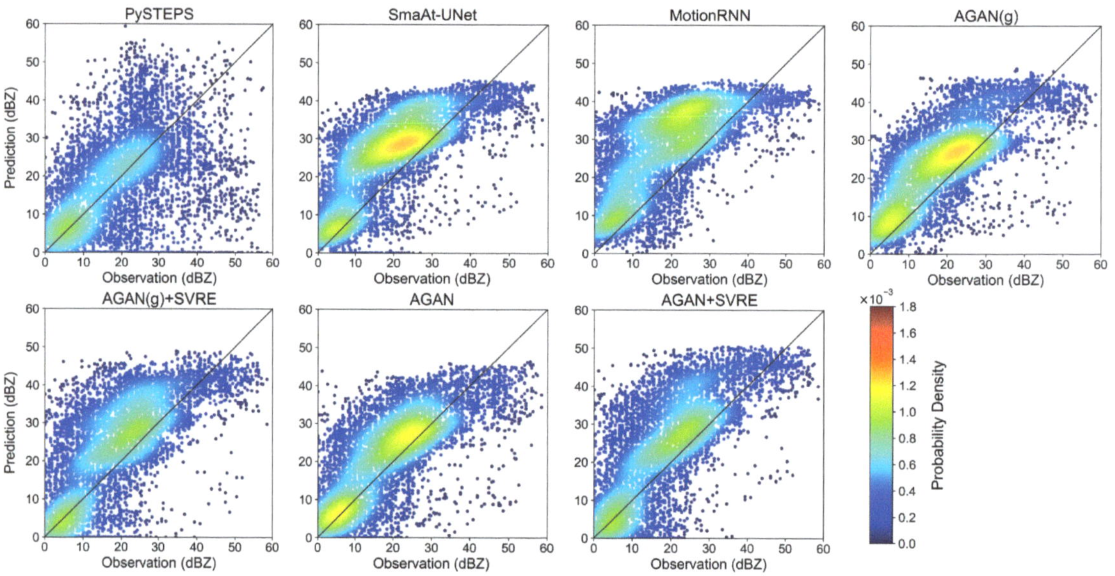

Figure 5. Contrast scatter plot of +60 min prediction and observation in case 1.

Table 4. The +60 min nowcasting performances in case 1. The up and down arrows in the heading indicate whether the highest or the lowest was the best for different metrics. A bold number indicates that the model in its row had the best performance, evaluated with the metric in its column.

Model	POD ↑	FAR ↓	CSI ↑	MBE ↓	MAE ↓	RMSE ↓	SSIM ↑	JSD ↓
PySTEPS	0.483	**0.455**	0.344	**−0.2**	**8.3**	**11.8**	0.231	0.672
SmaAt-UNet	**0.936**	0.556	0.431	10.4	11.1	15.1	0.304	0.664
MotionRNN	0.914	0.601	0.384	11.0	11.7	15.5	0.264	0.619
AGAN(g)	0.899	0.510	0.464	10.1	10.8	14.9	0.354	0.661
AGAN(g) + SVRE	0.881	0.554	0.421	9.1	10.5	14.4	0.317	0.650
AGAN	0.819	0.482	**0.464**	8.3	9.7	13.7	0.358	0.621
AGAN + SVRE	0.927	0.536	0.448	9.1	10.2	14.0	**0.368**	**0.616**

To further understand the results in Table 4, we plotted a Taylor diagram for the two experiments, which is shown in Figure 6. Taylor diagrams are widely used for the performance evaluation of meteorological models, since they can provide a concise statistical summary of the correlation coefficient (CC), the centered root mean squared error (RMSE′), and the variance ratio ($\sigma_{\hat{y}}/\sigma_y$) in a single diagram, based on the decomposition law proved by Taylor [64], which is explained in the following equation.

$$\frac{\mathrm{RMSE}'^2}{\sigma_y^2} = \frac{\mathrm{RMSE}^2 - \mathrm{MBE}^2}{\sigma_y^2} = \left(\frac{\sigma_{\hat{y}}}{\sigma_y}\right)^2 + 1 - 2\frac{\sigma_{\hat{y}}}{\sigma_y} \cdot \mathrm{CC} \tag{17}$$

In a Taylor diagram, the overall bias of a model can be attributed to the variability part measured by $\sigma_{\hat{y}}/\sigma_y$ and the correlation part measured by CC, which are denoted by the radial axis and the circumferential axis, respectively. The overall bias can be described by the RMSE′, which is depicted by an arc with the center lying on the horizontal radial axis. From Figure 6, it can be observed that the correlation coefficients of all the DL models were concentrated around 0.7. PySTEPS had the lowest CC of around 0.6 but the highest variance ratio of over 0.9, which coincided with the visualization results. It was also observed that $\sigma_{\hat{y}}/\sigma_y$ of the AGAN + SVRE was considerably higher than $\sigma_{\hat{y}}/\sigma_y$ of the AGAN(g), implying that the introduction of the GA strategy and the SVRE loss function could narrow the spatial variability distance between the prediction and the observation. It was also demonstrated that the AGAN + SVRE worked better in regard to the spatial variability representation than the DL-based models.

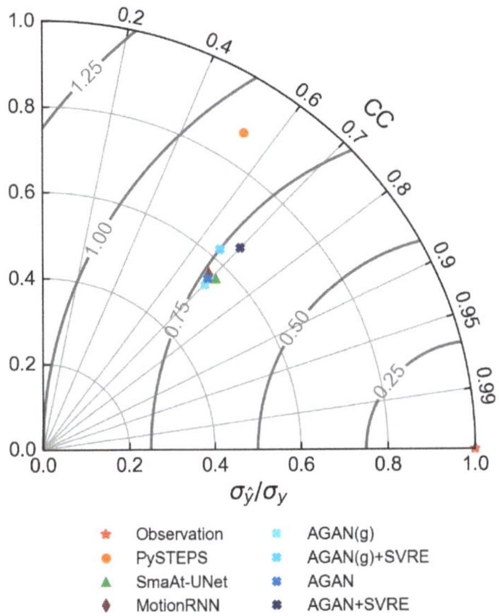

Figure 6. Taylor diagram of the +60 min prediction and the observation in case 1.

We also calculated the power spectral density (PSD) of both the X-axis and Y-axis of the radar maps to evaluate the models' abilities to capture local-scale weather patterns, and present the results in Figure 7. The horizontal axis of the PSD line plot was set to the logarithmic wavelength, instead of the frequency in the original concept of PSD, to more intuitively reflect the model's ability to capture weather patterns of different spatial scales [32]. If the prediction's PSD and the observation's PSD were close in a short-wavelength interval, we could say that the prediction had a similar local spatial pattern as that of the observation. The figure shows that, compared to the optical-flow-based PySTEPS method, all the DL models underestimated the PSD of the radar reflectivity along both the X-axis and Y-axis, especially for small-scale patterns (wavelength below 16 km), corresponding to the aforementioned blurry effect. It can be observed that the PSD of models trained with the GA strategy or the SVRE loss function was higher than the AGAN(g)'s PSD, indicating that the combination of the GA strategy and the SVRE loss function alleviated the systematic underestimation of spatial variability. Meanwhile, the AGAN + SVRE was ahead of the CNN-based SmaAt-UNet at all scales. The performance of the MotionRNN was the worst in small-scale local patterns with a spatial scale of fewer than 4 km, especially along the Y-axis. The results suggest that the AGAN + SVRE is more qualified for capturing local-scale patterns, which also verifies its ability to perform spatial variability representation from the side. Although the prediction of PySTEPS was the closest to the observation, the model's prediction was limited by its low correlation (CC) and forecasting accuracy (CSI).

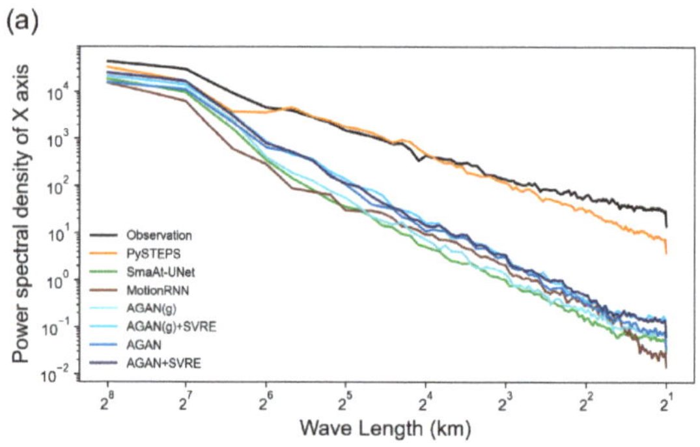

Figure 7. Power spectral density (PSD) of the +60 min prediction and observation in case 1. (**a**,**b**) are PSD of the X- and Y-axis, respectively.

4.2.2. Case 2

The second case was snipped from a growing local storm cell on 9 August 2019 UTC, lasting from 6:00 to 7:54. The +60 min predictions are visualized with the observations in Figure 8. The pixelwise nowcasting performances of case 2 were also evaluated with a contrast scatter plot, presented in Figure 9 The results show that the intensity predicted by PySTEPS at the storm center was very close to that of the observation, but the location of the storm center deviated from the observation. In contrast, DL models could successfully forecast the correct location of the storm center, whereas the underestimation of the peak reflectivity intensity and the exaggeration of the storm extent still existed, which could also be confirmed in the scatter plot. The last subfigures in Figure 8 show that the GA strategy could slightly alleviate this exaggeration effect and SVRE could increase the peak reflectivity intensity, pushing it closer to that of the observation. With the combination of the GA strategy and the SVRE loss function, the AGAN + SVRE gave the best prediction compared to the other baseline models.

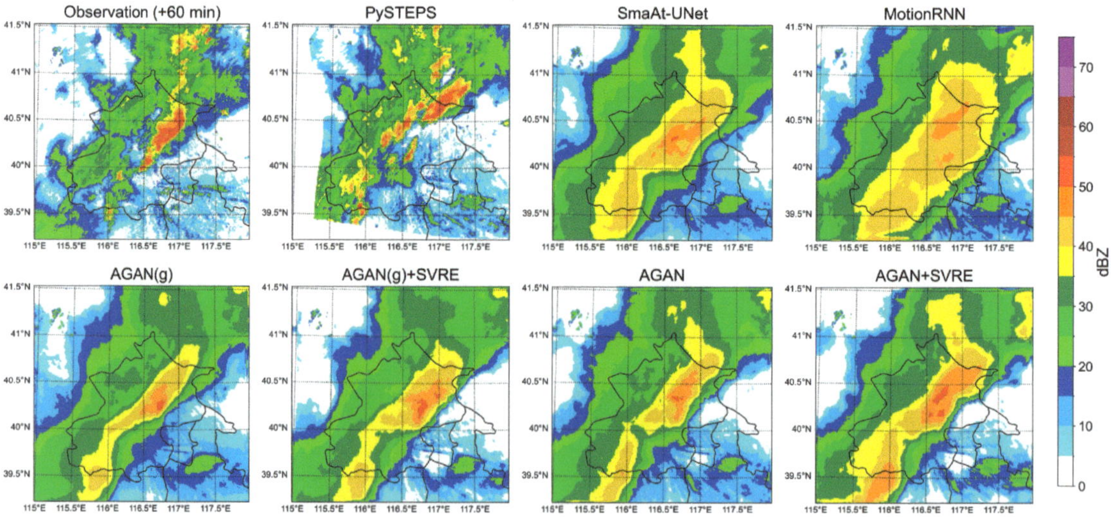

Figure 8. Visualization of the +60 min prediction and the observation in case 2.

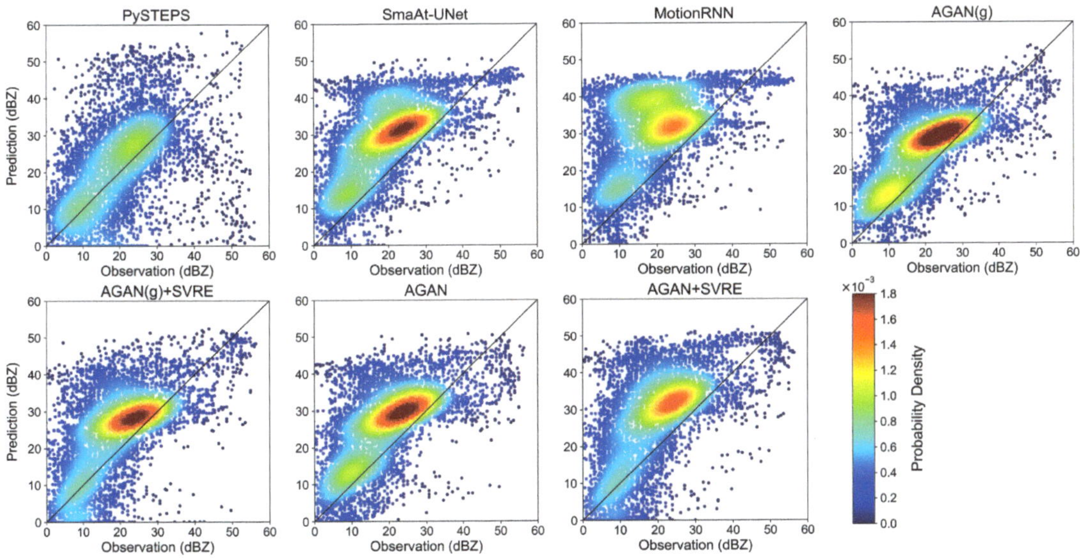

Figure 9. Contrast scatter plot of +60 min prediction and observation in case 2.

Table 5 enumerates the +60 min nowcasting performances of case 2. The GA strategy and the SVRE loss function improved the performance of the AGAN's generator in regard to the POD, MAE, RMSE, SSIM, and JSD, which was similar to case 1. However, the CSI of the model trained with the GA strategy and the SVRE loss function was reduced in both cases, which was different from the findings of the overall performances on the whole test set.

Table 5. The +60 min nowcasting performances in case 2. The up and down arrows in the heading indicate whether the highest or the lowest was the best for different metrics. A bold number indicates that the model in its row had the best performance, evaluated with the metric in its column.

Model	POD ↑	FAR ↓	CSI ↑	MBE ↓	MAE ↓	RMSE ↓	SSIM ↑	JSD ↓
PySTEPS	0.696	**0.476**	0.426	**0.8**	**7.5**	**10.6**	0.228	0.642
SmaAt-UNet	**0.981**	0.647	0.351	9.8	10.2	13.9	0.284	0.442
MotionRNN	0.936	0.659	0.333	10.7	11.3	15.3	0.249	0.298
AGAN(g)	0.933	0.519	**0.465**	8.3	8.8	13.2	0.306	0.331
AGAN(g) + SVRE	0.873	0.524	0.445	7.0	8.2	12.0	0.305	0.409
AGAN	0.951	0.604	0.388	8.5	9.1	12.8	0.327	0.341
AGAN + SVRE	0.960	0.617	0.378	8.3	9.1	12.8	**0.332**	**0.271**

We also plotted the Taylor diagram (Figure 10) and the PSD line chart (Figure 11) for case 2. In the Taylor diagram, it was found that both the GA strategy and the SVRE loss function could promote σ_g/σ_y of the model, and SVRE had an even bigger promotional effect than GA. The RMSE′, related to the overall bias, was consequently reduced. For the comparison experiment, the advantage of AGAN + SVRE over other baseline models was similar to case 1. It achieved the highest CC, the lowest RMSE′, and the second-highest variance ratio below PySTEPS. For the PSD of the predictions in Figure 11), the similarities and differences of the nowcasting performances between the models in case 2 were homogeneous to those in case 1. The PSD of radar images predicted by the AGAN + SVRE were the closest to that of the observation among all of the DL models.

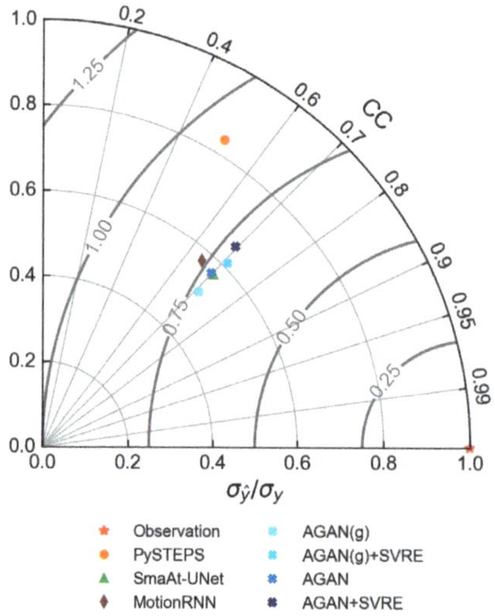

Figure 10. Taylor diagram of the +60 min prediction and the observation in case 2.

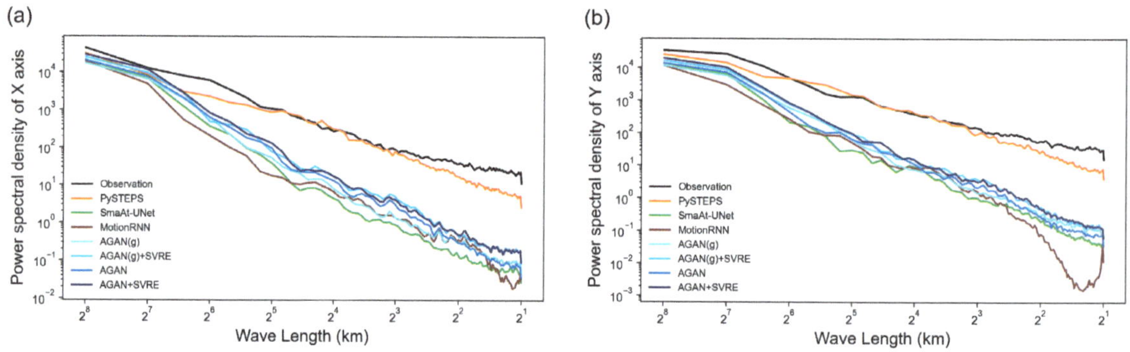

Figure 11. Power spectral density (PSD) of the +60 min prediction and observation in case 2. (**a**,**b**) are PSD of the X- and Y-axis, respectively.

4.3. Discussion

Thus far, we have evaluated the overall performances of different models and further analyzed their strengths and shortcomings with two storm cases. Generally, the results proved that the GA strategy and the SVRE loss function could alleviate the blurry effect of DL nowcasting models. The SVRE loss function and the GA strategy could boost DL nowcasting models by bridging the gaps between the predictions and the observations, particularly regarding spatial variability differences. As expected, the spatial variability-related metrics (SSIM, JSD, and PSD) of the test set and the two selected cases demonstrated that the GA strategy and the SVRE loss function could enhance the spatial variability representation of DL nowcasting models. More specifically, the enhancement of spatial variability representation derived from SVRE instead of GA. This might have been influenced by the hyperparameter selection in generative adversarial training. Since GANs are difficult to train, a larger weight has to be attached to the SVRE loss term and the reconstruction term to ensure convergence, leading to enhanced difference between the two components.

However, our methods also had several limitations. The first was that the forecasting accuracy of models trained with the SVRE loss function slightly reduced in heavy-rainfall cases. The AGAN(g)+SVRE and AGAN + SVRE could improve the CSI of the AGAN(g) and AGAN on the whole test set, but they failed in the two storm cases. This was probably because the goal of better spatial variability representation of extreme storm events can lead to more overestimated pixels or false alarms, which have a negative impact on forecasting accuracy. Another limitation came from the computational cost. The generative adversarial training process and the approximated integration over latent variables can significantly increase the convergence time for the training and inference processes of our model, making it difficult to apply to scenarios that require an extremely rapid response.

5. Conclusions

Previous deep learning models for radar nowcasting suffer from the systematic "blurry" problem and do not accurately represent the spatial variability of radar echo images. This study presented a Spatial Variability Representation Enhancement (SVRE) loss function and an Attentional Generative Adversarial Network (AGAN) to solve the problem, and evaluated them with a regional CINRAD dataset. An ablation experiment and a comparison experiment were implemented to verify the effects of the generative adversarial (GA) training strategy and SVRE loss and to compare the proposed model to current advanced radar nowcasting models. The performances of the models were validated on the whole test set and then inspected in two typical cases. Several metrics were selected to evaluate the forecasting accuracy and spatial variability representation. The results showed that both the GA strategy and the SVRE loss function could improve nowcasting performance by enhancing the spatial variability representation of the radar reflectivity. The GA strategy and SVRE also helped our model outperform other advanced baseline nowcasting models. The main contributions of this study are the following:

- We propose the SVRE loss function and the AGAN to alleviate the blurry effect of DL nowcasting models. Both of them can reduce this effect by enhancing the spatial variability of radar reflectivity.
- We attribute the blurry effect of DL nowcasting models to the deficiency in spatial variability representation of radar reflectivity or the precipitation field, which provides a new perspective for improving radar nowcasting.

Consequently, this study provides a feasible solution based on dense radar observations for high-resolution radar nowcasting applications. The limitations of our methods include reduced forecasting accuracy in high-intensity storm events and heavy computation costs. Future studies will focus on overcoming these limitations for high-intensity and small-scale storm events.

Author Contributions: Conceptualization, A.G.; methodology, A.G.; software, A.G.; validation, A.G.; formal analysis, A.G.; investigation, A.G. and R.L.; resources, B.P. and H.C.; data curation, M.C.; writing—original draft preparation, A.G. and R.L.; writing—review and editing, H.C., B.P. and G.N.; visualization, A.G. and R.L.; supervision, G.N.; project administration, G.N.; funding acquisition, G.N. All authors have read and agreed to the published version of the manuscript.

Funding: This work was supported by the National Key Research and Development Program of China (2018YFA0606002) and the Fund Program of State Key Laboratory of Hydroscience and Engineering (61010101221). The work of Haonan Chen was supported by Colorado State University.

Data Availability Statement: The codes for the experiments and the results can be found at https://github.com/THUGAF/SVRE-Nowcasting (accessed on 14 June 2023). The intermediate products, such as the feature maps and the pre-trained models, are available upon request.

Acknowledgments: The authors gratefully acknowledge the anonymous reviewers for providing careful reviews and comments on this article.

Conflicts of Interest: The authors declare no conflict of interest.

Abbreviations

The following abbreviations are used in this manuscript:

SVRE	Spatial varibility representation enhancement
AGAN	Attentional generative adversarial network
DL	Deep learning
GA	Generative adversarial
CNN	Convolutional neural network
RNN	Recurrent neural network
POD	Probability of detection
FAR	False alarm ratio
CSI	Critical sucess index
MBE	Mean bias error
MAE	Mean absolute error
RMSE	Root mean squared error
JSD	Jensen-Shannon divergence
SSIM	Structural similarity index measure
PSD	Power spectral density
CC	Correlation coefficient

References

1. WMO. *Guidelines for Nowcasting Techniques*; World Meteorological Organization: Geneva, Switzerland , 2017.
2. Wapler, K.; de Coning, E.; Buzzi, M. Nowcasting. In *Reference Module in Earth Systems and Environmental Sciences*; Elsevier Amsterdam, The Netherlands, 2019.
3. Han, L.; Zhao, Y.; Chen, H.; Chandrasekar, V. Advancing Radar Nowcasting Through Deep Transfer Learning. *IEEE Trans. Geosci Remote Sens.* **2022**, *60*, 4100609. [CrossRef]
4. Ma, Y.; Chen, H.; Ni, G.; Chandrasekar, V.; Gou, Y.; Zhang, W. Microphysical and polarimetric radar signatures of an epic flood event in Southern China. *Remote Sens.* **2020**, *12*, 2772. [CrossRef]
5. Ochoa-Rodriguez, S.; Wang, L.; Willems, P.; Onof, C. A review of radar-rain gauge data merging methods and their potential for urban hydrological applications. *Water Resour. Res.* **2019**, *55*, 6356–6391. [CrossRef]
6. Fulton, R.A.; Breidenbach, J.P.; Seo, D.J.; Miller, D.A.; O'Bannon, T. The WSR-88D rainfall algorithm. *Weather Forecast.* **1998** *13*, 377–395. [CrossRef]
7. Bringi, V.N.; Chandrasekar, V. *Polarimetric Doppler Weather Radar: Principles and Applications*; Cambridge University Press Cambridge, UK, 2001.
8. Overeem, A.; Buishand, T.A.; Holleman, I. Extreme rainfall analysis and estimation of depth-duration-frequency curves using weather radar. *Water Resour. Res.* **2009**, *45* , W10424. [CrossRef]
9. Dixon, M.; Wiener, G. TITAN: Thunderstorm identification, tracking, analysis, and nowcasting—A radar-based methodology *J. Atmos. Ocean. Technol.* **1993**, *10*, 785–797. [CrossRef]
10. Johnson, J.; MacKeen, P.L.; Witt, A.; Mitchell, E.D.W.; Stumpf, G.J.; Eilts, M.D.; Thomas, K.W. The storm cell identification and tracking algorithm: An enhanced WSR-88D algorithm. *Weather Forecast.* **1998**, *13*, 263–276. [CrossRef]
11. Tuttle, J.D.; Foote, G.B. Determination of the boundary layer airflow from a single Doppler radar. *J. Atmos. Ocean. Technol.* **1990** *7*, 218–232. [CrossRef]
12. Germann, U.; Zawadzki, I. Scale-dependence of the predictability of precipitation from continental radar images. Part I Description of the methodology. *Mon. Weather Rev.* **2002**, *130*, 2859–2873. [CrossRef]
13. Ruzanski, E.; Chandrasekar, V.; Wang, Y. The CASA nowcasting system. *J. Atmos. Ocean. Technol.* **2011**, *28*, 640–655. [CrossRef]
14. Woo, W.c.; Wong, W.k. Operational application of optical flow techniques to radar-based rainfall nowcasting. *Atmosphere* **2017** *8*, 48. [CrossRef]
15. Bowler, N.E.; Pierce, C.E.; Seed, A. Development of a precipitation nowcasting algorithm based upon optical flow techniques *J. Hydrol.* **2004**, *288*, 74–91. [CrossRef]
16. Radhakrishna, B.; Zawadzki, I.; Fabry, F. Predictability of precipitation from continental radar images. Part V: Growth and decay *J. Atmos. Sci.* **2012**, *69*, 3336–3349. [CrossRef]
17. Han, L.; Sun, J.; Zhang, W. Convolutional neural network for convective storm nowcasting using 3-D Doppler weather radar data. *IEEE Trans. Geosci. Remote Sens.* **2019**, *58*, 1487–1495. [CrossRef]
18. Sun, D.; Roth, S.; Lewis, J.; Black, M.J. Learning optical flow. In Proceedings of the European Conference on Computer Vision Marseille, France, 12–18 October 2008; Springer: Berlin/Heidelberg, Germany, 2008; pp. 83–97.
19. Novák, P.; Březková, L.; Frolík, P. Quantitative precipitation forecast using radar echo extrapolation. *Atmos. Res.* **2009**, *93*, 328–334 [CrossRef]
20. Voulodimos, A.; Doulamis, N.; Doulamis, A.; Protopapadakis, E. Deep learning for computer vision: A brief review. *Comput Intell. Neurosci.* **2018**, *2018*, 7068349. [CrossRef] [PubMed]

21. Deng, L.; Liu, Y. *Deep Learning in Natural Language Processing*; Springer: Berlin/Heidelberg, Germany, 2018.
22. Reichstein, M.; Camps-Valls, G.; Stevens, B.; Jung, M.; Denzler, J.; Carvalhais, N. Deep learning and process understanding for data-driven Earth system science. *Nature* **2019**, *566*, 195–204. [CrossRef]
23. Prudden, R.; Adams, S.; Kangin, D.; Robinson, N.; Ravuri, S.; Mohamed, S.; Arribas, A. A review of radar-based nowcasting of precipitation and applicable machine learning techniques. *arXiv* **2020**, arXiv:2005.04988.
24. Jing, J.; Li, Q.; Peng, X. MLC-LSTM: Exploiting the Spatiotemporal Correlation between Multi-Level Weather Radar Echoes for Echo Sequence Extrapolation. *Sensors* **2019**, *19*, 3988. [CrossRef]
25. Camps-Valls, G.; Tuia, D.; Zhu, X.X.; Reichstein, M. *Deep Learning for the Earth Sciences: A Comprehensive Approach to Remote Sensing, Climate Science and Geosciences*; John Wiley & Sons: Hoboken, NJ, USA, 2021; pp. 218–219.
26. Ji, S.; Xu, W.; Yang, M.; Yu, K. 3D convolutional neural networks for human action recognition. *IEEE Trans. Pattern Anal. Mach. Intell.* **2012**, *35*, 221–231. [CrossRef]
27. Qiu, Z.; Yao, T.; Mei, T. Learning spatio-temporal representation with pseudo-3d residual networks. In Proceedings of the IEEE International Conference on Computer Vision, Venice, Italy, 22–29 October 2017; pp. 5533–5541.
28. Tran, D.; Bourdev, L.; Fergus, R.; Torresani, L.; Paluri, M. Learning spatiotemporal features with 3d convolutional networks. In Proceedings of the IEEE International Conference on Computer Vision, Santiago, Chile, 7–13 December 2015; pp. 4489–4497.
29. Klein, B.; Wolf, L.; Afek, Y. A dynamic convolutional layer for short range weather prediction. In Proceedings of the IEEE Conference on Computer Vision and Pattern Recognition, Boston, MA, USA, 7–12 June 2015; pp. 4840–4848.
30. Ayzel, G.; Heistermann, M.; Sorokin, A.; Nikitin, O.; Lukyanova, O. All convolutional neural networks for radar-based precipitation nowcasting. *Procedia Comput. Sci.* **2019**, *150*, 186–192. [CrossRef]
31. Ronneberger, O.; Fischer, P.; Brox, T. U-net: Convolutional networks for biomedical image segmentation. In Proceedings of the International Conference on Medical Image Computing and Computer-Assisted Intervention, Munich, Germany, 5–9 October 2015; Springer: Berlin/Heidelberg, Germany, 2015; pp. 234–241.
32. Ayzel, G.; Scheffer, T.; Heistermann, M. RainNet v1. 0: A convolutional neural network for radar-based precipitation nowcasting. *Geosci. Model Dev.* **2020**, *13*, 2631–2644. [CrossRef]
33. Trebing, K.; Stanczyk, T.; Mehrkanoon, S. Smaat-unet: Precipitation nowcasting using a small attention-unet architecture. *Pattern Recognit. Lett.* **2021**, *145*, 178–186. [CrossRef]
34. Shi, X.; Chen, Z.; Wang, H.; Yeung, D.Y.; Wong, W.K.; Woo, W.C. Convolutional LSTM network: A machine learning approach for precipitation nowcasting. In Proceedings of the Advances in Neural Information Processing Systems, Montreal, QC, Canada, 7–12 December 2015; pp. 802–810.
35. Hochreiter, S.; Schmidhuber, J. Long short-term memory. *Neural Comput.* **1997**, *9*, 1735–1780. [CrossRef] [PubMed]
36. Shi, X.; Gao, Z.; Lausen, L.; Wang, H.; Yeung, D.Y.; Wong, W.k.; Woo, W.C. Deep learning for precipitation nowcasting: A benchmark and a new model. In Proceedings of the Advances in Neural Information Processing Systems, Long Beach, CA, USA, 4–9 December 2017; pp. 5617–5627.
37. Wang, Y.; Long, M.; Wang, J.; Gao, Z.; Philip, S.Y. Predrnn: Recurrent neural networks for predictive learning using spatiotemporal lstms. In Proceedings of the Advances in Neural Information Processing Systems, Long Beach, CA, USA, 4–9 December 2017; pp. 879–888.
38. Wang, Y.; Gao, Z.; Long, M.; Wang, J.; Yu, P.S. Predrnn++: Towards a resolution of the deep-in-time dilemma in spatiotemporal predictive learning. *arXiv* **2018**, arXiv:1804.06300.
39. Wu, H.; Yao, Z.; Wang, J.; Long, M. MotionRNN: A flexible model for video prediction with spacetime-varying motions. In Proceedings of the IEEE/CVF Conference on Computer Vision and Pattern Recognition, Nashville, TN, USA, 20–25 June 2021; pp. 15435–15444.
40. Jing, J.; Li, Q.; Ding, X.; Sun, N.; Tang, R.; Cai, Y. Aenn: A Generative Adversarial Neural Network for Weather Radar Echo Extrapolation. *Int. Arch. Photogramm. Remote Sens. Spat. Inf. Sci.* **2019**, *42*, 89–94. [CrossRef]
41. Tian, L.; Li, X.; Ye, Y.; Xie, P.; Li, Y. A Generative Adversarial Gated Recurrent Unit Model for Precipitation Nowcasting. *IEEE Geosci. Remote Sens. Lett.* **2020**, *17*, 601–605. [CrossRef]
42. Fang, W.; Pang, L.; Yi, W.; Sheng, V.S. AttEF: Convolutional LSTM Encoder-Forecaster with Attention Module for Precipitation Nowcasting. *Intell. Autom. Soft Comput.* **2021**, *29*, 453–466. [CrossRef]
43. Huang, Q.; Chen, S.; Tan, J. TSRC: A Deep Learning Model for Precipitation Short-Term Forecasting over China Using Radar Echo Data. *Remote Sens.* **2022**, *15*, 142. [CrossRef]
44. Arnaud, P.; Bouvier, C.; Cisneros, L.; Dominguez, R. Influence of rainfall spatial variability on flood prediction. *J. Hydrol.* **2002**, *260*, 216–230. [CrossRef]
45. Courty, L.G.; Rico-Ramirez, M.A.; Pedrozo-Acuna, A. The significance of the spatial variability of rainfall on the numerical simulation of urban floods. *Water* **2018**, *10*, 207. [CrossRef]
46. Goodfellow, I.; Bengio, Y.; Courville, A. *Deep Learning*; MIT Press: Cambridge, MA, USA, 2016.
47. Xie, P.; Li, X.; Ji, X.; Chen, X.; Chen, Y.; Liu, J.; Ye, Y. An Energy-Based Generative Adversarial Forecaster for Radar Echo Map Extrapolation. *IEEE Geosci. Remote Sens. Lett.* **2020**, *19*, 1–5. [CrossRef]
48. Ravuri, S.; Lenc, K.; Willson, M.; Kangin, D.; Lam, R.; Mirowski, P.; Fitzsimons, M.; Athanassiadou, M.; Kashem, S.; Madge, S.; et al. Skilful precipitation nowcasting using deep generative models of radar. *Nature* **2021**, *597*, 672–677. [CrossRef]

49. Fan, Y.; Lu, X.; Li, D.; Liu, Y. Video-based emotion recognition using CNN-RNN and C3D hybrid networks. In Proceedings of the 18th ACM International Conference on Multimodal Interaction, Tokyo, Japan, 12–16 November 2016; pp. 445–450.
50. Pulkkinen, S.; Nerini, D.; Pérez Hortal, A.A.; Velasco-Forero, C.; Seed, A.; Germann, U.; Foresti, L. Pysteps: An open-source Python library for probabilistic precipitation nowcasting (v1. 0). *Geosci. Model Dev.* **2019**, *12*, 4185–4219. [CrossRef]
51. Harshvardhan, G.; Gourisaria, M.K.; Pandey, M.; Rautaray, S.S. A comprehensive survey and analysis of generative models in machine learning. *Comput. Sci. Rev.* **2020**, *38*, 100285.
52. Ng, A.; Jordan, M. On discriminative vs. generative classifiers: A comparison of logistic regression and naive bayes. *Adv. Neural Inf. Process. Syst.* **2001**, *14*.
53. Goodfellow, I.; Pouget-Abadie, J.; Mirza, M.; Xu, B.; Warde-Farley, D.; Ozair, S.; Courville, A.; Bengio, Y. Generative adversarial nets. In Proceedings of the Advances in Neural Information Processing Systems, Montreal, QC, Canada, 8–13 December 2014; pp. 2672–2680.
54. Mao, X.; Li, Q.; Xie, H.; Lau, R.Y.; Wang, Z.; Paul Smolley, S. Least squares generative adversarial networks. In Proceedings of the IEEE International Conference on Computer Vision, Venice, Italy, 22–29 October 2017; pp. 2794–2802.
55. Isola, P.; Zhu, J.Y.; Zhou, T.; Efros, A.A. Image-to-image translation with conditional adversarial networks. In Proceedings of the IEEE Conference on Computer Vision and Pattern Recognition, Honolulu, HI, USA, 21–26 July 2017; pp. 1125–1134.
56. Thangjai, W.; Niwitpong, S.A.; Niwitpong, S. Confidence intervals for the common coefficient of variation of rainfall in Thailand. *PeerJ* **2020**, *8*, e10004. [CrossRef] [PubMed]
57. Woo, S.; Park, J.; Lee, J.Y.; Kweon, I.S. Cbam: Convolutional block attention module. In Proceedings of the European conference on Computer Vision (ECCV), Munich, Germany, 8–14 September 2018; pp. 3–19.
58. Chen, M.; Gao, F.; Kong, R.; Wang, Y.; Wang, J.; Tan, X.; Xiao, X.; Zhang, W.; Wang, L.; Ding, Q. Introduction of auto-nowcasting system for convective storm and its performance in Beijing Olympics meteorological service. *J. Appl. Meteorol. Sci.* **2010**, *21*, 395–404.
59. Stanski, H.; Wilson, L.; Burrows, W. Survey of common verification methods in meteorology. In *World Weather Watch Technical Report*; WMO: Geneva, Switzerland,1989.
60. Han, L.; Liang, H.; Chen, H.; Zhang, W.; Ge, Y. Convective precipitation nowcasting using U-Net Model. *IEEE Trans. Geosci. Remote Sens.* **2022**, *60*, 4103508. [CrossRef]
61. Chen, L.; Cao, Y.; Ma, L.; Zhang, J. A deep learning-based methodology for precipitation nowcasting with radar. *Earth Space Sci.* **2020**, *7*, e2019EA000812. [CrossRef]
62. Wang, Z.; Bovik, A.C.; Sheikh, H.R.; Simoncelli, E.P. Image quality assessment: From error visibility to structural similarity. *IEEE Trans. Image Process.* **2004**, *13*, 600–612. [CrossRef] [PubMed]
63. Heusel, M.; Ramsauer, H.; Unterthiner, T.; Nessler, B.; Hochreiter, S. Gans trained by a two time-scale update rule converge to a local nash equilibrium. *Adv. Neural Inf. Process. Syst.* **2017**, *30*, 6629–6640.
64. Taylor, K.E. Summarizing multiple aspects of model performance in a single diagram. *J. Geophys. Res. Atmos.* **2001**, *106*, 7183–7192. [CrossRef]

remote sensing

MDPI

Article

A Nonlinear Grid Transformation Method for Extrapolating and Predicting the Convective Echo of Weather Radar

Yue Sun [1], Hui Xiao [1,2,*], Ye Tian [3] and Huiling Yang [1,2]

1 Key Laboratory of Cloud-Precipitation Physics and Severe Storms (LACS), Institute of Atmospheric Physics, Chinese Academy of Sciences, Beijing 100029, China
2 College of Earth Sciences, University of Chinese Academy of Sciences (UCAS), Beijing 100049, China
3 Beijing Meteorological Observation Center, Beijing Meteorological Service, Beijing 100089, China
* Correspondence: hxiao@mail.iap.ac.cn; Tel.: +86-10-82995318

Abstract: A nonlinear grid transformation (NGT) method is proposed for weather radar convective echo extrapolation prediction. The change in continuous echo images is regarded as a nonlinear transformation process of the grid. This process can be reproduced by defining and solving a 2×6 transformation matrix, and this approach can be applied to image prediction. In ideal experiments with numerical and path changes of the target, NGT produces a prediction result closer to the target than does a conventional optical flow (OF) method. In the presence of convection lines in real cases, NGT is superior to OF: the critical success index (CSI) for 40 dBZ of the echo prediction at 60 min is approximately 0.2 higher. This is due to the better estimation of the movement of the whole cloud system in the NGT results since it reflects the continuous change in the historical images. For the case with a mesoscale convective complex, the NGT results are better than the OF results, and a deep learning result is cited from a previous study for the same case for 20 and 30 dBZ. However, the result is the opposite for 40 dBZ, where the deep learning method may produce an overestimation of the stronger echo.

Keywords: weather radar nowcasting; echo image prediction; grid transformation

1. Introduction

A weather radar is an important means to monitor and warn of severe convective weather [1–3]. Although data assimilation of weather radar helps to improve numerical forecasting [4,5], extrapolation prediction based on weather radar echoes remains the most direct method for nowcasting a convective event, especially for mesoscale convective weather [6,7] with rapid development and dramatic changes. By means of the current observed radar data, the movement and variation of the echo are obtained, and the position and intensity of the echo in the next few minutes or longer period are estimated.

The methods of extracting radar echo moving features from horizontal two-dimensional radar images can be divided into two categories according to the target. One focuses on identifying and tracking a whole cloud, such as the Thunderstorm Identification, Tracking, Analysis, and Nowcasting algorithm (TITAN) [8]. The cloud is determined by the maximum circumscribed ellipse under a given reflectivity threshold to obtain its trajectory and variation characteristics. Another category of methods focuses on obtaining the vector field of echo movement under a uniform grid. The Tracking Radar Echo by Correlation algorithm (TREC) [9], a common algorithm in the past, determines the echo moving vector on the specified grid point based on the cross-correlation between the subunits around adjacent time points. Both TITAN and TREC contribute to a classical auto-nowcast system [3]. The Optical Flow (OF) method can also produce a vector field based on images around adjacent time points. By assuming the target gray level is constant and introducing the global smoothing condition (known as Horn-Schunk OF, [10]) or the local consistency condition (known as Lucas-Kanade OF, [11]), moving vectors can be solved based on the local change and spatial gradient of the images. A recent study shows that different nowcasting schemes based on OF can both produce better results than TREC to some extent [12]. In addition, an improved prediction method combines OF with an assimilation system and considers water content and principal component analysis [13].

However, some challenges remain in the feature extraction and extrapolation of convective clouds in the above methods. The alternating development of convective cells often occurs in convective systems. The moving direction of the whole cloud system typically depends on the combination of individual cell movement and the direction of new cell emergence, which may not be correctly reproduced by a method based on a few images at adjacent times, such as TREC and OF. Although TITAN can provide the echo movement trajectory based on longer historical echo image samples, each convective cell in a cloud system may not only move in different directions but also merge, resulting in complex topological relationships that are difficult to extrapolate. In addition, the shape and intensity of convective echoes can vary locally and rapidly, which makes it difficult to capture classical local image features such as Scale-Invariant Feature Transform (SIFT) [14] and Harris corner detectors [15]. Therefore, other image recognition and analysis algorithms based on those features cannot be effectively applied to the extrapolation of weather radar echoes. In general, there is a lack of algorithms that consider multiple historical times and do not rely on identifying individual targets.

There are also some learning methods that have been applied to radar echo and precipitation nowcasting [16–23] in recent years. These methods allow multiple observed image inputs and predict the image in the future, which appears to address the limitations mentioned above. However, since the model parameters and modeling process are implicit and there is a requirement for a large number of samples for training, the reliability of these methods requires further verification in application.

In this paper, a new method is proposed for the extrapolation and prediction of weather radar echo images. The change in continuous echo images is regarded as a nonlinear transformation process of the grid by a transformation matrix. By defining and solving this matrix, the extrapolation and prediction of the echo image can be realized. The process of establishing the new method is described in Section 2. The echo prediction experiments are shown in Section 3, including a set of ideal cases and three real cases. A summary and discussion are provided in Section 4.

2. Nonlinear Grid Transformation Method

2.1. Matrix for Grid Transformation

The concept of the transformation matrix is first introduced in this section, including the conventional transformation matrix and the extended transformation matrix used in this paper. The connection between the transformation matrix and the echo image prediction is introduced in Sections 2.2 and 2.3.

The general form of the transformation matrix used in this paper is a matrix $M_{p\times p}$, as shown in Equation (1):

$$M_{p\times p}=\begin{bmatrix} m_{1,1} & m_{1,2} & \ldots m_{1,p-1} & m_{1,p} \\ m_{2,1} & m_{2,2} & \ldots m_{2,p-1} & m_{2,p} \\ & & \ldots & \\ m_{p-1,1} & m_{p-1,2} & \ldots m_{p-1,p-1} & m_{p-1,p} \\ 0 & 0 & \ldots & 0 & 1 \end{bmatrix} \tag{1}$$

where the first $p-1$ rows are independent elements, the value in row p and column p is 1, and the other values in row p are 0. When p is 3, a two-dimensional point set $[X_1, Y_1]$ can be transformed to $[X_2, Y_2]$ as follows:

$$[X_2, Y_2, 1]^T = M_{3\times 3}[X_1, Y_1, 1]^T \tag{2}$$

where the "1" at the third location in the square bracket provides a nonhomogeneous term when performing linear transformation of the point coordinates. Equation (2) is known as "affine transformation" [24], which is a common algorithm of image geometric transformation. With this matrix, the original point set can be translated, rotated, scaled, and deformed (sheared, stretched, etc.). For the convenience of observation, $[X_1, Y_1]$ is set to an orthogonal grid, and Figure 1a shows a comprehensive example, which includes changes such as moving to the right, magnification, counterclockwise rotation, and stretching the upper right corner. This transformation is repeatable. After N-1 times left multiplication by $M_{3\times3}$, the transformation result $[X_N, Y_N]$ is obtained in the same way:

$$[X_N, Y_N, 1]^T = (M_{3\times3})^{N-1}[X_1, Y_1, 1]^T \tag{3}$$

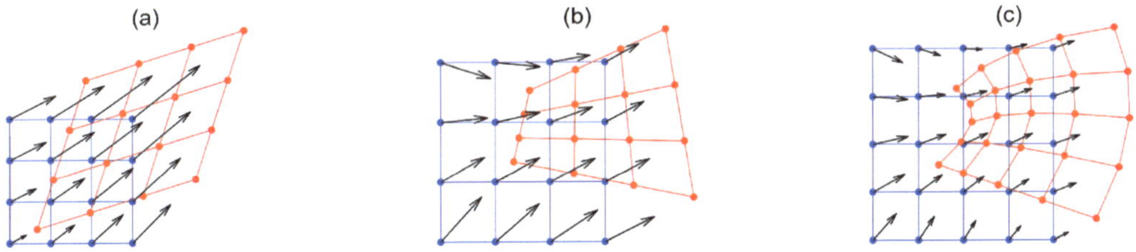

Figure 1. Example of grid transformation from $[X_1, Y_1]$ (blue grids) to $[X_2, Y_2]$ (red grids) using (**a**) $M_{3\times3}$, (**b**) $M_{4\times4}$, and (**c**) $M_{6\times6}$. The black arrows represent the transformation direction of a grid.

Related principles and examples of such common 2D image transformation are seen in [24]. Note that only the first two rows of $M_{3\times3}$ can actually be taken if only one transformation is needed. In addition, the inverse matrix of $M_{3\times3}$ represents the opposite transformation. Usually, when at least three groups of X and Y before and after the transformation are known, the six parameters in $M_{3\times3}$ can be solved to achieve a similar transformation for other point sets, or the expected transformation can be achieved by directly specifying the 6 parameters.

However, this conventional affine transformation has a limited ability to characterize image deformation. Note that in Figure 1a, although the shape has changed to some extent, the opposite edges are still parallel, so it is obviously not suitable for characterizing the changes in weather echo images. The transformation could be expanded by adding more parameters. It is easy to extend M in the form of Equation (1). When expanding the formula with coordinates multiplied by M, the newly added variables must be known or knowable and cannot be completely linearly related to the point sequence of the original X_1 and Y_1. A nonlinear XY term is proposed to be added, and the transformation matrix M is 4×4 as follows:

$$[X_2, Y_2, X_2Y_2, 1]^T = M_{4\times4}[X_1, Y_1, X_1Y_1, 1]^T \tag{4}$$

An example of the transformation of Equation (4) is given in Figure 1b, which shows that it can already characterize the case where the contrast is not parallel. However, every edge of the orthogonal grid after transformation is still a straight line. To further improve the deformation ability of the transformation, the nonlinear terms X^2 and Y^2 are added, and a 6×6 transformation matrix is used:

$$\left[X_2, Y_2, X_2Y_2, X_2^2, Y_2^2, 1\right]^T = M_{6\times6}\left[X_1, Y_1, X_1Y_1, X_1^2, Y_1^2, 1\right]^T \tag{5}$$

Now, the transformation can characterize not only the translation of the grid points but also the flexible bending deformation of the original orthogonal grid (Figure 1c), which has the potential to characterize the movement and shape change of the weather echo. Therefore, Equation (5) is used as the form of nonlinear grid transformation discussed in the following sections.

2.2. Estimation of the Nonlinear Grid Transformation Matrix

As mentioned in Section 1, it is difficult to extract the target feature points that do not change with time in the weather echo image, so a pair of known and corresponding $[X_1, Y_1]$ and $[X_2, Y_2]$ is hard to find. Thus, a major challenge is how to solve Equation (5) to obtain the transformation matrix $M_{6\times6}$ when given two or more echo images.

Assuming the original image values (reflectivity values) at two adjacent times are Q_1 and Q_2, $[X_1, Y_1]$ is the original orthogonal grid of the image, and $[X_2, Y_2]$ is the point set after transformation by $M_{6\times6}$. Since $[X_2, Y_2]$ probably does not coincide with the original orthogonal grid, let Q_{2*} be the estimated value on $[X_2, Y_2]$ of Q_2. Here, Figure 1c can be taken as a simple schematic diagram, where the blue grids are $[X_1, Y_1]$ and the red grids are $[X_2, Y_2]$. The pixels of Q_1 and Q_2 are 2D fields corresponding to the blue grids, while the pixels of Q_{2*} correspond to the red grids. The goal of the expected transformation is to make Q_1 look like Q_2, where the difference between Q_1 and Q_{2*} is as small as possible after moving the image values of Q_1 from $[X_1, Y_1]$ to $[X_2, Y_2]$. The above objective can be expressed as minimizing Equation (6):

$$J = \sum (Q_1 - Q_{2*})^2 \tag{6}$$

where the summation is for all grids of the image. However, Q_{2*} cannot be obtained by interpolation since the transformation and $[X_2, Y_2]$ are unknown in advance. Thus, an estimate of Q_{2*} is proposed using a Taylor first-order expansion as follows:

$$Q_{2*} \approx Q_2 + \frac{\partial Q_2}{\partial x}(X_2 - X_1) + \frac{\partial Q_2}{\partial y}(Y_2 - Y_1) \tag{7}$$

The approximation of Equation (7) is clearly rough and does not always hold, so some additional conditions are needed. First, the grid spacing should be larger than the displacement of the object in the images. Considering that the jet velocity in convective weather is often larger than 20 m s^{-1}, as is the outflow formed by convective precipitation, clouds may move 4.8~7.2 km or more with the wind within the interval of an operational weather radar, whose volume scan often takes 4~6 min. Therefore, the grid spacing selected for echo extrapolation should be greater than this order of magnitude, for example, at least 10 km $\times$ 10 km. Then, the echo image should be smoothed to some extent, such as via small-scale two-dimensional Gaussian filtering, to smooth the local jagged texture that may exist in the image so that the first-order partial derivative of the space in Equation (7) is representative.

After substituting Equation (7) into Equation (6), the problem of minimizing (6) can be transformed into solving a system of equations on the original orthogonal grid, where each equation is:

$$\Delta Q_{2-1} + \frac{\partial Q_2}{\partial x}(X_2 - X_1) + \frac{\partial Q_2}{\partial y}(Y_2 - Y_1) = 0 \quad (\Delta Q_{2-1} \neq 0) \tag{8}$$

where ΔQ_{2-1} is the difference of Q_2 minus Q_1 on each grid. The expressions of X_2 and Y_2 can be obtained by expanding Equation (5) as follows:

$$X_2 = m_{1,1}X_1 + m_{1,2}Y_1 + m_{1,3}X_1Y_1 + m_{1,4}X_1^2 + m_{1,5}Y_1^2 + m_{1,6} \tag{9}$$

$$Y_2 = m_{2,1}X_1 + m_{2,2}Y_1 + m_{2,3}X_1Y_1 + m_{2,4}X_1^2 + m_{2,5}Y_1^2 + m_{2,6} \tag{10}$$

After substituting Equations (9) and (10) into Equation (8), the nonhomogeneous linear equations of the parameters to be solved can be obtained:

$$B\,M^* = \frac{\partial Q_2}{\partial x}X_1 + \frac{\partial Q_2}{\partial y}Y_1 - \Delta Q_{2-1} \tag{11}$$

where the known quantity B and the parameter M^* to be solved are shown in Equations (12) and (13) as follows:

$$B = \left[\frac{\partial Q_2}{\partial x}X_1, \frac{\partial Q_2}{\partial x}Y_1, \frac{\partial Q_2}{\partial x}X_1Y_1, \frac{\partial Q_2}{\partial x}X_1^2, \frac{\partial Q_2}{\partial x}Y_1^2, \frac{\partial Q_2}{\partial x}, \frac{\partial Q_2}{\partial y}X_1, \frac{\partial Q_2}{\partial y}Y_1, \frac{\partial Q_2}{\partial y}X_1Y_1, \frac{\partial Q_2}{\partial y}X_1^2, \frac{\partial Q_2}{\partial y}Y_1^2, \frac{\partial Q_2}{\partial y}\right] \quad (12)$$

$$M^* = [m_{1,1}, m_{1,2}, m_{1,3}, m_{1,4}, m_{1,5}, m_{1,6}, m_{2,1}, m_{2,2}, m_{2,3}, m_{2,4}, m_{2,5}, m_{2,6}]^T \quad (13)$$

As long as more than 6 grids are satisfied, the system of equations consisting of Equation (11) can be solved by Least Squares Estimation (LSE). The obtained M^* contains the elements of the first two rows of $M_{6\times6}$, and $[X_2, Y_2]$ can then be obtained by Equation (14):

$$[X_2, Y_2]^T = \begin{bmatrix} m_{1,1}, & m_{1,2}, & m_{1,3}, & m_{1,4}, & m_{1,5}, & m_{1,6} \\ m_{2,1}, & m_{2,2}, & m_{2,3}, & m_{2,4}, & m_{2,5}, & m_{2,6} \end{bmatrix} \left[X_1, Y_1, X_1Y_1, X_1^2, Y_1^2, 1\right]^T \quad (14)$$

For a single transformation, the 2 × 6 transformation matrix in Equation (14) is sufficient. For the remaining 3~5 rows in $M_{6\times6}$, each row can be obtained by LSE after substituting $[X_2, Y_2]$ into Equation (5). Thus, a method for solving the nonlinear transformation matrix is obtained when two images are given. This approach is also applicable to multiple images. For example, for the case of three images, Equation (11) on each grid using the first and second images is first listed in the system of equations, then that of the second and third images are appended to the system of equations, and finally, they are solved together via LSE. This process enables the final transformation matrix to reflect the continuous movement and deformation characteristics of multiple images.

There are four more points to note:

(1) Quadratic coordinates and distance derivatives are involved in the operation of Equation (11), which makes it possible for the value of the coefficient to span many orders of magnitude. Therefore, all the variables involved in the actual calculation should be at least double-precision floating-point numbers.

(2) An underestimation of image change will occur if points of $\Delta Q_{2-1} = 0$ are considered when preparing Equation (11) because when only points of $\Delta Q_{2-1} = 0$ are used, the final obtained $M_{6\times6}$ will be a unit matrix, that is, the image is unchanged.

(3) Although the time complexity for solving Equation (11) appears to increase quadratically with the number of grids, the size of the original image does not become a problem because the method is not based on detailed texture and rather requires a certain degree of coarse grid spacing. A current personal computer can complete the calculation in 1–10 s when the historical sequence length is approximately 10, and the number of grid points is less than 100 × 100.

(4) After obtaining the transformation matrix and before extrapolating using Equation (14), the image and coordinates are replaceable. For example, the filtered coarse-resolution image can be replaced by the original fine-resolution image. However, the unit of coordinates cannot be replaced because the transformation matrix is based on the coordinates used. For example, the coordinates cannot be changed from rectangular coordinate distance to latitude and longitude once the transformation matrix is obtained or from the latitude and longitude of a region to another different region.

2.3. Practical Steps for Extrapolation Prediction

An ideal extrapolation prediction scheme is to first find $M_{6\times6}$ and obtain the transformation matrix after multiplication according to the time N that needs to be extrapolated, which is similar to Equation (3). Then, the inverse matrix is produced, and the discrete point set of reverse extrapolation is obtained and used for backward interpolation. Finally, the predicted image is obtained directly by linear interpolation. However, during the experiment for constructing the new method in this paper, some anomalies caused by nonlinear effects when N was large (such as greater than 10) were found. For example, the image does not further deform or move as expected but rolls back. Therefore, a more reliable extrapolation method is proposed.

First, only $M_{2\times6}$ is obtained using Equation (11). Then, $[X_2, Y_2]$ are obtained using Equation (14), and $[X_1, Y_1]$ are subtracted from $[X_2, Y_2]$ to find a vector field u and v to represent the one-time grid transformation. According to the time N to be extrapolated, the original grid points are moved in the opposite direction of u and v in N steps, and u and v on the new discrete points are interpolated in each step, which is known as backward interpolation. The values of the final backward positions are obtained by linear interpolation and are deemed to appear on the orthogonal grid point after N steps; thus, the predicted image is obtained. Although this process is somewhat different from a direct transformation similar to Equation (3), it can reflect the grid movement and deformation characteristics generated by the solved transformation matrix, and there is no abnormality caused by nonlinear effects.

The above introduces the image extrapolation method proposed in this paper, called the Nonlinear Grid Transformation (NGT) method. The main steps of echo extrapolation prediction are supplemented and summarized as follows:

(1) Prepare a set of time-continuous historical echo images prior to the start time of the prediction. Perform resolution reduction and two-dimensional Gaussian filtering if necessary.
(2) For each pair of adjacent images, each Equation (11) from a grid in which $\Delta Q_{2-1} \neq 0$ is selected continuously to form a system of linear equations.
(3) Solve the system of linear equations in the last step to obtain the transformation matrix $M_{2\times6}$. Obtain $[X_2, Y_2]$ according to Equation (14), and subtract the origin orthogonal grid $[X_1, Y_1]$ to obtain grid transformation vector fields u and v. Here, the u and v are generated one-time.
(4) If a resolution reduction is applied in the first step, u and v are linearly interpolated to the original resolution.
(5) Based on the original image at the start time of the prediction, u and v are used to conduct backward interpolation, and the extrapolation image at N steps is obtained.

3. Weather Radar Echo Prediction Experiments

3.1. Experimental Setting

3.1.1. Case Setting and Statistics

To examine the proposed NGT method, a set of ideal cases and three real cases are used. In the ideal experiment, a simple two-dimensional normal distribution image with an extreme value region is used to simulate the echo of a precipitation cloud. Some simple changes, such as moving or numerically changing, are added to the images to test whether NGT can achieve the expected results. The three experiments using real radar observations are based on different radars and convective weather events. The original input echo images have a range of 600 km × 600 km and a resolution of 1 km × 1 km. The data form of these images is Constant Altitude Plan Position Indicator (CAPPI) or composite reflectivity, whose details are provided in Sections 3.3–3.5.

For real cases, three commonly used skill scores, namely, the critical success index (CSI, also known as the threat score), the probability of detection (POD), and the false alarm ratio (FAR), are used to examine the prediction results [25]. These skill scores are calculated as follows:

$$\mathrm{CSI} = \frac{\mathrm{N_A}}{\mathrm{N_A} + \mathrm{N_B} + \mathrm{N_C}} \tag{15}$$

$$\mathrm{POD} = \frac{\mathrm{N_A}}{\mathrm{N_A} + \mathrm{N_C}} \tag{16}$$

$$\mathrm{FAR} = \frac{\mathrm{N_B}}{\mathrm{N_A} + \mathrm{N_B}} \tag{17}$$

where N_A is the number of grid points with successful predictions (the hit number) for which both the observation and prediction are greater than a given threshold; N_B is the number of grid points with false alarms for which the observation is less than the given threshold, and the prediction is greater than the given threshold; and N_C is the number of grid points with unsuccessful predictions, for which the condition is opposite to N_B. Referring to previous studies [18,19], the thresholds of radar reflectivity (Z) on the echo image are set to 20, 30, and 40 dBZ. A perfect prediction result has a CSI and POD close to 1 and a FAR close to 0. CSI is the overall statistic, while POD and FAR help to further explain, for example, whether a lower CSI is due mainly to a low hit rate or a high false alarm rate.

When comparing the CSI of different methods, a static reference value is additionally calculated, where the CSI of this static reference is obtained without processing the image at the starting time of the prediction. The purpose of doing so is to test the effectiveness of the CSI in the prediction results. Specifically, for clouds with a large range or inconspicuous movement, higher CSI may be obtained even without any echo extrapolation. Therefore, an effective prediction result should have a CSI larger than the static reference at the same time; otherwise, it should be regarded as a quantitative result with a negative effect.

3.1.2. The Method for Comparison

A method that can also output the image moving vector field is to be selected for comparison to examine whether the image prediction is improved by the newly proposed NGT method. Note that TREC and OF can both fulfill that. However, OF has lower computational complexity than TREC at the same resolution, and some recent studies indicate that OF makes better predictions [12]. Therefore, a simple and feasible OF scheme is adopted. A global smoothing constraint condition of Horn-Schunk OF [10,26] is implemented, and the minimizing function is Equation (18):

$$J = \int_{\Omega} \left[\left(I_x u + I_y v + I_t \right)^2 + \alpha^2 \left(|\nabla u|^2 + |\nabla v|^2 \right) \right] d\Omega \tag{18}$$

where Ω is the area domain of grid points, I_t, I_x, and I_y are the temporal and spatial partial derivatives that synthesize two adjacent times, and α^2 is a smoothing parameter. The iterative scheme to solve the optical flow vector fields u and v is as follows:

$$u^{n+1} = \overline{u}^n - I_x \frac{I_x \overline{u}^n + I_y \overline{v}^n + I_t}{\alpha^2 + I_x^2 + I_y^2} \tag{19}$$

$$v^{n+1} = \overline{v}^n - I_y \frac{I_x \overline{u}^n + I_y \overline{v}^n + I_t}{\alpha^2 + I_x^2 + I_y^2} \tag{20}$$

where $\overline{u}$ and $\overline{v}$ are the regional average values of u and v, respectively, and the superscripts n and $n + 1$ represent the order of iterations. Here, α^2 is set to the global average of ${I_x}^2 + {I_y}^2$. The number of iterations is 1000 to ensure the solution instead of setting a breakdown condition. More details about the calculation scheme of I_t, I_x, I_y, $\overline{u}$, $\overline{v}$ and the boundary are provided in [26].

The OF adopts a similar method as the NGT, where the vector fields u and v for extrapolation are obtained first, and then backward interpolation is used on the image at the predicted time. For real cases, the solving of u and v is performed under an average grid of 10 km × 10 km. An additional two-dimensional Gaussian filter with a standard deviation of 1 grid and a radius of 10 grid is applied to NGT. The prediction is under the resolution of the original 1 km × 1 km grid. Images of the prediction start time and the previous time are used as historical samples in OF, while NGT uses historical samples one hour ahead of the prediction start time. The impact of historical sample selection on OF is briefly discussed in Section 3.2, and the impact of historical sample length on NGT is briefly discussed at the end of Section 3.3.

3.2. Ideal Cases

As shown in Figure 2a, both NGT and OF can make basically accurate predictions when only a simple translation exists. This also shows that the OF extrapolation scheme used for comparison in this paper is reliable for simple prediction.

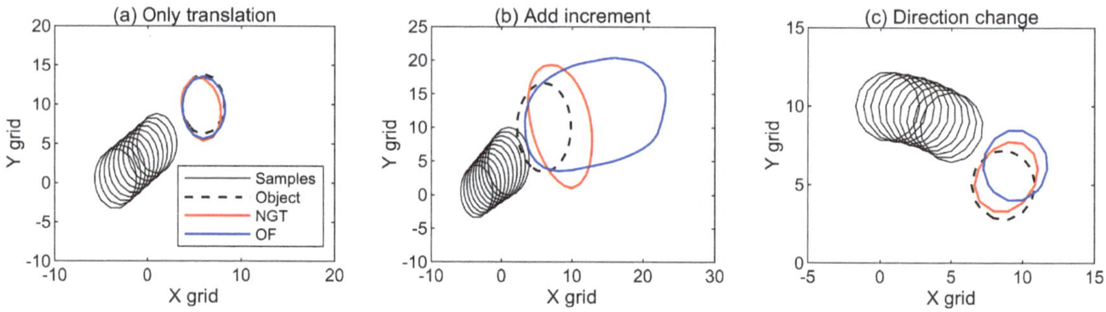

Figure 2. Example of extrapolation results in ideal experiments based on simple images: (**a**) only translation, (**b**) add increment, and (**c**) direction change. Samples are two-dimensional normal distribution images with an initial value range of 0~1, and contour lines for 0.5 are shown. The translation speed in (**a**) and (**b**) is 0.5 grids per time in both the X and Y directions. A 0.01 increment over time is added in (**b**). The rotation speed of the extreme center in (**c**) is 3 degrees per time. The historical samples shown are t = 1~10, the predictions start from t = 10, and the object is at t = 20. NGT is modeled using all ten historical samples, while OF uses only t = 9~10. The X and Y axes are distances with arbitrary units.

When the image value has a time increment (Figure 2b), although the results of the two methods are both biased, the result of NGT is closer to the target object, while OF has almost more than twice the deviation, possibly due to the time increment, which does not conform to the basic assumption of the OF method itself.

When the moving path of the target object has an arc with rotation characteristics (Figure 2c), the result of NGT is closer to the target, while the result of OF seems to be only a linear extrapolation of the historical sample trajectory, resulting in an obvious deviation. In addition, it can be expected that the change in the historical sample selection will produce different OF results, where the predicted red circle will appear at the extension lines of the central positions of the two historical samples, which is still farther from the target. That is, the OF method is weak at extrapolating this path turning, regardless of how historical samples are selected. Note that the use of more than two historical samples for OF in [12] is expected also not helpful under this condition since it does not change the algorithm feature. In summary, compared with the conventional OF method, NGT can perceive more image changes from historical samples and can obtain more accurate extrapolation prediction results when the target value and path change.

The deviation of the NGT should also be noted. It is seen from Section 2.2 that only the spatial derivatives and coordinates of the image are involved in NGT, while the simple target used in Figure 2 has similar spatial derivatives in all directions, which might bring confusion and cause deviation. However, since this defect does not exist in real cases, it is no longer discussed below.

3.3. Real Case 1: A Convective Line with Shape Evolution

A convective weather event with obvious movement and shape change in radar echo images is selected first. A squall line event occurred in the U.S. on 20 May 2011, which was analyzed in a previous study [27]. The stage is selected in which a convective line is detected by a NEXRAD S-band radar (the station code is KTLX, located at 97.28°W, 35.33°N) when this convective line with shape-shifting moves to the radar. This radar performs a volume scan consisting of 15 elevations in 255 s. The CAPPI at a height of 3 km is obtained by triple linear interpolation from the volume scan data, which can not only reflect the main movement of the echo but also automatically shield most of the clutter in the boundary layer.

As shown in Figure 3, the convection line and its strong echo front are first in an inverted L-shape. As the cloud system moves northeastward, the convection line gradually evolves into a bow shape. Due to the north side of the cloud body moving out of the radar detection range, the original inverted L-shaped east–west echo band also gradually disappears. The visual moving directions of cells, new cells emerging and the whole cloud system are marked in Figure 3a. The horizontal wind U and V from ERA5 reanalysis data (Figure 4) are cited as a physical wind field for comparison with NGT and OF to analyze their features. The direction of new cell emergence is toward the southeast side, which is the windward direction of the relatively low level (e.g., 950 hPa), while the direction of the cell movement at a height of 3 km is along the north and slightly eastward, basically the same as that of the wind field at a similar height (from 850 hPa to 700 hPa). The combination of the above two directions eventually makes the whole cloud system move northeast.

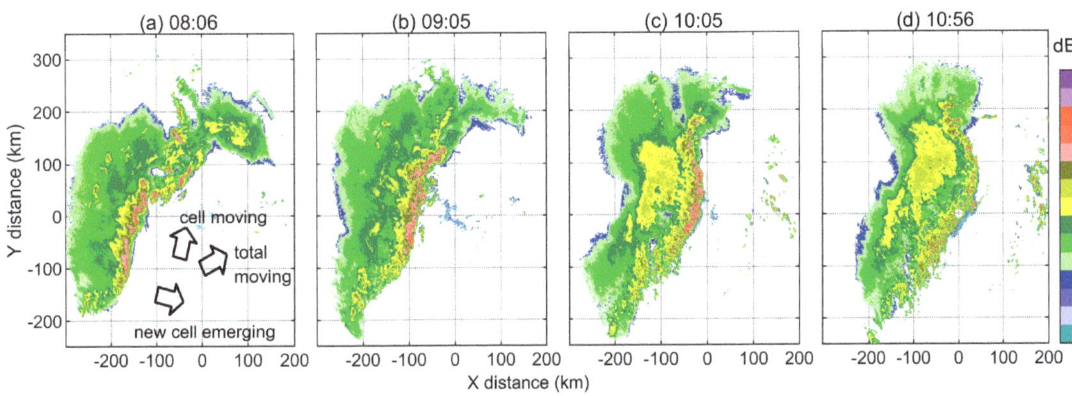

Figure 3. CAPPI at a height of 3 km observed by KTLX radar on 20 May 2011. X and Y represent west–east and south–north distances relative to the radar site. (**a**) 08:06; (**b**) 09:05; (**c**) 10:05; (**d**) 10:56. The time is in UTC. The black arrows in (**a**) represent the visual directions of cell movement, new cell emergence, and total movement of the whole system.

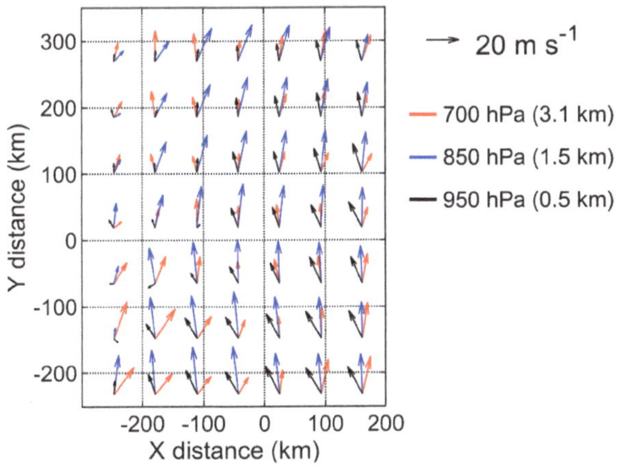

Figure 4. Wind field from ERA5 reanalysis data at 09:00 UTC.

Taking a 60 min prediction as an example, the data in Figure 3b are taken as the starting time, and Figure 3c is taken as the prediction target where the convection line moves and deforms. The results of NGT are basically consistent with the observed front edge of the convective line (Figure 5a). The grid transformation vectors mainly point to the northeast, which is consistent with the overall movement direction of the cloud system. In addition, these vectors point to the west at the northernmost point, which is also consistent with the image change characteristics of the weakening and disappearance of the east–west echo band on the north side in Figure 3. On the other hand, the OF results obviously underestimate the movement of the convection line (Figure 5b). The vector value of OF on the southeast side is small, which may be related to the interference of small-scale scattered clouds in the clear sky region at the front. More importantly, although the vector field of OF is similar to the wind field at the front of the convective line, pointing north and similar to the direction of monomer movement, this is not the direction of the overall cloud movement, which may be the main reason for the underestimation of movement.

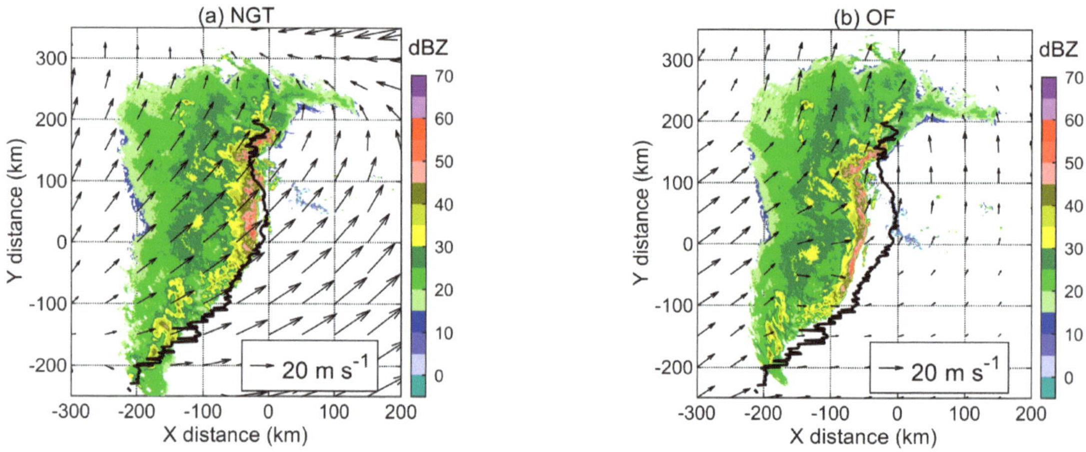

Figure 5. Sixty-minute prediction results. The target time is 10:05 UTC on 20 May 2011. The shading is the predicted CAPPI at a height of 3 km. The black solid lines are the 30 dBZ leading edge of the convective echo observed at the target time (Figure 3c). The vector fields are obtained by NGT and OF for backward interpolation. (**a**) NGT; (**b**) OF.

Although the prediction effect decreases with time, there are differences between the two methods (Figure 6). The CSI of NGT is lower than that of OF only in the first 1–3 times of prediction and is better than that of OF in other prediction periods. For the higher 40 dBZ echo, NGT has the most obvious advantage over OF, and the CSI can be 0.2 higher than that of OF in the 40~60 min prediction period. The CSI of OF for 30 and 40 dBZ is lower than the static references within 60 min, which means that the prediction accuracy loses its value, while the CSI of NGT decreases more smoothly, indicating that it has the capability to perform echo prediction over a longer prediction period. In addition, due to the overall deviation of the position of the convective line predicted by OF, its prediction skill scores are all lower than those of NGT (Table 1).

Figure 7 briefly shows the effect of historical sample length on NGT. The CSI is lower when only two historical samples (less than 5 min in this case) are taken. With an increase in the length of the historical sample, the CSI rises in oscillation first and then rises slowly when the sample is longer than 20~30 min, indicating that a longer sample length leads to a better NGT prediction effect. However, since the duration of convective weather is usually only a few hours or less, it is not practical to use historical samples longer than 60 min. Therefore, only 60 min of the historical sample for NGT is used and discussed in most parts of this paper.

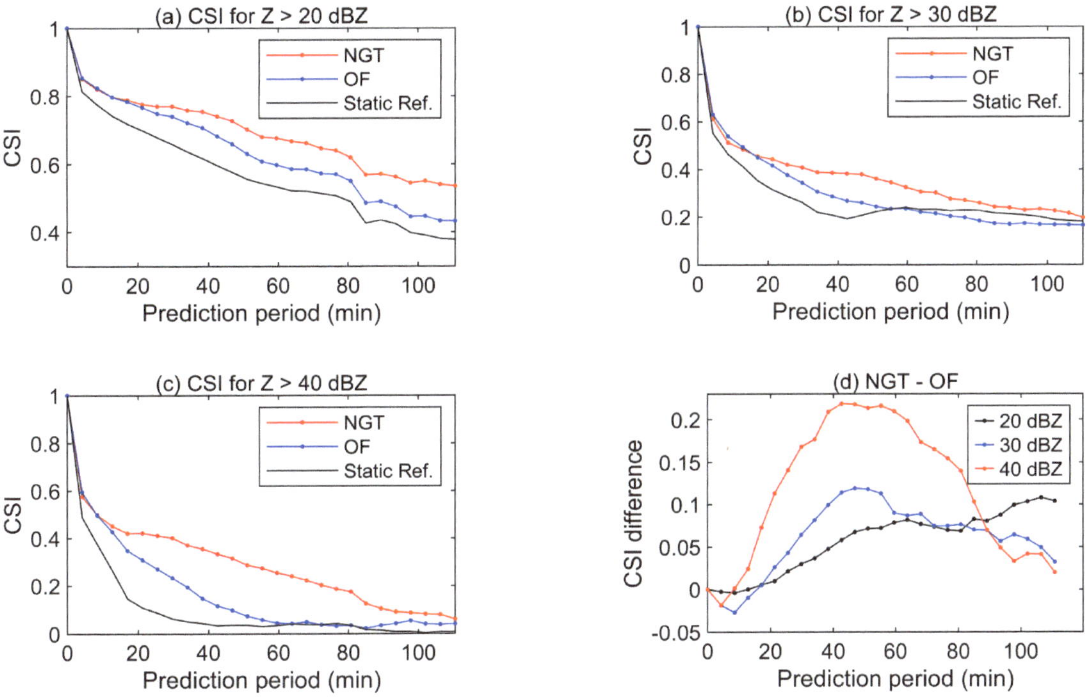

Figure 6. CSI of the prediction by NGT and OF for (**a**) Z > 20 dBZ, (**b**) Z > 30 dBZ, (**c**) Z > 40 dBZ, and (**d**) NGT—OF. The prediction starts at 9:05 UTC on 20 May 2011. The "static ref." lines mean a reference without any extrapolation on the data of the starting time.

Table 1. Skill scores of the two methods for 60 min prediction. The target time is 10:05 UTC on 20 May 2011.

Threshold (dBZ)	POD		FAR		CSI	
	NGT	OF	NGT	OF	NGT	OF
20	**0.84**	0.75	**0.23**	0.26	**0.68**	0.60
30	**0.41**	0.29	**0.38**	0.44	**0.33**	0.23
40	**0.42**	0.08	**0.61**	0.90	**0.25**	0.04

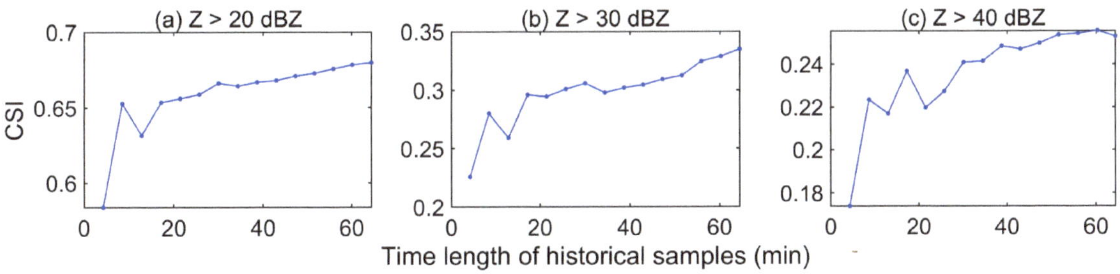

Figure 7. The CSI of 60 min prediction by NGT using different time lengths of historical samples for (**a**) Z > 20 dBZ, (**b**) Z > 30 dBZ, and (**c**) Z > 40 dBZ. The prediction starts at 9:05 UTC on 20 May 2011.

3.4. Real Case 2: A Convective Event with Convective Line Formation

To examine whether the NGT is always better than OF in different stages of a convective event, it is necessary to select a case that has been detected by radar for a more complete life cycle. The selected case is a convective event that occurred in Shandong Province, China, on 17 May 2020. The radar data are from a CINRAD-SA type S-band radar (the station number is Z9532, located at 120.23°E, 35.99°N). This radar performs a volume scan consisting of 9 elevations within 342 s. Other information is provided in [28]. The CAPPI at 3 km height is used for analysis, similar to the last case in Section 3.3.

As shown in Figure 8, scattered convective clouds appear at the northwest side of the radar first, develop into a severe convective cell group, and then form a convective line, passing over the radar origin and moving southeastward. In the following time, the stratiform cloud area at the back of the convection line gradually expands, and the strong echo at the front gradually weakens, which is no longer displayed together here.

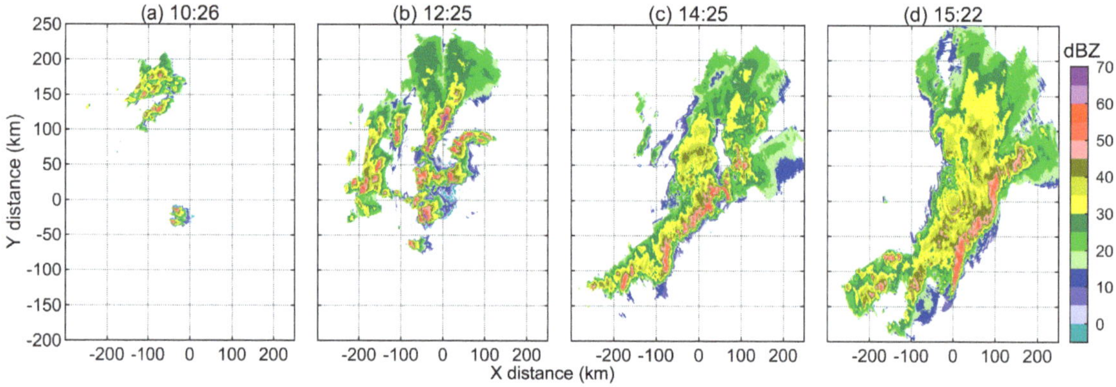

Figure 8. CAPPI at 3 km height observed by Z9532 radar on 17 May 2020. X and Y represent west–east and south–north distances relative to the radar site. (**a**) 10:26; (**b**) 12:25; (**c**) 14:25; (**d**) 16:59. The time is in UTC.

Figure 9 shows the 60 min prediction skill scores of almost the entire weather process. Clearly, NGT is not always superior to OF in echo prediction for 20 and 30 dBZ. However, the CSI of NGT for 40 dBZ is higher than that of OF at the target time of 14~16 UTC, which is the stage where the convective line exists.

A one-hour prediction is taken as an example for analysis similar to Section 3.3, where the data in Figure 8c are taken as the starting time, and Figure 8d is taken as the prediction target when the convection line persists. Note that 57 min is used here to represent a one-hour period since there is no data point exactly at the next 60 min. The NGT results show that the predicted convective lines are basically consistent with the observations (Figure 10a). Note that there is a large grid transformation vector over the normal wind speed in the clear sky area on the southeast side of the convective line pointing to the southeast, which indicates that the image is greatly stretched southeastward in the prediction. However, in the OF results (Figure 10b), an underestimation of the movement of the convective line remains. This underestimation is similar to that in Section 3.3 and is thus also due to the deviation of the moving direction of the whole cloud system and cells. In this 60 min prediction example, the CSI of NGT is better than that of OF (Figure 11), and the advantage is most obvious for 40 dBZ. Regarding the other skill scores, only the FAR of NGT is inferior to that of OF for 20 and 30 dBZ (Table 2). This may be due to the increase in the false alarm rate of NGT caused by a displacement greater than the expected echo at X = 0~100 km and Y = −200~−100 km in Figure 10a, but this does not affect the overall advantage of NGT.

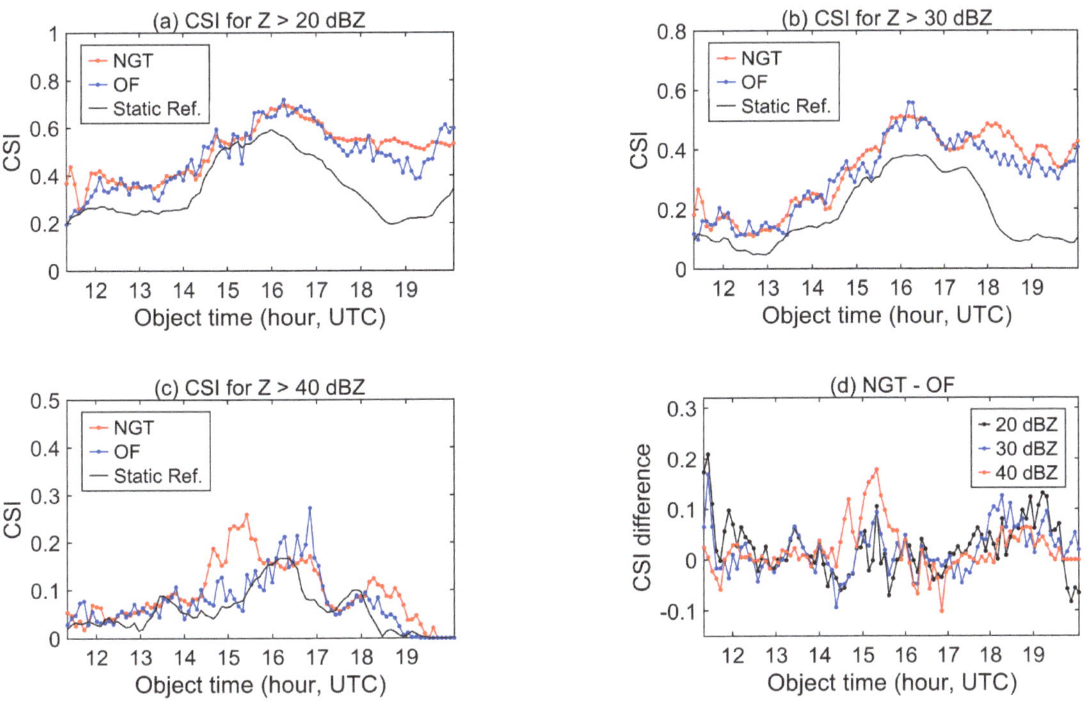

Figure 9. CSI of the 60 min extrapolating prediction by NGT and OF for (**a**) Z > 20 dBZ, (**b**) Z > 30 dBZ, (**c**) Z > 40 dBZ, and (**d**) NGT—OF on 17 May 2020. The prediction starts at 9:05 UTC on 20 May 2011.

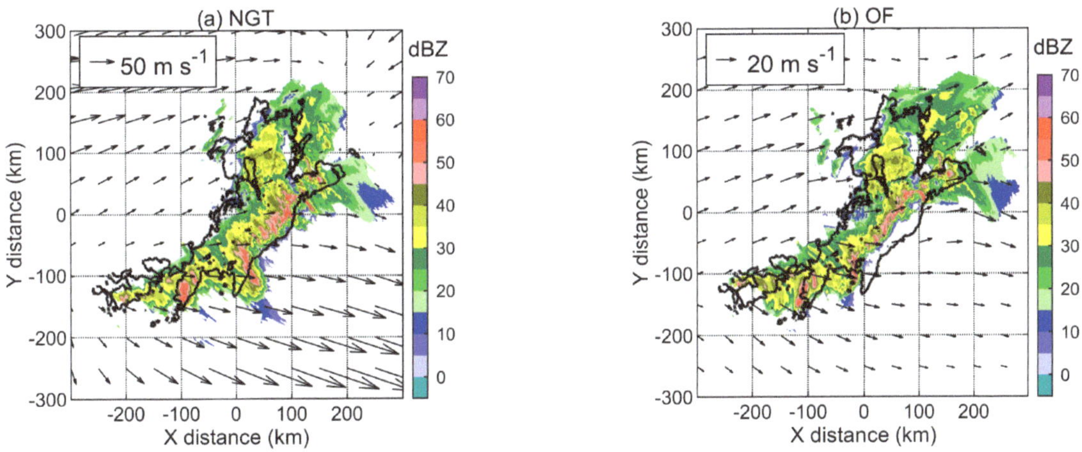

Figure 10. Fifty-seven-minute prediction results. The target time is 15:22 UTC on 17 May 2020. The shading is the predicted CAPPI at a height of 3 km. The black solid lines are the 30 dBZ contours of the convective echo observed at the target time (Figure 8c). The vector fields are obtained by NGT and OF for backward interpolation. (**a**) NGT; (**b**) OF.

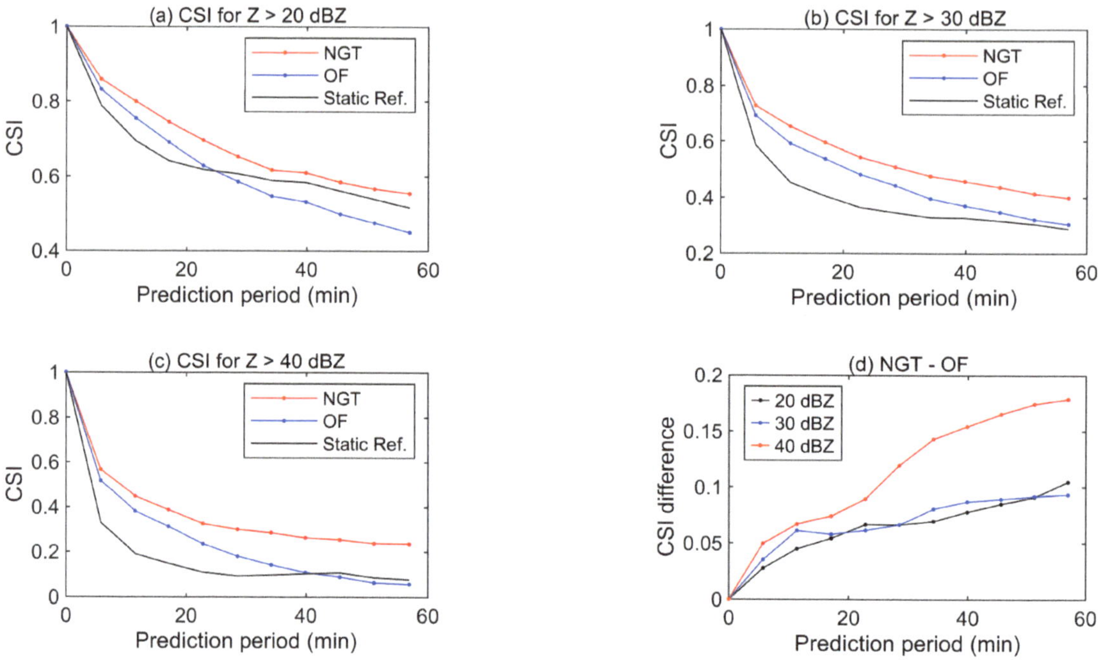

Figure 11. CSI of the prediction by NGT and OF for (**a**) Z > 20 dBZ, (**b**) Z > 30 dBZ, (**c**) Z > 40 dBZ, and (**d**) NGT—OF. The prediction starts at 14:25 UTC on 17 May 2020.

Table 2. Skill scores of the two methods for 60 min prediction. The target time is 15:22 UTC on 17 May 2020.

Threshold (dBZ)	POD		FAR		CSI	
	NGT	OF	NGT	OF	NGT	OF
20	**0.61**	0.47	0.15	**0.08**	**0.55**	0.45
30	**0.47**	0.34	0.27	**0.25**	**0.40**	0.31
40	**0.33**	0.08	**0.55**	0.83	**0.24**	0.06

3.5. Real Case 3: A Mesoscale Convective Complex and Comparison with Previous Deep Learning Results

As a newly proposed method, NGT must be tested in different types of convective weather, and it is best to compare it with other newer popular methods. We have identified a case that can approximately meet this goal. A convective weather process dominated by a Mesoscale Convective Complex (MCC) occurred in Hebei Province, China, on 21 June 2017. A 30 min prediction result was presented in previous deep learning studies by Liang et al. [18] and Han et al. [19], where 8 years of data were used for training and 30 min of historical data were inputted to generate predictions. They mainly used a U-Net model, which is a convolutional neural network and is constructed by stacking downsampling and upsampling convolutional modules yielding a unique U-shape network architecture. The table of skill scores was given in [18] for this prediction example. Therefore, it is possible to compare NGT with the previous deep learning method by collecting the same radar data and repeating only the 30-min predictions during the same prediction period instead of collecting a large number of training datasets to reproduce the deep learning result or directly citing the resulting images, which is not applicable. The radar echo image used in [18] is composite reflectivity, which is a combination of six operational radars. The station numbers are Z9010, Z9220, Z9311, Z9313, Z9314, and Z9335, where Z9313 and Z9314 are CINRAD-CB type, and others are CINRAD-SA type. The time interval of the image is 6 min.

The observation at the target time and the prediction results of NGT and OF are shown in Figure 12, while the results of the deep learning method under the same prediction conditions are seen in [18]. The differences between the observation and the prediction results from both NGT and OF seem no longer as obvious as shown in the above two sections. For the CSI (Figure 13), the results of NGT and OF are very close, where NGT is slightly higher for 20 and 30 dBZ. In addition, the CSIs of NGT and OF are only slightly higher than the static reference, indicating that the prediction of echo movement is limited. Considering that the OF method is sensitive to the instantaneous movement of the cloud, these results indicate that the movement itself of MCC is not as obvious as the squall lines referred to in the last sections. More importantly, the deep learning method appears to be a more advanced method with results better than those of NGT and OF for 40 dBZ; however, its CSI for 20 and 30 dBZ is not only worse than that of NGT and OF but also worse than the static reference, which indicates that it may not correctly reflect the movement of the echo.

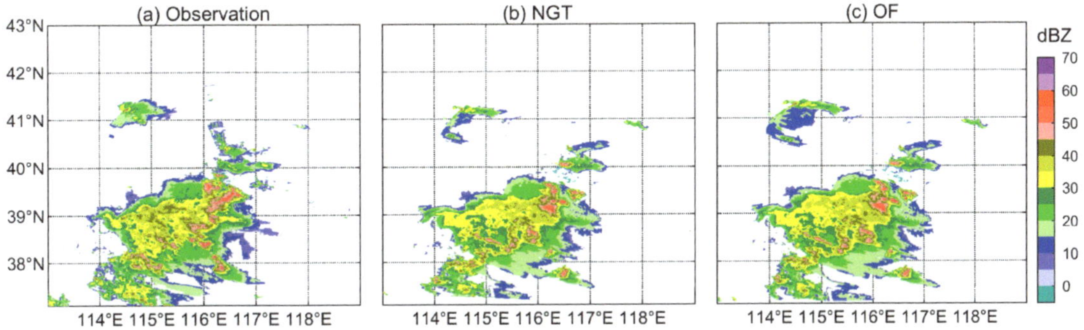

Figure 12. (**a**) Composite reflectivity obtained by multiple radars at 11:30 UTC (target time) on 21 June 2017 and 30 min prediction results by (**b**) NGT and (**c**) OF.

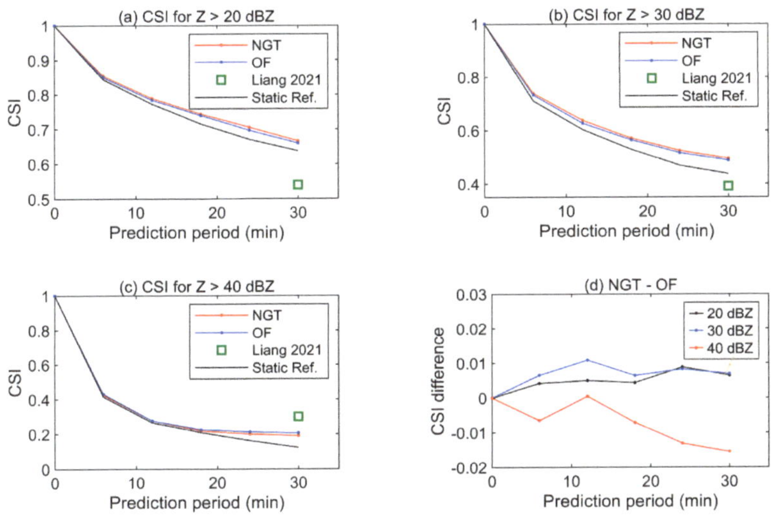

Figure 13. CSI of the prediction by NGT and OF for (**a**) Z > 20 dBZ, (**b**) Z > 30 dBZ, (**c**) Z > 40 dBZ, and (**d**) NGT—OF. The prediction starts at 11:00 UTC on 21 June 2017. Liang 2021 is cited from [18] as a deep learning method for comparison.

As shown in the skill scores (Table 3), although the POD of this deep learning method at 20 and 30 dBZ is the highest, the FAR is also higher than that of the other two methods. That is, both the hit rate and false rate are high, indicating an overestimation of the reflectivity of the echo images. This overestimation is seen in the image in Liang et al. (2021) and might be one of the reasons that the deep learning method obtains a better skill score for a larger echo threshold.

Table 3. Skill scores of two methods for 30 min prediction and the comparison with results from a previous deep learning study. The target time is 11:30 UTC on 21 June 2017.

Threshold (dBZ)	POD			FAR			CSI		
	NGT	OF	Liang 2021 *	NGT	OF	Liang 2021 *	NGT	OF	Liang 2021 *
20	0.75	0.77	**0.84**	**0.13**	0.18	0.33	0.67	0.66	**0.54**
30	0.61	0.63	**0.67**	**0.27**	0.32	0.49	0.49	0.49	**0.39**
40	0.28	0.31	**0.51**	0.62	0.62	**0.57**	0.19	0.21	**0.30**

* The results of "Liang 2021" are cited and reorganized from a previous study [18] as a deep learning method for comparison based on the best scores derived from the U-Net model and TrajGRU model.

4. Conclusions

The NGT method is proposed for weather radar convective echo extrapolation prediction. A set of ideal experiments and three real echo extrapolation experiments were performed to demonstrate the performance of the new method. The results are compared with those from an optical flow (OF) method, and a brief comparison with a previous study based on a deep learning method is conducted in a real case. The main conclusions are as follows.

(1) In the ideal experiments for simple targets, NGT can achieve better prediction results than OF when there are changes in the value and path of the image target.
(2) For cases with a convective line, NGT is superior to OF, where the CSI for 40 dBZ of the echo prediction at 60 min is approximately 0.2 higher.
(3) For the case of MCC, NGT is better than OF and a deep learning method for 20 and 30 dBZ, while the deep learning method produces the best skill score for 40 dBZ. However, the good performance of the deep learning method may be due to the overestimation of the stronger echo.

The conventional OF method is based on a few historical data samples, tending to present the instantaneous movement direction of convective cells. Therefore, although the results of OF are closer to the physical motion of a cloud, there is a deviation compared with the movement of the overall cloud system. On the other hand, NGT can be modeled based on historical 60 min data. Although the grid transformation vector of the NGT is generally different from the physical wind field, the results are closer to the continuous variation characteristics of the echo image. Therefore, the movement and variation pattern of the whole convection line can be predicted more approximately.

In general, NGT, as a new method, improves the prediction effect of at least one type of convective system. This improvement in nowcasting capability can help to respond to severe convective weather, such as making corresponding changes to outdoor activities and transportation plans and better arranging artificial hail suppression operations.

However, many studies must be conducted in the future. The sensitivity of the prediction effect to the parameters in the prediction method needs to be further studied. More importantly, the CSIs of NGT and OF for stronger echo are still low and are markedly less than those for weaker echoes, which may be caused by the spatial nonuniformity and the strength variation of convective echoes. Therefore, an extrapolation of the echo strength may be needed to improve the result. Moreover, since deep learning methods have the potential to better predict the increase in echo strength, more operational observation cases are needed for further study to take full advantage of different prediction methods.

Author Contributions: Conceptualization, Y.S. and Y.T.; methodology, Y.S. and H.X.; software, Y.S.; validation, Y.S.; formal analysis, Y.S. and H.X.; investigation, Y.S.; resources, H.X., Y.T. and H.Y.; data curation, Y.T. and H.Y.; writing—original draft preparation, Y.S.; writing—review and editing, H.X., Y.T. and H.Y.; visualization, Y.S.; supervision, H.X.; project administration, H.X.; funding acquisition, Y.S., H.X. and H.Y. All authors have read and agreed to the published version of the manuscript.

Funding: This research was supported in part by the National Key Research and Development Plan of China (grant no. 2019YFC1510304), the National Natural Science Foundation of China (grant no. 42105127), the Research and Experiment on the Construction Project of Weather Modification Ability in Central China (grant no. ZQC-T22254), and the Special Research Assistant Project of Chinese Academy of Sciences.

Data Availability Statement: Data are available in a publicly accessible repository that does not issue DOIs. Publicly available datasets were analyzed in this study. The NEXRAD radar base data used in Section 3.3 can be found at https://www.ncdc.noaa.gov/nexradinv/map.jsp, accessed on 7 February 2023. Other data sharing is not applicable.

Conflicts of Interest: The authors declare no conflict of interest.

References

1. Wilson, J.W.; Crook, N.A.; Mueller, C.K.; Sun, J.Z.; Dixon, M. Nowcasting Thunderstorms: A Status Report. *Bull. Am. Meteorol. Soc.* **1998**, *79*, 2079–2099. [CrossRef]
2. Schumacher, R.S.; Rasmussen, K.L. The formation, character and changing nature of mesoscale convective systems. *Nat. Rev. Earth Environ.* **2020**, *1*, 300–314. [CrossRef]
3. Mueller, C.; Saxen, T.; Roberts, R.; Wilson, J.; Betancourt, T.; Dettling, S.; Oien, N.; Yee, J. NCAR Auto-Nowcast System. *Wea. Forecast.* **2003**, *18*, 545–561. [CrossRef]
4. Sun, J.Z.; Wang, H.L. WRF-ARW Variational Storm-Scale Data Assimilation: Current Capabilities and Future Developments. *Adv. Meteorol.* **2013**, *2013*, 13. [CrossRef]
5. Dance, S.L.; Ballard, S.P.; Bannister, R.N.; Clark, P.; Cloke, H.L.; Darlington, T.; Flack, D.L.A.; Gray, S.L.; Hawkness-Smith, L.; Husnoo, N.; et al. Improvements in Forecasting Intense Rainfall: Results from the FRANC (Forecasting Rainfall Exploiting New Data Assimilation Techniques and Novel Observations of Convection) Project. *Atmosphere* **2019**, *10*, 125. [CrossRef]
6. Roy, S.S.; Mohapatra, M.; Tyagi, A.; Bhowmik, S.K.R. A review of Nowcasting of convective weather over the Indian region. *Mausam* **2019**, *70*, 465–484. [CrossRef]
7. Sokol, Z.; Szturc, J.; Orellana-Alvear, J.; Popová, J.; Jurczyk, A.; Célleri, R. The Role of Weather Radar in Rainfall Estimation and Its Application in Meteorological and Hydrological Modelling—A Review. *Remote Sens.* **2021**, *13*, 351. [CrossRef]
8. Dixon, M.; Wiener, G. Titan-thunderstorm identification, tracking, analysis, and nowcasting-a radar-based methodology. *J. Atmos. Oceanic. Technol.* **1993**, *10*, 785–797. [CrossRef]
9. Tuttle, J.D.; Foote, G.B. Determination of the boundary-layer air-flow from a single doppler radar. *J. Atmos. Oceanic. Technol.* **1990**, *7*, 218–232. [CrossRef]
10. Horn, B.K.P.; Schunck, B.G. Determining optical flow. *Artif. Intell.* **1981**, *17*, 185–203. [CrossRef]
11. Lucas, B.D.; Kanade, T. An Iterative Image Registration Technique with an Application to Stereo Vision. In Proceedings of the 7th International Joint Conference on Artificial Intelligence, Vancouver, BC, Canada, 24 August 1981; pp. 674–679.
12. Woo, W.-C.; Wong, W.-K. Operational Application of Optical Flow Techniques to Radar-Based Rainfall Nowcasting. *Atmosphere* **2017**, *8*, 48. [CrossRef]
13. Bechini, R.; Chandrasekar, V. An Enhanced Optical Flow Technique for Radar Nowcasting of Precipitation and Winds. *J. Atmos. Ocean. Technol.* **2017**, *34*, 2637–2658. [CrossRef]
14. Munich, M.; Pirjanian, P.; Di Bernardo, E.; Goncalves, L.; Karlsson, N.; Lowe, D. SIFT-ing through features with ViPR-Application of visual pattern recognition to robotics and automation. *IEEE Robot. Autom. Mag.* **2006**, *13*, 72–77. [CrossRef]
15. Sánchez, J.; Monzón, N.; Salgado, A. An Analysis and Implementation of the Harris Corner Detector. *Image Process. Line* **2018**, *8*, 305–328. [CrossRef]
16. Ravuri, S.; Lenc, K.; Willson, M.; Kangin, D.; Lam, R.; Mirowski, P.; Fitzsimons, M.; Athanassiadou, M.; Kashem, S.; Madge, S.; et al. Skilful precipitation nowcasting using deep generative models of radar. *Nature* **2021**, *597*, 672–677. [CrossRef]
17. Pan, X.; Lu, Y.H.; Zhao, K.; Huang, H.; Wang, M.J.; Chen, H.A. Improving Nowcasting of Convective Development by Incorporating Polarimetric Radar Variables Into a Deep-Learning Model. *Geophys. Res. Lett.* **2021**, *48*, 10. [CrossRef]
18. Liang, H.; Chen, H.; Zhang, W.; Ge, Y.; Han, L. Convective Precipitation Nowcasting Using U-Net Model. In Proceedings of the 2021 IEEE International Geoscience and Remote Sensing Symposium IGARSS, Brussels, Belgium, 11–16 July 2021; pp. 7134–7137. [CrossRef]
19. Han, L.; Liang, H.; Chen, H.; Zhang, W.; Ge, Y. Convective Precipitation Nowcasting Using U-Net Model. *IEEE Trans. Geosci. Remote Sens.* **2022**, *60*, 8. [CrossRef]

20. Yao, S.; Chen, H.N.; Thompson, E.J.; Cifelli, R. An Improved Deep Learning Model for High-Impact Weather Nowcasting. *IEEE J. Sel. Top. Appl. Earth Obs. Remote Sens.* **2022**, *15*, 7400–7413. [CrossRef]
21. Luo, C.; Li, X.; Wen, Y.; Ye, Y.; Zhang, X. A Novel LSTM Model with Interaction Dual Attention for Radar Echo Extrapolation. *Remote Sens.* **2021**, *13*, 164. [CrossRef]
22. Bouget, V.; Béréziat, D.; Brajard, J.; Charantonis, A.; Filoche, A. Fusion of Rain Radar Images and Wind Forecasts in a Deep Learning Model Applied to Rain Nowcasting. *Remote Sens.* **2021**, *13*, 246. [CrossRef]
23. Sun, N.; Zhou, Z.; Li, Q.; Jing, J. Three-Dimensional Gridded Radar Echo Extrapolation for Convective Storm Nowcasting Based on 3D-ConvLSTM Model. *Remote Sens.* **2022**, *14*, 4256. [CrossRef]
24. Szeliski, R. *Computer Vision: Algorithms and Applications*; Springer: Berlin/Heidelberg, Germany, 2010.
25. Roebber, P.J. Visualizing Multiple Measures of Forecast Quality. *Weather. Forecast.* **2009**, *24*, 601–608. [CrossRef]
26. Meinhardt-Llopis, E.; Pérez, J.S.; Kondermann, D. Horn-Schunck Optical Flow with a Multi-Scale Strategy. *Image Process. Line* **2013**, *20*, 151–172. [CrossRef]
27. Xue, L.L.; Fan, J.W.; Lebo, Z.J.; Wu, W.; Morrison, H.; Grabowski, W.W.; Chu, X.; Geresdi, I.; North, K.; Stenz, R.; et al. Idealized Simulations of a Squall Line from the MC3E Field Campaign Applying Three Bin Microphysics Schemes: Dynamic and Thermodynamic Structure. *Mon. Weather. Rev.* **2017**, *145*, 4789–4812. [CrossRef]
28. Sun, Y.; Xiao, H.; Yang, H.; Chen, H.; Feng, L.; Shu, W.; Yao, H. A Uniformity Index for Precipitation Particle Axis Ratios Derived from Radar Polarimetric Parameters for the Identification and Analysis of Raindrop Areas. *Remote Sens.* **2023**, *15*, 534. [CrossRef]

remote sensing

Article

Correction for the Attenuation Due to Atmospheric Gas and Stratiform Clouds in Triple-Frequency Radar Observations of the Microphysical Properties of Snowfall

Yue Chang [1,2,3], Hongbin Chen [1,3,*], Xiaosong Huang [1], Yongheng Bi [1], Shu Duan [1], Pucai Wang [1,3] and Jie Liu [4]

[1] Key Laboratory of Middle Atmosphere and Global Environment Observation, Institute of Atmospheric Physics, Chinese Academy of Sciences, Beijing 100029, China; chyue@cuit.edu.cn (Y.C.); hxs@mail.iap.ac.cn (X.H.); pcwang@mail.iap.ac.cn (P.W.)
[2] College of Electronic Engineering, Chengdu University of Information Technology, Chengdu 610225, China
[3] University of Chinese Academy of Sciences, Beijing 100049, China
[4] Meteorological Observation Center of China Meteorological Administration, Beijing 100081, China
* Correspondence: chb@mail.iap.ac.cn

Abstract: For triple-frequency radar, the attenuation attributed to atmospheric gases and stratiform clouds is diverse due to different snowfall microphysical properties, particularly in regions far from the radar. When using triple-frequency ground-based radar measurements, evaluating the attenuation of the three radars at different heights is common to derive attenuation-corrected effective reflectivity. Therefore, this study proposes a novel quality-controlled approach to identify radar attenuation due to gases and stratiform clouds that can be neglected due to varying snowfall microphysical properties and assess attenuation along the radar observation path. The key issue lies in the lack of information about vertical hydrometeor and cloud distribution. Therefore, European Centre for Medium-Range Weather Forecasts (ECMWF) reanalysis data are employed. The Self-Similar-Rayleigh-Gans Approximation (SSRGA) for the nonspherical scattering model in the Passive and Active Microwave TRAnsfer model 2 (PAMTRA2) is compared and analyzed against other scattering models to obtain the optimal triple-frequency radar attenuation correction strategies for stratiform cloud meteorological conditions with varying snowfall microphysical properties. This methodology paves the way for understanding differential attenuation attributed to gas and stratiform clouds with snowfall microphysical properties. Simultaneously, the bin-by-bin approximation method is used to perform the attenuation correction. The two-way attenuation correction increased up to 4.71 dB for heights above 6 km, remaining minimal for regions with heights below 6 km. These values, attributable to gases and stratiform clouds' two-way attenuation, are nonnegligible, especially at distances far from the W-band radar at heights above 6 km. Both values are relatively small for the X- and Ka-band radars and can be neglected for the varying snowfall microphysical properties. The attenuation correction of triple-frequency radar reflectivity is validated using the cross-calibration and dual-frequency reflectivity ratios. The results show that the method is valid and feasible.

Keywords: triple-frequency radar; gas and stratiform clouds attenuation; attenuation correction; SSRGA; snowfall microphysical properties

1. Introduction

Real-time measurement of the three-dimensional (3D) structure of clouds and precipitation remains challenging in atmospheric-sounding technology. Radar remains a vital tool for obtaining macro and microstructural characteristics of clouds. Ground-based triple-frequency (X, Ka, and W bands) cloud and precipitation radars are critical components of observational systems designed for detailed cloud structure characterization. These radars provide deeper insights into microphysical processes and retrieve more cloud and precipitation parameters than single-frequency radar, thanks to measurements at multiple frequency bands. However, the triple-frequency method, a powerful technique in microphysical research, demands precise radar calibration and reliable attenuation correction [1,2].

Within the radar reflectivity range, frequency-dependent hydrometeor scattering and absorption properties are governed by microphysical characteristics. Numerical and observational results have indicated that the triple-frequency radar signature has the potential for retrieving morphological parameters constrained by triple-frequency radar measurements [3,4]. Furthermore, when combined with other remote sensors, triple-frequency radar has significantly improved retrievals for ice [3,5,6] and rain [7–9]. Due to frequency-dependent properties, different hydrometeors exhibit varying attenuation or backscattering at each band of a triple-frequency radar. For instance, snowflakes tend to produce noticeable attenuation at and above the W band [5], although attenuation corrections at higher frequencies have been elusive in previous studies.

Distinct reflectivity signals result from differential scattering or attenuation, which are influenced by particle microphysical properties, including particle size distribution (PSD), type, density, and phase state. Numerous studies have investigated triple-frequency radar measurements with other ground-based equipment for actual snowfall observations [3,5,10–17]. While most of these studies rely on Rayleigh and non-Rayleigh scattering methodologies, the irregular and complex nature of most ice crystals and snowflakes can introduce errors when calculating backscattering and attenuation using the Mie or T-Matrix scattering models. Hogan proposed the SSRGA scattering model [18,19], providing a faster method for computing the scattering properties of aggregated ice particles and snowflakes than the Discrete Dipole Approximation (DDA). Furthermore, it offers more accurate backscattering calculations for irregular snowflakes at higher frequencies compared to the Mie solution or Rayleigh approximation.

Moreover, due to the lack of information about hydrometeor vertical distribution, it is essential to derive hydrometeor vertical profiles. In this study, we utilize ECMWF's ERA5 hourly reanalysis data to obtain atmospheric microphysics information, focusing on continuous and relatively homogeneous stratiform clouds to calculate attenuation contributions using the SSRGA approach. This approach appears promising in representing snowflakes more accurately according to their actual habits when combined with ERA5 hourly reanalysis profile data interpolated vertically.

In reality, various attenuations caused by gas and stratiform clouds are non-negligible, especially when at least one of the frequencies is affected by attenuation [20]. Therefore, this study utilizes triple-frequency radar observations (X, Ka, and W bands) with vertical pointing to quantify differential attenuation due to specific gases (e.g., oxygen and water vapor) and stratiform clouds using the SSRGA model in PAMTRA2 [21]. This study aims to develop an attenuation correction method for triple-frequency radar based on the first triple-frequency (X, Ka, and W bands) radar observations collected during winter in Zhangshanying, Yanqing, Beijing.

In past attenuation studies, there was no information about vertical hydrometeor and cloud distribution; the only information was path-integrated attenuation in snowstorms from actual observations. This technique provides bin-by-bin vertical distribution information to calculate the attenuation due to gases and stratiform clouds with snowfall microphysical properties. The influence of neglecting the effect of attenuation produced by different stratiform layers is vital for triple-frequency radar and needs to be assessed. The methodology combines PAMTRA2 with the SSRGA model and ECMWF reanalysis data for vertical profiles of gases and hydrometeor layers to calculate attenuation under different layer conditions. The ECMWF reanalysis data grid has a 0.25-degree hourly resolution, and vertical interpolation is performed for the calculation to match radar range bins. This approach provides more information than 12 h of sounding data.

The paper is organized as follows: Section 2 provides an overview of the triple-frequency radar instruments and ground-based auxiliary observation equipment, specifically the micro-rain radar deployed during the Zhangshanying Experiment. Section 3 details the attenuation correction method used, focusing on the SSRGA scattering model for extinction calculation, followed by the bin-by-bin approximation method for attenuation correction. In the first part of Section 4, we compare atmospheric gases and hydrometeors attenuation at X, Ka, and W band radars and calculate attenuation attributed to stratiform clouds using different scattering models. Section 5 presents the vertical effective reflectivity profile of the attenuation correction effect of the triple-frequency radar using different scattering models. We validated the attenuation correction of triple-frequency radar reflectivity using X- and W-band reflectivity and their dual-frequency reflectivity ratios (DFR). We also evaluate scattering models suitable for triple-frequency radar with varying snowfall microphysical properties. Then, Section 6 compares results with a micro-rain radar to further validate W-band radar attenuation correction. The final part of this section presents the W-band radar attenuation correction results. Section 7 presents the results of the attenuation correction comparison. Finally, Section 8 summarizes the main discussions of this study.

2. The Triple-Frequency Radar Field Measurement Campaign Site and Observational System and the Micro-Rain Radar

This study focuses on stratiform clouds with snowfall microphysical properties, and all the data used in this work were obtained from China's first triple-frequency radar and the ERA5 ECMWF reanalysis data. This triple-frequency radar experiment for winter precipitation was conducted by the Key Laboratory of Middle Atmosphere and Global Environment Observation at the Institute of Atmospheric Physics of the Chinese Academy of Sciences. The experimental site was located in Zhangshanying, Yanqing, Beijing (40.4898°N, 115.8596°E, 494 m above mean sea level) during two periods, from 11 November 2020, to 30 March 2021, and 11 November 2021, to 30 March 2022.

The triple-frequency cloud and precipitation radar system comprises X, Ka, and W band radars with multi-antenna coaxial scanning and Doppler signal observational capabilities. These radars were vertically pointed, providing a comprehensive view of hydrometeors at three frequencies during the experiment.

Before and during the field campaign, these three radars underwent meticulous metal sphere calibration. The reflectivity factor measurements of the Ka and W band radars were independently adjusted. The corrected values for the Ka-band radar reflectivity were 11.3 dB, and for the W-band radar reflectivity, it was 13.5 dB. Additionally, the range bin data of the W-band radar were shifted forward by two range bins. Figure 1 illustrates the schematic diagram of the triple-frequency cloud and precipitation radar system, showcasing its flexibility and mobility, including the main radar components, shelters, comprehensive data processing, and observation software.

The Ka-band (millimeter wave) and the X-band (centimeter wave) are dual-transmitting and dual-receiving all-solid-state frequency-modulated continuous-wave radars. In contrast, the W-band (millimeter wave) is a single-transmitting and dual-receiving combined pulsed Doppler radar. In the dual-transmitting and dual-receiving mode, two transmitters transmit horizontal and vertical polarization waves simultaneously, while the two receivers collect data returned by these two polarization waves simultaneously. Because a dual-polarization operation mode is in trial operation, spectrum accumulation cannot be performed in this mode, and the observation data of the dual-polarization mode is not yet available. The hardware employs dual-frequency coplanar antennas and integrated coaxial design techniques to synchronize devices with different frequencies for coordinated real-time observations with multiple frequency radars targeting the same hydrometeor.

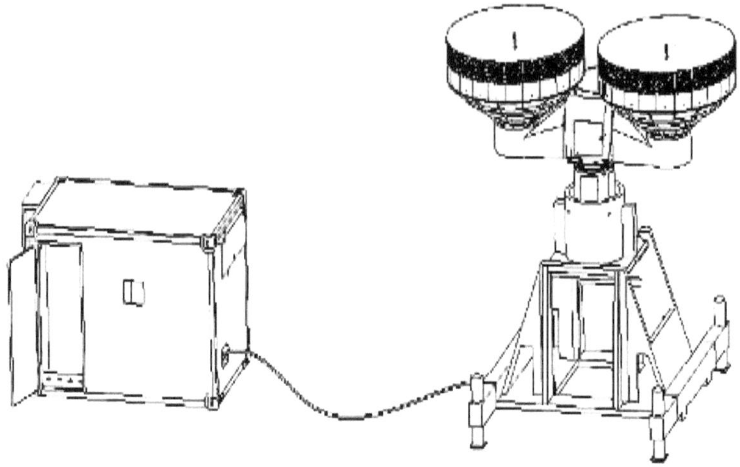

(a) (b)

Figure 1. The schematic diagram of the triple-frequency radar system of the Institute of Atmospheric Physics (square cabin system (**a**), Triple-frequency multiantenna coaxial scanning dual-polarization Doppler radar equipment (**b**)).

The technical specifications and settings of the three vertically pointing radars at the Zhangshanying site in Yanqing are detailed in Table 1. All three radars share a temporal resolution of 1.5 s. The range resolution of the W-band radar was set at 120 m to enhance the sensitivity of the W-band radar, while the X- and Ka-band radars were set at 30 m. Only zenith observation datasets are considered for the attenuation correction, although scanning data were collected during the campaign. It is important to note that we assume continuous and homogeneous stratiform clouds within observation volumes at the same layers are relatively homogeneous, and we interpolate the data for the W-band radar to match the triple-frequency radar's scattering properties in the attenuation correction process.

Table 1. Technical specifications and settings of the three vertically pointing radars operated at Zhangshanying of Yanqing site.

Specifications	X	Ka	W
Frequency/GHz	9.375	35.015	94.5
Pulse repetition period/μs	833–1667	833–1667	100–500
Detection capability/dBz	−21@5 km	−34@5 km	−22@5 km
Antenna gain/dB	43.6	54.57	56.4
3 dB beam width/°	1	0.33	0.24
Average transmit power/W	118	55.5	16.3
Noise figure/dB	2.9	3.8	5.1
Range resolution/m	10, 30	10, 30	30, 120
Pulse repetition frequency/Hz	2500	2500	5000
Wavelength/mm	32.017	8.566	3.171

The ground-based auxiliary equipment micro-rain radar (METEK MRR-2) was deployed 10 m from the triple-frequency radar. The micro-rain radar operates at a frequency of 24.23 GHz. It provides height detection capabilities within a range from 0 to 6 km. The radar system offers a height resolution of 300 m. The rainfall rates, droplet size distribution, radar reflectivity, falling velocity of precipitation particles, and the reflectivity factor attenuation are provided. The METEK MRR-2 micro-rain radar yields valuable data related to rainfall rates, droplet size distribution, radar reflectivity, falling velocity of precipitation particles, and the attenuation of the reflectivity factor.

3. Attenuation Correction Method Based on the SSRGA Scattering Model in PAMTRA2

3.1. The SSRGA Scattering Model in PAMTRA2

PAMTRA2 offers SSRGA for radar applications, which is crucial for researching frozen hydrometeors [21]. Tyynelä [22] proposed that SSRGA can accurately calculate the scattering properties of realistic snowflakes, aligning with other uncertainties inherent in radar measurements. Leinonen [23] further demonstrated a direct relationship between the backward scattering properties of snowflakes and the mass distribution within the snowflake, confirming the applicability of this approach. Hogan [18] introduced a simple analytical solution based on the self-similarity of snowflakes and validated the SSRGA approach's efficiency. SSRGA scales effectively to particles larger than the radar wavelength, offering faster calculations compared to Discrete Dipole Approximation (DDA). Additionally, SSRGA accommodates in-situ measurements of hydrometeor properties and provides more precise scattering calculations for snowflakes compared to the Mie solution or Rayleigh approximation, especially at higher frequencies.

In this study, we focus on the SSRGA scattering model in PAMTRA2 to compute attenuation along the triple-frequency radar path, accounting for hydrometeor vertical distribution and utilizing ECMWF reanalysis data for stratiform clouds with varying snowfall microphysical properties. Simultaneously, different hydrometeor properties are applied in various layers.

3.2. Calculation of Attenuation Coefficients of Different Cloud Layers for Triple-Frequency Radar Based on PAMTRA2

Ice water content (IWC) is a crucial cloud microphysical parameter for precipitation formation, defined as the cloud mass of ice per unit volume of atmospheric air. It is expressed as

$$\mathrm{IWC} = \int_0^\infty \mathrm{m}\,(\mathrm{D})\mathrm{N}\,(\mathrm{D})\mathrm{d}\,\mathrm{D}, \tag{1}$$

where m(D) represents the mass of ice crystal diameter D. Due to the irregular characteristics of frozen snowflakes and ice crystals, in nature, the definition of D has no standard form. Moreover, it is significant as it constitutes the bulk cloud properties. N(D) follows the exponential distribution given by

$$\mathrm{N(D)} = \mathrm{N_0}\exp(-\Lambda \mathrm{D}), \tag{2}$$

with the mass of various ice snowflakes represented by a power law

$$\mathrm{m} = \frac{\pi}{6}\,\rho_\mathrm{e}\,\mathrm{D}^3, \tag{3}$$

Hence, the equation for IWC can be simplified to

$$\mathrm{IWC} = \frac{\pi}{6}\int \rho_\mathrm{s}\,(\,\mathrm{D}\,)\,\mathrm{D}^3\,\mathrm{N}\,(\,\mathrm{D}\,)\,\mathrm{d}\,\mathrm{D}, \tag{4}$$

where the unit of IWC is g m^{-3} and D is in mm.

With the extinction cross section $\sigma_\mathrm{e}(\mathrm{D})$, the attenuation can be expressed as

$$\mathrm{K} = 4.34 \times 10^3 \int_0^\infty \mathrm{N}\,(\mathrm{D})\sigma_\mathrm{e}\,(\mathrm{D})\,\mathrm{d}\,\mathrm{D}, \tag{5}$$

where K is in dB km^{-1}. PAMTRA2 is used to calculate the attenuation coefficient.

The attenuation due to atmospheric propagation in different radar range bins is attributed to atmospheric gases and frozen particles during snowfall. This two-way extinction thickness is defined as

$$\tau(R) = \exp\left(-2\int_0^R K(R)\,dR\right), \tag{6}$$

where $K = K_a + K_c$, K_a represent the absorption and scattering coefficients (i.e., extinction or attenuation) due to atmospheric gas (primarily oxygen and water vapor) and clouds, respectively. K_c represents the absorption and scattering coefficient due to clouds. R (km) represents the distance or range from the radar. The attenuation correction equation can be expressed as

$$Z_r(R) = \frac{Z_m(R)}{\tau(R)}, \tag{7}$$

where Z_m ($mm^6\ m^{-3}$) is the measured radar effective reflectivity and Z_r represents the attenuated correction value of the radar effective reflectivity.

3.3. Attenuation Correction Method with Bin-by-Bin Approximation

The attenuation correction method with bin-by-bin approximation [24] is defined as

$$Z(i) = \frac{Z_m(i)}{\tau(i-1)} \exp\{K(R) \times \Delta R\}, \tag{8}$$

where $\tau(i) = \begin{cases} 1 & i = 0 \\ \exp\{-2\sum_{i=1}^{i} K(i) \times \Delta R\} & i > 1 \end{cases}$ is the extinction thickness at each range bin. Z(i) is determined from the measured value of the $Z_m(i)$ radar echo. This method calculates the attenuation correction for each range bin individually along the radial direction. It leverages the actual vertical stratification information of gases and hydrometeors, enhancing the stability and accuracy of the attenuation correction process.

4. Analysis of the Triple-Frequency Radar Attenuation

The attenuation experienced by each particular radar band used in triple-frequency radars varies due to atmospheric gases and frozen snowflakes during snowfall. Thus, it is crucial to estimate attenuation values for all three radars. Attenuation from water vapor and oxygen has been computed to correct radar-effective reflectivity measurements. This correction utilizes the Rosenkranz98 [25] and Liebe93 [26] models integrated into PAMTRA2 and atmospheric profiles obtained from the ECMWF reanalyzed data during the triple-frequency radar measurements. It is worth noting that the triple-frequency radars were meticulously calibrated using a metal sphere before and during the field campaign.

4.1. Attenuation Due to Atmospheric Gases and Hydrometeors of Triple-Frequency Radar

Hydrometeors and atmospheric gases contribute differently to attenuation in triple-frequency radars. In this study, we calculate the attenuation caused by atmospheric gases, mainly oxygen, and water vapor, using the gas absorption models Rosenkranz98 and Liebe93 via PAMTRA2. We adapted the oxygen model to the US standard atmospheric profiles. Given the lack of real-time vertical atmospheric profiles, we rely on ECMWF reanalysis data, which provide temperature, pressure, and relative humidity information.

Figure 2 presents the attenuation contributions of hydrometeors and gases at the triple-frequency radar for the ECMWF profile data of 21 January 2022 within the 0~10 km range bins. The results from the two algorithms, Rosenkranz98 (black) and Liebe93 (blue), demonstrate good agreement. The two-way attenuation values vary for the three radars, ranging from 0.001 to 0.01 dB km^{-1} for the X-band, 0.01 to 0.05 dB km^{-1} for the Ka-band, and 0.01 to 3 dB km^{-1} for the W-band. The maximum attenuation occurs at the W-band radar, consistent with the gas attenuation (2.675 dB km^{-1}) calculated by Neto [27]. Therefore, atmospheric gas and hydrometeor attenuation at the X- and Ka-band radars are negligible compared to their observed values, while attenuation at the W-band radar is non-negligible and requires correction.

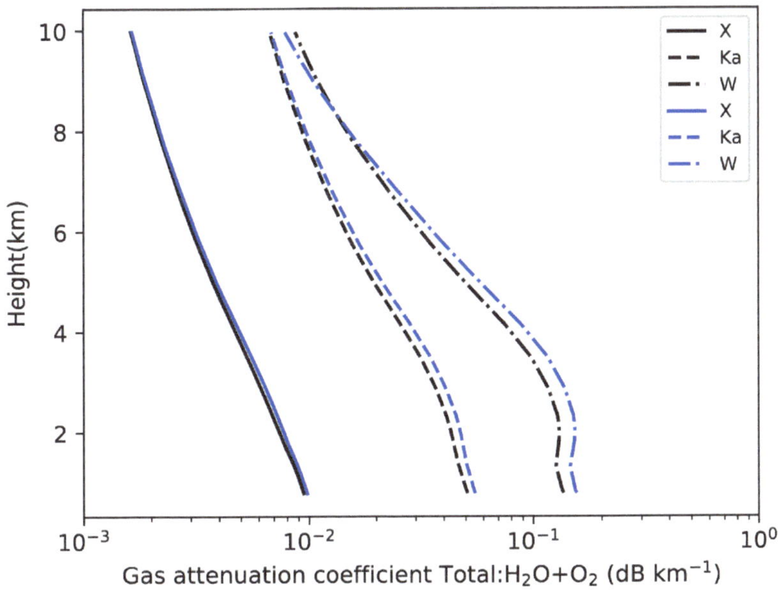

Figure 2. The gas and hydrometeor attenuation coefficient at the triple-frequency radar at 06:00 UTC on 21 January 2022. (Black represents the Rosenkranz98 method, blue represents the Liebe93 method).

4.2. The Attenuation Coefficient for Snowflake Particles of Triple-Frequency Radar

To achieve more accurate attenuation correction, we simulate and compare different scattering models before correction, including Rayleigh (Ray), Mie, T-Matrix (TMM), and SSRGA, offering more realistic scattering properties compared to sphere/spheroid models. Our simulations assume a temperature of -10 °C and a particle diameter range from 0.01 to 10 mm. Figure 3 illustrates the two-way attenuation coefficient signatures for the triple-frequency radars. When the particle size is less than 0.2 mm, the two-way attenuation for the triple-frequency radar can be neglected. However, as particle size increases, different scattering models yield significantly varying results. For the Rayleigh and SSRGA scattering models (Figure 3a,d), the maximum attenuation value at the W-band radar reaches up to 0.05 and 1.5 dB, while the attenuation values at the X- and Ka-band radars remain minimal and can be considered negligible. The Mie and TMM scattering models (Figure 3b,c) produce nearly identical attenuation values at the X-band radar and the Ka and W-band radars, which is inconsistent with actual measurements. Therefore, due to the nonspherical and complex nature of ice crystals and snowflake-like particles, the SSRGA scattering model, which closely approximates the realistic shape of snowflakes and offers high computational efficiency, is selected for this study. It implies that attenuation caused by snowfall with varying microphysical properties increases with frequency, constraining the maximum range of measurement for the W-band radar.

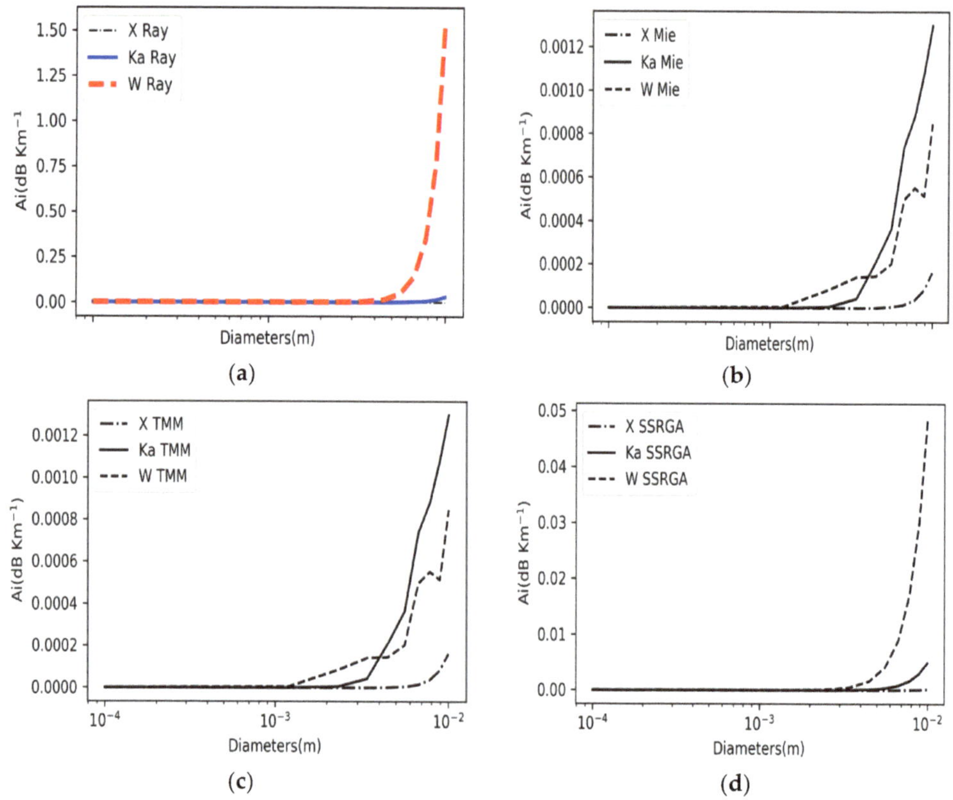

Figure 3. The attenuation coefficient for the triple-frequency radar at different snowflake diameters at a temperature −10 °C: (**a**) Rayleigh scattering, (**b**) Mie scattering, (**c**) T matrix scattering, (**d**) SSRGA scattering.

5. Comparison of the Attenuation Correction of Gas and Stratiform Clouds for Triple-Frequency Radar Effective Reflectivity Profiles

In the following sections, we utilize the different scattering models to perform the gas and hydrometeors attenuation corrections for the triple-frequency radar. The ECMWF reanalysis data are used in this study for the atmospheric vertical profile information. For calculating the attenuation, we assume that the snowflake particle size ranges from 0.001 to 10 mm and that the ice density is 0.9167 g cm^{-3}.

5.1. Case Study of the Triple-Frequency Radar Profile

The two-way gas and stratiform cloud attenuation correction at the X-band radar is performed with the Rayleigh, Mie, TMM, and SSRGA scattering models in PAMTRA2. The attenuation correction with different scattering models and the non-correction reflectivity profile at the X-band radar are shown in Figure 4 at 8:35 UTC on 21 January 2022. The attenuation correction and the non-correction are nearly coincidental. The signature indicates that the attenuation due to the gas and stratiform clouds with snowfall microphysical properties is relatively tiny compared to their X-band radar measurements. Therefore, the attenuation owing to gas and stratiform clouds with snowfall microphysical properties is negligible for the X-band radar.

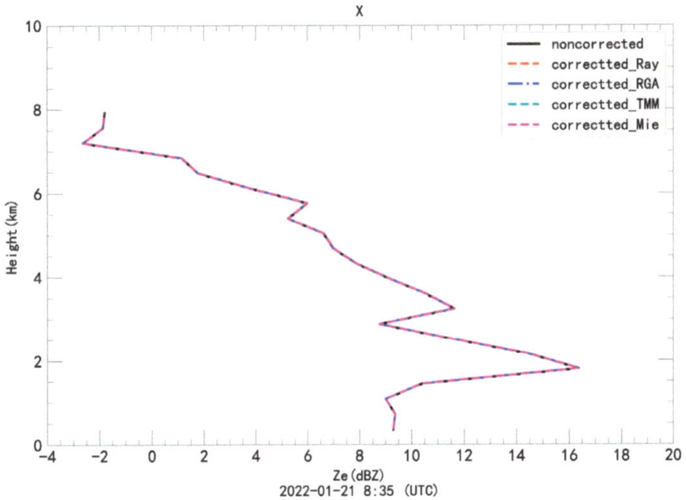

Figure 4. Comparison of gas and stratiform clouds with snowfall microphysical properties attenuation correction using Rayleigh, Mie, T matrix and SSRGA models for the effective reflectivity profile of the X-band radar at 8:35 UTC of 21 January 2022.

In the same way, we performed attenuation correction of gas and stratiform clouds with snowfall microphysical properties at the Ka-band radar. The attenuation correction and non-corrected effective reflectivity results are displayed in Figure 5. Only the Mie scattering model slightly increases after attenuation correction for the regions with heights above 6 km. The maximum value reaches 1.45 dB, while the attenuation correction values using the other models are minimal. Consequently, the attenuation corrections for the Ka-band radar are not needed, which are produced by gas and stratiform clouds with snowfall microphysical properties.

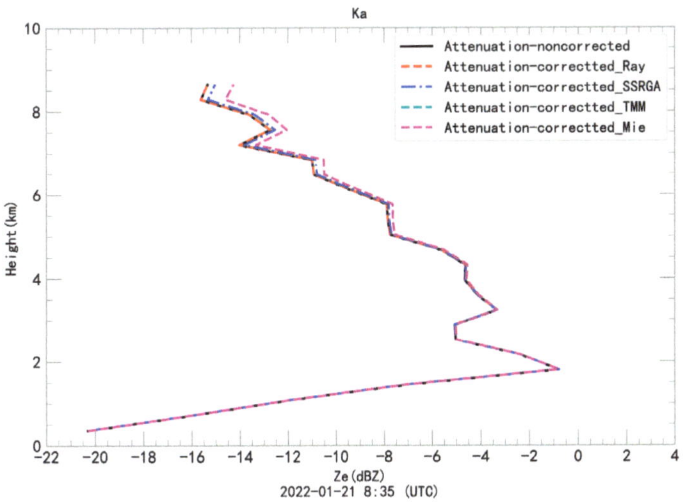

Figure 5. *Cont.*

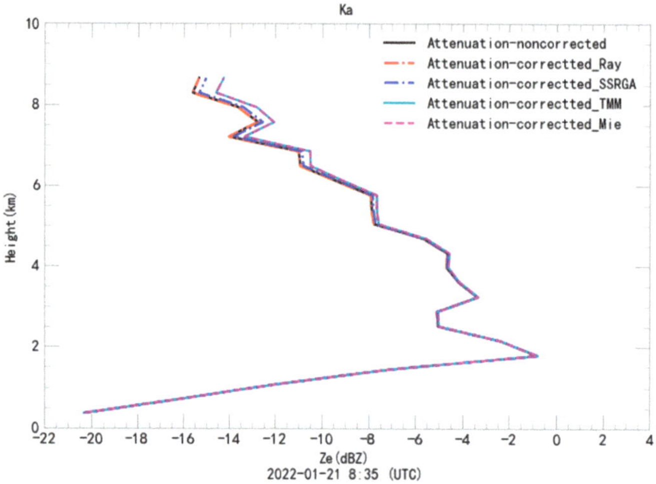

Figure 5. Comparison of gas and stratiform clouds with snowfall microphysical properties attenuation correction using Rayleigh, Mie, T matrix and SSRGA models for the effective reflectivity profile of the Ka-band radar at 8:35 UTC of 21 January 2022.

However, due to gas and stratiform clouds with snowfall microphysical properties, the attenuation correction for the W-band radar indicated in Figure 6 with the different scattering models is significant at 8:35 UTC on 21 January 2022. The Mie and TMM scattering models coincided. Nevertheless, these models and the Rayleigh scattering model are far from the range of the measured reflectivity for regions above 3 km, so these scattering models are considered unsuitable. Only the gas and stratiform clouds with snowfall microphysical properties attenuation correction of the SSRGA scattering model are more appropriate than other models. The maximum attenuation correction value reaches 4.38 dB for regions above 6 km, and the values are small for regions below 6 km. Therefore, the SSRGA model was adopted to make the attenuation correction for the W-band radar in this study.

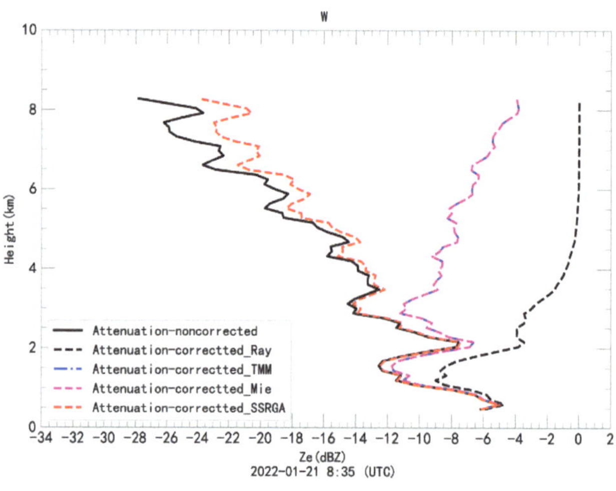

Figure 6. Comparison of gas and stratiform clouds with snowfall microphysical properties attenuation correction using Rayleigh, Mie, T matrix and SSRGA model for the effective reflectivity profile of the W-band radar at 8:35 UTC of 21 January 2022.

5.2. Case Study of the Attenuation Correction for the W Band Radar Effective Reflectivity

In these case studies, all observational data were obtained from the triple-frequency radar at Zhangshanying, the Yanqing site in Beijing. For the snowfall period from 8:04 to 9:25 UTC on 21 January 2022 and from 19:54 to 5:43 UTC on 17–18 March 2022, the attenuation due to gas was estimated and utilized to correct the radar reflectivity measurements via the Rosenkranz98 model and the atmospheric profiles collected from the ECMWF reanalysis data. The SSRGA scattering model is applied to calculate the W-band radar attenuation caused by stratiform clouds with snowfall microphysical properties in PAMTRA2. The reflectivities of the triple-frequency radar, the W-band radar reflectivity after the attenuation correction, and their dual-frequency reflectivity ratios are presented in Figures 7 and 8. It must be noted that these correction values were minimal for the regions with heights below 6 km; as expected, they increased obviously for the regions with heights above 6 km.

The comparison of the results before and after the W-band effective reflectivity attenuation correction and the effective reflectivities of Ka-band and X-band radar from 16:04 to 17:25 UTC on 21 January 2022 is illustrated in Figure 7. The −15 °C isotherm (dashed line in the time–height plots) is 4.4 km, and the −10 °C isotherm (continuous line in the time–height plots) is approximately 1.0 km. The attenuation correction of the W band radar effective reflectivity for the regions with heights above 6 km is increased to some degree in Figure 7b compared with Figure 7a. The maximum correction value reaches 4.42 dB, and the minimum correction value is around zero during the observational period. In the regions above 6 km, the effective reflectivity improved to some extent. Attenuation is found to be negligible on average for the regions with heights below 6 km.

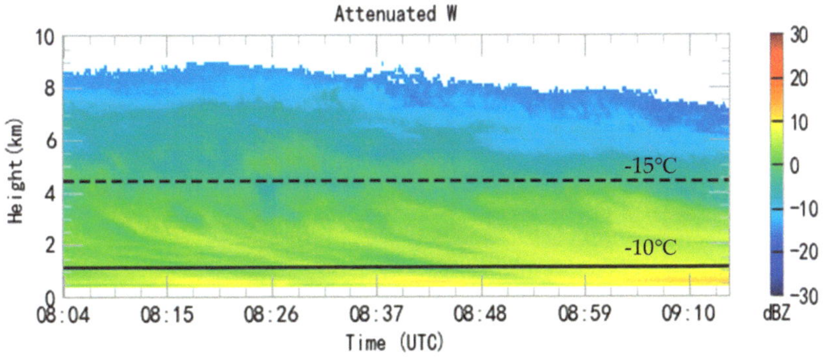

(a)

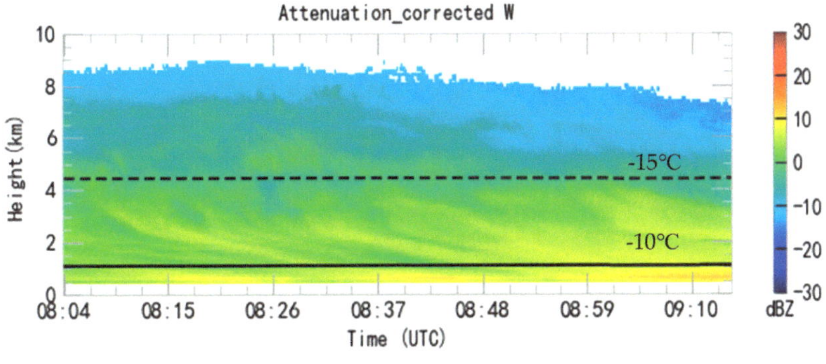

(b)

Figure 7. *Cont.*

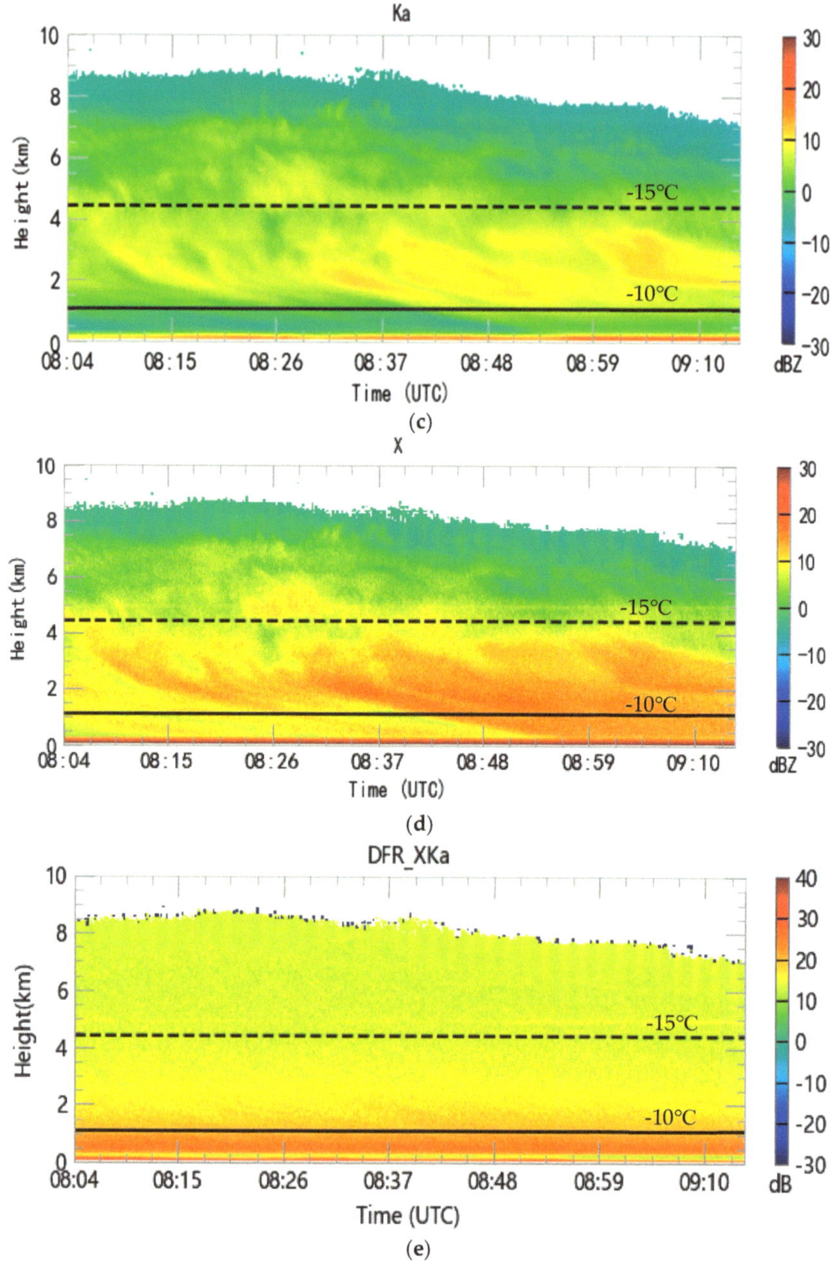

Figure 7. *Cont.*

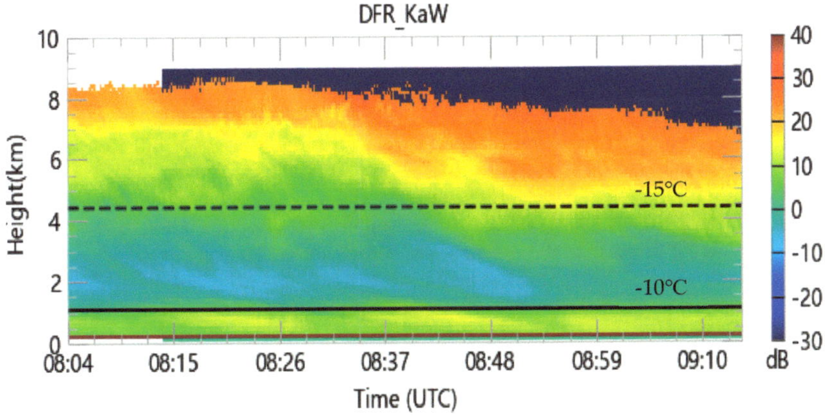

(f)

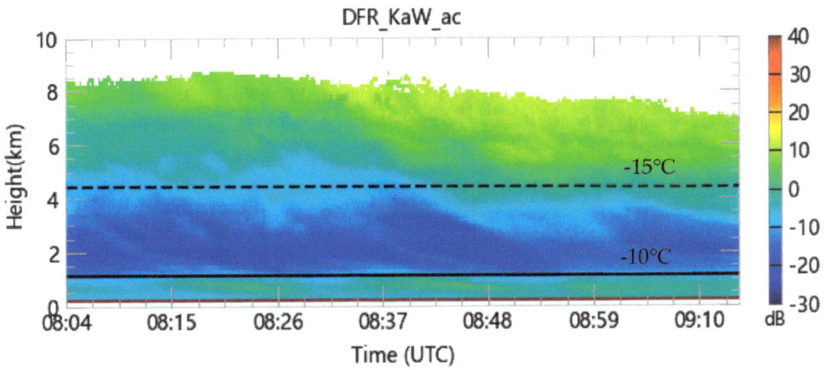

(g)

Figure 7. Comparison of attenuations due to gas and stratiform clouds with snowfall microphysical properties before (**a**) and after (**b**) the correction of effective reflectivity for W-band radar; the effective reflectivity of Ka-band (**c**), and X-band radar (**d**); the dual-frequency ratio of X-Ka (**e**), Ka-W (**f**), and Ka-W (the W-band radar reflectivity is after attenuation correction) (**g**) band radar of 21 January 2022 between 8:04 and 9:25 (UTC). The continuous line and dashed line are the −10 °C and −15 °C isotherms (provided by the ECMWF reanalysis data), respectively.

Figure 7c,d show the features of the effective reflectivities of Ka-band and X-band radar during the snowfall period from 8:04 to 9:25 UTC on 21 January 2022 and from 19:54 to 5:43 UTC on 17–18 March 2022. The Ka-band and X-band radar undergo minimal attenuations relative to the W-band radar. We found that the attenuations of the W-band radar increase with heights in the regions above 5 km. The signatures of the effective reflectivity of the triple-frequency radar indicate that it has the effect of cross-calibration to use triple-frequency radar measurements and help correct for the attenuation of the W-band radar measurements. The features also demonstrate that most attenuation results from the upper-lever clouds, far from the radar.

Figure 7e–g show the dual-frequency reflectivity ratios of X-Ka band radar, Ka-W band radar, and Ka-W_ac (the W-band radar reflectivity is after attenuation correction). In Figure 7e, the dual-frequency reflectivity ratios of X-Ka are divided into two regions with heights below and above 2 km, but they are both close to a constant value. The features indicate the consistency of X- and Ka-band radar reflectivity measurements during snowfall. In Figure 7f, the dual-frequency reflectivity ratios of Ka-W are significant for the regions with heights above 5 km, though they are at the same altitude. However, the W-band radar reflectivity is after attenuation correction, and the dual-frequency reflectivity ratios of Ka-W_ac remain consistent for the regions with heights above 5 km. It is confirmed that the attenuation correction for the W-band radar used in this study is rational.

The other case is illustrated in Figure 8 with before and after the W-band effective reflectivity attenuation correction, the effective reflectivity of the Ka-band, and the X-band radar of 17 March 2022 between 19:54 and 18 March 2022 5:25 UTC, and it is used to validate the effect of attenuation correction. The −15 °C isotherm (dashed line in the time–height plots) is around 4.4 km, and the −5 °C isotherm (continuous line in the time–height plots) is about 0.7 km due to the high ground temperature. As expected, the comparison between Figure 8a,b show that the attenuation corrections of the W-band radar effective reflectivity for regions with heights above 5.5 km were also observed to be distinctly improved. The maximum correction value is up to 4.71 dB, and the minimum is zero during the observational period. For regions with heights above 5.5 km, the effective reflectivity is optimized to a certain degree. Based on this feature, the attenuation from gas and stratiform clouds with snowfall microphysical properties can be negligible on average below 5.5 km. In this case, the maximum attenuation value exceeds the above case. The reason for this difference is that the snowfall intensity in the second case is more significant than in the first. The snow particle sizes are different in the two cases, as well. In sum, the attenuation systematically increases along the radar path. These signatures in Figures 7 and 8 indicate that using the SSRGA scattering model in PAMTRA2 is practical and feasible to calculate the attenuation of the W-band radar simulation.

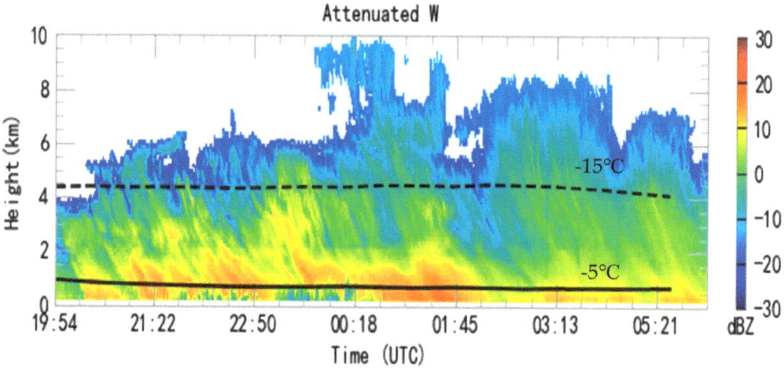

(a)

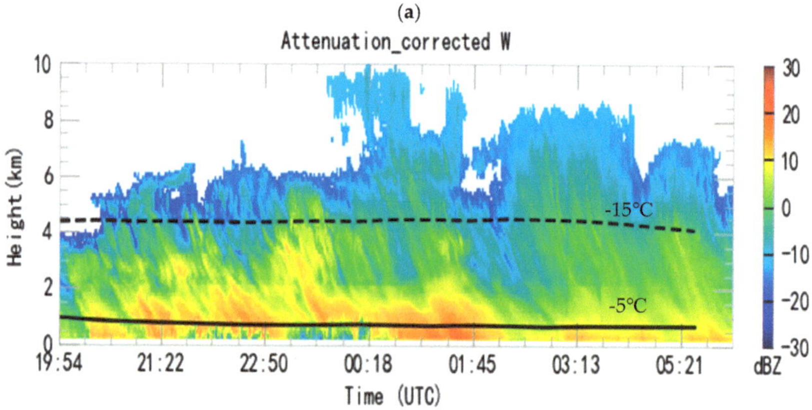

(b)

Figure 8. *Cont.*

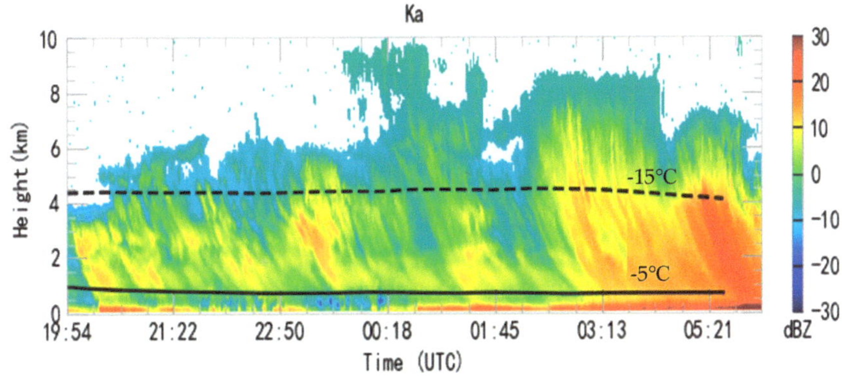

(c)

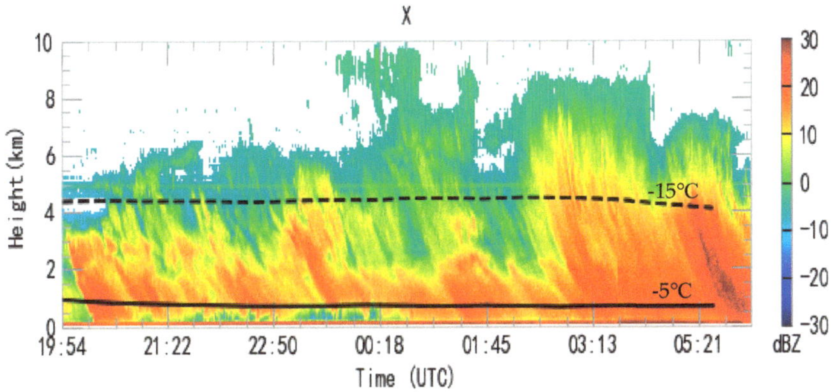

(d)

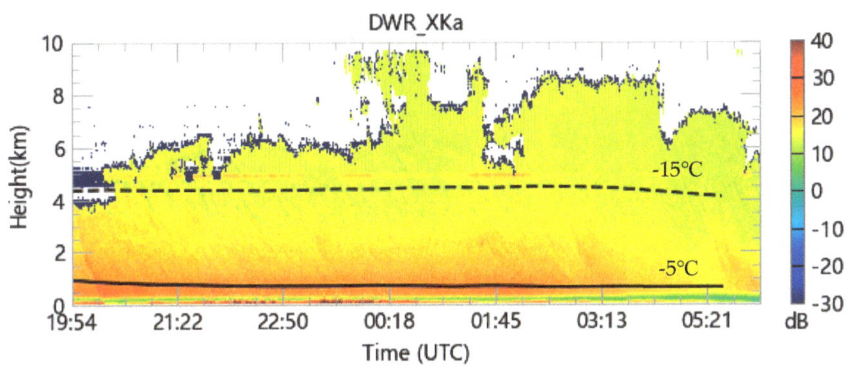

(e)

Figure 8. *Cont*.

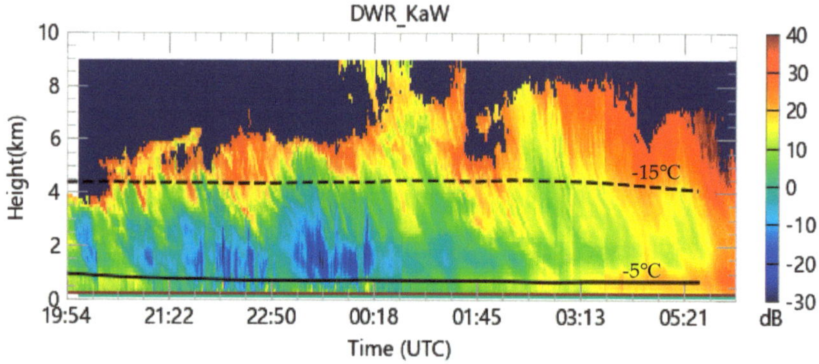

(f)

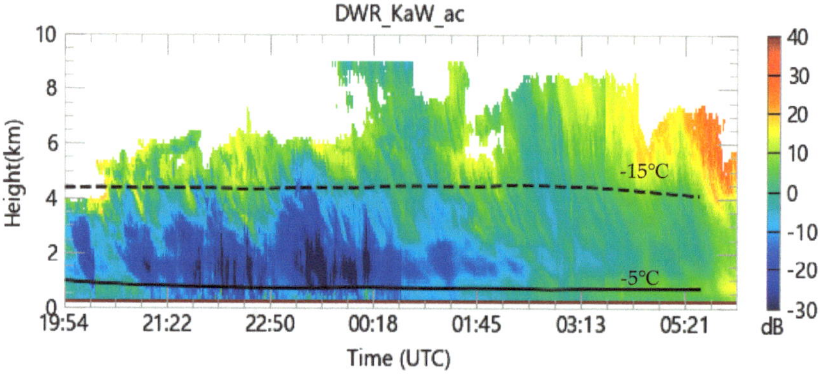

(g)

Figure 8. Comparison of attenuations due to gas and stratiform clouds with snowfall microphysical properties before (**a**) and after (**b**) the correction of effective reflectivity for W-band radar, the effective reflectivity of Ka-band (**c**) and X-band radar (**d**); the dual-frequency ratio of X-Ka (**e**), Ka-W (**f**) and Ka-W (the W-band radar reflectivity is after attenuation correction) (**g**) band radar of 17 March 2022 between 19:54 and 18 March 2022 5:25 (UTC). The continuous line and dashed line are the −5 °C and −15 °C isotherms (provided by the ECMWF reanalysis data), respectively.

Figure 8c,d show the features of the effective reflectivity of Ka-band and X-band radar during the snowfall period from 9:54 to 5:43 UTC of 17~18 March 2022. Same as in the first case, the Ka-band and X-band radar have tiny attenuations compared to the W-band radar. The attenuations of the W-band radar increase with heights in the regions above 5 Km, especially in the regions above 8 Km. The characteristics of the effective reflectivity of the triple-frequency radar suggest that it has the effect of cross-calibration to use triple-frequency radar measurements. We can better analyze the features of the triple-frequency radar. There are bigger attenuations far from the radar and from the upper-lever clouds.

Figure 8e–g show the dual-frequency reflectivity ratios of X-Ka band radar, Ka-W band radar, and Ka-W_ac (the W-band radar reflectivity is after attenuation correction). In Figure 8e, the dual-frequency reflectivity ratios of X-Ka are divided into two regions with heights below and above 2 km. However, each has a proximate constant value except for the regions below 500 m. The features suggest that the X- and Ka-band radar reflectivity measurements are consistent during snowfall and attenuate less. In Figure 8f, the dual-frequency reflectivity ratios of Ka-W are significant differences for the regions with heights above 5 km, while the differences are relatively small. However, the W-band radar reflectivity is after attenuation correction, and the dual-frequency reflectivity ratios of Ka-

W_ac have a significant improvement for the regions with heights above 5 km. Therefore, the method of attenuation correction for the W-band radar works here. In addition, the clouds are inhomogeneous in this case, resulting in biases after the attenuation correction of the W-band radar.

6. Comparison Analysis with Micro-Rain Radar Attenuation

To test the effectiveness of the W-band radar's attenuation correction, we compare it with the attenuation observed by the ground-based micro-rain radar of 20 January 2022. It is important to note that the micro-rain radar operates at a wavelength of 12.5 mm, distinct from the W-band radar's wavelength of 3.171 mm. Although the attenuations caused by atmospheric gases and stratiform clouds with snowfall microphysical properties differ for the two radars, we can assess the strength of their attenuations vertically.

Figure 9 displays the attenuated radar reflectivity of the micro-rain radar. It is worth mentioning that the maximum measurable distance of the micro-rain radar is only 6.0 km. As seen in the figure, the attenuation of the MRR-2 during the snowfall period is negligible below 5 km, which aligns with the results of the attenuation correction for the W-band radar using PAMTRA2. However, around 9:00, there is a notable attenuation at a height of approximately 2 km. It likely indicates that attenuation increases with snowfall intensity during this period.

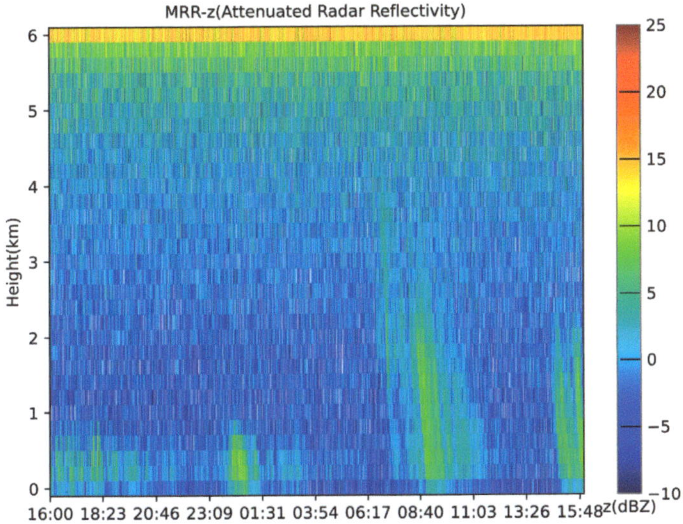

Figure 9. The attenuated radar reflectivity at the MRR-2 of 20 January 2022.

7. Discussion

In this study, we proposed a two-way gas and cloud attenuation correction for the W-band radar method based on the SSRGA scattering model in PAMTRA2. It is challenging to do it due to the need for more information about hydrometeor vertical distribution. This approach utilizes the SSRGA scattering model, which closely approximates the realistic snowflake structure and atmospheric vertical profiles from ECMWF reanalysis data. Compared to traditional methods like DDA, the method offers high computational efficiency for attenuation correction in triple-frequency radar. It is worth noting that the information about hydrometeor vertical distribution varies with the spatial and temporal conditions during the snowfall. Thus, it can introduce errors in the attenuation correction. In order to improve the accuracy of the attenuation correction, we should make more vertical observations of meteorological parameters in future experiments.

In future work, we plan to systematically apply this methodology to expand the dataset for triple-frequency radar observations, helping us better understand snow attenuation at W-band and improve data quality for identifying triple-frequency radar characteristics during snowfall. Additionally, we intend to explore the application of this attenuation correction strategy in scanning modes based on profile information at various levels. Optimizing the observation scheme for the triple-frequency radar and collecting more data will allow us to gain insights into the relationship between attenuation in triple-frequency radar and clouds with snowfall microphysical properties. When dual-polarization data become available, we will explore polarization parameter attenuation correction to enhance attenuation correction effectiveness further when combined with this method.

8. Conclusions

By analyzing the results from the triple-frequency radar reflectivity case studies, we found that the two-way attenuation correction of the W-band radar increased by up to 4.71 dB for regions with heights above 6 km. At the same time, it remained minimal for regions with heights below 6 km. The attenuation values, attributed to gases and stratiform clouds with snowfall microphysical properties, resulted in non-negligible attenuation far from the W-band radar for regions with heights above 6 km. Based on the results shown in Figures 7–9, we validated this attenuation by comparing it to the attenuation observed by the ground-based micro-rain radar, the cross-calibrating among the triple-frequency radar, and their dual-frequency reflectivity ratios. It proved that the attenuation correction for the W-band radar is effective. We can obtain better attenuation correction when more information about vertical hydrometeor distribution is available in the future.

Unlike previous attenuation studies that relied only on path-integrated attenuation observations during snowstorms [28,29], this technique provides bin-by-bin vertical distribution information to calculate attenuation due to gases and stratiform clouds with varying snowfall microphysical properties. Assessing the impact of neglecting attenuation effects from different stratiform layers is crucial for triple-frequency radar. The quality of the W-band radar data was improved to help further analyze the characteristics of triple-frequency radar during snowfall.

Author Contributions: Conceptualization, Y.C. and H.C.; methodology, X.H.; formal analysis, Y.B.; investigation, S.D.; data curation, J.L.; writing—original draft preparation, Y.C.; writing—review and editing, H.C. and P.W. All authors have read and agreed to the published version of the manuscript.

Funding: This research was funded by the National Key Research and Development Program of China, "Fine-scale observations and data inversion of land-air interactions in Northern China" (Grant 2022YFF0801301) and the National Key Research and Development Program of China, "Research and Experiment of Vertical Comprehensive Meteorological Observation Technology in Megacities" (Grant 2017YFC1501700).

Data Availability Statement: The data that support the findings of this study are available from the corresponding author upon reasonable request.

Conflicts of Interest: The authors declare no conflict of interest.

References

1. Oue, M.; Kollias, P.; Matrosov, S.Y.; Battaglia, A.; Ryzhkov, A.V. Analysis of the microphysical properties of snowfall using scanning polarimetric and vertically pointing mul-ti-frequency Doppler radars. *Atmos. Meas. Tech.* **2021**, *14*, 4893–4913. [CrossRef]
2. Falconi, M.T.; Von Lerber, A.; Ori, D.; Marzano, F.S.; Moisseev, D. Snowfall retrieval at X, Ka and W bands: Consistency of backscattering and micro-physical properties using BAECC ground-based measurements. *Atmos. Meas. Tech.* **2018**, *11*, 3059–3079. [CrossRef]
3. Leinonen, J.; Lebsock, M.D.; Tanelli, S.; Sy, O.O.; Dolan, B.; Chase, R.J.; Finlon, J.A.; von Lerber, A.; Moisseev, D. Retrieval of snowflake microphysical properties from multifrequency radar observations. *Atmos. Meas. Tech.* **2018**, *11*, 5471–5488. [CrossRef]
4. Tridon, F.; Battaglia, A.; Kneifel, S. Estimating total attenuation using Rayleigh targets at cloud top: Applications in multi-layer and mixed-phase clouds observed by ground-based multifrequency radars. *Atmos. Meas. Tech.* **2020**, *13*, 5065–5085. [CrossRef]
5. Mason, S.L.; Chiu, C.J.; Hogan, R.J.; Moisseev, D.; Kneifel, S. Retrievals of Riming and Snow Density from Vertically Pointing Doppler Radars. *J. Geophys. Res. Atmos.* **2018**, 123. [CrossRef]

6. Gorgucci, E.; Baldini, L.; Adirosi, E.; Montopoli, M. Ice water content assessment in the single-, dual-, and triple-frequency radar scenarios. *Remote. Sens. Environ.* **2020**, *254*, 112242. [CrossRef]
7. Tridon, F.; Battaglia, A.; Luke, E.; Kollias, P. Rain retrieval from dual-frequency radar Doppler spectra: Validation and potential for a midlatitude precipi-tating case-study. *Q. J. Roy. Meteor. Soc.* **2017**, *143*, 364–1380. [CrossRef]
8. Battaglia, A.; Kollias, P.; Dhillon, R.; Roy, R.; Tanelli, S.; Lamer, K.; Grecu, M.; Lebsock, M.; Watters, D.; Mroz, K.; et al. Spaceborne Cloud and Precipitation Radars: Status, Challenges, and Ways Forward. *Rev. Geophys.* **2020**, *58*, e2019RG000686. [CrossRef]
9. Mróz, K.; Battaglia, A.; Kneifel, S.; D'Adderio, L.P.; Neto, J.D. Triple-Frequency Doppler Retrieval of Characteristic Raindrop Size. *Earth Space Sci.* **2020**, *7*, e2019EA000789. [CrossRef]
10. Hogan, R.J.; Illingworth, A.J. The Potential of Spaceborne Dual-Wavelength Radar to Make Global Measurements of Cirrus Clouds. *J. Atmos. Ocean. Technol.* **1999**, *16*, 518–531. [CrossRef]
11. Liao, L.; Meneghini, R. A Study on the Feasibility of Dual-Wavelength Radar for Identification of Hydrometeor Phases. *J. Appl. Meteorol. Clim.* **2011**, *50*, 449–456. [CrossRef]
12. Matrosov, S.Y. Feasibility of using radar differential Doppler velocity and dual-frequency ratio for sizing particles in thick ice clouds. *J. Geophys. Res. Atmos.* **2011**, *116*, 1–13. [CrossRef]
13. Kneifel, S.; Kollias, P.; Battaglia, A.; Leinonen, J.; Maahn, M.; Kalesse, H.; Tridon, F. First observations of triple-frequency radar Doppler spectra in snowfall: Interpretation and ap-plications. *Geophy. Res. Lett.* **2016**, *43*, 2225–2233. [CrossRef]
14. Grecu, M.; Tian, L.; Heymsfield, G.M.; Tokay, A.; Olson, W.S.; Heymsfield, A.J.; Bansemer, A. Nonparametric methodology to estimate precipitating ice from multiple-frequency radar reflectivity observations. *J. Appl. Meteorol. Clim.* **2018**, *57*, 2605–2622. [CrossRef]
15. Mason, S.L.; Hogan, R.J.; Westbrook, C.D.; Kneifel, S.; Moisseev, D.; von Terzi, L. The importance of particle size distribution and internal structure for triple-frequency radar re-trievals of the morphology of snow. *Atmos. Meas. Tech.* **2019**, *12*, 4993–5018. [CrossRef]
16. Barrett, A.I.; Westbrook, C.D.; Nicol, J.C.; Stein, T.H.M. Rapid ice aggregation process revealed through triple-wavelength Doppler spectrum radar analysis. *Atmos. Meas. Tech.* **2019**, *19*, 5753–5769. [CrossRef]
17. Tridon, F.; Battaglia, A.; Chase, R.J.; Turk, F.J.; Leinonen, J.; Kneifel, S.; Mroz, K.; Finlon, J.; Bansemer, A.; Tanelli, S.; et al. The Microphysics of Stratiform Precipitation During OLYMPEX: Compatibility Between Tri-ple-Frequency Radar and Airborne In Situ Observations. *J. Geophys. Res. Atmos.* **2019**, *124*, 8764–8792. [CrossRef]
18. Hogan, R.J.; Westbrook, C.D. Equation for the Microwave Backscatter Cross Section of Aggregate Snowflakes Using the Self-Similar Rayleigh–Gans Approximation. *J. Atmos. Sci.* **2014**, *71*, 3292–3301. [CrossRef]
19. Hogan, R.J.; Honeyager, R.; Tyynelä, J.; Kneifel, S. Calculating the millimetre-wave scattering phase function of snowflakes using the self-similar Rayleigh–Gans Approximation. *Q. J. R. Meteorol. Soc.* **2017**, *143*, 834–844. [CrossRef]
20. L'Ecuyer, T.S.; Stephens, G.L. An Estimation-Based Precipitation Retrieval Algorithm for Attenuating Radars. *J. Appl. Meteorol.* **2002**, *41*, 272–285. [CrossRef]
21. Mech, M.; Maahn, M.; Kneifel, S.; Ori, D.; Orlandi, E.; Kollias, P.; Schemann, V.; Crewell, S. PAMTRA 1.0: The Passive and Active Microwave radiative TRAnsfer tool for simulating radiometer and radar measurements of the cloudy atmosphere. *Geosci. Model. Dev.* **2020**, *13*, 4229–4251. [CrossRef]
22. Tyynelä, J.; Leinonen, J.; Westbrook, C.D.; Moisseev, D.; Nousiainen, T. Applicability of the Rayleigh-Gans approximation for scattering by snowflakes at microwave frequencies in vertical incidence. *J. Geophys. Res. Atmos.* **2013**, *118*, 1826–1839. [CrossRef]
23. Leinonen, J.; Moisseev, D.; Nousiainen, T. Linking snowflake microstructure to multi-frequency radar observations. *J. Geophys. Res. Atmos.* **2013**, *118*, 3259–3270. [CrossRef]
24. Zhang, P.C.; Wang, Z.H. A Study on Algorithm to Make Attenuation Correction to Radar Observations of Radar Reflectivity Factor (I): Theoretical Analysis. *Plateau Meteorol.* **2001**, *20*, 1–5.
25. Rosenkranz, P.W. Water vapor microwave continuum absorption: A comparison of measurements and models. *Radio Sci.* **1998**, *33*, 919–928, reprinted in *Radio Sci.* **1999**, *34*, 1025. [CrossRef]
26. Liebe, H.J.; Hufford, G.A.; Cotton, M.G. Propagation Modeling of Moist Air and Suspended Water/Ice Particles below 1000GHz. In Proceedings of the Special Meeting of the Electromagnetic Wave Propagation Panel Symposium, Palma de Mal/orca, Spain, 17–20 May 1993.
27. Neto, J.D.; Kneifel, S.; Ori, D.; Trömel, S.; Handwerker, J.; Bohn, B.; Hermes, N.; Mühlbauer, K.; Lenefer, M.; Simmer, C. The TRIple-frequency and Polarimetric radar Experiment for improving process observations of winter precipitation. *Earth Syst. Sci. Data* **2019**, *11*, 845–863. [CrossRef]
28. Liao, L.; Meneghini, R. Physical Evaluation of GPM DPR Single- and Dual-Wavelength Algorithms. *J. Atmos. Ocean. Technol.* **2019**, *36*, 883–902. [CrossRef]
29. Kalogeras, P.; Battaglia, A. Improving Millimeter Radar Attenuation Corrections in High-Latitude Mixed-Phase Clouds via Radio Soundings and a Suite of Active and Passive Instruments. *IEEE Trans. Geosci. Remote. Sens.* **2022**, *60*, 4303611. [CrossRef]

remote sensing

MDPI

Communication

Statistical Analysis of Mesovortices during the First Rainy Season in Guangdong, South China

Ying Tang [1,2], Xin Xu [2,3,*], Yuanyuan Ju [2,3], Zhenyu Wu [1], Shushi Zhang [4], Xunlai Chen [5] and Qi Xu [6]

1 Nanjing Marine Radar Institute, Nanjing 210023, China
2 Key Laboratory of Mesoscale Severe Weather, Ministry of Education, School of Atmospheric Sciences, Nanjing University, Nanjing 210023, China
3 Key Laboratory of Radar Meteorology, China Meteorology Administration, Nanjing 210023, China
4 Key Laboratory of Transportation Meteorology of China Meteorological Administration, Nanjing Joint Institute for Atmospheric Sciences, Nanjing 210023, China
5 Shenzhen Meteorology Bureau, Shenzhen 518040, China
6 Jiangsu Air Traffic Management Branch Bureau of Civil Aviation Administration of China, Nanjing 211113, China
* Correspondence: xinxu@nju.edu.cn

Abstract: Based on Doppler radar observation and reanalysis data, the statistical characteristics of mesovortices (MVs) during the first rainy season (April–June) in Guangdong, South China, from 2017 to 2019 are studied, including their spatiotemporal distributions, structural features and favorable environmental conditions. The results show that the MVs usually exhibit short lifetimes; about 70% last for less than 30 min. The intensity and horizontal scale of the MVs are proportional to their lifetime. Long-lived MVs have larger horizontal scales and stronger intensities than short-lived ones. The MVs are mainly observed over the Pearl River Delta region, followed by western Guangdong Province, but relatively fewer in both eastern and northern Guangdong Province. The uneven spatial distribution of the MVs is closely related to the differences in the topographical features and the environment conditions over South China. MVs are prone to form over flat regions. The Pearl River Delta and eastern Guangdong regions are relatively flat compared to the more mountainous western and northern Guangdong regions. Moreover, the monsoonal south-westerlies, water vapor flux, atmospheric instability and vertical wind shear over southwest Guangdong are significantly larger than those in other regions and are favorable for the formation of MVs. The occurrence frequencies of MVs in central and southern parts of Guangdong display similar diurnal variations, reaching the peak during the late afternoon and early evening while dropping to the minimum overnight. However, the situation is opposite in northern Guangdong, with the peak overnight and the minimum during the late afternoon and early evening. The regional differences in diurnal variation are likely related to the moving direction of mesoscale convective systems (MCSs) in Guangdong.

Keywords: mesovortices; Doppler radar; first rainy season; Pearl River Delta region

1. Introduction

South China is an area with high frequency of severe convective weather where disastrous weather events such as short-term heavy rainfall, tornadoes and gales frequently occur [1,2]. Severe convective weather often causes serious casualties and property losses. For example, on 13 April 2016, a strong squall line caused gusts in most of Guangdong Province, with the wind speed in most areas reaching 13.9 m s^{-1} (even up to 41.5–46.1 m s^{-1} in some areas). Meanwhile, it was also accompanied by short-term heavy rainfall, causing great economic losses [3]. On 4 March 2018, a rarely seen strong squall line occurred over the northern and central parts of Jiangxi Province. During this process, high winds exceeding 17.2 m s^{-1} were found at 521 observational stations, and the wind speeds at 172 stations were above 24.5 m s^{-1}. This severe squall line affected more than 267,000 people and caused great damages to a large number of houses and crops, resulting in a direct economic loss of CNY 410 million [4].

The genesis of severe convective weather and parent mesoscale convective systems (MCSs) in South China has been studied extensively in the past. For example, Du et al. [5] investigated the effects of topography, land–sea contrast and cold pool on convection initiation in the coastal area of South China through a series of sensitivity experiments. The moving direction of cold pool outflow changed due to terrain blocking, thereby affecting the development and movement of the MCS. Based on Doppler radar observation, Liu et al. [6] analyzed an extreme precipitation event caused by two MCSs in South China on 11 May 2014. Results showed that the rapid splitting and reconstructing of bow echoes within the MCSs are conducive to the occurrence of extreme rainfall. By using an objective identification algorithm of convection initiation, Bai et al. [7] conducted a statistical study on the radar climatology of convections in the coastal area of South China. Convection initiation exhibited three peaks, i.e., a late-night-to-morning peak on the windward coast and offshore, a noon-to-late-afternoon peak on the coastal land, and an evening-to-early-morning peak over the northwestern highland, respectively.

Past studies have revealed the relationship between severe convective weather and meso-β-scale convective systems (with horizontal scales ranging from 20 to 200 km). However, the forecasting of severe convective weather is still one of the key challenging tasks in current weather forecasts due to its small horizontal scale and abruptness [8]. With the progress of observation techniques, especially the widespread application of Doppler weather radar, and the continuous development of numerical models, increasing attention has been paid to the meso-γ-scale (with horizontal scales from 2–20 km) or even smaller-scale processes. In particular, severe convective weather (e.g., tornadoes and straight-line winds) can be directly caused by meso-γ-scale vortices. For example, Trapp et al. [9] found that ~26% of meso-γ-scale vortices can develop into tornadoes, in accordance with the radar analysis of tornadic mesocyclones. Based upon single Doppler radar analysis, Wakimoto et al. [10] examined the formation of derechos (i.e., high straight-line winds) produced by bow echoes. They found that the near-surface high winds were due to the superposition of the ambient flow with the rotational flow of low-level meso-γ-scale vortices formed at the bow echo leading convective line. This was also confirmed by Atkins and Laurent [11] and Xu et al. [12] by virtue of the Weather Research and Forecasting Model simulations.

In general, meso-γ-scale vortices can be classified into two types according to their parent convective systems. One is mesocyclones caused by isolated cells [13], and the other is the so-called mesovortices (MVs) generated in the low level of organized MCSs [14]. Previous studies of meso-γ-scale vortices in China focused on mesocyclones [15–17]. Taking a squall line that occurred in Guangdong on 20 March 2013 for example [18], there were many supercells within this squall line. Most of the gales were generated by the supercellular mesocyclones. However, there is still a lack of thorough studies of MVs in China. Recently, Tang et al. [19] (hereafter T20) conducted a statistical analysis of MVs that occurred in the warm season of the Yangtze–Huai River Basin (YHRB) using 3-year radar observations. Their results showed that most of the MVs were short-lived, with an average lifetime of 26 min. Long-lived MVs were prone to having a larger size and stronger intensity than those of short-lived MVs. In this work, the characteristics of MVs that occurred in the first rainy season of South China (i.e., April to June) are studied by using 3-year radar observations during 2017–2019, and these characteristics are then compared to those in the YHRB.

The rest of this paper is organized as follows: Data and methods are introduced in Section 2. The main results are presented in Section 3. Finally, a discussion and conclusions are provided in Section 4.

2. Data and Methods

2.1. Data

In this study, observations from nine Doppler weather radars in Guangdong from April to June in 2017–2019 were used. The distribution of these radar stations is shown in Figure 1, covering almost the whole Guangdong Province. As in T20, the 88D2ARPS program of the Advanced Regional Prediction System (ARPS) from the Center for Analysis and Prediction of Storms (CAPS) at the University of Oklahoma [20] was used for the raw radar data quality control, which removes non-meteorological echoes and performs the radial velocity de-aliasing [21]. The processed radar data were further used to identify and statistically analyze the MVs. Finally, the ERA5 reanalysis data [22] with a high horizontal resolution of $0.25° \times 0.25°$ from the European Centre for Medium-Range Weather Forecasts (ECMWF) were utilized to study the environmental conditions for MVs in Guangdong from April to June.

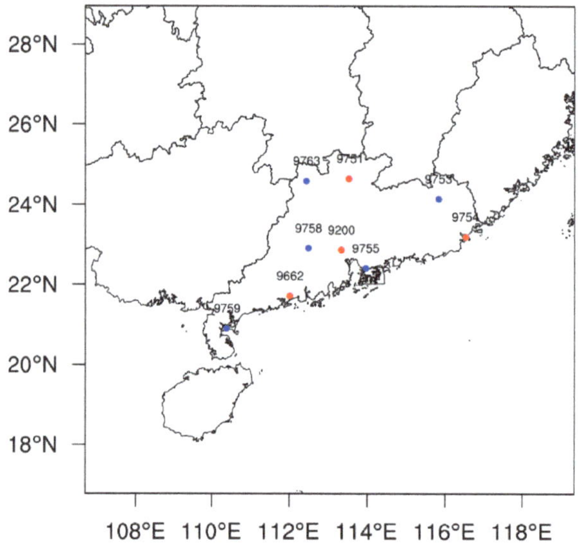

Figure 1. Distribution of nine Doppler weather radars in Guangdong: Guangzhou (9200), Yangjiang (9662), Shaoguan (9751), Meizhou (9753), Shantou (9754), Shenzhen (9755), Zhaoqing (9758), Zhanjiang (9759) and Lianzhou (9763) radar stations. Red dots represent radars used to identify MVs.

2.2. Identification of MVs

Since MVs are mainly generated in organized MCSs, it was necessary to identify the MCSs in advance. Firstly, the composite reflectivity from the nine Doppler radars (Figure 1) was used to identify convective zones (CZs) of >35 dBZ. Then, the area of each CZ was calculated, and a CZ with an area greater than 1000 km^2 was considered as an MCS. More details about the detection of MCSs can be found in T20.

In the past, the identification of meso-γ-scale vortices (e.g., mesocyclones) or even smaller-scale circulations (e.g., tornadoes) mainly depended on the appearance of a positive–negative velocity pair in the Doppler radar radial velocity field [23]. The rotational strength of a vortex is measured by the difference between the maximum inbound and outbound radial velocities; however, this measurement is susceptible to velocity noises. When the environmental wind speed is strong, the positive–negative velocity cannot be identified because it is easily submerged in the background wind field. Moreover, when the vortex is far from the radar station, the azimuthal resolution becomes too low to well resolve the velocity pair. In this study, a more advanced linear least square derivative method [24] was used to calculate the azimuthal shear of an MV. This method considers the contribution of all radial velocities in a given range to the azimuthal shear at the center point, thus reducing the noise errors caused by the radial velocities.

In this study, we only calculated the azimuthal shear at the 0.5° elevation within 150 km of the radar because the MVs are mainly generated in the lower troposphere below 3 km. As in T20, a lower limit of 10^{-3} s^{-1} was adopted for the azimuthal shear of MVs, which is one order greater than the Coriolis parameter in mid-latitudes. The duration of the MV must exceed 18 min, i.e., three volume scans of the radar. On this basis, a storm cell identification technology similar to WSR-88D was adopted for the forward and backward tracking of MVs to obtain their life cycles [25].

Because the radar radial velocity cannot be composed, the identification of MVs can only be performed for each individual radar. Despite the dense distribution of Doppler radars in Guangdong Province, only four radars covering almost the whole province were selected to identify the MVs. These four radars are the Shaoguan (9751), Shantou (9754), Yangjiang (9662) and Guangzhou (9200) radar stations, representing the northern, eastern and western Guangdong Province and the Pearl River Delta region, respectively.

3. Characteristics of MVs during the First Rainy Season in Guangdong, South China

3.1. Spatiotemporal Distribution

During the first rainy season (April–June) in Guangdong, South China, from 2017 to 2019, a total of 7965 MVs were identified using the four radars in Guangdong Province (Table 1). Figure 2 shows the uneven spatial distribution of MVs in Guangdong Province. The MVs are most concentrated in the Pearl River Delta region, accounting for about 41% of the total, followed by western Guangdong where the MVs account for about 27%. There are far fewer MVs generated in eastern and northern Guangdong, which together account for less than half of the total. In particular, the MVs are sparsely distributed near the Nanling mountains in northern Guangdong. The uneven spatial distribution of the MVs can be attributed to the following two reasons: The first is the different topographical features of Guangdong. Laing and Fritsch [26] have pointed out that MVs are prone to form over flat regions, such as the America Great Plains. Purpura et al. [27] documented supercell activity (i.e., meso-γ-scale mesovortex activity) across the Appalachian Mountains in America and also found that there are fewer supercells in the more mountainous areas. The formation of MVs is therefore affected by the topography of the Nanling mountains in northern Guangdong and the Lianhua, Luofu and Jiulian Mountains in eastern Guangdong. In contrast, the relatively flat terrain in western Guangdong and Pearl River Delta region is conducive to the formation of the MVs. Secondly, the uneven spatial distribution of MVs is intimately related to the differences in environmental conditions in different regions of Guangdong, which will be discussed in Section 3.3. In terms of lifetime, most of the MVs (about 67%) last less than 30 min, and about 30% last more than half an hour, while the ones lasting more than one hour only account for about 5% (Table 1). The statistical characteristics of MV lifetime in Guangdong are similar to those in the YHRB [19].

Table 1. Statistics for the number and lifetime of the MVs identified by four radars in Guangdong from April to June during 2017–2019. Percentages of the total MVs with different lifetimes are shown in parentheses.

	Guangzhou (9200)	Shaoguan (9751)	Yangjiang (9662)	Shantou (9754)	Total
Number of MVs	3267	794	2182	1722	7965
18–30 min	2153 (66%)	551 (69%)	1466 (67%)	1172 (68%)	5342 (67%)
30–60 min	992 (30%)	223 (28%)	626 (29%)	479 (28%)	2320 (29%)
>60 min	122 (4%)	20 (3%)	90 (4%)	71 (4%)	303 (4%)

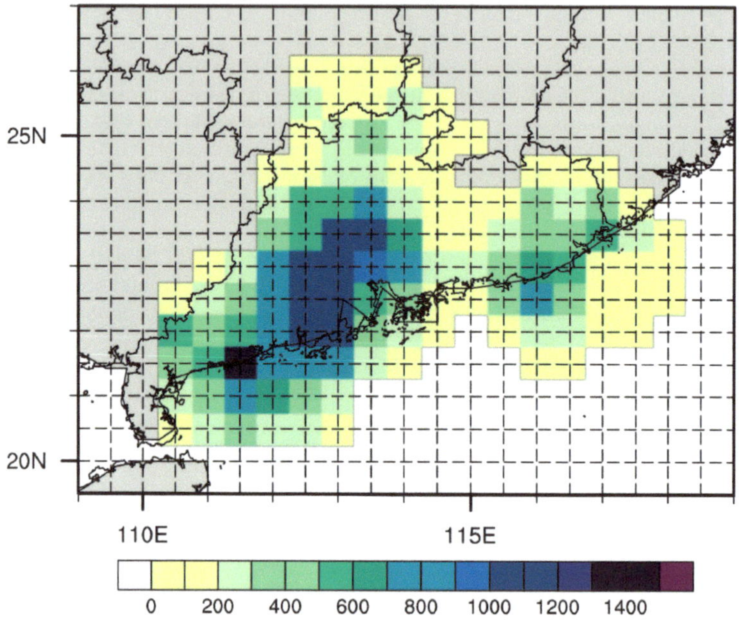

Figure 2. Spatial distribution of occurrence frequency of MVs during the first rainy season in Guangdong, South China, from 2017 to 2019.

Figure 3 shows the monthly and diurnal variations of the MVs in Guangdong. The number of MVs generated in June is significantly larger than that in April and May, accounting for about half of the total (Figure 3a). In addition, the MVs display notable diurnal variations (Figure 3b). MV occurrence is most frequent from the late morning to the afternoon (0900 BJT to 1800 BJT, i.e., Beijing time) but drops to the minimum overnight (2100 BJT to 0300 BJT). This diurnal variation is just opposite to that in the warm season in the YHRB [19]. Specifically, the MVs in the YHRB present slight diurnal variations (figures omitted), with two weak peaks at night (1800 BJT to 2100 BJT) and in the morning (0600 BJT to 0900 BJT), respectively. In addition, there is a minimum at night (0300 BJT to 0600 BJT).

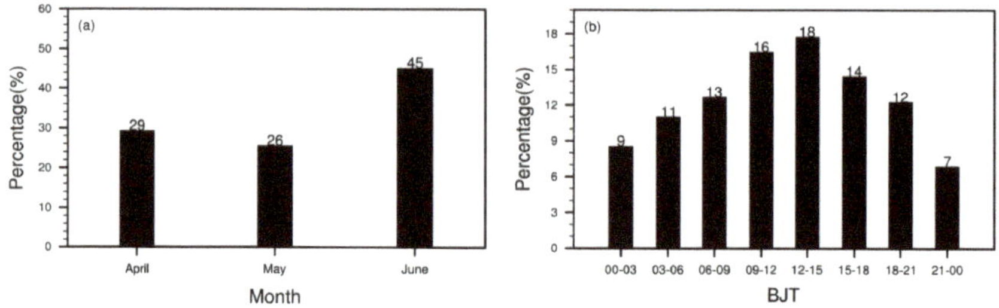

Figure 3. (**a**) Monthly and (**b**) diurnal variations of the MVs in Guangdong from April to June during 2017–2019.

Figure 4 shows the distributions of the diameters and intensities (in terms of azimuthal shear) of the MVs. It can be seen that the MV diameters are mostly between 4 km and 12 km (about 80%), while only a few are larger than 12 km or smaller than 4 km (Figure 4a). Most of the MVs (about 94%) are weak, with intensities between 0.001 s^{-1} and 0.004 s^{-1}, while only about 6% are stronger than 0.004 s^{-1} (Figure 4b). It is indicated that the diameter of the MV is proportional to its lifetime (Table 2). The MVs presenting a short lifetime of

18–30 min have a mean diameter of 6.96 km, the MVs lasting for 30–60 min have a mean diameter of 8.35 km, and the MVs lasting for more than one hour have a mean diameter of 10.3 km. Similarly, the azimuthal shear of the MVs is also proportional to their lifetime. The MVs with a short lifetime of 18–30 min have an average intensity of 0.002 s^{-1}, while the MVs with a long lifetime of exceeding 60 min have an average intensity of 0.003 s^{-1}. Thus, a longer lifetime is accompanied by greater diameter and stronger intensity, which is similar to the characteristics of the MVs in the YHRB [19].

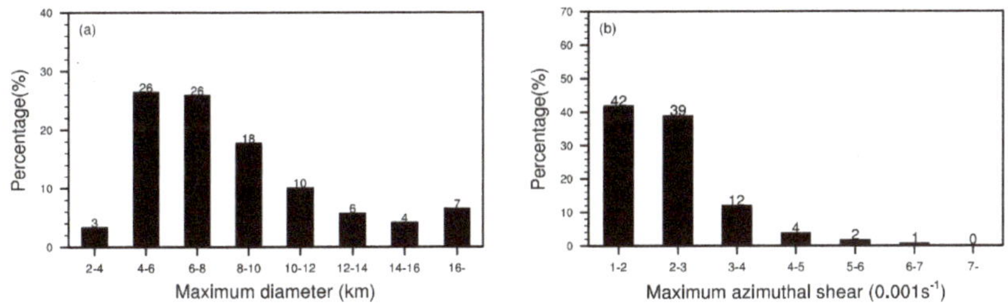

Figure 4. Distributions of the (**a**) diameter (km) and (**b**) intensity ($10^{-3} \cdot s^{-1}$) of the MVs in Guangdong from April to June during 2017–2019.

Table 2. Average diameter, azimuthal shear intensity and lifetime of the MVs in Guangdong from April to June during 2017–2019.

	Diameter (km)	Azimuthal Shear ($10^{-3} \cdot s^{-1}$)	Lifetime (Minutes)
All MVs	7.49	2.15	26
Short-lived MVs	6.96	2.03	20
Medium-lived MVs	8.35	2.33	36
Long-lived MVs	10.30	2.88	75

3.2. Comparison of MVs in Different Regions of Guangdong

As mentioned above, most MVs during the first rainy season in Guangdong appear in the Pearl River Delta region and western Guangdong. In this subsection, the temporal variations of MVs in different regions of Guangdong are compared. About half of the MVs in the Pearl River Delta region occur in June, which is significantly more than that in April and May. Similar monthly variations of MVs are found in eastern Guangdong. However, there are less significant monthly variations in northern and western Guangdong. The percentages of MVs generated in each month (April to June) are almost the same (Figure 5a).

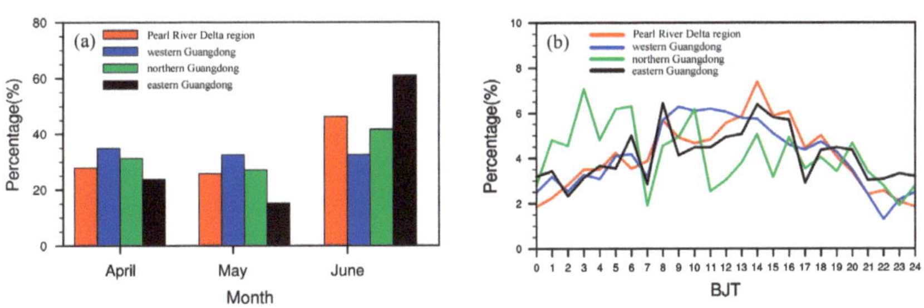

Figure 5. (**a**) Monthly variations and (**b**) diurnal variations of MVs in the four different regions of Guangdong. Red bar (solid line) represents the Pearl River Delta region, with blue bar (solid line) for western Guangdong, green bar (solid line) for northern Guangdong and black bar (solid line) for eastern Guangdong.

The diurnal variations are quite similar for the MVs in the River Delta region and eastern Guangdong (Figure 5b), which show an afternoon-to-night peak and night-to-early-morning minimum. However, the MVs in northern and western Guangdong present a different diurnal variation. The MVs in northern Guangdong show a night-to-early-morning peak and afternoon-to-evening minimum. The MVs in western Guangdong have a midday peak and night-to-early-morning minimum. This may be explained by the moving direction of MCSs in Guangdong. On the one hand, during April and May in South China, there is still weak cold air frequently affecting Guangdong from the north. Under the combined influence of dry and cold air from the north and warm and moist air from the sea, there is usually an MCS accompanied by a cold front, affecting Guangdong from the north to the south. The MCS tends to be triggered in northern Guangdong during the night-to-morning period and then moves eastward and southward to southern Guangdong from the afternoon to evening [28]. Strong urban heat island effects over the River Delta region could also play some role in the development of MCSs. The afternoon peak of MCSs is partially attributed to the solar heating-induced increase in CAPE. Therefore, the MVs generated in the MCS present diurnal variations consistent with those of the MCSs.

Moreover, the statistical characteristics of MV lifetime in the Pearl River Delta region are similar to those in the other three regions, with an average lifetime of 25–30 min. The azimuthal shear intensity and horizontal scale of the MVs are also similar in the Pearl River Delta region and northwest Guangdong, with an average diameter of about 7 km and an average intensity of about 0.002 s^{-1}. The MVs in eastern Guangdong are slightly weaker and smaller than those in the other three regions (Table 3).

Table 3. Average diameter, azimuthal shear intensity and lifetime of the MVs in different regions of Guangdong from April to June during 2017–2019.

	Diameter (km)	Azimuthal Shear ($10^{-3}\cdot s^{-1}$)	Lifetime (Minutes)
Pearl River Delta	7.64	2.16	27
Western Guangdong	7.54	2.23	26
Northern Guangdong	7.34	2.28	25
Eastern Guangdong	7.22	1.98	26

3.3. Environmental Conditions

As noted above, the spatial distribution of the MVs during the first rainy season in South China is affected by environmental conditions. During this period, the South Asia high center is located in the upper troposphere (200 hPa) over the Indochina Peninsula and adjacent areas (Figure 6a). South China is on the southern side of the upper-level jet axis at its entrance region. In the middle troposphere (500 hPa), the subtropical high lies between 10°N and 20°N, and South China lies on the northern side of the subtropical high and in the westerlies at the bottom of the East Asia trough (Figure 6b). In the lower troposphere (850 hPa), with the onset of the southwest monsoon, the equatorial westerlies from the Indian Ocean and the cross-equatorial flow from near Kalimantan Island entering the South China Sea converge and blow over South China (Figure 6c,d). Meanwhile, the low-level southwesterly jets transport abundant warm and moist air to South China, especially to the west of the Pearl River Estuary (the Pearl River Delta region and western Guangdong), resulting in the higher potential pseudo-equivalent temperature and water vapor flux in this region than in northern and eastern Guangdong.

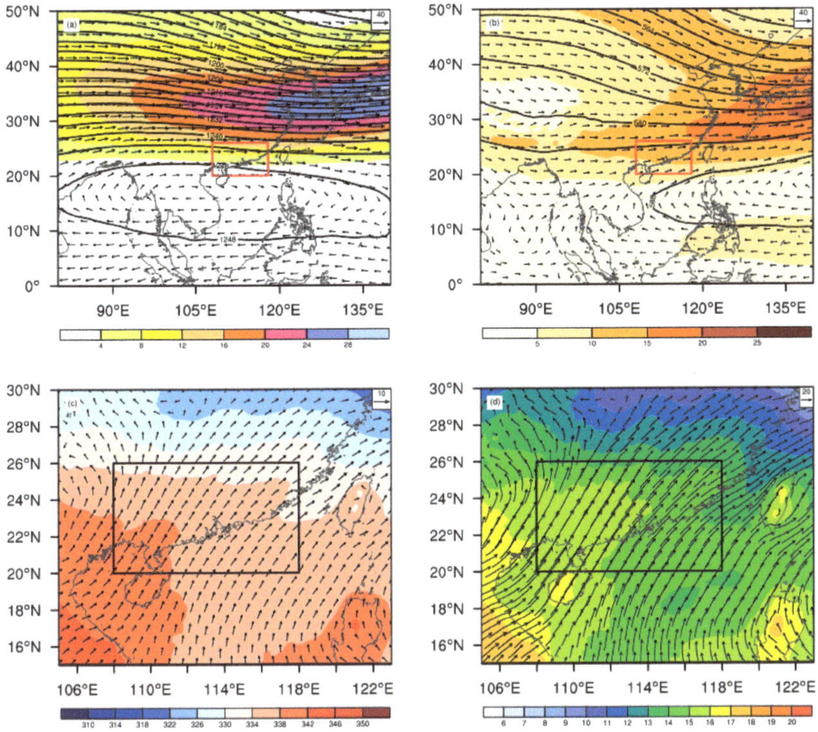

Figure 6. Wind fields (arrows and shaded, unit: m·s^{-1}) and geopotential height fields (black contours, unit: m) during April–June in (**a**) East Asia at 200 hPa and (**b**) Pan-South China at 500 hPa; (**c**) wind field (arrows, unit: m·s^{-1}) and potential pseudo-equivalent temperature (shaded, unit: K) in Pan-South China at 850 hPa; (**d**) water vapor flux (arrows and shaded, unit: g·cm^{-1}·hPa^{-1}·s^{-1}) in Pan-South China at 850 hPa. Both the red and black boxes represent Guangdong Province.

Due to the warm and moist airflow transported by the southwest monsoon, the mean low-level instability during April–June is relatively higher across the whole Guangdong Province (Figure 7a). The potential pseudo-equivalent temperature in southwest Guangdong (30–32 °C) is significantly higher than that in other regions (28–30 °C). In addition, the existence of the low-level jet causes stronger low-level wind shear over southwest Guangdong than in other regions (Figure 7b). In short, the relatively higher instability and stronger low-level vertical wind shear are conducive to the formation of MVs and are similar to the environmental conditions for the formation of MVs in the YHRB. Weisman and Trapp [29] studied the effects of vertical wind shear and CAPE on the genesis of MVs using idealized simulations. The results show that the stronger the vertical wind shear is, especially for the low-level wind shear, the stronger the MVs will be. High CAPE is favorable for the formation of MVs. Atkins and Laurent [11] obtained a similar result by using the Advanced Research Weather Research and Forecasting (ARW-WRF) model to study the relationship between the development of MVs and these two parameters.

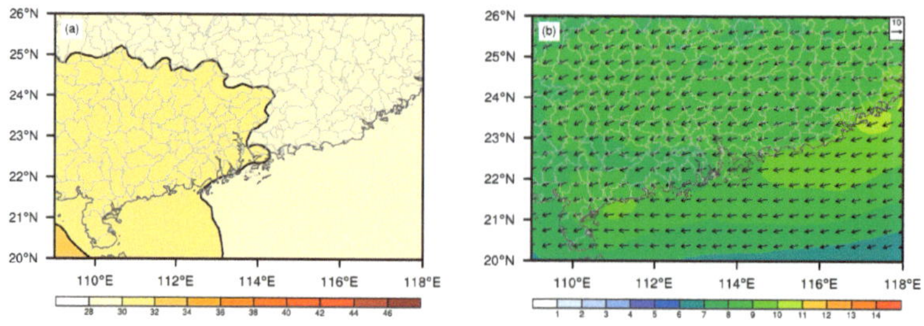

Figure 7. (**a**) Potential pseudo-equivalent temperature (shaded and black contours, unit: °C) and (**b**) vertical wind shear (shaded and arrows, unit: m·s^{-1}) at 1000–700 hPa in Guangdong from April to June.

4. Conclusions

In this paper, the mesovortices (MVs) that occurred during the first rainy season (April to June) in South China during 2017–2019 were analyzed using Doppler radar observations as well as ERA5 reanalysis. The spatiotemporal distributions and structural features of MVs were examined, as were their favorable environmental conditions. The main conclusions are as follows:

The MVs are mainly observed in the Pearl River Delta region, followed by western Guangdong, and relatively fewer in both eastern and northern Guangdong. Similar to the Yangtze–Huaihe River Basin (YHRB), the MVs in South China present a very short lifetime, with about 70% lasting less than 30 min. The intensity and horizontal scale of the MVs are proportional to their lifetime; i.e., long-lived MVs have larger horizontal scales and stronger intensities than short-lived ones. Different from the MVs in the YHRB, the MVs in South China display more significant diurnal variations, which occur most frequently from late morning toward evening (1100 BJT to 1700 BJT) but drop to the minimum overnight (2000 BJT to 0200 BJT on the next day). In contrast, the MVs in the YRHB present slight diurnal variations, with two weak peaks at night (1800 BJT to 2100 BJT) and in the morning (0600 BJC to 0900 BJT) as well as a minimum in the early morning (0300 BJT to 0600 BJT).

The diurnal variations are quite similar for the MVs in the Pearl River Delta region, western and eastern Guangdong, with an afternoon-to-evening peak and a night-to-early-morning minimum. On the contrary, the diurnal variation of MVs in northern Guangdong is opposite to that mentioned above. This is because during the first rainy season in South China, under the combined influence of dry and cold air from the north and the warm and moist air from the sea, there is usually an MCS accompanied by a cold front, which thereby affects South China from north to south. The MCS tends to be triggered in northern Guangdong from night to morning and then moves to the southern coastal areas from afternoon to evening. As a result, MVs generated within MCSs have a similar diurnal variation to their parent systems.

The uneven spatial distribution of MVs is closely related to the environmental differences in South China. Affected by the southwest monsoon, the water vapor flux, low-level instability and vertical wind shear in southwest Guangdong are significantly greater than those in other regions during the first rainy season and are favorable to the formation of MVs.

This study contributes to the understanding of MVs in South China and thus has a great significance for the operational nowcasting and warning of MVs. However, the relationship between the MVs and severe weather (such as gales and heavy rainfall [30,31]) in South China is still unclear and will be studied in the future.

Author Contributions: Conceptualization, Y.T. and X.X.; formal analysis, Y.T. and X.X.; methodology, Y.T. and X.X.; resources, X.C. and Q.X.; writing—original draft preparation, Y.T. and X.X.; writing—review and editing, Y.T., X.X., Y.J., Z.W. and S.Z. All authors have read and agreed to the published version of the manuscript.

Funding: This work is supported by the National Natural Science Foundation of China (42122036, 42230607, 41750965) and the Second Tibetan Plateau Scientific Expedition and Research (STEP) program (2019QZKK0105).

Data Availability Statement: The data that support the findings of this study are available from the first author, upon reasonable request.

Acknowledgments: The authors would like to thank the anonymous reviewers for helpful comments and suggestions.

Conflicts of Interest: The authors declare no conflict of interest.

References

1. Fan, L.M.; Yu, X.D. Characteristic analyses on environmental parameters in short-term severe convective weather in China. *Plateau. Meteorol.* **2013**, *32*, 156–165.
2. Xue, X.Y.; Ren, G.Y.; Sun, X.B.; Ren, Y.Y.; Yu, Y. Climatological characteristics of meso-scale and micro-scale strong convective weather events in China. *Clim. Environ. Res.* **2019**, *24*, 199. (In Chinese)
3. Wang, L.; Shen, X.; Wang, Y.; Zhang, C.; Wang, Y.; Li, X. Mechanism analysis of a squall line upscale growing process in South China. *Plateau. Meteorol.* **2021**, *40*, 145–158.
4. Zhang, C.; Zhi, S.L.; Xu, A.H. Analysis of radar echo characteristics of rare force 10 gale with a strong squall line event in Jiangxi. *Torrential Rain Disasters* **2019**, *38*, 135–143. (In Chinese)
5. Du, Y.; Chen, G.X.; Han, B.; Bai, L.Q.; Li, M.H. Convection initiation and growth at the coast of South China. Part II: Effects of the terrain, coastline, and cold pools. *Mon. Weather Rev.* **2020**, *148*, 3871–3892. [CrossRef]
6. Liu, X.; Luo, Y.L.; Guan, Z.Y.; Zhang, D.L. An extreme rainfall event in coastal South China during SCMREX-2014: Formation and roles of rainband and echo trainings. *J. Geophys. Res. Atmos.* **2018**, *123*, 9256–9278. [CrossRef]
7. Bai, L.Q.; Chen, G.X.; Huang, L. Image processing of radar mosaics for the climatology of convection initiation in South China. *J. Appl. Meteorol. Climatol.* **2020**, *59*, 65–81. [CrossRef]
8. Yu, X.D.; Zhou, X.G.; Wang, X.M. The advances in the nowcasting techniques on thunderstorms and severe convection. *Acta Meteorol. Sin.* **2012**, *70*, 311–337.
9. Trapp, R.J.; Tessendorf, S.A.; Godfrey, E.S.; Brooks, H.E. Tornadoes from squall lines and bow echoes. Part I: Climatological distribution. *Weather. Forecast.* **2005**, *20*, 23–34. [CrossRef]
10. Wakimoto, R.M.; Murphey, H.V.; Davis, C.A.; Atkins, N.T. High winds generated by bow echoes. Part II: The relationship between the mesovortices and damaging straight-line winds. *Mon. Weather Rev.* **2006**, *134*, 2813–2829. [CrossRef]
11. Atkins, N.T.; St. Laurent, M. Bow echo mesovortices. Part I: Processes that influence their damaging potential. *Mon. Weather Rev.* **2009**, *137*, 1497–1513. [CrossRef]
12. Xu, X.; Xue, M.; Wang, Y. The genesis of mesovortices within a real-data simulation of a bow echo system. *J. Atmos. Sci.* **2015**, *72*, 1963–1986. [CrossRef]
13. Burgess, D.W.; Donaldson, R.J., Jr.; Desrochers, P.R. *The Tornado: Its Structure, Dynamics, Prediction, and Hazards*; American Geophysical Union: Washington, DC, USA, 1993. [CrossRef]
14. Funk, T.W.; Darmofal, K.E.; Kirkpatrick, J.D.; Dewald, V.L.; Przybylinski, R.W.; Schmocker, G.K.; Lin, Y.-J. Storm reflectivity and mesocyclone evolution associated with the 15 April 1994 squall line over Kentucky and Southern Indiana. *Weather Forecast.* **1999**, *14*, 976–993. [CrossRef]
15. Yu, X.D.; Zheng, Y.Y.; Zhang, A.M.; Yao, Y.Q.; Fang, C. The detection of a severe tornado event in Anhui with China new generation weather radar. *Plateau. Meteorol.* **2006**, *25*, 914–924.
16. Feng, J.Q.; Tang, D.Z.; Yu, X.D.; Zhang, H.M.; Chen, Q.P. The accuracy statistics of mesocyclone identification products from CINRAD/SA. *Meteorol. Mon.* **2010**, *36*, 47–52.
17. Fang, C.; Zheng, Y. The analysis of mesocyclone product from the doppler weather radar. *Meteorol. Mon.* **2007**, *33*, 16–20.
18. Fang, C.; Yu, X.D.; Zhu, W.J.; Yin, Z.H.; Zhou, K.H. Characteristics of the thunderstorm gale process in Hunan and Guangdong on 20 March 2013. *Meteorol. Mon.* **2015**, *41*, 1305–1314.
19. Tang, Y.; Xu, X.; Xue, M.; Tang, J.; Wang, Y. Characteristics of low-level meso-γ-scale vortices in the warm season over East China. *Atmos. Res.* **2020**, *235*, 104768. [CrossRef]
20. Xue, M.; Droegemeier, K.K.; Wong, V.; Shapiro, A.; Brewster, K.; Carr, F.; Weber, D.; Liu, Y.; Wang, D. The Advanced Regional Prediction System (ARPS)—A multi-scale nonhydrostatic atmospheric simulation and prediction tool. Part II: Model physics and applications. *Meteor. Atmos. Phys.* **2000**, *76*, 143–165. [CrossRef]
21. Brewster, K.; Hu, M.; Xue, M.; Gao, J. Efficient assimilation of radar data at high resolution for short-range numerical weather prediction. In Proceedings of the World Weather Research Program Symposium on Nowcasting and Very Short-range Forecasting, Toulouse, France, 6 September 2005.
22. Hersbach, H.; Dee, D. ERA5 Reanalysis is in Production. *ECMWF Newsl.* **2016**, *147*, 7.
23. Mitchell, E.D.; Vasiloff, S.V.; Stumpf, G.J.; Witt, A.; Eilts, M.D.; Johnson, J.T.; Thomas, K.W. The national severe storms laboratory tornado detection algorithm. *Weather Forecast.* **1998**, *13*, 352–366. [CrossRef]

24. Smith, T.M.; Elmore, K.L. The use of radial velocity derivatives to diagnose rotation and divergence. In Proceedings of the 11th Conf. on Aviation, Range, and Aerospace, Hyannis, MA, USA, 5 October 2004.
25. Johnson, J.T.; Mackeen, P.L.; Witt, A.; Mitchell, E.D.; Stumpf, G.J.; Eilts, M.D.; Thomas, K.W. The storm cell identification and tracking algorithm: An enhanced WSR-88D algorithm. *Weather. Forecast.* **1998**, *13*, 263–276. [CrossRef]
26. Laing, A.G.; Fritsch, J.M. The global population of mesoscale convective complexes. *Q. J. R. Meteorol. Soc.* **1997**, *123*, 389–405 [CrossRef]
27. Purpura, S.M.E.; Davenport, M.D.; McKeown, K.E.; Riggin, R.R. Environmental evolution of supercell thunderstorms interacting with the Appalachian Mountains. *Weather. Forecast.* **2023**, *38*, 179–198. [CrossRef]
28. Wu, N.G.; Ding, X.; Wen, Z.P.; Chen, G.X.; Meng, Z.Y.; Lin, L.X.; Min, J.Z. Contrasting frontal and warm-sector heavy rainfalls over South China during the early-summer rainy season. *Atmos. Res.* **2020**, *235*, 104693. [CrossRef]
29. Weisman, M.L.; Trapp, R.J. Low-level mesovortices within squall lines and bow echoes. Part I: Overview and dependence on environmental shear. *Mon. Weather Rev.* **2003**, *131*, 2779–2803.
30. Wei, P.; Xu, X.; Xue, M.; Zhang, C.; Wang, Y.; Zhao, K.; Zhou, A.; Zhang, S.; Zhu, K. On the Key Dynamical Processes Supporting the 21.7 Zhengzhou Record-breaking Hourly Rainfall in China. *Adv. Atmos. Sci.* **2023**, *40*, 337–349. [CrossRef]
31. Liu, Q.; Xu, X.; Zhao, K.; Zhou, A. A merger-formation bow echo caused by low-level mesovortex. *J. Geophy. Res. Atmos.* **2023**, *128*, 8. [CrossRef]

remote sensing

MDPI

Article

Statistical Bias Correction of Precipitation Forecasts Based on Quantile Mapping on the Sub-Seasonal to Seasonal Scale

Xiaomeng Li [1,2], Huan Wu [1,2,*], Nergui Nanding [3], Sirong Chen [4], Ying Hu [1,2] and Lingfeng Li [1,2]

1 School of Atmospheric Sciences, and Southern Marine Science and Engineering Guangdong Laboratory (Zhuhai), Sun Yat-sen University, Zhuhai 519082, China
2 Guangdong Province Key Laboratory for Climate Change and Natural Disaster Studies, Sun Yat-sen University, Zhuhai 519082, China
3 School of Earth Sciences, Yunnan University, Kunming 650032, China
4 Guangxi Climate Center, Nanning 530022, China
* Correspondence: wuhuan3@mail.sysu.edu.cn

Abstract: Accurate precipitation forecasting is challenging, especially on the sub-seasonal to seasonal scale (14–90 days) which mandates the bias correction. Quantile mapping (QM) has been employed as a universal method of precipitation bias correction as it is effective in correcting the distribution attributes of mean and variance, but neglects the correlation between the model and observation data and has computing inefficiency in large-scale applications. In this study, a quantile mapping of matching precipitation threshold by time series (MPTT-QM) method was proposed to tackle these problems. The MPTT-QM method was applied to correct the FGOALS precipitation forecasts on the 14-day to 90-day lead times for the Pearl River Basin (PRB), taking the IMERG-final product as the observation. MPTT-QM was justified by comparing it with the original QM method in terms of precipitation accumulation and hydrological simulations. The results show that MPTT-QM not only improves the spatial distribution of precipitation but also effectively preserves the temporal change, with a better precipitation detection ability. Moreover, the MPTT-QM-corrected hydrological modeling has better performance in runoff simulations than the QM-corrected modeling, with significantly increased KGE metrics ranging from 0.050 to 0.693. MPTT-QM shows promising values in improving the hydrological utilities of various lead time precipitation forecasts.

Keywords: precipitation; bias correction; Quantile Mapping; sub-seasonal to seasonal forecast

1. Introduction

Meteorological disasters represent one of the most serious types of natural disaster in the world. Among the different kinds of meteorological disasters, a flood disaster induced by heavy precipitation has a wide range of influence, a long duration, and causes significant property loss and casualties [1]. In light of the global warming environment, it is expected that the frequency and intensity of flood disasters will continue to increase [2]. Therefore, there is an urgent need to detect and monitor flood events. Precipitation forecasting is one of the most important and effective tools for obtaining information in flood monitoring [3]. Therefore, if more accurate precipitation forecast information were to be provided before the occurrence of heavy precipitation, this would mark a great contribution to flood forecasting and monitoring and disaster prevention [4,5].

Numerical weather forecast technology has undergone unprecedented development, and the quality of precipitation forecasting has also significantly improved [6], particularly on the short-and-medium-term scale (0–10 days). The effective lead time for a disastrous weather forecast needs to be extended to 14 days through the development of certain skills so as to ensure the significant value of forecasting for decision making [7,8]. However, as the atmosphere is a nonlinear system with inherent randomness [9], there are deviations between the numerical models and the observed data. At the same time, the predictable lead time of a model has a certain range [10]. For example, the predictable lead time of a daily weather forecast is generally around two weeks [11]. In recent years, a series of new random physical process experiments [12] and an updated parameterization scheme [13,14] were designed to improve the model ensemble predictions. However, weather forecasting on the sub-seasonal to seasonal scale is still challenging [15].

Therefore, to obtain accurate and reliable precipitation forecast information and provide a solid foundation for flood forecasting and monitoring, the bias correction of the model precipitation forecast is a significant step. Bias correction is an essential link in the process for obtaining a medium-and-long-term forecast, especially for lead times beyond 14 days. In recent decades, scholars have proved that statistical post-processing methods can effectively reduce or eliminate the systematic errors in the original model data. A variety of bias correction models have been developed based on statistical methods, such as analogs [16–19], QM, and other non-parametric methods [20–25] that are easy to implement and fast to calculate. There are various parametric methods based on complex mathematical and physical models, such as the non-homogeneous Gaussian regression model [26], logistic regression model [27], Bayesian model averaging model (BMA) [28,29], Bayesian joint probability (BJP) [30–32], Kalman filtering [33], etc. In recent years, with the development of machine learning technology, this kind of method has been widely used for the bias correction of model data. For example, random forest [34], artificial neural network (ANN) [35,36], convolutional neural network (CNN) [37], and other neural-network-based composite methods [38] have been employed.

QM, as the most efficient method, has been widely used to correct satellite precipitation products and ensemble numerical forecast and general circulation model (GCM) climate forecast data. At the same time, QM can directly calibrate runoff simulations using the hydrological model [39–41]. In addition, studies show that assimilating transformed precipitation into the NWP model using QM can also improve the typhoon forecast [42,43]. In the procedure of the QM method, the cumulative distribution function (CDF) of the model data and the observation data are established, respectively. Then, the transfer function (TF) between the two types of data is established for correction, or the model data are directly mapped to the CDF of the observation data [44] to correct the model data. The QM correction method can capture the average evolution of, and variability in, precipitation while adjusting all statistical moments. Many different test schemes based on QM have been successfully applied for bias correction. For example, Terink et al. [45] adjusted the daily RCM simulation precipitation and temperature data of the Rhine River Basin and found that the QM method operated relatively well under normal and extreme conditions. Bennett et al. [20] used QM to correct the annual and seasonal RCM rainfall bias in Australia, and they highlighted that the spatial distribution was improved after bias correction. Similarly, Themeßl et al. [25] found that the QM method effectively corrected the modeled daily precipitation in Alpine areas by analyzing seven bias correction methods. Huang et al. [22] established a five-parameter gamma Gaussian model on the basis of QM, which was successfully used to calibrate the monthly and seasonal precipitation forecasts of GCMS. Although the QM method is effective in correcting distribution attributes such as the mean and variance, the performance of the QM method in optimizing the spatial distribution of forecasting precipitation and detecting the occurrence of precipitation events is not satisfactory. Moreover, the QM method ignores the correlation between prediction data and observation data [41].

Therefore, if the QM method is directly used to correct the forecast precipitation on the sub-seasonal to seasonal scale (14–90 days) with a high temporal resolution (e.g., 3 h), the detailed temporal and spatial characteristics of precipitation will be blurred. In this study, a new bias correction method based on the QM was proposed to calibrate forecasting precipitation on the sub-seasonal to seasonal scale with a high temporal resolution. The new method firstly matches the precipitation threshold according to the time series and then corrects the model precipitation data by QM. The performance of the new bias correction method was analyzed against the observation data and the original QM method in terms of both precipitation accumulation and hydrological simulations.

The remainder of the paper is structured as follows. The datasets and study areas used in this study are described in Section 2. A detailed description of the proposed bias correction scheme is provided in Section 3. The results are presented in Section 4. Section 5 is the discussion and Section 6 is the conclusion.

2. Study Area and Data

2.1. Datasets

The problem of an insufficient spatial distribution of precipitation observed by surface rainfall stations is overcome by satellite remote sensing. Satellite remote sensing precipitation is an important source of precipitation data in many remote areas, particularly in the case of ungauged basins. In 2014, the Global Precipitation Measurement (GPM) was jointly developed by the National Aeronautics and Space Administration (NASA) and the Japan Aerospace Exploration Agency (JAXA) in order to provide high-resolution precipitation data globally. GPM is the inheritance and improvement of the Tropical Rainfall Measuring Mission (TRMM) satellites. On the one hand, the spatial resolution of the precipitation products ranges from 0.25 degree to 0.1 degrees, and the time resolution is increased from 3 h to 30 min. On the other hand, GPM's dual-band (Ku, Ka) radar system and high-performance microwave radiometer significantly enhance the detection ability for weak rainfall (<0.5 mm/h) and solid precipitation. The core observation platform of GPM is composed of dual-frequency precipitation radar (DPR) and the 13-channel GPM microwave imager (GMI) carried by GPM. DPR is the first type of active spaceborne remote-sensing and dual-frequency rain radar in the world, which is composed of Ku-band radar (13.6 GHz) and Ka-band radar (35.5 GHz). Ku-band radar has a better detection effect for medium-intensive precipitation, and Ka-band radar is more sensitive to small precipitation particles due to its shorter detection band. The GPM IMERG-final product is used as the reference data for analysis in this study. The original temporal resolution of IMERG-final is half-hourly, and the spatial resolution is 0.1 degrees [46]. In this study, IMERG-final was resampled to 0.125 degrees using the arithmetic mean method and accumulated to a three-hourly resolution.

The Flexible Global Ocean–Atmosphere–Land System model (FGOALS) was developed by The Institute of Atmospheric Physics (IAP), Chinese Academy of Sciences (CAS), and the Laboratory of Numerical Modelling for Atmospheric Sciences and Geophysical Fluid Dynamics (LASG) [47]. FGOALS is also one of the coupled models for China's participation in the 6th International Coupled Model Comparison Program (CMIP6). The output data from the historical climate simulation experiment (2001–2020), based on the updated version of the FGOALS model [48], i.e., FGOALS-f3-L, was used as the model prediction data to be corrected using the new bias correction method on the sub-seasonal to seasonal scale in this study. The FGOALS-f3-L data were interpolated using the first-order conservation interpolation method into 0.125 degrees, consistent with the IMERG-final. The research period of this study was between 2001 and 2020, of which 2001–2015 was the historical period for the experimental data, and the period of 2016–2020 was set as the verification period.

2.2. Study Area

The Pearl River is one of the seven major rivers in China. The Pearl River flows through Yunnan, Guizhou, Guangxi, Guangdong, Hunan, Jiangxi, and other provinces (autonomous regions) and the northeast of Vietnam, with a total length of 2214 km and a total drainage area of 453,690 square kilometers. Of this, the PRB in China covers an area of 442,100 square kilometers, and the basin in Vietnam covers an area of 11,590 square kilometers. The PRB is composed of four water systems, including Xijiang River, Beijiang River, Dongjiang River, and the rivers in the Pearl River Delta. The PRB is located in the inland and subtropical climate zone. The average precipitation from April to September is between 600 and 1900 mm, and the runoff of the PRB from April to September accounts for approximately 80% of the annual runoff [49]. The PRB was the main research area of this study, extending eastward to eastern Guangdong and southward to the coastal areas of western Guangdong and southern Guangxi (Figure 1).

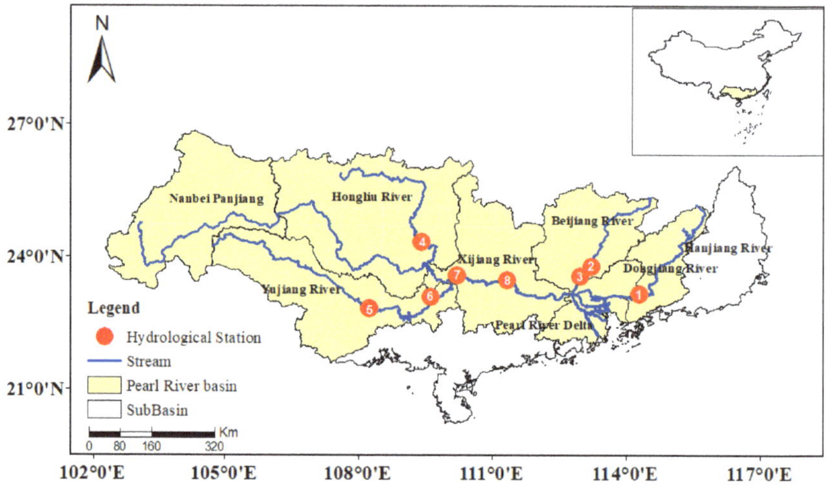

Figure 1. Research area of PRB in China.

The runoff data of eight hydrological stations in the PRB were selected for this study (Table 1). The time series of the observed runoff data is 2016–2020, and the source of the data is the Pearl River Administration of Navigational Affairs, https://zjhy.mot.gov.cn/zhuhangsj/shuiqingxx/, accessed on 9 May 2021.

Table 1. Information of main hydrological stations in the PRB.

Num.	Name	Longitude	Latitude	Drainage Area (Km^2)
1	Boluo	114.3	23.167	25,325
2	Feilaixia	113.236	23.786	34,000
3	Shijiao	112.963	23.554	38,363
4	Liuzhou	109.397	24.329	45,413
5	Nanning	108.236	22.833	72,656
6	Guigang	109.613	23.089	85,148
7	Dahuangjiangkou	110.204	23.582	288,544
8	Wuzhou	111.329	23.465	327,006

3. Method

The QM method uses a single transfer function to map the model simulation data to the CDF distribution of the observed data. When the simulated and observed values are relatively close, the revision is better; however, when the difference between the two values is large, the QM method can-not improve the model data significantly and may even introduce new biases. The QM method tends to reproduce the average precipitation from the observation, but the reproducing is not based on the one-by-one mapping between the observation and the model, let alone the correction of the modeled number of wet days [50,51]. In general, the QM method maps all the same precipitation amounts simulated by the model at different times to the same percentile value of observed precipitation, causing exactly the same revised precipitation values.

Manolis et al. [52] used different instances of gamma function that are fitted on multiple discrete segments of the precipitation CDF, instead of the common quantile–

quantile approach that uses one theoretical distribution to fit the entire CDF. This allows to better transfer the observed precipitation statistics to the raw model data. However, new uncertainties may be introduced in the CDF fitting using the Gamma-theoretic distribution.

The bias correction methods proposed in this study are described in this section, and the technical workflow is shown in Figure 2a. First, the new bias correction method, which is called the quantile mapping of matching precipitation threshold by time series (MPTT-QM) method, was proposed in this study. A threshold segmentation was performed using different percentiles of historical observed precipitation data [52]. The CDF distribution function of the observed and model data for each interval was then calculated using the nonparametric method of empirical distribution. Four discrimination factors were established according to the threshold distribution characteristics of the observed precipitation in historical periods and the error relationship between the observed and model precipitation. The weighted results of the four discrimination factors were used to determine the threshold intervals of the forecast precipitation, and then the CDF matching method (Figure 2b) was used for further correction within the determined interval.

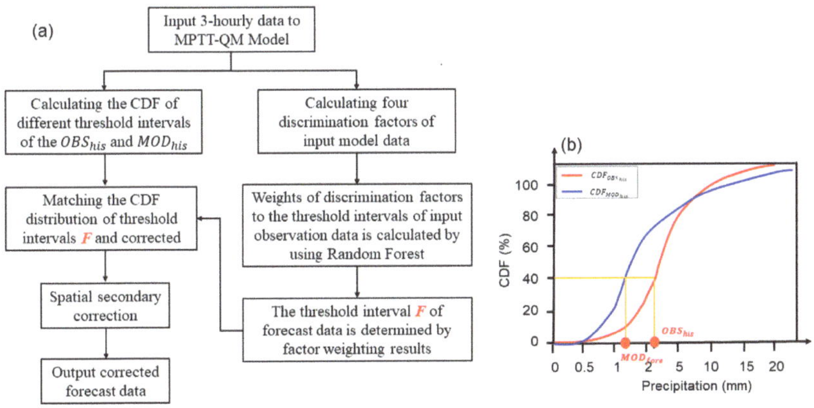

Figure 2. (**a**) The workflow of the MPTT-QM method; and (**b**) the schematic diagram of the QM method. MOD_{fore} represents the forecast model data, OBS_{his} represents the mapped observation data by CDF.

3.1. Quantile Mapping of the Matching Precipitation Threshold by Time

The QM method is effective in correcting distribution attributes such as the mean and variance but neglects the correlation between the model and observation data. In this study, the MPTT-QM bias correction model was proposed to solve this problem while retaining the advantages of QM. The percentile method, which is commonly used in precipitation research, was used to determine the threshold interval for this study [53,54]. Specifically, a set of 12 percentile threshold values (i.e., the 10th, 20th, ... , 80th, 85th, 90th, 95th, and 98th percentiles) was used to classify all the observed 3-h precipitation data into 13 intervals, similar to those proposed in previous studies [55,56]. For each interval, a list of date–time stamps (referred to as $T_{OBS_{his(i)}}$) for all the observed values was derived as follows:

$$T_{OBS_{his(i)}} = (time_1, time_2, time_3 \cdots, time_k) \quad (1)$$

where i ($i = 1, 2, \cdots n, n = 13$) indicates each of the intervals, and $time_k$ indicates the occurrence time when the precipitation intensity falls within the ith interval. The CDF distribution of the observation data in each interval, hereafter referred to as $CDF_{OBS_{his(i)}}$, is calculated. According to $T_{OBS_{his(i)}}$, the corresponding model precipitation values (referred to as $MOD_{his(i)}$) are derived to calculate the model CDF (referred to as $CDF_{MOD_{his(i)}}$). An initial bias correction is performed using the QM method (Figure 2b):

$$MOD_{BC_k} = CDF^{-1}_{OBS_{his(i)}}\left(CDF_{MOD_{his(i)}}\left(MOD_{fore_k}\right)\right) \tag{2}$$

where MOD_{fore_k} represents the forecast model data at a certain time k to be corrected on the sub-seasonal to seasonal scale, and MOD_{BC_k} is the corrected results. The problem with the original QM method is that the same precipitation value modeled at different times is mapped onto the same value by the original QM method. To overcome this issue, instead of using the identical CDF function for the bias correction, we propose four discriminant factors in MPTT-QM to adjust the forecast value so as to determine which interval's CDF matching function should be chosen for the correction. The details of the discriminant factors are described below.

3.2. *Discrimination Factors*

3.2.1. Discrimination Factor One

The threshold distribution of the observation data (i.e., IMERG-final) is an important reference for estimating the distribution interval of the forecasting threshold. F_{time_t} represents the threshold interval with the maximum probability of precipitation distribution at the t_{th} time of a year according to long-term observation. Specifically,

$$F_{time_t} = Max\left(k_{(i,t)}/Sum_i\right) \tag{3}$$

where $k_{(i,t)}$ denotes the number of times when the precipitation values fall in each threshold interval at the same time of the year (i.e., the same month, day, and hour), i denotes the threshold intervals ($i = 1, 2 \cdots n, n = 13$), and t denotes the 3-h interval time in a year ($t = 1, 2 \cdots m; m = 366 \times 8$). Sum_i is the total number of OBS_{his} values falling in the ith interval for all the observation times.

3.2.2. Discrimination Factor Two

The error relationship between historical observation and historical model data can be used as an important reference for estimating the distribution interval of the forecasting threshold [23]. The historical precipitation data are extracted at the same time (the same month, day, and hour) as the forecasting precipitation, so that there are fifteen groups in total. The average precipitation in the historical period is defined as the sixteenth group of data. The series of data are written as $MOD_j(j = 1, 2, \cdots, n, n+1, n = 15)$. The correlation coefficient between the MOD_{fore} and $MOD_j(j = 1, 2, \cdots, n, n+1, n = 15)$ is calculated, respectively, and the *j-group* with the largest correlation coefficient is recorded as MOD_{Max}. One extracts the observation data of the corresponding year of MOD_{Max}, which is written as OBS_{Max}. The linear fitting method is used to simulate the error relationship between MOD_{Max} and OBS_{Max} [24]:

$$OBS_{Max} = a + b \times MOD_{Max} \tag{4}$$

where a and b are the linear fitting parameters calculated by the ordinary least squares method. It is assumed that the same error relationship is also followed for future forecast data:

$$MOD_{fore-BC} = a + b \times MOD_{fore} \tag{5}$$

where $MOD_{fore-BC}$ is the forecast data corrected using the error relationship. The distribution interval of $MOD_{fore-BC}$ is taken as F_2.

3.2.3. Discrimination Factors Three and Four

The removal of multiplicative errors is a convenient method for bias correction on sub-seasonal to seasonal scales [57]. $F_{MOD_{fore_t}}$ $(t = 1, 2 \cdots m; m = 366 \times 8)$ represents the forecast threshold interval, with multiplicative error being adjusted as follows:

$$F_{MOD_{fore_t}} = MOD_{fore_t} \times \overline{OBS_{his_t}} / \overline{MOD_{his_t}} \tag{6}$$

where MOD_{fore_t} is the forecast value to be corrected, $\overline{OBS_{his_t}}$ is the average value of the observation in the historical period for the same time as the forecast time, and $\overline{MOD_{his_t}}$ is the average value of the forecast period. The distribution interval of $F_{MOD_{fore_t}}$ is taken as F_3. The distribution interval of $\overline{OBS_{his_t}}$ is taken as F_4.

3.2.4. Random Forest

Random forest is a classification algorithm based on a tree classifier, which was first proposed by Breiman [58]. There are many advantages to random forest, for example, it relieves the overfitting problem that often occurs in machine learning. At the same time, the selection of characteristic genes can be carried out. A large number of theoretical and applied studies have proved the accuracy of the random forest model from different angles [59,60]. At present, random forest is considered as one of the best machine learning models due to its tolerance of outliers and noise in the dataset. The contribution weights of different factors to a group of data can be obtained through the random forest model, which is also one of the characteristics of random forest.

In this study, four discriminant factors of each time step in the historical period were used as input data, and the true threshold interval of the historical observation data was used as the target data. Then, the contribution weight of each factor to the true threshold interval of the observed data was calculated through the random forest model.

$$F = \sum_t (F_t \times Weight_t) \tag{7}$$

Above, F_t is the t_{th} factor, and the weight coefficient of F_t is $Weight_t (t = 1, 2, \ldots, 4)$. The weighted result F is the estimated distribution interval of the forecasting threshold. This means that the forecasting data should be matched with the distribution interval F, and the CDF function of the history observation and history model in interval F should be adopted for further correction.

It is critical issue to address the 0 value in the forecast. The method proposed by Tian et al. [61] is adopted in this study. When the forecast precipitation is 0, the OBS_{his} and MOD_{his} of the first eight timesteps (one day) of MOD_{fore} are extracted to calculate the number of missed (m) and false alarms (f) of MOD_{his} according to OBS_{his}. If $m <= f$, it is determined that the forecast precipitation is 0. If $m > f$, the mean value of the first eight timesteps is used to replace the forecast.

3.3. Additional Spatial Correction

In order to maintain better smoothness and continuity of the spatial distribution of the corrected model precipitation data, the outliers need to be removed from the data. Dixon and Dean [62] proposed a simplified outlier test method for smaller sample sizes ($n < 10$), namely the Q-test (or Dixon's Q-test). This method has been widely used in many scientific research fields, such as international analytical chemistry and materials, for a long time. The calculation of the statistic Q value is very simple; the difference between the suspicious value and its nearest value is divided by the range. The calculation formula is written as follows:

$$Q_1 = \frac{x_2 - x_1}{x_n - x_1} \text{ or } Q_n = \frac{x_n - x_{n-1}}{x_n - x_1} \tag{8}$$

According to the measured sample number and the given confidence, one can check the critical value table to obtain the value $Q_p(n)$. If Q_1 (or Q_n) > $Q_p(n)$, there are outliers in the sample data. Otherwise, the data are without any outlier. In this study, we select the spatial window of 3 × 3 grid cells (the number of samples $n = 9$) and gradually slide it to identify outliers. We replace the outlier with the arithmetic mean of the data of the nearest eight non-outlier and non-missing grid cells. That is, a simple interpolation and replacement calculation is performed on the outlier. The outliers among the corrected data are removed by the additional spatial correction. The spatial continuity distribution of the corrected data is better maintained.

3.4. The DRIVE Model

The hydrologic model has been demonstrated to be an effective and efficient tool for monitoring, simulating, and forecasting floods [63,64]. The hydrological simulations were conducted using the Dominant river Routing Integrated with VIC Environment (DRIVE) model, which was developed by Wu [65] through coupling the DRTR (Dominant River Tracing, DRT-based runoff-Routing) model with the VIC (Variable Infiltration Capacity) land surface model. To be applied for spatially distributed and real-time runoff prediction, the VIC model has further been significantly modified (in particular, from its original point-based model structure to a grid-based model structure) so that the modified VIC as a runoff generation component of the DRIVE model is capable of simulating spatially distributed runoff at each time step (i.e., computing all the grid boxes at each time step) [65]. The DRTR model includes a package of hydrographic upscaling (from fine spatial resolution to coarse resolution) algorithms and resulting global datasets (flow direction, river network, drainage area, flow distance, slope, etc.) especially designed for large-scale hydrologic modeling. The DRTR model is grid based and very convenient for simulating spatially distributed streamflow by coupling with the modified VIC model. More details about the DRIVE model can be found in Wu et al. [66]. The DRIVE model has been used routinely for global flood forecasting and monitoring [66], implementing TRMM global satellite precipitation products [67].

3.5. Evaluation Methods

Six precipitation products were used in this study to evaluate precipitation and hydrological performance. They are IMERG-final (IMERG, observation) and FGOALS (model). The FGOALS model precipitation data are corrected by QM and MPTT-QM for 14-day and 90-day lead time forecasts, respectively, which are called QM-14day, QM-90day, MPTT-QM-14day, and MPTT-90day. This also means that the revised calculation is repeated every 14 (90) days. After the completion of each correction process, the observations for these 14 (90) days were summarized into the historical phase dataset, and the initial conditions were recalculated for the next revision.

Firstly, the 3-hourly precipitation products of 2016–2020 were accumulated to daily, 5 days, 15 days, and monthly. Then, the precipitation accuracy was evaluated for each time scale. The model data were assessed through three widely used statistical evaluation metrics: the correlation coefficient (R), root mean square error (RMSE), and mean bias (MB). A higher R, lower RMSE, and absolute MB indicate better agreement between the estimations and observations. The formulas are provided in Table 2. In addition, three indicators were selected in this study to evaluate the precipitation detection capabilities, including the probability of detection (POD), critical success index (CSI), and false alarm ratio (FAR). POD represents the ratio of correct estimates to the number of precipitation occurrences based on observations. FAR denotes the proportion of precipitation occurrences that were erroneously detected. CSI indicates the overall performance in terms of detection capability by integrating POD and FAR. The values of these indicators range from 0 to 1, and a higher POD and CSI and lower FAR indicate a better performance.

Therefore, six different types of precipitation data were used in this study to run the DRIVE model on the 3-hourly time scale and the 0.125-degree spatial scale. The simulated runoff was compared with the runoff observed at eight hydrological stations in the PRB. The Kling–Gupta efficiency coefficient (KGE) was selected as the hydrological assessment indicator. The KGE coefficient is a comprehensive evaluation index integrating the correlation coefficient (R), bias ratio (β), and variability ratio (γ). KGE can comprehensively evaluate the performance of simulation data, and the optimum score is 1.

Table 2. Statistical metrics used for evaluating precipitation and runoff estimates. P is precipitation estimate; P_{obs} is observation data; Q is runoff estimate; Q_{obs} is observation runoff from the gauge; Cov is the covariance; σ is the standard deviation and μ is the mean value; n is the number of data pairs; H is the number of observed precipitation events detected correctly by the products; F is the number of precipitation events detected by the products but not observed; and M is the number of precipitation events that the products cannot detect.

Statistical Metrics	Formulas	Optimal Score
Correlation coefficient (R)	$R = \frac{Cov(P,P_{obs})}{\sqrt{\sigma(P)\sigma(P_{obs})}}$	1
Root mean square error ($RMSE$)	$RMSE = \sqrt{\frac{1}{n}\sum(P - P_{obs})^2}$	0
Mean bias (MB)	$MB = \frac{1}{n}\sum(P - P_{obs})$	0
Probability of detection (POD)	$POD = \frac{H}{H+M}$	1
Critical success index (CSI)	$CSI = \frac{H}{H+M+F}$	1
False alarm ratio (FAR)	$CSI = \frac{F}{H+F}$	0
Bias ratio (β)	$\beta = \frac{\mu_P}{\mu_{P_{obs}}}$	1
Variability ratio (γ)	$\gamma = \frac{\sigma_P/\mu_P}{\sigma_{P_{obs}}/\mu_{P_{obs}}}$	1
Kling–Gupta efficiency (KGE)	$KGE = 1 - \sqrt{(R-1)^2 + (\beta-1)^2 + (\gamma-1)^2}$	1

4. Results

4.1. Precipitation Assessment Results

Figure 3 illustrates the spatial patterns of daily average precipitation from 2016 to 2020, derived from the IMERG, FGOALS, QM, and MPTT-QM precipitation products. It can be seen that the IMERG precipitation in the PRB shows a decreasing trend from east to west. The maximum precipitation area is concentrated in the east of the PRB. The FGOALS model data show a higher value in the eastern area and lower value in the western area. However, the overall precipitation value is smaller than that of IMERG, and there are also great differences in the spatial details. The precipitation products corrected by QM effectively improve the precipitation in the eastern and central areas of the PRB, and QM is more similar to IMERG in terms of spatial distribution. However, the figure shows that the maximum daily rainfall area of IMERG precipitation reaches more than 5 mm/day in the east of the PRB, and there are certain differences between QM-14day and QM-90day, on the one hand, and IMERG, on the other. The precipitation corrected by MPTT-QM is more consistent with the overall spatial distribution of IMERG. It clearly shows four rainbands with decreasing precipitation from east to west. The distribution of MPTT-QM-14day is better than that of MPTT-QM-90day. MPTT-QM-90day has an overestimation trend in the central part of the Pearl River Basin relative to MPTT-QM-14day.

To further evaluate the spatial distribution consistency between the corrected precipitation and IMERG precipitation, Figure 4 shows a density scatter diagram of the daily average precipitation distribution of each grid cell in the study area. It can be seen from Figure 4a that the FGOALS data show an overall small trend for IMERG. In particular, when the daily average precipitation of IMERG is distributed in the range of 4–5 mm, the FGOALS is as small as 50%. However, for some precipitation maxima, FGOALS shows a higher estimation. The QM-corrected data can effectively solve the problem of the serious underestimation of FGOALS. However, when IMERG precipitation is in the range of 4–6 mm, QM-14day and QM-90day are still slightly low. No matter how high or low the precipitation value is, the scattered points marking the spatial distribution of MPTT-QM and IMERG are basically situated around the y = x baseline, and the performance is clearly better than that of the QM method. As can be seen in Figure 4, there is a more pronounced overestimation trend in MPTT-QM-90day than MPTT-QM-14day when the precipitation level is in the 5 mm interval, which is consistent with Figure 3. In terms of spatial correlation coefficient values and root mean square error values, MPTT-QM-90day and MPTT-QM-14day are very similar, but the spatial mean bias is smaller for MPTT-QM-14day.

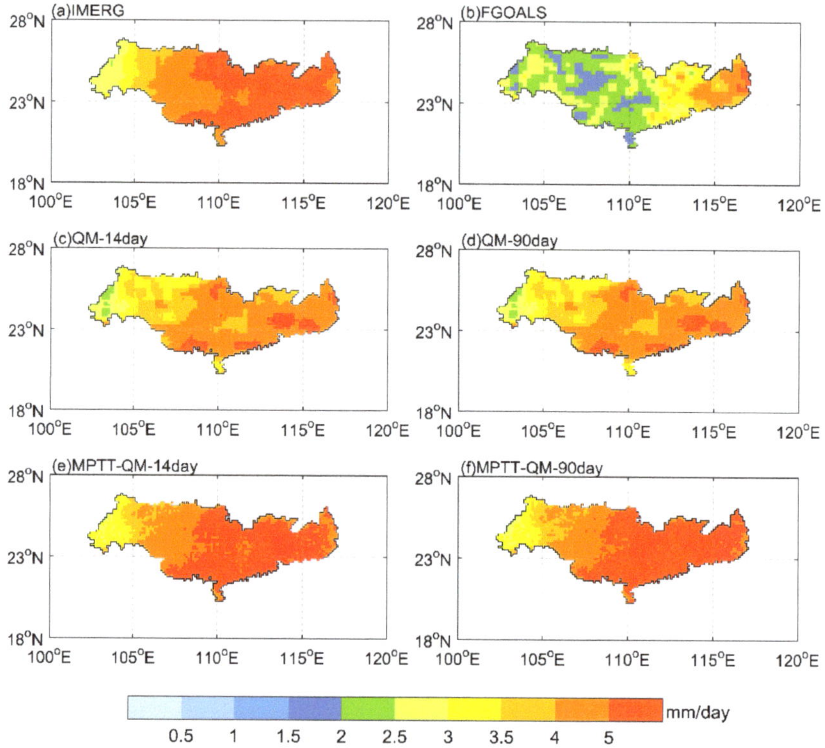

Figure 3. Spatial patterns of daily average precipitation from 2016 to 2020 in the PRB: (**a**–**f**) represent the IMERG, FGOALS, QM-14day, QM-90day, MPTT-QM-14day, and MPTT-QM-90day, respectively.

To verify the time series trend of precipitation on the basin scale, the basin mean values of different precipitation products in the PRB were calculated on the 15-day scale and the monthly scale in this study (Figure 5). The figure shows that the basin mean precipitation of IMERG at 15 days (red line in Figure 5a–e) has changed steadily over the past five years, and the maximum precipitation period is mainly from May to September each year. The three wet years are 2016, 2019, and 2020. Among these, the wet years of 2016 and 2020 were caused by short-term heavy precipitation events, and that of 2019 was caused by continuous precipitation events.

Regarding the IMERG data, the basin mean precipitation values in the first half of June and August in 2016 and the first half of June and September in 2020 reached 200 mm. The basin mean precipitation in the first half of June 2020 reached 242.24 mm, which is the highest value within the past five years. In Figure 5a, the precipitation of the FGOALS model (blue line) shows a time lag trend and lower precipitation relative to IMERG. Figure 5b,c illustrates that the QM does not have the ability to change the trend of model precipitation. It can only adjust the precipitation value at each time step to make it larger and, therefore, closer to the distribution of IMERG. Moreover, the MPTT-QM model can change the trend of precipitation in the time series (Figure 5d,e). Comparing the MPTT-QM with the QM and FGOALS data, it can be found that MPTT-QM has better consistency with IMERG, especially for 2017, 2018, and 2019. However, for the extreme precipitation events in 2016 and 2020, although MPTT-QM can effectively improve the original data, the correction performance of MPTT-QM for extreme precipitation events still needs to be further improved.

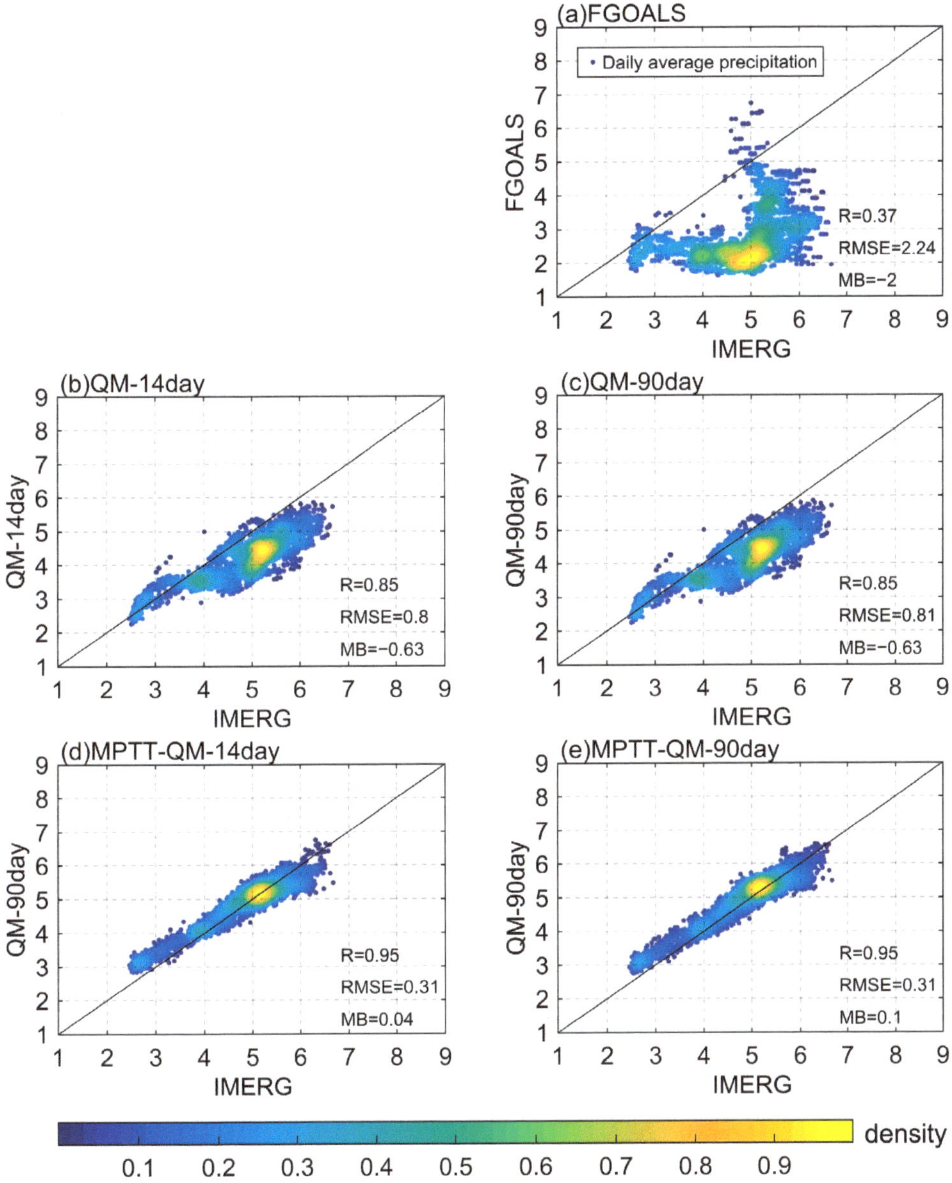

Figure 4. Density scatter diagram of daily average precipitation of each grid cell relative to IMERG in the PRB from 2016 to 2020. Each scatter point represents the daily average precipitation value of each grid cell in the study area, the color bar represents the density of precipitation values. and (**a**–**e**) represent the FGOALS, QM-14day, QM-90day, MPTT-QM-14day, and MPTT-QM-90day, respectively (unit: mm/day).

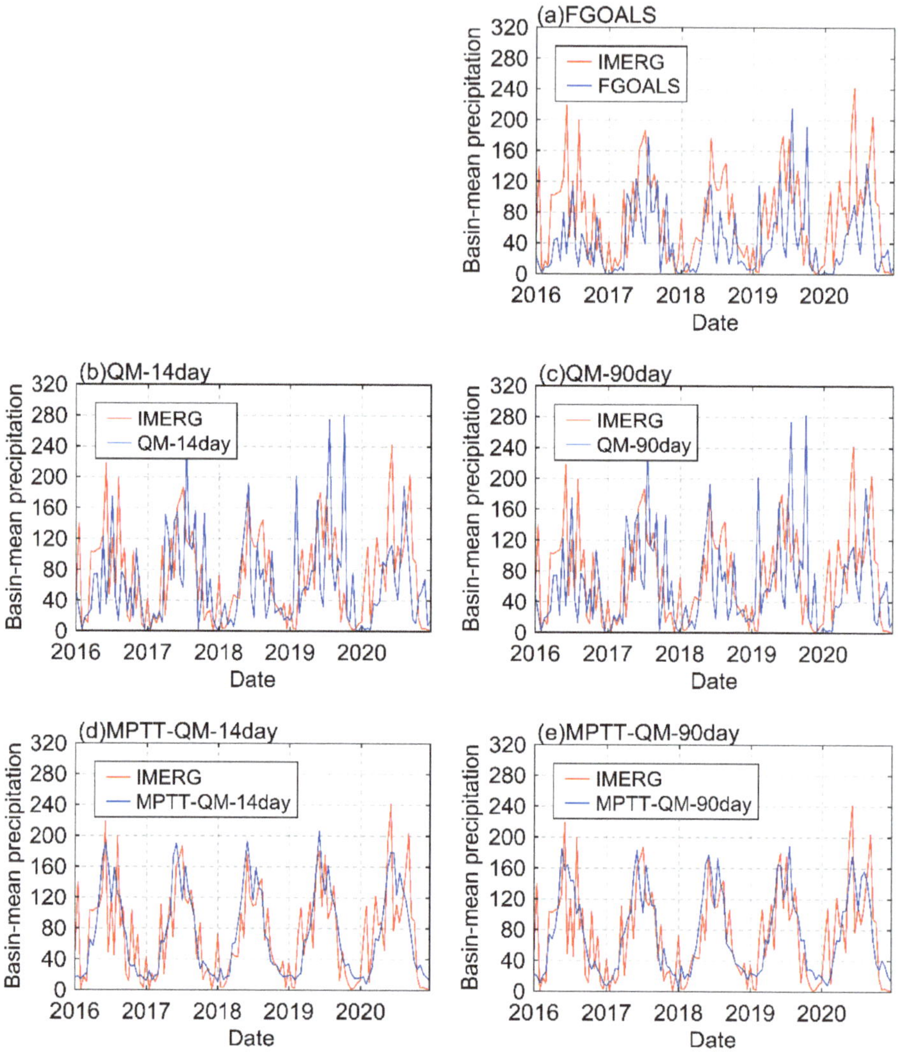

Figure 5. Time series distribution of average precipitation in the PRB on a half-month scale from 2016 to 2020. The red line represents the IMEGR, and the blue line in figure (**a**–**e**) represents the FGOALS, QM-14day, QM-90day, MPTT-QM-14day, and MPTT-QM-90day, respectively (unit: mm).

Figure 6 demonstrates the boxplots of R, RMSE, and MB for the MPTT-QM model and other precipitation products over four different time scales. The MPTT-QM model performs better than the other products, with a higher R, lower RMSE, and lower MB. With the increase in the time scale from days to months, the R value becomes higher. However, the daily scale is enhanced to a stronger degree than the monthly scale. For example, for the correlation coefficient R, the median value of FGOALS on the daily scale is 0.02, and the values of MPTT-QM-14day and MPTT-QM-90day are 0.15 and 0.13, which are 650% and 550% higher than the original model data. The monthly FGOALS median value is 0.33, and the MPTT-QM-14day and MPTT-QM-90day are 0.78 and 0.76, respectively, which are 136% and 130% higher than the value for FGOALS. For the QM precipitation products, the performance of R and RMSE is equivalent to that of FGOALS, and there is no significant improvement. On the daily scale, the performance of R and RMSE are slightly decreased. MB is effectively improved using the QM method. The median value of MB is increased from −2.19 to −0.66 (QM-14day) and −0.66 (QM-90day), respectively, on the daily scale. Meanwhile, the median of MB of the MPTT-QM model performed better, with values of 0.07 (MPTT-QM-14day) and 0.13 (MPTT-QM-90day) on the daily scale, respectively. Overall, in terms of time distribution, MPTT-QM-14day outperform MPTT-QM-90day significantly.

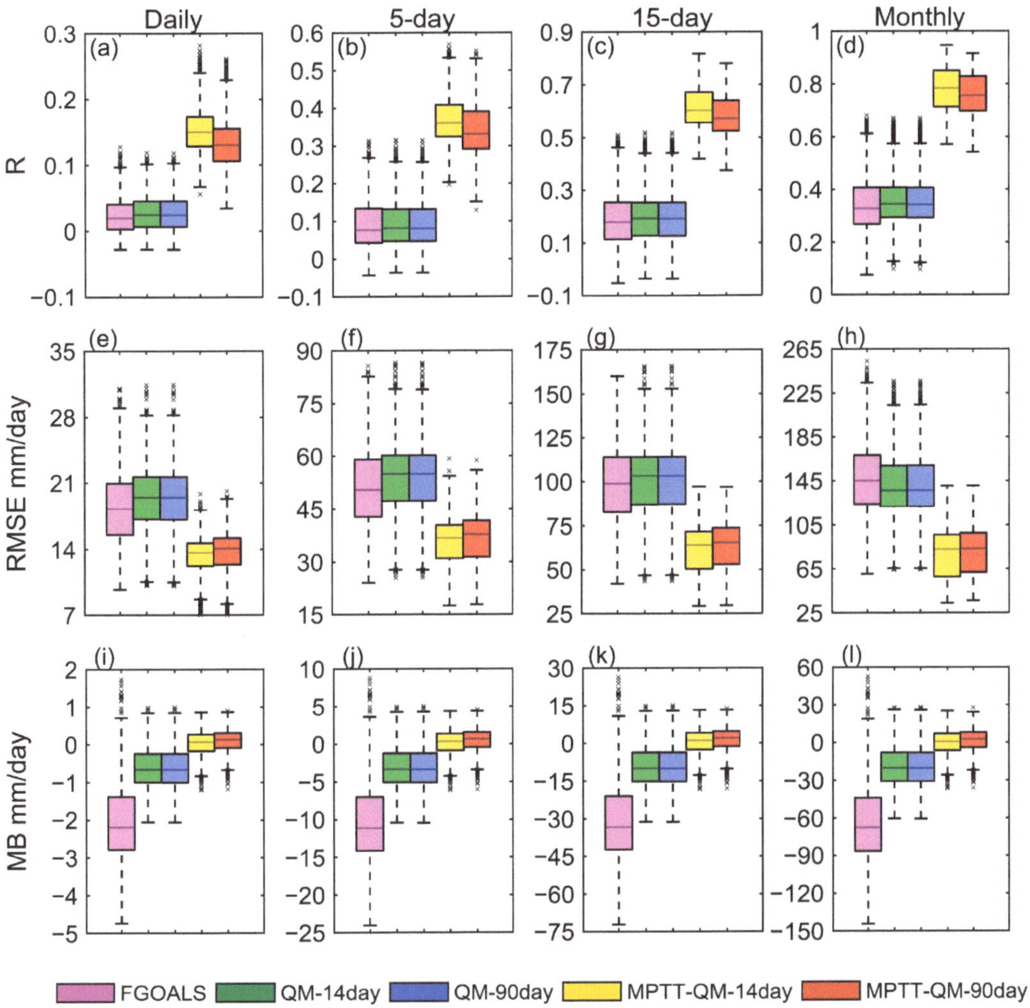

Figure 6. Boxplots of correlation coefficient (R), root mean square error (RMSE), and mean bias (MB) for five precipitation products over four different time scales: daily, 5-day, 15-day, and monthly. (**a–d**) represent the boxplot of R, (**e–h**) represent the boxplot of RMSE, and (**i–l**) represent the MB.

Table 3 summarizes the statistical metrics of the five precipitation products. The values in the table are the basin average values of the statistical metrics for the PRB on a daily scale. Although the FAR values of MPTT-QM-14day and MPTT-QM-90day are slightly higher, other indicators should be comprehensively considered. Combining the results of several assessment indicators in Figures 4 and 6 in terms of temporal and spatial distribution, the performance of the revised product decreases significantly as the prediction time increases. Overall, the MPTT-QM-14day shows the best performance among the five precipitation products, with POD of 0.950. The second-best-performing product is MPTT-QM-90day, which has the highest CSI value.

Table 3. Summary of performance statistical metrics for the FGOALS, QM-14day, QM-90day, MPTT-QM-14day, and MPTT-QM-90day precipitation products. The precipitation threshold is set to 0.1 mm/day.

Names	CSI	POD	FAR
FGOALS	0.302	0.412	0.461
QM-14day	0.322	0.444	0.462
QM-90day	0.322	0.443	0.462
MPTT-QM-14day	0.486	0.950	0.501
MPTT-QM-90day	0.488	0.946	0.498

It is worth mentioning that the POD values of MPTT-QM-14day and MPTT-QM-90day are greater than 90% for all grid cells in the PRB. For MPTT-QM-14day, 49.6% of the total grid cells in the PRB have POD values greater than 0.95, and 40% have POD values greater than 0.95 for MPTT-QM-90day. The MPTT-QM-14day and MPTT-QM-90day methods effectively improve the value of CSI by more than 0.4, increasing from 3.6% to 95.3% and 95.1%, respectively.

4.2. Hydrological Assessment Results

First, six precipitation products were used to run the DRIVE model. The KGE coefficient results obtained by comparing the runoff observation data of the hydrological stations on the monthly scale are shown in Figure 7. The runoff simulation results show that the calibrated DRIVE model is efficient. The runoff results of IMERG-DRIVE show that the KGE coefficients of the eight hydrological stations in the PRB are more than 0.48, and there are five stations with monthly KGE values greater than 0.60 (Figure 7). The average of the monthly KGE coefficient is 0.59. Among the stations, the Nanning station data are simulated best, and the monthly KGE coefficient reaches 0.66. The simulation effect of FGOALS-DRIVE on the PRB is unsatisfactory. The average of the monthly KGE coefficient is 0.01. These unsatisfactory results are related to the poor self-quality of the FGOALS precipitation. After QM bias correction, the effect of the runoff simulation was slightly improved. The monthly KGE average increased to 0.16 (QM-14day) and 0.15 (QM-90day), respectively. However, the runoff simulation results of QM still fall short of the credible standard. According to the results of the MPTT-QM-14day runoff simulation, there are five hydrological stations with monthly KGE values greater than 0.4 in the PRB, and the average value reaches 0.45. The performance of MPTT-QM-90day is slightly worse, the average value reaches 0.40. The performance of the MPTT-QM method in hydrology is more related to the self-quality of the observed precipitation.

On the daily scale, the average correlation between the output runoff data of IMERG-DRIVE and the observed runoff data is 0.74, indicating that the data are highly correlated (Figure 8a). The average correlation coefficient of FGOALS-DRIVE is 0.19, and Nanning station has the highest correlation, which is 0.23. The average correlation coefficients of both QM-14day and QM-90day are 0.16 on the daily scale, a value which is slightly lower than that of FGOALS. The runoff correlation was significantly improved by the MPTT-QM method. After MPTT-QM-14day correction, the correlation coefficients of seven hydrological stations were greater than 0.4. Wuzhou station, located in the middle of the PRB, has the highest correlation coefficient, which reaches 0.60. On the monthly scale, MPTT-QM-14day performed better, raising the average correlation coefficient of FGOALS on the monthly scale from 0.37 to 0.71. Therefore, by comprehensively comparing the KGE coefficient and correlation coefficient of QM and MPTT-QM relative to the observed runoff, it can be seen that the MPTT-QM method is more effective than the QM method.

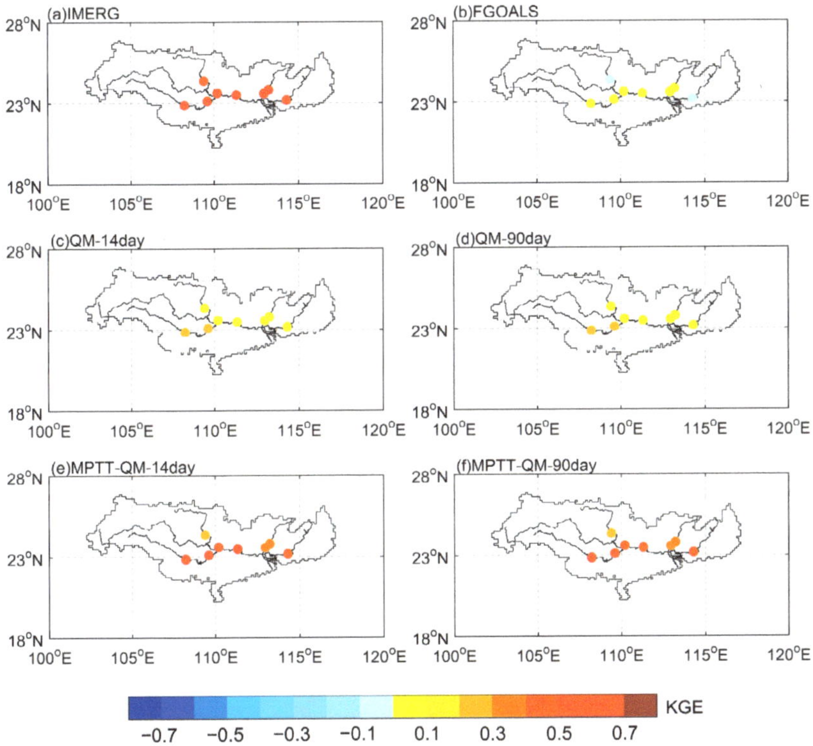

Figure 7. Spatial distribution of the monthly KGE coefficients of hydrological stations in the PRB: (**a**–**f**) represent the IMERG, FGOALS, QM-14day, QM-90day, MPTT-QM-14day, and MPTT-QM-90day, respectively.

The KGE coefficients of area rainfall of the upstream basin and simulated runoff for the five precipitation products, compared with IMERG, are shown in Table 4. In general, the performance of MPTT-QM in estimating precipitation and runoff of the eight stations in the PRB is significantly better than that of QM method. The MPTT-QM-14day is the best-performing model. For the MPTT-QM-14day model, the average KGE coefficient of the area rainfall compared with FGOALS is increased by nearly 3.82 times, and the average KGE coefficient of the runoff is increased by more than 12.94 times for the eight hydrological stations in the PRB. For the MPTT-QM-90day model, the average KGE coefficients of area rainfall of the upstream basin and runoff are increased by 3.78 times and nearly 12.60 times, respectively. The same conclusion is obtained using the QM method. According to the KGE coefficient of QM-14days, the average rainfall value of the eight stations is doubled, and the average runoff value is increased by nearly 5.03 times. The KGE coefficient of QM-90day is nearly doubled and increased by 4.98 times. It can be seen that both QM and MPTT-QM perform better for the 14-day lead time forecast than 90-day one. Although the KGE coefficient of the area rainfall is higher, both the MPTT-QM and QM bias correction methods are clearly more efficient in improving the runoff simulation.

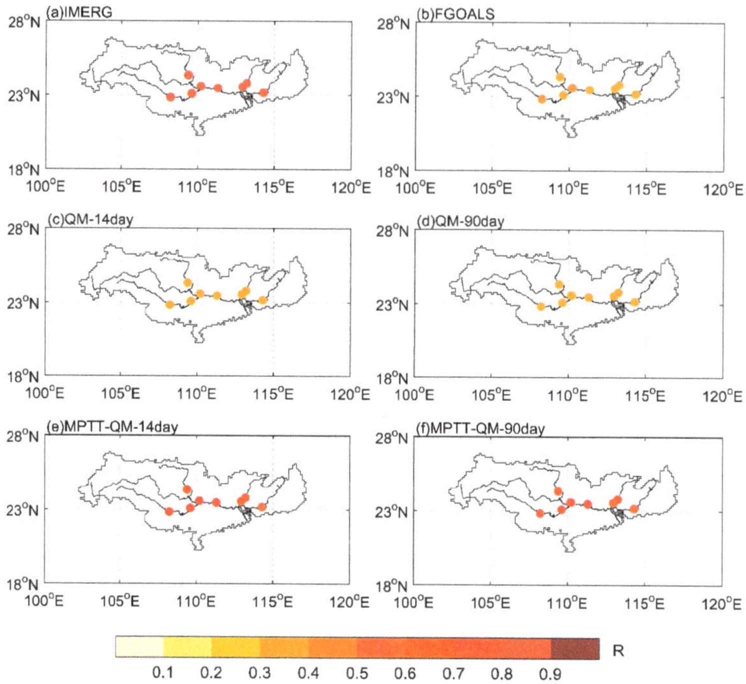

Figure 8. Spatial distribution of daily correlation coefficients (R) of hydrological stations in the PRB: (**a**–**f**) represent the IMERG, FGOALS, QM-14day, QM-90day, MPTT-QM-14day, and MPTT-QM-90day, respectively.

Table 4. Summary of performance based on the monthly KGE coefficients of hydrological stations in the PRB for the FGOALS, QM-14day, QM-90day, MPTT-QM-14day, and MPTT-QM-90day. For a hydrological station, *P* represents the KGE coefficient of model-simulated area rainfall of the upstream basin compared with IMERG. *Q*. represents the KGE coefficient of model-simulated runoff compared with the IMERG-DRIVE-simulated runoff.

Names		Boluo	Feilaixia	Shijiao	Liuzhou	Nanning	Guigang	Dahuangjiangkou	Wuzhou
FGOAIS	P	0.291	0.006	0.018	0.053	0.220	0.201	0.300	0.293
	Q	0.428	0.012	−0.014	−0.033	−0.085	−0.107	0.103	0.093
QM-14day	P	0.365	0.188	0.204	0.352	0.383	0.369	0.478	0.480
	Q	0.439	0.194	0.186	0.242	0.364	0.313	0.344	0.318
QM-90day	P	0.364	0.186	0.202	0.348	0.381	0.366	0.475	0.476
	Q	0.437	0.191	0.184	0.240	0.360	0.309	0.342	0.316
MPTT-QM-14day	P	0.761	0.798	0.804	0.793	0.892	0.886	0.876	0.863
	Q	0.701	0.573	0.617	0.550	0.793	0.768	0.777	0.766
MPTT-QM-90day	P	0.731	0.780	0.788	0.774	0.882	0.881	0.890	0.883
	Q	0.701	0.573	0.594	0.524	0.802	0.769	0.724	0.720

Figure 9 shows the monthly runoff intensity–time curve after the removal of the missing values of the Guigang, Nanning, Wuzhou, and Liuzhou stations in the PRB from 2016 to 2020. The figure shows that the runoff curve (red) simulated by IMERG-DRIVE is closely matched with the distribution of the observation data for most time periods. The runoff value of FGOALS-DRIVE is very small for all the hydrological stations. The QM-DRIVE-simulated runoff data are similar to the precipitation, which is corrected by QM. Only the numerical value can be changed, and it is difficult to change the trend of the runoff. The runoff data simulated by MPTT-QM-DRIVE at four stations show a good performance, which is basically consistent with the runoff intensity–time curve of IMERG-DRIVE.

Figure 10 illustrates the intensity–time curve of the monthly area rainfall of the upstream basin for the Guigang, Nanning, Wuzhou, and Liuzhou stations. The figure demonstrates that the variation in area rainfall is more complex relative to the runoff. The area rainfall of the upstream basin for the four stations obtained by the MPTT-QM method also indicates a good performance.

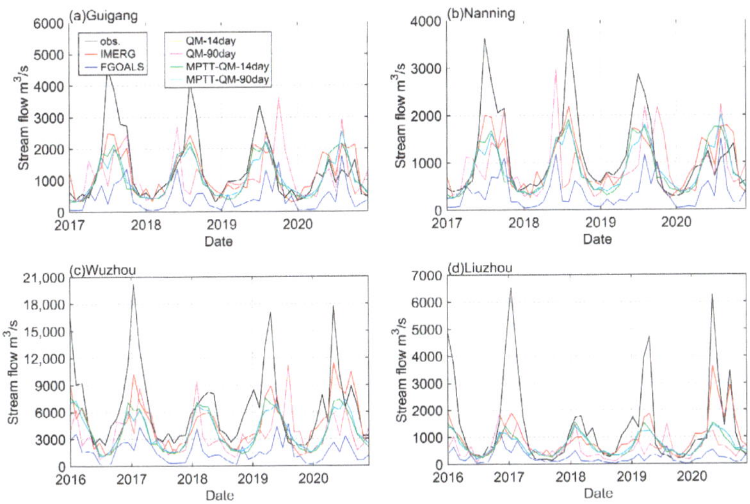

Figure 9. Runoff intensity–time curve of the PRB on the monthly scale from 2016 to 2020: (**a**–**d**) represent the Guigang, Nanning, Wuzhou, and Liuzhou stations, respectively.

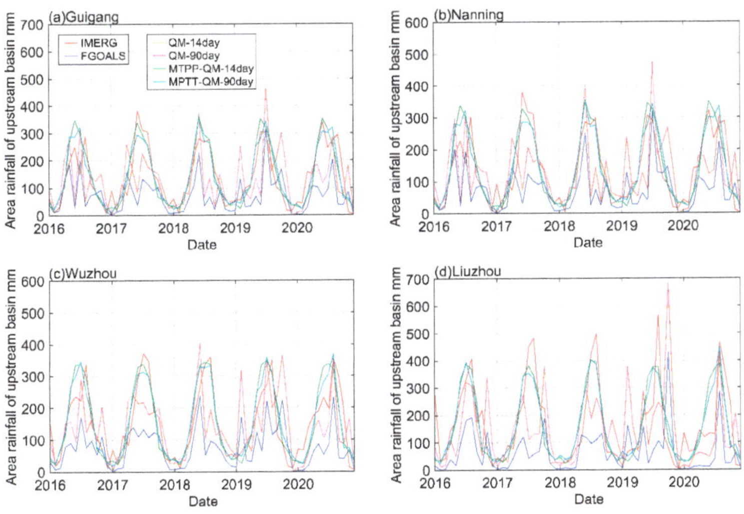

Figure 10. Area rainfall of the upstream basin intensity–time curve of the PRB on the monthly scale from 2016 to 2020: (**a**–**d**) represent the Guigang, Nanning, Wuzhou, and Liuzhou stations, respectively (unit: mm).

5. Discussion

In general, the MPTT-QM-corrected precipitation data indicate a significantly better performance than that of the original QM method in terms of the consistency of the temporal and spatial distribution on the sub-seasonal to seasonal scale. This is because MPTT-QM overcomes the problem that the same precipitation value modeled at different times is mapped onto the same value using the original QM method by correction at the precipitation threshold intervals. In order to more accurately identify the threshold intervals of the forecast precipitation, the four discrimination factors are selected to comprehensively consider the threshold distribution characteristics of the observed precipitation in the historical period and the error relationship between the observed precipitation and the model precipitation.

At the same time, due to the fact that precipitation errors can be transmitted using the hydrological model, more accurate precipitation data will also lead to an improvement of the hydrological simulation performance. Therefore, the MPTT-QM method also has an excellent performance in hydrological simulations.

When the MPTT-QM model is applied in practice, it will provide a solid foundation for the prediction and early warning of flood disasters. At the same time, the MPTT-QM model also requires further improvement. For example, the performance of the MPTT-QM model will decrease slightly with the lead time increase in the bias correction. Moreover, the question of how to predict and correct the occurrence of extreme precipitation events using the MPTT-QM model is the primary problem to be solved in the next stage of model research and development.

6. Conclusions

In this paper, we proposed a new precipitation bias correction method based on QM to match precipitation thresholds by time series, which is called the MPTT-QM model. FGOALS model data were used to estimate 3-h precipitation in the PRB at a spatial resolution of 0.125 degrees. The model performance and retrieval results are summarized as follows:

1. The MPTT-QM model has better consistency with IMERG than the original QM model in terms of spatial distribution. The MPTT-QM model excelled in terms of the RMSE and MB;
2. MPTT-QM can effectively optimize the change in the precipitation series and improve the correlation coefficient between the model and observation data, which the QM method cannot achieve to any meaningful extent. For a 14-day lead time forecast, MPTT-QM increases the average correlation coefficient of the PRB by nearly six times compared to the original FGOALS model on the daily scale;
3. MPTT-QM also shows a stable performance in terms of the POD and CSI. MPTT-QM shows a good precipitation detection ability for the 14-day to 90-day lead time forecasts;
4. Based on the hydrological performance evaluation, the KGE coefficients of the eight hydrological stations are improved significantly using the MPTT-QM-DRIVE model compared to the QM-DRIVE model.

Author Contributions: X.L. and H.W. conceived and designed the methods and frameworks; X.L. completed the experiments and wrote the paper; N.N. and H.W. helped to edit the paper. Y.H. and L.L. helped analyze the results; S.C. provided funding support. All authors have read and agreed to the published version of the manuscript.

Funding: This study was supported by the National Natural Science Foundation of China (Grants: 42088101, 42275019, U1811464), and also partially supported by the Program for Guangdong Introducing Innovative and Entrepreneurial Teams (Grants: 2017ZT07X355, 2020B1212060025), the Key R&D Program of Guangxi (Guike AB21075008) and Hainan R&D Program (CXFZ2022J074). The authors would like to thank the anonymous reviewers for their constructive suggestions.

Data Availability Statement: Please contact the corresponding author for the method code and data referenced in this publication.

Acknowledgments: The authors thank the China Meteorological Administration (CMA), Institute of Atmospheric Physics, Chinese Academy of Science, and the National Oceanic and Atmospheric Administration (NOAA) for providing the datasets.

Conflicts of Interest: The authors declare no conflict of interest.

References

1. Cross, R. International federation of red cross and Red Crescent Societies. *Personnel* **2003**, *1*, 754.
2. Pachauri, R.K.; Allen, M.R.; Barros, V.R.; Broome, J.; Cramer, W.; Christ, R.; Church, J.A.; Clarke, L.; Dahe, Q.; Dasgupta, P. Climate change 2014: Synthesis report. In *Contribution of Working Groups I, II and III to the Fifth Assessment Report of the Intergovernmental Panel on Climate Change*; IPCC: Geneva, Szwitzerland, 2014.
3. Pappenberger, F.; Bartholmes, J.; Thielen, J.; Cloke, H.L.; Buizza, R.; de Roo, A. New dimensions in early flood warning across the globe using grand-ensemble weather predictions. *Geophys. Res. Lett.* **2008**, *35*, L10404. [CrossRef]
4. Fritsch, J.M.; Carbone, R. Improving quantitative precipitation forecasts in the warm season: A USWRP research and development strategy. *Bull. Am. Meteorol. Soc.* **2004**, *85*, 955–966. [CrossRef]
5. Schumacher, R.S. The studies of precipitation, flooding, and rainfall extremes across disciplines (SPREAD) workshop: An interdisciplinary research and education initiative. *Bull. Am. Meteorol. Soc.* **2016**, *97*, 1791–1796. [CrossRef]
6. Vitart, F. Monthly forecasting at ECMWF. *Mon. Weather Rev.* **2004**, *132*, 2761–2779. [CrossRef]
7. Keenan, T.; Joe, P.; Wilson, J.; Collier, C.; Golding, B.; Burgess, D.; May, P.; Pierce, C.; Bally, J.; Crook, A. The Sydney 2000 world weather research programme forecast demonstration project: Overview and current status: Overview and current status. *Bull. Am. Meteorol. Soc.* **2003**, *84*, 1041–1054. [CrossRef]
8. Miyakoda, K.; Smagorinsky, J.; Strickler, R.F.; Hembree, G. Experimental extended predictions with a nine-level hemispheric model. *Mon. Weather Rev.* **1969**, *97*, 1–76. [CrossRef]
9. Lorenz, E.N. Deterministic nonperiodic flow. *J. Atmos. Sci.* **1963**, *20*, 130–141. [CrossRef]
10. Lorenz, E.N. Atmospheric predictability as revealed by naturally occurring analogues. *J. Atmos. Sci.* **1969**, *26*, 636–646. [CrossRef]
11. Li, J.; Ding, R. Temporal-spatial distributions of predictability limit of short-term climate. *Chin. J. Atmos. Sci* **2008**, *32*, 975–986.
12. Li, W.; Zhu, Y.; Zhou, X.; Hou, D.; Sinsky, E.; Melhauser, C.; Peña, M.; Guan, H.; Wobus, R. Evaluating the MJO prediction skill from different configurations of NCEP GEFS extended forecast. *Clim. Dyn.* **2019**, *52*, 4923–4936. [CrossRef]
13. Vitart, F. Impact of the Madden Julian Oscillation on tropical storms and risk of landfall in the ECMWF forecast system. *Geophys. Res. Lett.* **2009**, *36*, L15802. [CrossRef]
14. Zhu, Y.; Zhou, X.; Li, W.; Hou, D.; Melhauser, C.; Sinsky, E.; Peña, M.; Fu, B.; Guan, H.; Kolczynski, W. Toward the improvement of subseasonal prediction in the National Centers for environmental prediction global ensemble forecast system. *J. Geophys. Res. Atmos.* **2018**, *123*, 6732–6745. [CrossRef]
15. Guan, H.; Zhu, Y.; Sinsky, E.; Li, W.; Zhou, X.; Hou, D.; Melhauser, C.; Wobus, R. Systematic error analysis and calibration of 2-m temperature for the NCEP GEFS reforecast of the Subseasonal Experiment (SubX) Project. *Weather Forecast.* **2019**, *34*, 361–376. [CrossRef]
16. Delle Monache, L.; Eckel, F.A.; Rife, D.L.; Nagarajan, B.; Searight, K. Probabilistic weather prediction with an analog ensemble. *Mon. Weather Rev.* **2013**, *141*, 3498–3516. [CrossRef]
17. Hagedorn, R.; Buizza, R.; Hamill, T.M.; Leutbecher, M.; Palmer, T. Comparing TIGGE multimodel forecasts with reforecast-calibrated ECMWF ensemble forecasts. *Q. J. R. Meteorol. Soc.* **2012**, *138*, 1814–1827. [CrossRef]
18. Hamill, T.M.; Bates, G.T.; Whitaker, J.S.; Murray, D.R.; Fiorino, M.; Galarneau, T.J.; Zhu, Y.; Lapenta, W. NOAA's second-generation global medium-range ensemble reforecast dataset. *Bull. Am. Meteorol. Soc.* **2013**, *94*, 1553–1565. [CrossRef]
19. Hamill, T.M.; Whitaker, J.S. Ensemble calibration of 500-hPa geopotential height and 850-hPa and 2-m temperatures using reforecasts. *Mon. Weather Rev.* **2007**, *135*, 3273–3280. [CrossRef]
20. Bennett, J.; Grose, M.; Post, D.; Ling, F.; Corney, S.; Bindoff, N. Performance of quantile-quantile bias-correction for use in hydroclimatological projections. In Proceedings of the MODSIM 2011-19th International Congress on Modelling and Simulation-Sustaining Our Future: Understanding and Living with Uncertainty, Perth, Australia, 12–16 December 2011; pp. 2668–2675.
21. Devi, U.; Shekhar, M.; Singh, G. Correction of mesoscale model daily precipitation data over Northwestern Himalaya. *Theor. Appl. Climatol.* **2021**, *143*, 51–60. [CrossRef]
22. Huang, Z.; Zhao, T.; Zhang, Y.; Cai, H.; Hou, A.; Chen, X. A five-parameter Gamma-Gaussian model to calibrate monthly and seasonal GCM precipitation forecasts. *J. Hydrol.* **2021**, *603*, 126893. [CrossRef]
23. Li, H.; Sheffield, J.; Wood, E.F. Bias correction of monthly precipitation and temperature fields from Intergovernmental Panel on Climate Change AR4 models using equidistant quantile matching. *J. Geophys. Res. Atmos.* **2010**, *115*, D10101. [CrossRef]
24. Piani, C.; Haerter, J.; Coppola, E. Statistical bias correction for daily precipitation in regional climate models over Europe. *Theor. Appl. Climatol.* **2010**, *99*, 187–192. [CrossRef]
25. Themeßl, M.J.; Gobiet, A.; Heinrich, G. Empirical-statistical downscaling and error correction of regional climate models and its impact on the climate change signal. *Clim. Chang.* **2012**, *112*, 449–468. [CrossRef]

26. Gneiting, T.; Raftery, A.E.; Westveld, A.H.; Goldman, T. Calibrated probabilistic forecasting using ensemble model output statistics and minimum CRPS estimation. *Mon. Weather Rev.* **2005**, *133*, 1098–1118. [CrossRef]
27. Wilks, D.S.; Hamill, T.M. Comparison of ensemble-MOS methods using GFS reforecasts. *Mon. Weather Rev.* **2007**, *135*, 2379–2390. [CrossRef]
28. Raftery, A.E.; Gneiting, T.; Balabdaoui, F.; Polakowski, M. Using Bayesian model averaging to calibrate forecast ensembles. *Mon Weather Rev.* **2005**, *133*, 1155–1174. [CrossRef]
29. Wilson, L.J.; Beauregard, S.; Raftery, A.E.; Verret, R. Calibrated surface temperature forecasts from the Canadian ensemble prediction system using Bayesian model averaging. *Mon. Weather Rev.* **2007**, *135*, 1364–1385. [CrossRef]
30. Robertson, D.; Shrestha, D.; Wang, Q. Post-processing rainfall forecasts from numerical weather prediction models for short-term streamflow forecasting. *Hydrol. Earth Syst. Sci.* **2013**, *17*, 3587–3603. [CrossRef]
31. Shrestha, D.L.; Robertson, D.E.; Bennett, J.C.; Wang, Q. Improving precipitation forecasts by generating ensembles through postprocessing. *Mon. Weather Rev.* **2015**, *143*, 3642–3663. [CrossRef]
32. Wang, Q.; Robertson, D.; Chiew, F. A Bayesian joint probability modeling approach for seasonal forecasting of streamflows at multiple sites. *Water Resour. Res.* **2009**, *45*, W05407. [CrossRef]
33. Cheng, W.Y.; Steenburgh, W.J. Strengths and weaknesses of MOS, running-mean bias removal, and Kalman filter techniques for improving model forecasts over the western United States. *Weather Forecast.* **2007**, *22*, 1304–1318. [CrossRef]
34. Taillardat, M.; Fougères, A.-L.; Naveau, P.; Mestre, O. Forest-based and semiparametric methods for the postprocessing of rainfall ensemble forecasting. *Weather Forecast.* **2019**, *34*, 617–634. [CrossRef]
35. Herman, G.R.; Schumacher, R.S. Money doesn't grow on trees, but forecasts do: Forecasting extreme precipitation with random forests. *Mon. Weather Rev.* **2018**, *146*, 1571–1600. [CrossRef]
36. Li, H.; Yu, C.; Xia, J.; Wang, Y.; Zhu, J.; Zhang, P. A model output machine learning method for grid temperature forecasts in the Beijing area. *Adv. Atmos. Sci.* **2019**, *36*, 1156–1170. [CrossRef]
37. Li, W.; Pan, B.; Xia, J.; Duan, Q. Convolutional neural network-based statistical post-processing of ensemble precipitation forecasts *J. Hydrol.* **2022**, *605*, 127301. [CrossRef]
38. Zhou, S.; Wang, Y.; Yuan, Q.; Yue, L.; Zhang, L. Spatiotemporal estimation of 6-hour high-resolution precipitation across China based on Himawari-8 using a stacking ensemble machine learning model. *J. Hydrol.* **2022**, *609*, 127718. [CrossRef]
39. Liu, L.; Gao, C.; Xuan, W.; Xu, Y.-P. Evaluation of medium-range ensemble flood forecasting based on calibration strategies and ensemble methods in Lanjiang Basin, Southeast China. *J. Hydrol.* **2017**, *554*, 233–250. [CrossRef]
40. Rajczak, J.; Kotlarski, S.; Schär, C. Does quantile mapping of simulated precipitation correct for biases in transition probabilities and spell lengths? *J. Clim.* **2016**, *29*, 1605–1615. [CrossRef]
41. Zhao, T.; Bennett, J.C.; Wang, Q.; Schepen, A.; Wood, A.W.; Robertson, D.E.; Ramos, M.-H. How suitable is quantile mapping for postprocessing GCM precipitation forecasts? *J. Clim.* **2017**, *30*, 3185–3196. [CrossRef]
42. Lien, G.-Y.; Miyoshi, T.; Kalnay, E. Assimilation of TRMM multisatellite precipitation analysis with a low-resolution NCEP global forecast system. *Mon. Weather Rev.* **2016**, *144*, 643–661. [CrossRef]
43. Da, C.; Kalnay, E. Improving Tropical Cyclone Predictions by Assimilation of Satellite-Retrieved Precipitation with Gaussian Transformation. In Proceedings of the AGU Fall Meeting Abstracts, Washington, DC, USA, 10–14 December 2018; p. NG23A–02
44. Wood, A.W.; Maurer, E.P.; Kumar, A.; Lettenmaier, D.P. Long-range experimental hydrologic forecasting for the eastern United States. *J. Geophys. Res. Atmos.* **2002**, *107*, ACL 6-1–ACL 6-15. [CrossRef]
45. Terink, W.; Hurkmans, R.; Torfs, P.; Uijlenhoet, R. Bias correction of temperature and precipitation data for regional climate model application to the Rhine basin. *Hydrol. Earth Syst. Sci. Discuss.* **2009**, *6*, 5377–5413.
46. Hou, A.Y.; Kakar, R.K.; Neeck, S.; Azarbarzin, A.A.; Kummerow, C.D.; Kojima, M.; Oki, R.; Nakamura, K.; Iguchi, T. The global precipitation measurement mission. *Bull. Am. Meteorol. Soc.* **2014**, *95*, 701–722. [CrossRef]
47. He, B.; Bao, Q.; Wang, X.; Zhou, L.; Wu, X.; Liu, Y.; Wu, G.; Chen, K.; He, S.; Hu, W. CAS FGOALS-f3-L model datasets for CMIP6 historical atmospheric model intercomparison project simulation. *Adv. Atmos. Sci.* **2019**, *36*, 771–778. [CrossRef]
48. He, B.; Liu, Y.; Wu, G.; Bao, Q.; Zhou, T.; Wu, X.; Wang, L.; Li, J.; Wang, X.; Li, J. CAS FGOALS-f3-L model datasets for CMIP6 GMMIP Tier-1 and Tier-3 experiments. *Adv. Atmos. Sci.* **2020**, *37*, 18–28. [CrossRef]
49. Zhao, Y.; Zou, X.; Cao, L.; Xu, X. Changes in precipitation extremes over the Pearl River Basin, southern China, during 1960–2012 *Quat. Int.* **2014**, *333*, 26–39. [CrossRef]
50. Minville, M.; Brissette, F.; Leconte, R. Uncertainty of the impact of climate change on the hydrology of a nordic watershed *J. Hydrol.* **2008**, *358*, 70–83. [CrossRef]
51. Wilby, R.L.; Hay, L.E.; Leavesley, G.H. A comparison of downscaled and raw GCM output: Implications for climate change scenarios in the San Juan River basin, Colorado. *J. Hydrol.* **1999**, *225*, 67–91. [CrossRef]
52. Grillakis, M.G.; Koutroulis, A.G.; Tsanis, I.K. Multisegment statistical bias correction of daily GCM precipitation output *J. Geophys. Res. Atmos.* **2013**, *118*, 3150–3162. [CrossRef]
53. Hyndman, R.J.; Fan, Y. Sample quantiles in statistical packages. *Am. Stat.* **1996**, *50*, 361–365.
54. Zhang, H.; Zhai, P. Temporal and spatial characteristics of extreme hourly precipitation over eastern China in the warm season *Adv. Atmos. Sci.* **2011**, *28*, 1177–1183. [CrossRef]
55. Bharti, V.; Singh, C.; Ettema, J.; Turkington, T. Spatiotemporal characteristics of extreme rainfall events over the Northwest Himalaya using satellite data. *Int. J. Climatol.* **2016**, *36*, 3949–3962. [CrossRef]

56. Martius, O.; Pfahl, S.; Chevalier, C. A global quantification of compound precipitation and wind extremes. *Geophys. Res. Lett.* **2016**, *43*, 7709–7717. [CrossRef]
57. Navarro-Racines, C.; Tarapues, J.; Thornton, P.; Jarvis, A.; Ramirez-Villegas, J. High-resolution and bias-corrected CMIP5 projections for climate change impact assessments. *Sci. Data* **2020**, *7*, 7. [CrossRef]
58. Breiman, L. Random forests. *Mach. Learn.* **2001**, *45*, 5–32. [CrossRef]
59. Chen, C.; Liaw, A.; Breiman, L. Using random forest to learn imbalanced data. *Univ. Calif. Berkeley* **2004**, *110*, 24.
60. Primajaya, A.; Sari, B.N. Random forest algorithm for prediction of precipitation. *Indones. J. Artif. Intell. Data Min.* **2018**, *1*, 27–31. [CrossRef]
61. Tian, Y.; Peters-Lidard, C.D.; Eylander, J.B.; Joyce, R.J.; Huffman, G.J.; Adler, R.F.; Hsu, K.l.; Turk, F.J.; Garcia, M.; Zeng, J. Component analysis of errors in satellite-based precipitation estimates. *J. Geophys. Res. Atmos.* **2009**, *114*, D24101. [CrossRef]
62. Dixon, W.J. Analysis of extreme values. *Ann. Math. Stat.* **1950**, *21*, 488–506. [CrossRef]
63. Reed, S.; Schaake, J.; Zhang, Z. A distributed hydrologic model and threshold frequency-based method for flash flood forecasting at ungauged locations. *J. Hydrol.* **2007**, *337*, 402–420. [CrossRef]
64. Yilmaz, K.K.; Adler, R.F.; Tian, Y.; Hong, Y.; Pierce, H.F. Evaluation of a satellite-based global flood monitoring system. *Int. J. Remote Sens.* **2010**, *31*, 3763–3782. [CrossRef]
65. Wu, H.; Adler, R.F.; Hong, Y.; Tian, Y.; Policelli, F. Evaluation of global flood detection using satellite-based rainfall and a hydrologic model. *J. Hydrometeorol.* **2012**, *13*, 1268–1284. [CrossRef]
66. Wu, H.; Adler, R.F.; Tian, Y.; Huffman, G.J.; Li, H.; Wang, J. Real-time global flood estimation using satellite-based precipitation and a coupled land surface and routing model. *Water Resour. Res.* **2014**, *50*, 2693–2717. [CrossRef]
67. Huffman, G.; Adler, R.; Bolvin, D.; Gu, G.; Nelkin, E.; Bowman, K.; Hong, Y.; Stocker, E.; Wolff, D. The TRMM multi-satellite precipitation analysis: Quasi-global, multi-year, combined sensor precipitation estimates at fine scales. *J. Hydrometeorol.* **2007**, *8*, 28–55. [CrossRef]

remote sensing

MDPI

Article

Machine Learning Model-Based Retrieval of Temperature and Relative Humidity Profiles Measured by Microwave Radiometer

Yuyan Luo [1], Hao Wu [2], Taofeng Gu [3], Zhenglin Wang [1,4], Haiyan Yue [5], Guangsheng Wu [3], Langfeng Zhu [2], Dongyang Pu [2], Pei Tang [6] and Mengjiao Jiang [1,*]

1 Plateau Atmospheres and Environment Key Laboratory of Sichuan Province, School of Atmospheric Sciences, Chengdu University of Information Technology, Chengdu 610225, China; 3210101022@stu.cuit.edu.cn (Y.L.)
2 Key Laboratory of China Meteorological Administration Atmospheric Sounding, School of Electrical Engineering, Chengdu University of Information Technology, Chengdu 610225, China; wuhao@cuit.edu.cn (H.W.); 3210307001@stu.cuit.edu.cn (D.P.)
3 Guangzhou Meteorological Observatory, Guangzhou 511430, China
4 Hainan International Commercial Aerospace Launch Co., Ltd., Wenchang 571300, China
5 Guangzhou Emergency Warning Information Release Center, Guangzhou 511430, China
6 Zhongshan Meteorological Service, Zhongshan 528400, China
* Correspondence: jiangmj@cuit.edu.cn

Abstract: The accuracy of temperature and relative humidity (RH) profiles retrieved by the ground-based microwave radiometer (MWR) is crucial for meteorological research. In this study, the four-year measurements of brightness temperature measured by the microwave radiometer from Huangpu meteorological station in Guangzhou, China, and the radiosonde data from the Qingyuan meteorological station (70 km northwest of Huangpu station) during the years from 2018 to 2021 are compared with the sonde data. To make a detailed comparison on the performance of machine learning models in retrieving the temperature and RH profiles, four machine learning algorithms, namely Deep Learning (DL), Gradient Boosting Machine (GBM), Extreme Gradient Boosting (XGBoost) and Random Forest (RF), are employed and verified. The results show that the DL model performs the best in temperature retrieval (with the root-mean-square error and the correlation coefficient of 2.36 and 0.98, respectively), while the RH of the four machine learning methods shows different excellence at different altitude levels. The integrated machine learning (ML) RH method is proposed here, in which a certain method with the minimum RMSE is selected from the four methods of DL, GBM, XGBoost and RF for a certain altitude level. Two cases on 29 January 2021 and on 10 February 2021 are used for illustration. The case on 29 January 2021 illustrates that the DL model is suitable for temperature retrieval and the ML model is suitable for RH retrieval in Guangzhou. The case on 10 February 2021 shows that the ML RH method reaches over 85% before precipitation, implying the application of the ML RH method in pre-precipitation warnings.

Keywords: microwave radiometer; radiosonde; temperature and humidity profiles; machine learning

1. Introduction

Atmospheric temperature and relative humidity (RH) are important parameters of the atmosphere and environment. Temperature and RH profiles with refined vertical resolution play an important role in urban meteorological forecasting [1–5]. The coastal city of Guangzhou is frequently hit by medium- and small-scale short-term weather events (such as torrential rains, typhoons and thunderstorms), which are extremely destructive and catastrophic despite the short activity time [6]. Accurate observation of the atmospheric vertical profile is fundamental for meteorological studies.

Although the traditional radiosonde data have high representativeness and reliability, the traditional observations are expensive and lack spatiotemporal resolution [7]. Ground-based microwave radiometers (MWRs) with passive remote sensing technology can overcome these shortcomings [8]. Since an MWR has the advantages of reliable calibration method, high resolution, unmanned continuous observation and simple operation, it is becoming an important instrument for remote sensing of atmospheric vertical profiles [9]. The MWR can continuously observe temperature, relative humidity and liquid water content within 0–10 km. In recent years, these data, combined with wind profile data, have gradually become an important reference for short-impending weather forecasting [9]. It is of high scientific importance and potential value to study the inversion of atmospheric temperature and humidity profile using microwave radiometer data [10]. However, the brightness temperature (BT) data from different channels of the MWR are disturbed by precipitation and cloud factors, resulting in abnormal values [11–13]. Meanwhile, when the sun is in the observation direction of the MWR, the BT data will be abnormally increased due to the influence of solar radiation, especially those used in low latitudes [14]. Therefore, it is essential to control the quality of the MWR observation data for a better forecast [15–17].

With the development of the ground-based MWR network, it has been widely applied to the detection of atmospheric vertical profiles in the boundary layer. Improving the reliability and accuracy of MWR observations is the priority for a refined atmospheric vertical profile. Meteorologists have proposed various methods to improve the accuracy of retrieval data, such as the linear statistical method [18], the best estimate method [19], neural networks [20,21] and machine learning [22,23]. Among these methods, the neural network performs well in solving the nonlinear relationship in the model. For example, Bao et al. [24,25] used the back-propagation neural network to retrieve the atmospheric temperature and RH profiles after the quality control of the first-level data. However, the traditional back-propagation neural network is time-consuming and requires a huge amount of data [26].

With the continuous development of artificial intelligence technology, the machine learning model has been increasingly applied in the field of microwave remote sensing, especially in atmospheric profile inversion. Gregori et al. [27] used the Gradient Boosting Machine (GBM) regression tree in a machine learning algorithm to estimate the boundary layer height using the MWR data and confirmed the excellent performance of machine learning in terms of training speed and retrieval accuracy. Jia [28] used the Extreme Gradient Boosting (XGBoost) machine learning model to predict the non-monsoonal winter precipitation over Eurasia. The results show that the XGBoost model performs significantly better than the traditional linear regression model. Liu [29] used the XGBoost model to correct the daily land surface temperature, where the rapidly increasing trend after the correction indicates an effective correction of the inhomogeneous land surface temperature in China. Recent studies have shown that the XGBoost model has great potential to improve climate prediction. The Random Forest (RF) algorithm has been applied to atmospheric environmental research in recent years [30,31]. Jiang et al. [32] used the RF machine learning model to establish an aerosol optical depth (AOD) dataset in the cloudy Sichuan Basin. GBM, XGBoost and RF all use boosting learning. The disadvantage of boosting learning is that there is a serial relationship between its base learners, and it is difficult to train data in parallel. The Deep Learning (DL) model is a machine learning algorithm that uses multi-layer artificial neural networks to achieve state-of-the-art accuracy in many tasks [33,34]. Similar to traditional machine learning algorithms, the DL model can model complex nonlinear systems. Moreover, it performs better in extracting the advantageous features with deeper network layers [35]. Recently, the performance of the DL model has been proven to be comparable to that of human experts [34].

In this study, four machine learning algorithms, namely the GBM, XGBoost algorithm, RF algorithm and DL algorithm, are used to compare with the MWR-derived first-level BT data. Based on this, the best machine learning method to improve the retrieval accuracy of profiles from the MWR data will be found. We try to give the evolution of RH profile transfer information, such as in which layer the water content could surge to a certain level, as an indicator of the coming precipitation. The techniques are only tested over a small region in Guangzhou, China. The rest of this paper is organized as follows. Section 2 describes the four machine learning algorithms, the datasets, the study region and the data preprocessing procedures. The comparison between the results of the four machine learning algorithms and the radiosonde data are presented in Section 3, where typical cases are also analyzed. Finally, the conclusion and discussion are presented in Section 4.

2. Data and Methods

The microwave radiometer BT data and the sonde data are observed from 2018 to 2021 in Huangpu and Qingyuan, respectively. Machine learning-based models are applied to retrieve temperature and relative humidity using the brightness temperature measured by the microwave radiometer based on the channel from 22.24 GHz to 31.4 GHz, 51.0 GHz to 58.0 GHz. The 2018–2020 dataset is used as the training sample, while the 2021 dataset is used as the validation sample.

2.1. Location of Observation Stations

The location of the observation stations is shown in Figure 1. The MWR data are observed from Huangpu station, while the radiosonde data are observed from the Qingyuan station. The distance between Huangpu station (113.29°N, 23.13°E) and Qingyuan station (113.05°N, 23.43°E) is about 70 km. The altitudes of Qingyuan station and Huangpu station are 79.2 m and 70.7 m, respectively. The two stations have similar underlying surface conditions, and there is no mountain barrier between them. The radiosonde data observed at Qingyuan station are well matched with the MWR BT data, which can be applied to train the machine learning algorithms in this study.

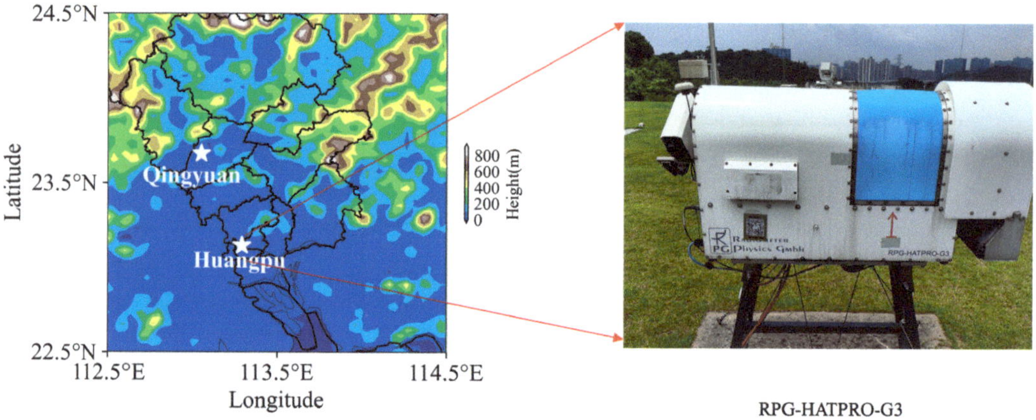

Figure 1. The location of observation stations. (The microwave radiometer data are observed from Huangpu station, while the radiosonde data are observed from the Qingyuan station).

2.2. Datasets

Two datasets, namely the MWR BT dataset and radiosonde dataset are used in this paper. The BT data are measured by an MWR located at the Huangpu station. The MWR uses the RPG-HATRPO-G3 from the Radiometer Physics GmbH in Germany, which is a 14-channel ground-based passive MWR with seven water vapor absorption channels (K-band) from 22.24 GHz to 31.40 GHz and seven oxygen absorption channels (V-band) from 51.00 GHz to 58.00 GHz [21,36]. The radiosonde data of temperature and RH are measured by an L-land GTS1 digital radiosonde at the Qingyuan station, which is launched twice daily at 11:00 and 23:00 UTC, respectively.

2.3. Data Preprocessing

The quality control of the BT data is performed before the training of four machine learning methods, in order to obtain better prediction results. The datasets during 2018–2021 are matched according to the principle of time consistency. There is a certain law of the time series variation in brightness temperature [20], and the transformation observed by each frequency channel of the MWR within 3 min should be continuous. First, the MWR BT samples are excluded if they do not meet the above conditions. Second, the MWR BT samples are further excluded if they are with large fluctuations, which may be errors of the instrument. After these two sample screening steps, the radiosonde data, temperature and RH are averaged by altitude. The interval altitude for a given level is set to be in the range of every 25 hPa and 50 hPa for below and above 700 hPa, respectively. In this way, the BT data are matched to the radiosonde level values. The matched samples are then classified into three categories based on RH values: clear sky, cloudy sky and rainy conditions. The inversion effect for MWR data is generally better under clear-sky conditions than under cloudy and rainy weather conditions [37–39]. The radiosonde data are processed according to the approach of Yan et al. [21] to determine the weather conditions and estimate cloud parameters. Theoretically, when a cloud forms, the RH at the corresponding height reaches 100%. However, due to factors such as condensation nuclei, the RH in the cloud layer is slightly lower than the theoretical value [40,41]. Therefore, 85% RH is used as the threshold value in the radiosonde data to determine the altitude level. The specific determination criteria are as follows:

(1) The measured data are classified as rainy-day data if the RH is greater than 85% from the ground to the height of 600 m.
(2) The data are classified as cloudy-sky data if the RH is less than 85% near the surface but greater than 85% in the upper atmosphere [20].
(3) The data are classified as clear-sky data if the RH is always less than 85% from the ground to any altitude level.

Thereafter, a total of 2461 quality-assured MWR data samples matched to the radiosonde sounding data from January 2018 to July 2020 are used for training, and 1321 quality-assured test samples during 2021 are used for validation. Due to the cloudy condition, 52% of the three years of data cannot be used.

2.4. Methods

2.4.1. Deep Learning (DL)

DL is an advanced machine learning architecture based on neural networks [42]. It aims to bridge the gap between machine learning and artificial intelligence by incorporating powerful learning capabilities and a wide range of applications. Compared to RF, GBM and XGBoost, DL shows superior adaptability. However, DL's performance is highly dependent on the available data, and it tends to excel when provided with a larger volume of data [42]. Recent research indicates that DL exhibits promising results in temperature retrieval [43,44]. The DL model in the study consists of two hidden layers, each containing 200 neurons. The activation function employed in the first hidden layer is relu, as it has been observed to provide superior results compared to other functions [45]. The second hidden layer and the output layer default to using linear activation functions by default. The entire model is trained using the Adam optimizer with the mean square error serving as the loss function.

2.4.2. Gradient Boosting Machine (GBM)

GBM is a boosting algorithm that uses different weights to linearly combine the base learners to reuse the learner with excellent performance [46]. The GBM algorithm calculates the pseudo-residuals according to the initial model. Then, it builds a base learner to interpret the pseudo-residuals, which can reduce the residuals in the gradient direction. Then, the base learner is multiplied by the weight coefficient and linearly combined with the original model to form a new model. The learning rate of the base learner is set to

0.1. The goal of the GBM is to find a model that minimizes the expectation of the loss function [47].

2.4.3. Extreme Gradient Boosting (XGBoost)

XGBoost is a gradient boosting-based integrated learning algorithm proposed by Chen and Guestrin [48]. The XGBoost [48] has been gradually applied in the atmospheric environment prediction. A second-order Taylor expansion is introduced in XGBoost, which increases the accuracy and enables loss functions to be customized via gradient descent. It adds the complexity of the tree model to the regularization term in order to prevent overfitting and, as a result, performs better in generalization [49]. However, the applications of the XGBoost machine learning method for the retrieval of meteorological profiles using MWRs are still few [21]. In this study, each tree is constructed using a learning rate of 0.3, a maximum tree depth of 6, a regularization weight of 10 and a total of 50 weak learners (trees).

2.4.4. Random Forest (RF)

RF is an integrated machine learning method that uses the random resampling technique bootstrap and the random node splitting technique to build multiple decision trees by Breiman [50]. A splitting technique with 50 random nodes is used to build the model. RF can be used for classification, clustering and regression data applications [51]. The RF model can analyze the classification features of complex interactions and has good robustness to data with noise and missing values. Meanwhile, it has a fast learning speed. The variable importance measure can be used as a feature selection tool for high-dimensional data [52]. Random forests are generally more effective at solving classification problems than regression problems. This is because random forests produce discrete outputs for classification tasks, rather than continuous outputs for regression tasks. In regression, the Random Forest model is limited in its ability to predict values beyond the range of the training set data. Therefore, when performing regression with a Random Forest, it is important to be aware that predictions may be limited within the range of the training data.

2.4.5. A 10-Fold Cross-Validation Method

Cross-validation is a common approach to model building and model parameter verification in machine learning, which is used to estimate the skill of a machine learning model [53]. In k-fold cross-validation [54,55], it is first randomly divided into k mutually exclusive subsets of similar size, i.e., $k-1$ is randomly selected as the training set each time, and the remaining 1 is used as the test set. When this round is completed, k copies are again randomly selected to train the data. After several rounds (less than k), the loss function is selected to evaluate the optimal model and parameters. In this study, k is set to 10. The four machine learning models are trained using 70% of the training samples, following standard training procedures, while the remaining 30% of the samples are used for validation. In addition, 10-fold cross-validation [32] is used for all four models.

3. Results and Case Illustration

The study first performs a 10-fold cross-verification analysis of training samples to verify the feasibility of the models. Second, the most appropriate methods of temperature and RH are found by analyzing the total scatter density, errors and RMSEs of different height layers of four machine learning methods with sonde data verification. Third, two cases are used to illustrate the results.

3.1. A 10-Fold Cross-Validation with Training Samples

In order to evaluate the performances of the four methods on the training dataset for temperature and RH, we use the 10-fold cross-validation method for verification, and the results are shown in Table 1. For temperature, the root-mean-square errors (RMSEs) of the DL, GBM, XGBoost and RF models are 2.32, 2.33, 2.49 and 3.07, respectively. The temperatures of the 10-fold cross-validation correlation coefficient (CV-R^2) of four methods is above 0.97. For RH, the RMSEs of the RF, XGBoost, GBM and DL models are 13.70, 13.72, 14.96 and 17.96, respectively. RH by the RF model shows the highest accuracy, with the sample-based 10-fold cross-validation CV-R^2 being 0.72. The results show the performance evaluation of the four machine methods, where the temperature performs better than RH. Compared with other methods, DL is suitable for MWR temperature and RF is better for RH retrieval. In general, the performance of the training result for RH is not so good.

Table 1. The 10-fold cross-validation results for four methods in terms of temperature and relative humidity (RH).

	Method	RMSE	CV-R^2	MAE
Temperature (°C)	DL	2.32	0.98	1.73
	GBM	2.33	0.98	1.80
	XGBoost	2.49	0.98	1.81
	RF	3.07	0.97	2.17
RH (%)	DL	17.96	0.53	14.04
	GBM	14.96	0.67	11.09
	XGBoost	13.72	0.72	9.49
	RF	13.70	0.72	9.92

RMSE: root-mean-square error; CV-R^2: cross-validation correlation coefficient; and MAE: mean average error.

3.2. Validation of Four Models with the Radiosonde Data

3.2.1. Scatter Density Variation

Radiosonde data are also used for comparison with the four model retrievals. The temperatures of the four models as a function of the radiosonde measurements from all 22 atmospheric vertical layers from 250 hPa to 1000 hPa are shown in Figure 2. The regression equations and coefficients of determination (R^2) are given, as well as the number of data points (N = 1321) and the RMSEs. Figure 2a shows that the linear regression relationship between the DL temperature and radiosonde temperature has a slope of 1.0, a y-intercept of 0.17 and minimal fluctuation around the regression line, with an R^2 of 0.98 and the lowest RMSE of 2.36 among the four models. As shown in Figure 2b, the linear regression relationship between the GBM temperature and the radiosonde temperature exhibits a slope of 0.99. The R^2 is 0.98, and the RMSE is 2.53. Figure 2c shows that the XGBoost model has a slope of 1.0, an R^2 of 0.97 and an RMSE of 3.07. The RF model has a slope of 0.95, an R^2 of 0.97 and an RMSE of 3.04, as shown in Figure 2d. The results show that the DL model has a high retrieval capability for temperature with an RMSE of 2.36 °C and an R^2 of 0.98.

Similarly, the RH of the four models as a function of the radiosonde measurements from all 22 atmospheric vertical layers is shown in Figure 3. The RMSEs of the DL, GBM, XGBoost and RF models are 20.08, 19.45, 19.72 and 19.07, respectively. The lack of independent cloud-related information may contribute to the deviations. The conditions with an RH less than 85% are considered clear-sky conditions in the study. However, in real atmospheric conditions, clouds may form due to the presence of cloud condensation nuclei when RH reaches around 85% [40]. In general, the RMSEs of RH are relatively greater compared with that of temperature.

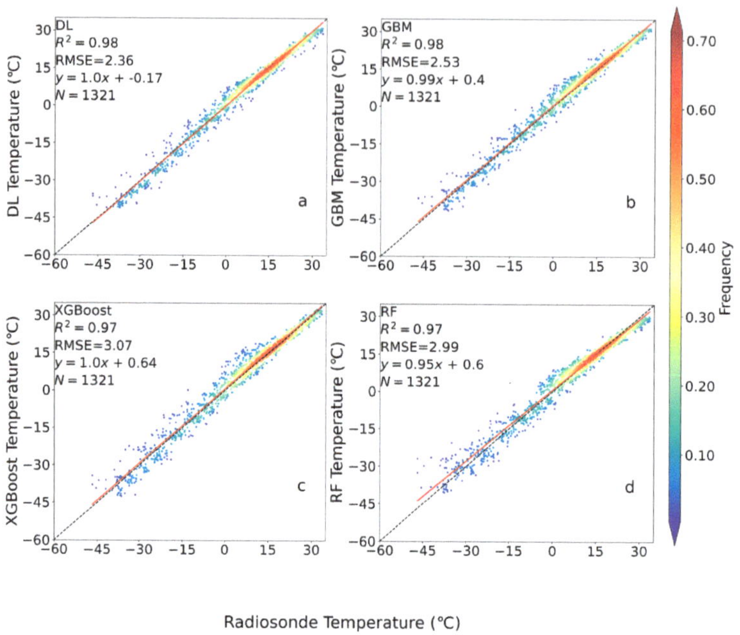

Figure 2. Retrievals for temperature as a function of radiosonde data by the (**a**) DL, (**b**) GBM, (**c**) XGBoost and (**d**) RF models. The red solid line is the line of best fit in linear regression.

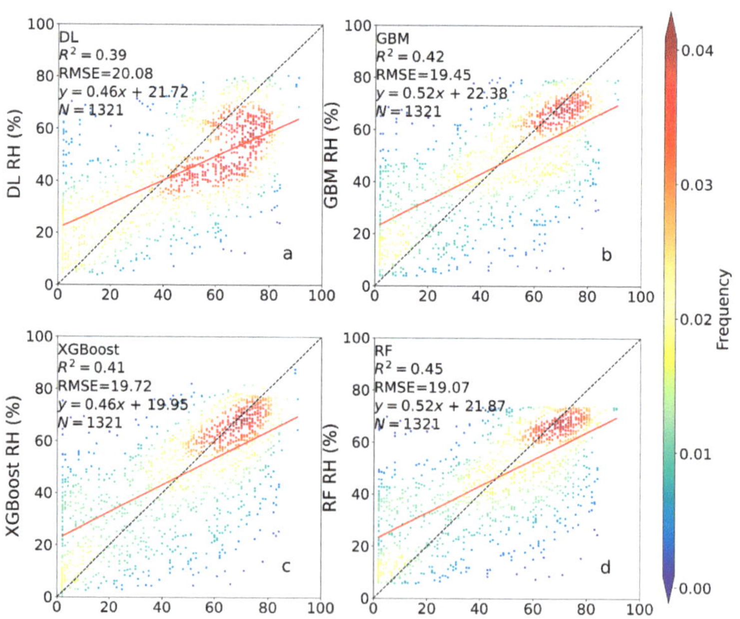

Figure 3. Retrievals for RH as a function of radiosonde data by the (**a**) DL, (**b**) GBM, (**c**) XGBoost and (**d**) RF models. The red solid line is the line of best fit in linear regression.

3.2.2. Bias and RMSEs Variation with Altitude

The temperature profile retrieval biases of the four models are shown in Figure 4. The red dotted lines represent the mean value of the biases, and the blue lines inside the box represent the median values. The blue shading indicates ±1 °C temperature biases. The left and right borders of the box contain the values from the first quartile to the third quartile. The blue dotted lines extend from each quartile to the minimum or maximum bias. In Figure 4a, the DL temperature bias is within ±1 °C for most of the pressure levels from 700 hPa to 1000 hPa, indicating high accuracy in retrieving temperature profiles in lower levels. From 250 hPa to 650 hPa, the box length for the DL model is shorter than that of the GBM, XGBoost and RF models, indicating that the temperature biases of the DL model are more concentrated around the mean and median values. Furthermore, most of the mean and median values of the temperature biases by the DL model are very close, indicating that the temperature biases of the DL model are more uniform and concentrated. The temperature bias from 250 hPa to 1000 hPa shows that the mean temperature bias measured by the DL model is negative near the surface and then becomes positive at 850 hPa with the increasing altitude. However, it turns negative again at 350 hPa with the increasing altitude. That is, the temperature bias from the DL model shows a distribution of "low at both ends and high in the middle". In contrast, the temperature biases by the RF model show a large fluctuation from left to right and are not stable enough at all levels (Figure 4d), which is similar to the trend of Yan et al. [21].

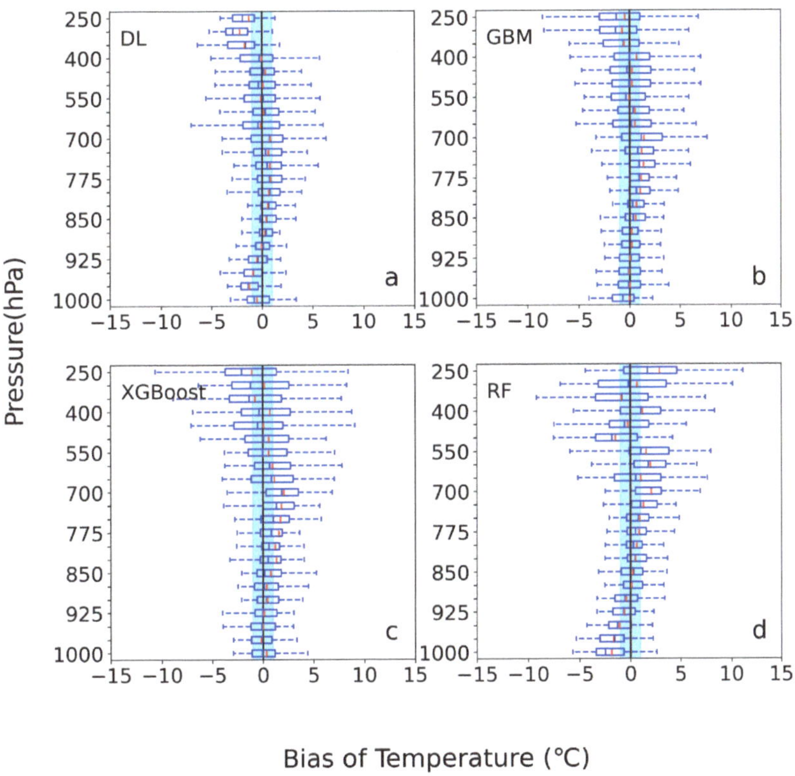

Figure 4. Temperature retrieval biases (the retrieval minus radiosonde) by the (**a**) DL, (**b**) GBM, (**c**) XGBoost and (**d**) RF models, respectively. Red dotted lines and blue lines within the boxes indicate the means and medians, respectively. The blue shadow means ±1 °C temperature biases.

The RH profile retrieval biases of the four models are shown in Figure 5. The red dotted lines and the blue lines are the same as in Figure 4, but the blue shadows show the biases ranging from −10% to +10%. In the four methods, the mean bias is usually close to the median in the lower-altitude layer, but deviates significantly from the median from 600 hPa to 800 hPa. The reason for the large deviation is the loss of the cloud information [18]. Figure 5d shows that the bias of the RF machine learning method remains within ±10% from the surface to 800 hPa, and its box length is almost the smallest among the four models from 800 hPa to 1000 hPa. As for GBM and XGBoost (Figure 5b,c), their bias also remains in the range of ±10% near the surface, but the interquartile range is much larger than the interquartile range of RF and the maximum bias exceeds 40%. Thus, the RF RH shows better retrieval near the surface than the other three methods.

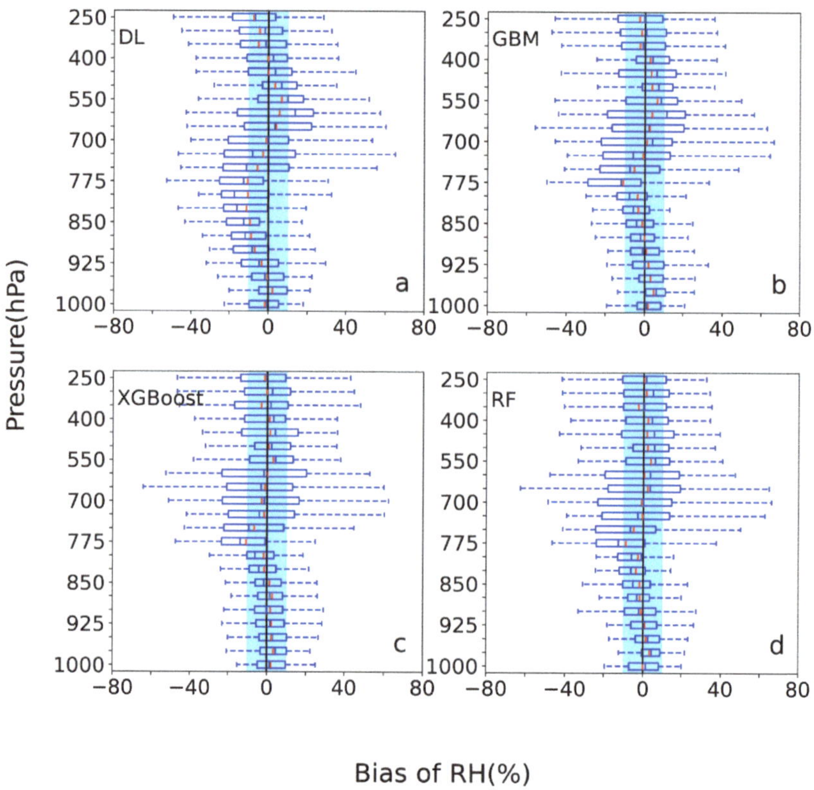

Figure 5. RH retrieval biases (the retrieval minus radiosonde) by the (**a**) DL, (**b**) GBM, (**c**) XGBoost and (**d**) RF models, respectively. Red dotted lines and blue lines within the boxes indicate the means and medians, respectively. The blue shadow means a bias of ±10%.

At the same time, we also compare and analyze the RMSEs of the four methods in different altitude layers in order to find the differences in the performance of the four methods on temperature and RH in different altitude layers. The RMSEs of the temperature and RH profiles of DL, GMB, XGBoost and RF at 22 height layers are shown in Figure 6. For temperature, the DL model shows a smaller RMSE than XGBoost, GBM and RF in the layers from 250 hPa to 1000 hPa. In particular, the DL RMSE is less than 1.5 °C in the layers from 775 hPa to 1000 hPa. For the RH, the RMSE of RH is larger than the RMSE of the temperature, and the RMSEs of the four machine learning methods from 600 hPa to 750 hPa are 20% to 30%; moreover, the RMSEs for all four methods generally increase with the altitude within this range, and the maximum deviation appears above the level of 700 hPa. This variation characteristic is similar to [20,21]. However, as we can see in the low-level RMSEs performance, the RMSEs of RF are between 10% and 15%. Therefore, from the overall characteristics of the RMSE of the two variables, the temperature RMSE is mainly concentrated in the upper layer, and the high RMSE of RH mainly occurs in the middle layer, which is in accordance with the situation found by Cimini et al. [8]. The results show that the DL temperature from 250 hPa to 1000 hPa performs better than the temperature of the other three methods, and the RF RH performs better in the low layers. In general, the performance of the training result for four machine learning methods is not so good.

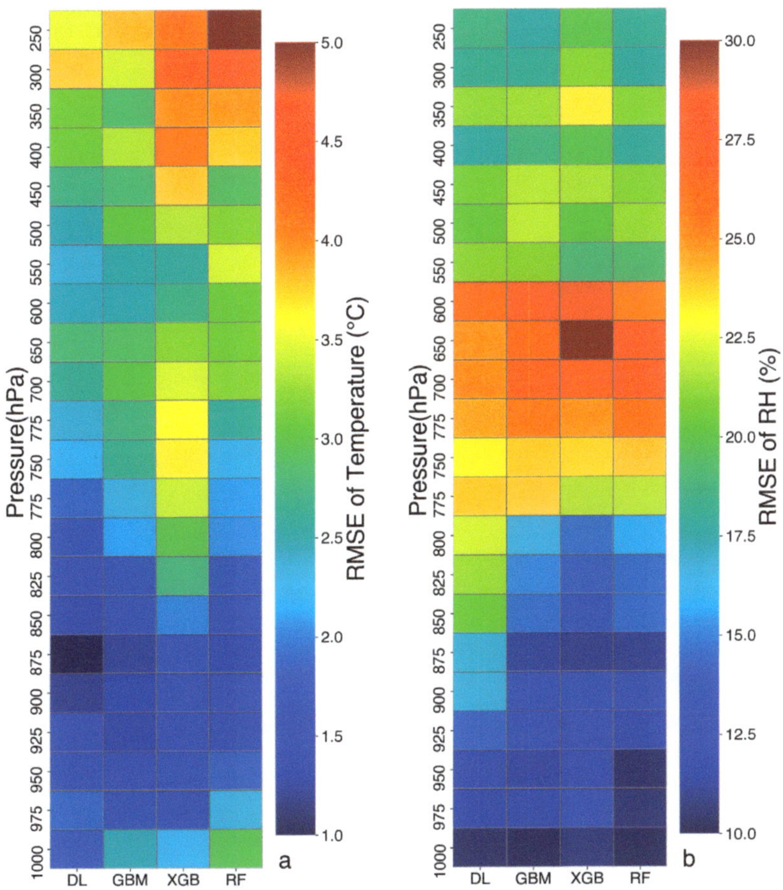

Figure 6. RMSEs for retrieval of temperature (**a**), and RH (**b**) profiles relative to the radiosonde data using the DL, RF, GBM and XGBoost methods.

The performance of the four models for RH retrieval at different levels is shown in Table 2. From 900 hPa to 1000 hPa, DL and RF demonstrate better accuracies in terms of R^2 (0.62 and 0.60), and the RF model performs better in terms of the RMSE (11.14) and the mean average error (8.83) compared with the DL model (12.93 and 8.84), indicating that the RF RH performs best from 900 hPa to 1000 hPa. From 775 hPa to 875 hPa, the XGBoost model shows better accuracies in terms of an RMSE of 14.97. From 600 hPa to 750 hPa, the RMSEs of the four methods are large. Above 550 hPa, RF performs well for RH retrieval, with an RMSE of 19.50. According to Table 2, the RF model achieves better retrieval ability from 900 hPa to 1000 hPa and above 550 hPa. From 775 hPa to 875 hPa, the XGBoost model achieves better retrieval ability. The DL model performs well for RH retrieval from 600 hPa to 750 hPa.

Table 2. Comparisons among different methods in terms of their performance at different levels.

Height	Method	RMSE	R^2	MAE
250–550 hPa	DL	19.69	0.34	15.09
	GBM	20.09	0.34	15.91
	XGBoost	20.71	0.29	16.20
	RF	19.50	0.36	15.44
700–750 hPa	DL	24.79	0.27	20.65
	GBM	26.07	0.18	21.48
	XGBoost	26.80	0.21	21.62
	RF	26.22	0.18	21.48
775–875 hPa	DL	20.99	0.18	17.73
	GBM	16.27	0.44	12.50
	XGBoost	14.97	0.53	11.38
	RF	15.55	0.47	11.99
900–1000 hPa	DL	12.93	0.62	8.84
	GBM	11.67	0.59	9.00
	XGBoost	12.19	0.56	9.43
	RF	11.14	0.60	8.83

Since the RHs of the four machine learning methods show different excellence at different height levels, a new integrated machine learning (ML) RH method is proposed here. The machine learning RH is to select the RH profiles by integrating the four methods of DL, GBM, XGBoost and RF, where the result of the method with the minimum RMSE for a certain level is adopted. The RMSE is 15.00 and R^2 is 0.64 by comparison of radiosonde RH and ML RH from all 22 atmospheric vertical levels.

3.3. *Case Illustration*

Based on the analysis in Section 3.2, DL (RMSE = 2.36, R^2 = 0.98) is the most suitable for temperature retrieval and ML (RMSE = 15.00, R^2 = 0.64) is the most suitable for RH retrieval. A case on 29 January 2021 is used for illustration. Another case with precipitation on 10 February 2021 is used to explain the changes in RH before the entire precipitation.

3.3.1. Case Analysis for DL Temperature and Machine Learning RH

Figure 7a shows the temperature profiles from the DL and radiosonde data. At 7:15 a.m. on 29 January 2021 (Beijing time, same below), the DL retrieval shows only a small difference with the radiosonde data from 775 hPa to 1000 hPa. In particular, the DL model agrees well with the radiosonde data at levels from 850 hPa to 900 hPa. However, the difference increases from 250 hPa to 700 hPa. Figure 7b shows the RH profiles by ML. Overall, the difference between the retrieved RH profile and the radiosonde data is greater than that of the temperature profile. The RH obtained from ML and radiosonde data show some agreement with the changing altitude. The integrated ML method performs better for RH from the layers 700 hPa to 875 hPa, with an RH bias lower than 10% compared with other altitudes.

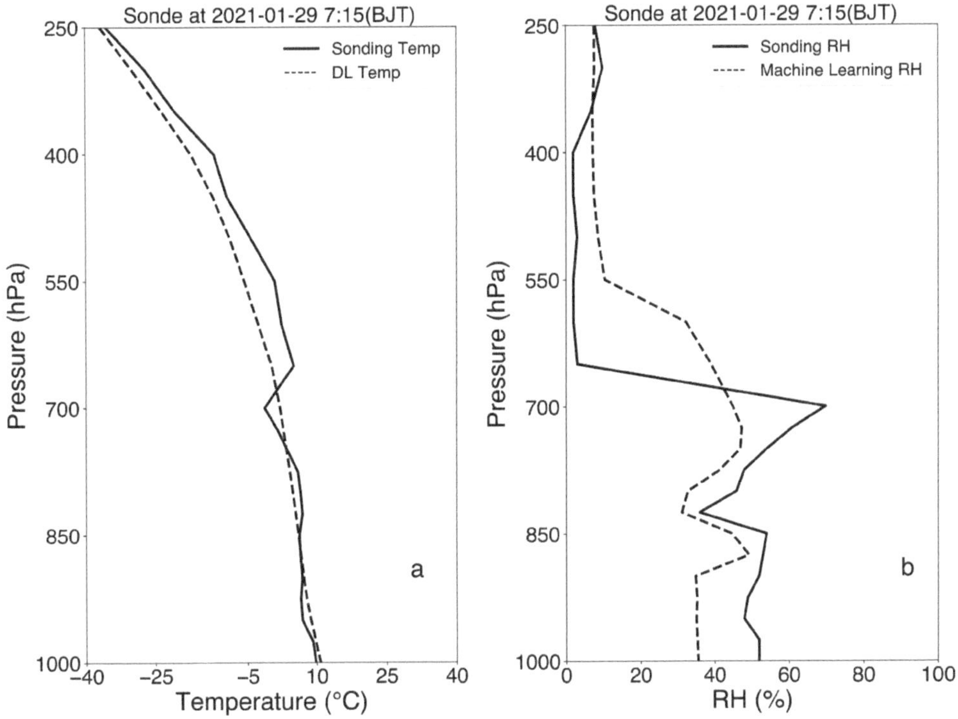

Figure 7. Comparison of (**a**) temperature profiles generated using the DL model and radiosonde data, and (**b**) RH profiles generated by integrating machine learning and the radiosonde data at 07:15 on 29 January 2021.

3.3.2. ML RH for Application before Precipitation

Another case on 10 February 2021 is used to explain the changes in ML RH before the entire precipitation process. The vertical profiles of RH predicted from 600 hPa to 1000 hPa levels and the observed data being the liquid water path (LWP) and the hourly surface precipitation histogram before and during a precipitation event are all shown in Figure 8. The precipitation started at 16:00 on 9 February 2021 and ended at 00:00 on 11 February 2021 (Figure 8a). The surface hourly precipitation reached the maximum value (6.9 mm) at 04:00 on 10 February 2021. Two phases of RH changes that occurred before 16:00 on 9 February 2021 are shown in Figure 8. In the first stage, Figure 8a shows the initial peak in the liquid water path (LWP) with a maximum value of 1319.60 g/m^2, while Figure 8b shows a gradual increase in RH at the lower levels. In the second stage, three consecutive peaks in the LWP (352.60 g/m^2, 1157.20 g/m^2 and 885.90 g/m^2) were observed, accompanied by an overall increase in RH at all levels. Notably, RH exceeded 85% from 750 hPa to 900 hPa prior to the onset of precipitation. The LWP has four peaks, indicating the continuous moisture transport and humidification process prior to precipitation. It also consists of the RH increase in Figure 8b. The RH increase obtained by the integrated ML method shows good agreement with the pre-precipitation LWP variation curve, indicating that the machine learning-based RH profiles successfully captured the significant increase in humidity before precipitation, which may provide some indication for precipitation forecasting.

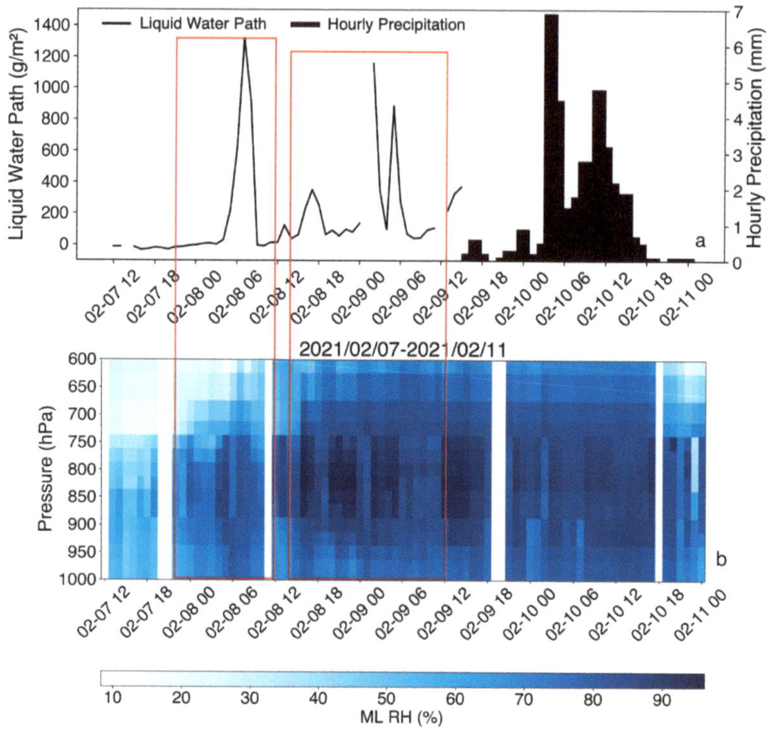

Figure 8. (**a**) Observed data of the liquid water path by MWR and the surface hourly precipitation histogram by Huangpu National Basic Meteorological Observation Station from 7 February 2021 to 11 February 2021, and (**b**) vertical profiles of ML RH from 600 hPa to 1000 hPa.

4. Discussion and Conclusions

The microwave radiometer (MWR) is widely used in meteorological observations, and the accuracy of temperature and relative humidity (RH) measurements can be affected by retrieval methods, weather conditions and environmental factors. In this study, we compared the Deep Learning (DL), Gradient Boosting Machine (GBM), Extreme Gradient Boosting (XGBoost) and Random Forest (RF) methods in retrieving temperature and RH profiles from 1000 hPa to 250 hPa using the MWR data and radiosonde data from 2018 to 2021, with the aim of improving the profile retrieval accuracy of the MWR.

Validation with radiosonde measurements shows that the DL model has better retrieval capability for temperature with a root-mean-square error (RMSE) of 2.07 °C and a correlation coefficient (R^2) of 0.98. Most of the temperature biases in the DL and XGBoost model are within ±1 °C from 700 hPa to 1000 hPa, and the RMSEs of the temperature profile using the DL model are less than 2.5 °C from 750 hPa to 1000 hPa. The RF model performs the best in retrieving the RH with the least bias near the surface, an RMSE around 12.5% and an interquartile range nearly the smallest among the four models.

A new integrated machine learning (ML) RH method is used to select the RH profiles by integrating the four models of DL, GBM, XGBoost and RF, where the result of the model with the minimum RMSE for a certain level is adopted. The RMSE is 15.00 and R^2 is 0.64 by comparison of radiosonde RH and ML RH from all 22 atmospheric vertical levels. We use the DL temperature and the ML RH to analyze two cases. A case on 29 January 2021 shows that DL is suitable for temperature retrieval and ML is suitable for RH retrieval. We apply the ML data to a precipitation case on 10 February 2021, and the results show that the change in ML RH shows a close correlation with the liquid water path before 16:00 on 9 February

2021. The ML RH reaches over 85% before 16:00 on February 9, indicating that the machine learning-based RH profiles successfully captured the significant increase in humidity prior to precipitation, which may provide some guidance for precipitation forecasting.

In conclusion, our study provides new insights into the performance of DL, GBM, XGBoost and RF in temperature and RH retrieval using MWR data. DL (RMSE = 2.36, R^2 = 0.98) shows superiority in temperature retrieval because the deep neural network architecture allows it to capture complex temperature patterns effectively [33,34]. Similar to traditional machine learning algorithms, DL can model complex nonlinear systems [35]. For the R^2 of the temperature, DL and GBM are both 0.98, which is 0.01 higher than that of the RF and XGBoost models. When comparing DL with GBM, the RMSE of temperature decreases from 2.53 to 2.36. The performance of the four models for RH retrieval at different levels is shown in Table 2. The RF model achieves better retrieval ability from 900 hPa to 1000 hPa and above 550 hPa. From 775 hPa to 875 hPa, the XGBoost model achieves better performance. The DL model performs well for RH retrieval from 600 hPa to 750 hPa. An integrated ML (RMSE = 15.00, R^2 = 0.64) approach improves RH retrieval because the ML method integrates the advantages of multiple methods.

It is important to note that our training datasets were obtained under clear-sky conditions without considering the data from cloudy conditions, which has certain limitations. Due to cloudy conditions, 52% of the three years of data cannot be used. A total of 2461 quality-assured MWR data samples matched to the radiosonde sounding data from January 2018 to July 2020 are used for training, and 1321 quality-assured test samples during 2021 are used for validation. Although 52% of the data cannot be used, the amount of available data based on observations is enough to represent the Guangzhou area. When applying these machine learning models in another region, it is critical to consider the region-specific characteristics and climatic conditions. The performance of the models may vary due to the differences in atmospheric dynamics, topographies and local weather patterns. Therefore, further investigation and validation specific to the target region would be necessary to assess the suitability and performance of the models. Our stations are only representative of the Guangzhou region. However, the models can be applied in other regions upon the observation available. Yan et al. [21] used microwave radiometer bright temperature data combined with DL, RF and XGBoost to invert the temperature and humidity profile under clear-sky conditions in Beijing. Bao et al. [24] used MWR BT data with a neural network to invert the temperature and humidity profile under clear-sky conditions in Nanjing. The lack of cloud information led to larger errors in RH in the middle layer from 700 hPa to 750 hPa. As shown in Li et al. [17] and Bao et al. [24], the correlation between the RH derived from the MWR and radiosonde data is much smaller than the correlation of temperature. This also proves our conclusion that the temperature inversion results are better than the RH inversion results. In addition, the distribution of the RMSE for all four models from 1000 hPa to 250 hPa is similar to the results of Che et al. [20], which showed that the RH RMSE tends to increase with height and the maximum deviation occurs in the middle atmosphere. Therefore, in this study, the ML method after the fusion of four models is proposed to reduce the influence of nonlinear relations on RH inversion. When the predicted hourly RH reaches the threshold of 85%, the warning information is provided to the forecasters. Consideration of region-specific characteristics is essential, and future research should explore cloud-related analysis. The suitability of DL for temperature retrieval and the effectiveness of ML for RH retrieval can be attributed to their respective model architectures and approaches. Future research will explore more in-depth cloud-associated analyses to address these limitations.

Author Contributions: Investigation, Y.L.; formal analysis, Y.L.; writing, Y.L.; visualization, Y.L.; editing, Y.L., H.W. and M.J.; conceptualization, H.W. and M.J.; supervision, H.W. and M.J.; data curation, T.G., H.Y., G.W. and L.Z.; writing—review and editing, Z.W., D.P. and P.T. All authors have read and agreed to the published version of the manuscript.

Funding: This study was funded by the Guangzhou Science and Technology Bureau (No. 202206010016), National Natural Science Foundation of China (Grant No. 42105073), Chengdu University of Information Technology Research Fund (KYTZ202217) and the China Scholarship Council.

Data Availability Statement: Not applicable.

Conflicts of Interest: The authors declare no conflict of interest.

References

1. Diao, M.H.; Jumbma, L.; Sheffield, J.; Wood, E.F.; Zondlo, M.A. Validation of AIRS/AMSU-A water vapor and temperature data with in situ aircraft observations from the surface to UT/LS from 87°N–67°S. *J. Geophys. Res. Atmos.* **2013**, *118*, 6816–6836. [CrossRef]
2. Osei, M.A.; Amekudzi, L.K.; Ferguson, C.R.; Danuor, S.K. Inter-comparison of AIRS temperature and relative humidity profiles with AMMA and DACCIWA radiosonde observations over West Africa. *Remote Sens.* **2020**, *12*, 2631. [CrossRef]
3. Liu, H.Y.; Li, J.; Cao, X.Y.; Xiong, B. Characteristics of the Atmosphere Remote Sensed by the Ground-Based 12-Channel Radiometer. *Remote Sens. Technol. Appl.* **2007**, *22*, 222–229.
4. Li, J.; Zeng, Q.C. Infrared Remote Sensing of Clear Atmosphere and Related Inversion Problem. Part II: Experimental Study. *Chin J. Atmos. Sci.* **1997**, *21*, 214–222. [CrossRef]
5. Wang, Y.; Huang, Y.; Huang, S.Y. A Preliminary Study of the Retrieval Methods for Atmosphere and Humidity profiles. *Chin Remote Sens. Resour.* **2008**, *20*, 23–26. [CrossRef]
6. Zeng, Q.; Qing, Z.P.; Zhu, M.; Zhang, F.G.; Wang, H.; Liu, Y.; Shi, Z.; Yu, Q. Application of Random Forest Algorithm on Tornado Detection. *Remote Sens.* **2022**, *14*, 2909. [CrossRef]
7. Ye, X.L.; Chen, Y.; Yang, S.; Yang, X.; Kan, Y.J. Quality Control method of single station surface air temperature data based on EEMD-CES. *Trans. Atmos. Sci.* **2019**, *42*, 390–398. [CrossRef]
8. Cimini, C.; Marzano, F.S.; Ciotti, P.; Cimini, D.; Westwater, E.R.; Han, Y.; Keihm, S.J.; Ware, R. Atmospheric Microwave Radiative Models Study Based on Ground-Based Multichannel Radiometer Observations in the 20–60 GHz Band. In Proceedings of the Fourteenth ARM Science Team Meeting Proceedings, Albuquerque, NM, USA, 22–26 March 2004.
9. Yao, Z.G.; Chen, H.B. Retrieval of Atmospheric Temperature Profiles with Neural Network Inversion of Microwave Radiometer Data in 6 Channels Near 118.75 GHz. *J. Meteorol. Sci.* **2006**, *26*, 252–259.
10. Liu, Y.Y.; Mao, J.T.; Liu, J.; Li, F. Research of BP Neural Network for Microwave Radiometer Remote Sensing Retrieval of Temperature, Relative Humidity, Cloud Liquid Water Profiles. *J. Plateau. Meteorol.* **2010**, *29*, 1514–1523.
11. Wang, Z.H.; Li, Q.; Chu, Y.L.; Zhu, Y.Y. Environmental Thermal Radiation Interference on Atmospheric Brightness Temperature Measurement with Ground-based K-band Microwave Radiometer. *J. Appl. Meteorol. Sci.* **2014**, *25*, 711–721. [CrossRef]
12. Zhao, L.; Ma, Y.F.; Zhang, G.X.; Yang, L.M. The Principle and Error Analysis of Microwave Radiometer MP-3000A. *Desert Oasis Meteorol.* **2009**, *3*, 54–57.
13. Che, Y.F.; Ma, S.Q.; Xing, F.H.; Li, S.T.; Dai, Y.R. Cloud Influence on Atmospheric Humidity Profile Retrieval by Ground-based Microwave Radiometer. *J. Appl. Meteorol. Sci.* **2015**, *26*, 193–202. [CrossRef]
14. Pan, Y.; Lei, L.F.; Li, Q.; Jiang, S.L.; Wang, Z.H. Analysis on the solar influence to brightness temperatures observed with a ground-based microwave radiometer. *J. Atmos. Sol. Terr. Phys.* **2021**, *222*, 105725. [CrossRef]
15. Qi, Y.J.; Fan, S.Y.; Mao, J.J.; Li, B.; Guo, C.W.; Zhang, S.T. Impact of Assimilating Ground-Based Microwave Radiometer Data on the Precipitation Bifurcation Forecast: A Case Study in Beijing. *Atmosphere* **2021**, *12*, 551. [CrossRef]
16. Qi, Y.; Fan, S.; Li, B.; Mao, J.; Lin, D. Assimilation of Ground-Based Microwave Radiometer on Heavy Rainfall Forecast in Beijing *Atmosphere* **2022**, *13*, 74. [CrossRef]
17. Li, Q.; Wei, M.; Wang, Z.; Jiang, S.; Chu, Y. Improving the Retrieval of Cloudy Atmospheric Profiles from Brightness Temperatures Observed with a Ground-Based Microwave Radiometer. *Atmosphere* **2021**, *12*, 648. [CrossRef]
18. Hogg, D.C.; Decker, M.T.; Guiraud, F.O.; Earnshaw, K.B.; Merritt, D.A.; Moran, K.P.; Sweezy, W.B.; Strauch, R.G.; Westwater, E.R.; Little, C.G. An Automatic Profiler of the Temperature, Wind and Humidity in the Troposphere. *J. Appl. Meteorol. Climatol.* **1983**, *22*, 807–831. [CrossRef]
19. Rodgers, C.D. *Inverse Methods for Atmospheric Sounding—Theory and Practice*; World Scientific: Singapore, 2008.
20. Che, Y.F.; Ma, S.Q.; Xing, F.H.; Li, S.T.; Dai, Y.R. An improvement of the retrieval of temperature and relative humidity profiles from a combination of active and passive remote sensing. *Meteorol. Atmos. Phys.* **2018**, *131*, 681–695. [CrossRef]
21. Yan, X.; Liang, C.; Jiang, Y.Z.; Luo, N.N.; Zang, Z.; Li, Z.Q. A Deep Learning Approach to Improve the Retrieval of Temperature and Humidity Profiles from a Ground-Based Microwave Radiometer. *IEEE Trans. Geosci. Remote Sens.* **2020**, *58*, 8427–8437 [CrossRef]
22. Chi, J.; Kim, H.C.; Lee, S.; Crawford, M.M. Deep learning based retrieval algorithm for Arctic sea ice concentration from AMSR2 passive microwave and MODIS optical data. *Remote Sens. Environ.* **2019**, *231*, 111204. [CrossRef]
23. Xue, Y.; Forman, B.A. Comparison of passive microwave brightness temperature prediction sensitivities over snow-covered land in North America using machine learning algorithms and the Advanced Microwave Scanning Radiometer. *Remote Sens. Environ.* **2015**, *170*, 153–165. [CrossRef]
24. Bao, Y.S.; Cai, X.; Qian, C.; Min, J.Z.; Lu, Q.F. 0–10 KM Temperature and Humidity Profiles Retrieval from Ground-based Microwave Radiometer. *J. Trop. Meteorol.* **2018**, *24*, 243–252. [CrossRef]

25. Zhang, T.H.; Bao, Y.S.; Qian, Y.Z.; Lin, L.B.; Liu, X.L.; Li, L.; Hou, Y.; Lei, H.Y.; Li, G.W.; Ma, J.; et al. Atmospheric temperature and humidity profile retrievals based on BP neural network and genetic algorithm. *J. Trop. Meteorol.* **2020**, *36*, 97–107. [CrossRef]
26. Zhao, Y.X.; Yan, H.L.; Wu, P.; Zhou, D. Linear correction method for improved atmospheric vertical profile retrieval based on ground-based microwave radiometer. *Atmos. Res.* **2020**, *232*, 104678. [CrossRef]
27. Gregori, D.A.M.; Guadalupe, S.H.; Juan, L.G.R.; Alberto, C.; Lucas, A.A. Estimating the urban atmospheric boundary layer height from remote sensing applying machine learning techniques. *Atmos. Res.* **2021**, *266*, 105962. [CrossRef]
28. Qian, Q.F.; Jia, X.J.; Lin, H.; Zhang, H.Z. Seasonal Forecast of Nonmonsoonal Winter Precipitation over the Eurasian Continent Using Machine-Learning Models. *J. Clim.* **2021**, *34*, 7113–7129. [CrossRef]
29. Liu, F.; Wang, X.M.; Sun, F.B.; Wang, H.; Wu, L.F.; Zhang, X.Z.; Liu, W.B.; Che, H.Z. Correction of Overestmation in Observed Land Surface Temperatures Baesd on Machine Learning Models. *J. Clim.* **2022**, *35*, 5359–5377. [CrossRef]
30. Chen, B.; Qin, H.L.; Chen, G.X.; Xue, H.J. Ocean Salinity as a Precursor of Summer Rainfall over the East Asian Monsoon Region. *J. Clim.* **2019**, *32*, 5659–5676. [CrossRef]
31. Tornow, F.; Domenech, C.; Fischer, J. On the Use of Geophysical Parameters for the Top-of-Atmosphere Shortwave Clear-Sky Radiance-to-Flux Conversion in EarthCARE. *J. Atmos. Ocean. Technol.* **2019**, *36*, 717–732. [CrossRef]
32. Jiang, M.J.; Chen, Z.H.; Yang, Y.S.; Ni, C.J.; Yang, Q. Establishment of aerosol optical depth dataset in the Sichuan Basin by the random forest approach. *Atmos. Pollut. Res.* **2022**, *13*, 101394. [CrossRef]
33. Bengio, Y. *Learning Deep Architectures for AI*; Now Publishers Inc.: Hanover, MA, USA, 2009; Volume 2, pp. 1–127. [CrossRef]
34. Schmidhuber, J. Deep learning in neural networks: An Overview. *Neural. Netw.* **2015**, *61*, 85–117. [CrossRef]
35. Zhang, W.; Peng, G.; Li, C.; Chen, Y.; Zhang, Z. A New Deep Learning Model for Fault Diagnosis with Good Anti-Noise and Domain Adaptation Ability on Raw Vibration Signals. *Sensors* **2017**, *17*, 425. [CrossRef]
36. Tan, H.B.; Mao, J.T.; Chen, H.H.; Chan, P.W.; Wu, D.; Li, F.; Deng, T. A Study of a Retrieval Method for Temperature and Humidity Profiles from Microwave Radiometer Observations Based on Principal Component Analysis and Stepwise Regression. *J. Atmos. Ocean. Technol.* **2011**, *28*, 378–389. [CrossRef]
37. Liu, J.Z.; Zhang, Q. Evaluation of microwave radiometer inversion products. *Meteorol. Sci. Technol.* **2010**, *38*, 325–331.
38. Li, J.Q.; Li, X.S.; Dong, W.X.; Lv, Q.Q. Applicability analysis of temperature and humidity data from RPG-HATPRO microwave Radiometer inversion. *J. Meteorol. Environ.* **2017**, *33*, 89–95.
39. Tian, J.B.; Zhang, Y.X.; Yan, H.Q.; Tang, W.T.; Kang, X.Y. Study on Neural Network Algorithm for Atmospheric Profile Based on Microwave Radiometer in Plateau Region. *Plateau. Mt. Meteorol. Res.* **2021**, *41*, 125–134. [CrossRef]
40. Wang, J.H.; Rossow, W.B. Determination of Cloud Vertical Structure from Upper-Air Observations. *J. Appl. Meteorol. Climatol.* **1995**, *34*, 2243–2258. [CrossRef]
41. Poore, K.D.; Wang, J.H.; Rossow, W.B. Cloud Layer Thicknesses from a Combination of Surface and Upper-Air Observations. *J. Clim.* **1995**, *8*, 550–568. [CrossRef]
42. LeCun, Y.; Bengio, Y.; Hinton, G.E. Deep Learning. *Nature* **2015**, *521*, 436–444. [CrossRef]
43. Choe, Y.J.; Yom, J.H. Improving accuracy of land surface temperature prediction model based on deep-learning. *Spat. Inf. Res.* **2020**, *28*, 377–382. [CrossRef]
44. Milad, A.; Adwan, I.; Majeed, S.A.; Yusoff, N.I.M.; Al-Ansari, N.; Yaseen, Z.M. Emerging Technologies of Deep Learning Models Development for Pavement Temperature Prediction. *IEEE Access* **2021**, *9*, 23840–23849. [CrossRef]
45. Hou, P.; Guo, P.; Wu, P.; Wang, J.; Gangopadhyay, A.; Zhang, Z. A Deep Learning Model for Detecting Dust in Earth's Atmosphere from Satellite Remote Sensing Data. In Proceedings of the IEEE International Conference on Smart Computing SMARTCOMP, Bologna, Italy, 14–17 September 2020; pp. 196–201. [CrossRef]
46. Friedman, J.H. Greedy function approximation: A gradient boosting machine. *Ann. Stat.* **2001**, *29*, 1189–1232. [CrossRef]
47. Friedman, J.H. Stochastic Gradient Boosting. *Comput. Stat. Data Anal.* **2002**, *38*, 367–378. [CrossRef]
48. Chen, T.; Guestrin, C. XGBoost: A Scalable Tree Boosting System. In Proceedings of the 22nd ACM SIGKDD International Conference on Knowledge Discovery and Data Mining KDD, San Francisco, CA, USA, 13–17 August 2016; pp. 785–794. [CrossRef]
49. Wang, H.; Liu, C.; Deng, L. Enhanced Prediction of Hot Spots at Protein-Protein Interfaces Using Extreme Gradient Boosting. *Sci. Rep.* **2018**, *8*, 14285–14297. [CrossRef] [PubMed]
50. Breiman, L. Random forest. *Mach. Learn.* **2001**, *45*, 5–32. [CrossRef]
51. Chen, X.; Ishwaran, H. Random Forests for Genomic Data Analysis. *Genomics* **2012**, *99*, 323–329. [CrossRef]
52. Wang, Q.; Nguyen, T.T.; Huang, J.Z.; Nguyen, T.T. An Efficient Random Forests Algorithm for High Dimensional Data Classification. *Adv. Data Anal. Classif.* **2018**, *12*, 953–972. [CrossRef]
53. Jung, Y.S. Multiple Predicting K-fold Cross-Validation for Model Selection. *J. Nonparametr. Stat.* **2018**, *30*, 197–215. [CrossRef]
54. Jung, Y.S.; Hu, J.H. A K-fold Averaging Cross-Validation Procedure. *J. Nonparametr. Stat.* **2015**, *27*, 167–179. [CrossRef]
55. Wang, T.T. Performance Evaluation of Classification Algorithms by K-fold and leave-one-out Cross Validation. *Pattern Recognit.* **2015**, *48*, 2839–2846. [CrossRef]

remote sensing

MDPI

Article

Study on the Vertical Structure and the Evolution of Precipitation Particle Spectrum Parameters of Stratocumulus Clouds over North China Based on Aircraft Observation

Jingyuan Xiong, Xiaoli Liu * and Jing Wang

Key Laboratory for Aerosol-Cloud-Precipitation of China Meteorological Administration, Nanjing University of Information Science and Technology, Nanjing 210044, China
* Correspondence: liuxiaoli2004y@nuist.edu.cn

Abstract: The understanding of the macro- and micro-structure, particle spectrum parameters, and their evolutions in different parts of stratocumulus clouds based on aircraft observation data, is important basic data for the development of cloud microphysical parameterization schemes and the quantitative retrieval of cloud-precipitation by radar and satellite detections. In this study, a total of ten vertical measurements during three aircraft observations were selected to analyze the vertical distribution of cloud microphysical properties in different parts of stratocumulus clouds in Hebei, North China. It was found that the downdraft in the cumulus cloud area was stronger than that in the stratiform cloud area, with the temperature at the same height higher than that in the stratiform cloud area, and the height of the 0 °C layers was correspondingly higher. In terms of particle spectrum parameters, the intercept and slope parameters of particle spectrum below melting levels in the cumulus part were higher than those in stratiform clouds area in the same weather process. In different vertical detection, it was found that the ice particles have begun to melt in the negative temperature layer near 0 °C level, and there might be sublimation, fragmentation, and aggregation in the melting process of ice phase particles. In addition, the melting process changed the spectral parameters greatly and also changed the correlation between the intercept and slope of the particle spectrum. The slope below the 0 °C level increased with the increase of intercept, which was greater than that above the 0 °C level. The relationship obtained between the intercept parameter of the particle's spectrum and temperature, and the correlation between the maximum diameter and slope parameter of the particle spectrum, have certain reference significance for cloud physical parameterization and the quantitative retrieval of cloud precipitation by radar and satellite in North China and similar climate background areas.

Keywords: aircraft observation; stratocumulus cloud; microphysical property; particle size distribution

1. Introduction

Under large-scale weather conditions, stratiform clouds associated with frontal systems are frequently seen in northern China. These clouds are usually multilayered mixed-phase clouds with durations ranging from a few hours to two days [1]. It was found that the microphysical properties of ice-phase particles and precipitation formation mechanisms vary widely among different cloud top temperatures, different locations in the cloud [2], and embedded convective cells in stratiform clouds [1]. Qi et al. found that in the convective cell with high supercooled water content, the growth of ice particles was mainly due to the aggregation and riming growth processes, while in the cloud area with low supercooled water content, the aggregation was the main growth process, and the formation of precipitation in the convective cell with high supercooled water content conforms to the "seeder—feeder" mechanism [3]. Kang et al. found that the occurrence and strengthen of convection could improve the growth rate of ice crystals. The low layer of cold cloud in the weak convective area had an explosive growth area of ice crystal concentration, while the middle and upper layers of cold cloud in the strong convective area had rapid growth areas of ice crystal concentration [4]. Yang et al. found that in stable stratus clouds, the content of supercooled water in the mixed phase layers was very low, ice particles grew mainly through the process of deposition and aggregation, and the ice crystallization of the cloud was relatively high. However, the content of supercooled water was higher in the stratus cloud area with more vigorous development, and the presence of a large number of liquid droplets also indicated that the conversion between the ice and liquid phase in the mixed layer was not sufficient [5]. Gao et al. showed that there were obvious differences in the shape and formation processes of ice particles in stratocumulus clouds. The shapes of ice particles in stratus cloud areas were complex, including needle, columnar, and dendritic particles, while cumulus cloud areas mainly consisted of dendritic particles, with obvious coalescence and riming processes [6].

Hu et al. found that Z_H increased and Z_{DR} decreased, as the height decreased above the melting layer. This predicts that the aggregation process has transformed the ice crystals from ellipsoidal to more nearly spherical aggregates [7]. Wei et al. found that small particles were more predominant both above and below the melt layer, with two peaks between −5 and 2 °C for ice-phase particles (50–300 μm) and snowflakes (>300 μm) [8]. In the early stage of precipitation development, the ice phase particles in clouds were dominated by graupels and line shapes, and in the mature stage of precipitation, the ice phase particles were dominated by graupels and aggregates [9]. The exponential distribution could better fit the ice phase particle spectral distribution pattern, and the power function could better fit the relationship between the two spectrum parameters [10,11].

The rate of melting of ice and snow crystals was an important factor in determining the thickness of the melting layer and the associated bright bands. It played an important role in weather forecasting and hydrological applications, and had important implications for snowfall under climate change [12–16]. Kain et al. found that the melting of snow cools the atmosphere. At sufficiently high precipitation rates, rain in clouds could convert into snow, which would allow the melt layer to disappear. Therefore, high precipitation rates may lead to an increase in snowfall intensity [17]. In addition to the properties of the ice crystals themselves [18], meteorological conditions such as air temperature and relative humidity also affected the melting rate of ice particles [19,20]. Heymsfield et al. found that the slope of the particle spectral distribution tends to decrease with melting at high relative humidity, and the maximum particle size of ice particles continues to increase during the melting process [21,22]. Heymsfield et al. defined the shape-sensitive parameter area ratio (Ar) of ice-phase particles and found that it was related to the position in the cloud and microphysical processes within the cloud, while there was a negative correlation between the area ratio and the particle size. The power function could fit the relationship between the two, well [21].

Generally speaking, aircraft observation data are important basic data for understanding the characteristics of cloud particle distribution and evolution, establishing cloud microphysical parameterization schemes, and inversion of cloud precipitation microphysical characteristics based on radar and satellite data. Although the distribution of particles, the morphology of ice particles, and water content in stratocumulus clouds have been well understood in previous studies, the evolution of microstructure and the vertical distribution of the precipitation particle spectrum in stratocumulus clouds based on the data of multiple flights observation, especially the spectrum evolutions of ice particles after they fall below the 0 °C layer, still lack inevitable discovery. In this paper, we explored the macro and microscopic characteristics of clouds by analyzing the aircraft observations of stratocumulus clouds in Hebei Province on 22 May 2017, 21 May 2018, and 24 August 2019, and analyzed the ice particle number concentrations, two-dimensional images, and the height distributions of the particle spectrum distribution. The relationships between spectrum parameters, temperature, and the maximum diameter of particles, were also studied. Our hope in undertaking this study was to further enrich the scientific understanding of stratocumulus cloud microstructure, and provide references for radar and satellite cloud property reversion and cloud microphysical parameterization in different parts of a stratocumulus cloud, so as to improve the cloud microphysical parameterization scheme in cloud simulations and improve the accuracy of stratocumulus cloud precipitation forecasts.

2. Data

2.1. Introduction of the Observation Data

The data in this paper were taken from the "13th Five-Year Plan" meteorological key project in Hebei Province's, "Experiment on artificial rainfall and hail prevention technology in the eastern foothills of Taihang Mountains" scientific field experiment. The data were taken from 3 observations, made on 22 May 2017, 21 May 2018, and 24 August 2019 (Figure 1). The measurements went through the negative temperature layer, the 0 °C layer, the melting layer, and the convective and stratiform cloud areas. The aircraft took off from Zhengding Airport, and the flight times were concentrated in the afternoon and night. The longest duration was 209 min and the shortest duration was 185 min. The aircraft carried out horizontal detection at different altitudes and carried out vertical detection in circling ascent and descent within the safe flight altitude limit. Based on the flight area and the abundance of ice-phase particle data, 10 vertical detections were selected for the study (Table 1).

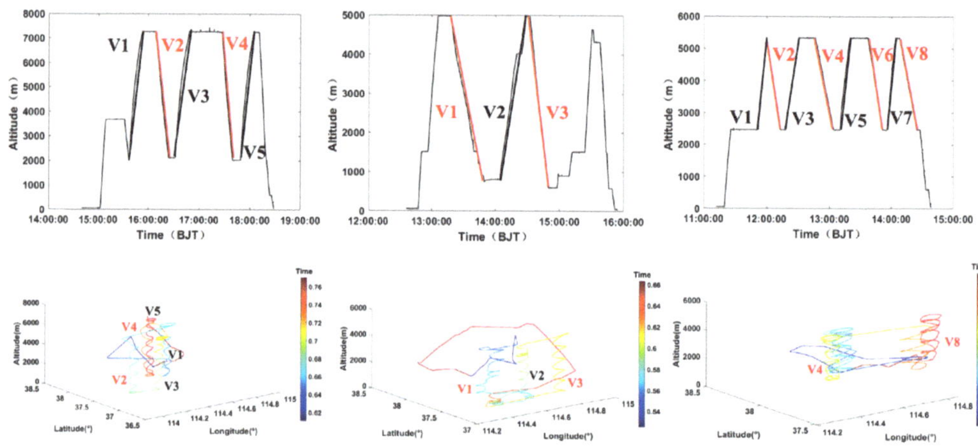

Figure 1. Flight track of the 3 aircraft (from left to right is 22 May 2017, 21 May 2018, and 24 August 2019).

Table 1. Details of the observation flights.

Flight		Detection Time (Beijing Time)	Range of Height (m)	Range of Temperature (°C)	Rate of Descent (m/s)	Width of Detection (km)
22 May 2017	V1	15:37:11–15:53:06	7263–1998	−15.4–8.1	5.51	10.72
	V2	16:08:32–16:23:38	7263–2121	−15.7–8.2	5.68	9.99
	V3	16:31:10–16:48:59	7267–2107	−15.6–8.6	4.83	9.40
	V4	17:27:07–17:39:54	7247–2009	−16.1–5.5	6.83	11.51
	V5	17:48:59–18:04:27	7256–2011	−15.9–4.9	5.65	7.90
21 May 2018	V1	13:17:43–13:54:07	4971–784	−1.2–12.4	1.92	16.18
	V2	14:04:17–14:30:00	4975–810	−1.8–10.3	2.70	13.39
	V3	14:32:32–14:49:50	4958–590	−1.6–11.3	4.21	7.35
24 August 2019	V4	12:46:39–13:04:59	5297–2463	−3–7.3	2.58	11.10
	V8	14:07:30–14:24:38	5328–2502	−2.3–7.3	2.75	10.47

2.2. Introduction of the Instrumentation

This paper adopted the observation data of King-air 350ER (No. 3523) of Hebei Artificial Weather Office, and the cloud physical detection system consisting of several probes on board the aircraft, which mainly includes the FCDP (The Fast Cloud Droplet Probe), the CDP (Cloud Droplet Probe), the CIP (Cloud Imaging Probe), the HVPS (The High Volume Precipitation Spectrometer), 2D-S combined probes, and the Airborne Integrated Meteorological Measurement System AIMMS-20 [23], which could measure in real-time 0.055–9075 μm of the spectral distribution of various particles; give 25–19,200 μm particle 2D images; give real-time measurements to obtain macroscopic information such as temperature, pressure, humidity, wind speed, wind direction, and vertical velocity of the atmosphere; and liquid water content, total water content, etc. in clouds and aircraft flight trajectory could also be detected in real time (Table 2).

Table 2. Particle probe and its parameters. (Reprinted/adapted with permission from Ref. [5]. 2023, Jiefan Yang).

Instrument Name	Equipment Manufacturer	Measuring Range	Resolution	Use
Passive Cavity Aerosol Spectrometer Probe	DMT	30 channels, 0.1~3 μm	0.1 μm	Used for the detection of an aerosol particle spectrum
Fast Cloud Droplet Probe	SPEC	21 channels, 2~50 μm	3 μm	Cloud particle spectrum
Cloud Droplet Probe (CDP)	DMT	30 channels, 2~50 μm		Cloud particle spectrum
Cloud Imaging Probe (CIP)	DMT	62 channels, 25~1500 μm	25 μm	Used to obtain a high-definition crystal grain spectrum and 2-dimensional particle image of ice, snow, cloud
Precipitation Imaging Probe (PIP)	DMT	62 channels, 100~6200 μm	100 μm	Used to obtain precipitation particle spectrum and image
Cloud Imaging Probe	SPEC	10~2000 μm	2.3 μm	Used for cloud droplets, snow and ice crystals, raindrop images
2D-S Optical	SPEC	10~1280 μm	100 μm	Used for cloud droplets, snow and ice crystals, raindrop images
High Volume Precipitation Spectrometer (HVPS)	SPEC	150~19,200 μm	150 μm	Used to obtain a clear precipitation particle spectrum and particle 2-dimensional image
LWC	DMT	0~3 g m^{-3}		Liquid water content
TWC	Nevzorov	0.005~3 g m^{-3}		Liquid water content, ice, and snow crystal water content
AIMMS-20	Aventech	Temperature: −50~50 °C Vertical velocity: 0~50 m s^{-1} Altitude: 0~13.7 km	Temperature: 0.3 °C Velocity: 0.75 m s^{-1} Altitude: 18.3 m	Used for measuring high temperature, pressure, humidity, wind, and aircraft motion parameters

The hourly observation data from the Chinese ground-based weather stations were used to calculate the cumulative precipitation over the detection period for the three aircraft observation areas. The detection area of 2017 was located in the northern part of the precipitation center, and the maximum intensity of precipitation was over 40 mm. In 2018, the detection area was located in the southern part of the precipitation center, and the maximum intensity of precipitation was 30 mm. The detection area of 2019 was located in the eastern part of the precipitation center, and the overall precipitation intensity was low, and the maximum intensity of precipitation was 5 mm (Figure 2).

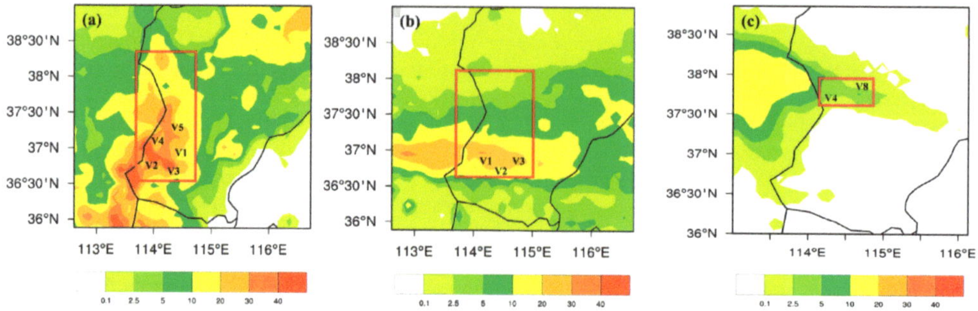

Figure 2. Cumulative precipitation: (**a**) is for 22 May 2017, (**b**) is for 21 May 2018, and (**c**) is for 24 August 2019.

2.3. Precipitation and Radar Data

In this paper, SA-band Doppler weather radar (38°21′7′′N, 114°42′43′′E) in Shijiazhuang and SA-band Doppler weather radar (36°36′11′′N, 114°28′59′′E) in Handan were mainly used. The radar wavelength was 10 cm, and the volume scan was completed every 6 min. As could be seen from the radar reflectivity of the 2.4° and 3.4° elevation angles, there was melting bright band in each stage. The observations in 2017 were in the southern parts of the cloud system, and the observations in 2018 and 2019 were in the northern part of the cloud system (Figures 3–5). From the vertical structure of radar echo, it could be observed that the V1, V2, and V3 detections on 22 May 2017 were in the strong echo region, while the rest were in the typical stratiform cloud region (Figure 6). From the radar echo image on 21 May 2018, it could be observed that there is a convective cell in the cloud in the vertical detection area of the V1 segment, and the other processes are layered echo area. The radar reflectivity of the V3 section weakens obviously at the height of 3 km (Figure 7). From the vertical radar reflectivity, it could be seen that on 24 August 2019, the detection in section V8 was located outside the convective cell. There were melting signatures based on the strengthening of the radar reflectivity below the 0 °C layer in both vertical detection areas, and the bottom of the melting layers were relatively low (Figure 8).

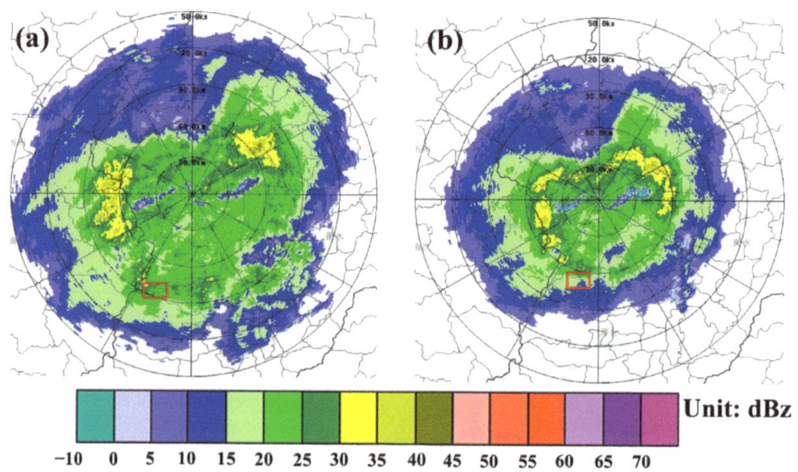

Figure 3. 22 May 2017 (**a**) 17:30, 2.4° elevation angle PPI, (**b**) 17:54, 3.4° elevation angle PPI. The red rectangle box is the aircraft detection area.

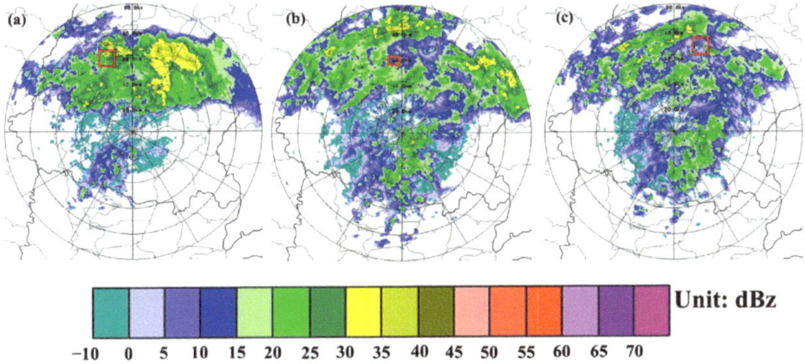

Figure 4. 21 May 2018 (**a**) 13:18, 2.4° elevation angle PPI, (**b**) 14:18, 2.4° elevation angle PPI, (**c**) 14:36, 3.4° elevation angle PPI. The red rectangle box is the aircraft detection area.

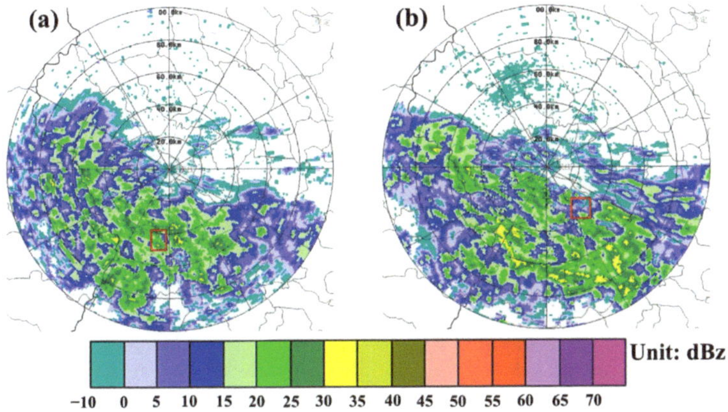

Figure 5. 24 August 2019, (**a**) 12:48, 3.4° elevation angle PPI, (**b**) 14:18, 2.4° elevation angle PPI. The red rectangle box is the aircraft detection area.

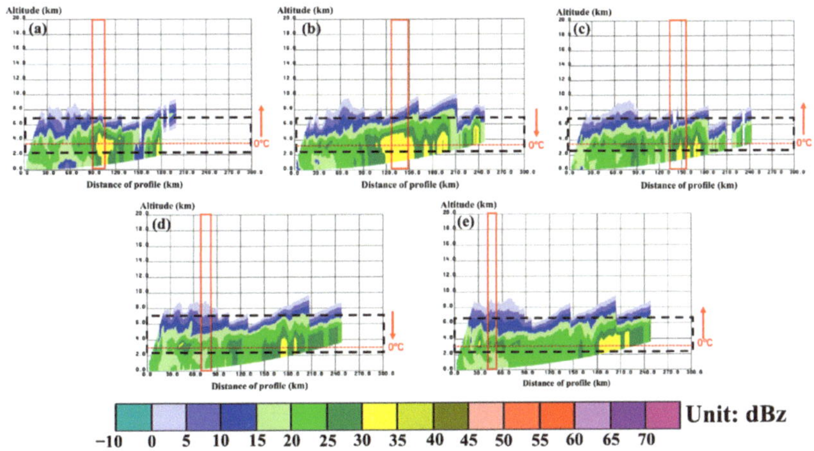

Figure 6. 22 May 2017, (**a**) 15:42 of V1, (**b**) 16:12 of V2, (**c**) 16:36 of V3, (**d**) 17:30 of V4, (**e**) 17:54 of V5. The red rectangle box is the aircraft detection area.

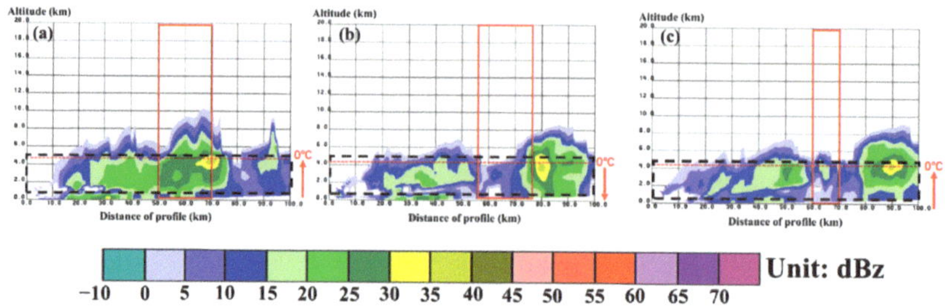

Figure 7. 21 May 2018, (**a**) 13:18 of V1, (**b**) 14:18 of V2, (**c**) 14:42 of V3. The red rectangle box is the aircraft detection area.

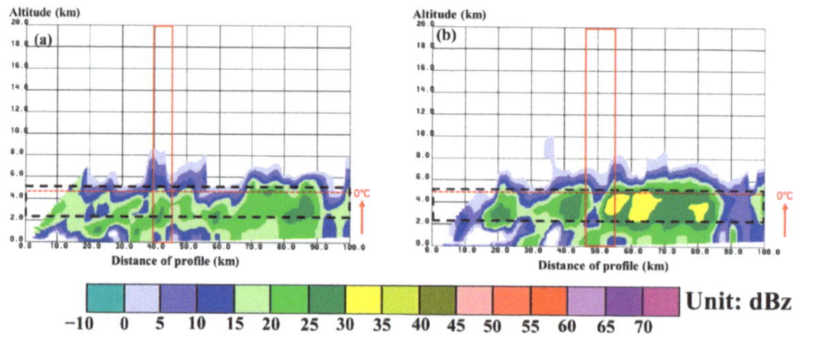

Figure 8. 24 August 2019, (**a**) 12:48 of V4, (**b**) 14:18 of V8. The red rectangle box is the aircraft detection area.

In the observation periods of 2017 and 2018, the cloud systems were in the stratocumulus development stage, there were strong convective echoes embedded in the stratiform cloud area at both of the two stages, and the precipitation in these two stages was strong. In 2019, the echo of the detected cloud system was weak and the precipitation was also weak, so the detected cloud system was stratocumulus in a mature stage.

3. Result

3.1. Vertical Distribution of Microphysical Characteristics in Clouds

On 22 May 2017, the detection of the V1 (Figure 9) and V2 (Figure S1 in Supplementary Materials) segments were in the convective cell region embedded in the stratiform cloud, the detection of the V3 segment (Figure S2 in Supplementary Materials) was in the convective cell edge, and the detections of the V4 (Figure 10) and V5 (Figure S3 in Supplementary Materials) segments were in the stratiform cloud regions. It was shown in Figure 9 that the peak concentrations of the HVPS particles are concentrated in small to medium sized particles. In the stratiform cloud region, the number concentration of particles was low, but with the decrease of height, the peak number concentration gradually tended toward the particles with medium diameter. The CDP particles presented a dominantly multi-modal distribution, and the HVPS particles presented a bimodal distribution (6700 m) in the convective cell, which was higher than that in the stratiform cloud area (3800–4000 m). The convective cell has higher liquid water content, the ice particles have better riming growth conditions. In the convective cell, the particle number concentrations detected by the CDP, the CIP and the HVPS were higher than that in the stratiform cloud area, and the particle spectrum was also wider. There were more larger particles in the convective cell, and the strong gravity drag of large particles made the downdraft in the convective cell stronger than that in the stratiform cloud area. It was found that the 0 °C in the convective cell area was higher than that in the stratiform cloud area (Table 3), and at the end of the detection stage, the temperatures of the same altitude level of V1 and V2 were higher than other stages (Figures 9 and 10).

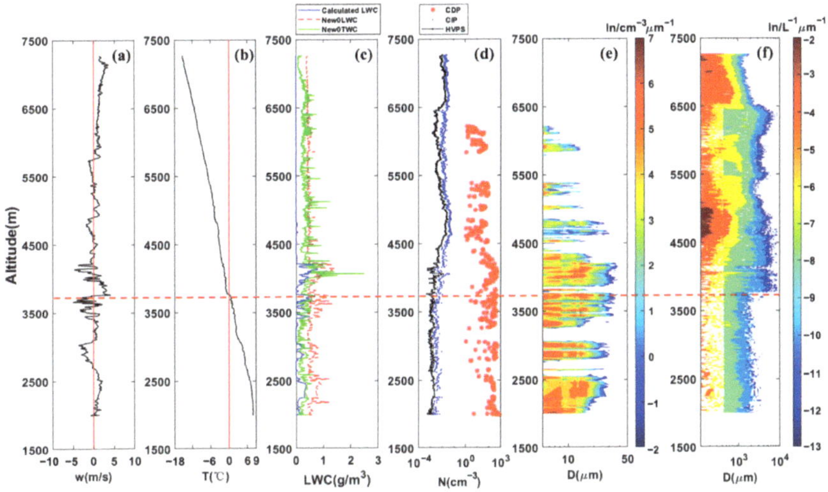

Figure 9. The aircraft observation data set of V1 on 22 May 2017 in the convective cell, (**a**) vertical velocity (the red line is 0, units: m s^{-1}, °C); (**b**) temperature (units: °C, the red line is 0); (**c**) calculated liquid water content (calculated LWC), probe detected liquid water content (New0LWC), probe detected total water content (New0TWC) (units: g m^{-3}); (**d**) particle number concentration of the CDP, the CIP, and the HVPS (units: cm^{-3}); (**e**) particle spectrum of CDP (units: cm^{-3}μm^{-1}); (**f**) particle spectrum of the CIP (100–400 μm) and the HVPS (400–8700 μm) (units: L^{-1}μm^{-1}). The red dotted line is the 0 °C layer.

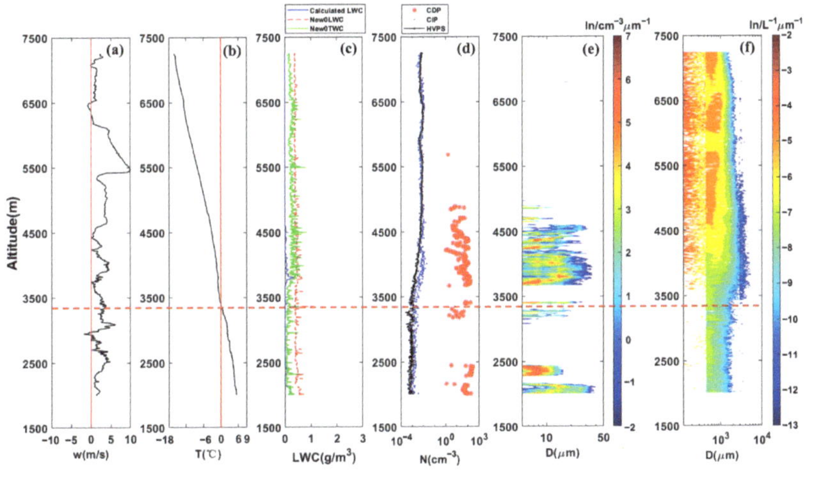

Figure 10. The aircraft observation data set of V4 on 22 May 2017 in the stratiform cloud area, (**a**) vertical velocity (the red line is 0, units: m s^{-1}, °C); (**b**) temperature (units: °C, the red line is 0); (**c**) calculated liquid water content (calculated LWC), probe detected liquid water content (New0LWC), probe detected total water content (New0TWC) (units: g m^{-3}); (**d**) particle number concentration of the CDP, the CIP, and the HVPS (units: cm^{-3}); (**e**) particle spectrum of CDP (units: cm^{-3}μm^{-1}); (**f**) particle spectrum of the CIP (100–400 μm) and the HVPS (400–8700 μm) (units: L^{-1}μm^{-1}). The red dotted line is the 0 °C layer.

Table 3. Comparison of microphysical quantities at different stages.

Detection		Height of 0 °C Layer (m)	CDP (cm^{-3}) (Max/Average)	CIP (cm^{-3}) (Max/Average)	HVPS (cm^{-3}) (Max/Average)	w (m/s)	
						Updraft (Max/Average)	Downdraft (Max/Average)
2017	V1	3764	725/55	19.17/0.54	$3.12 \times 10^{-2}/7.87 \times 10^{-3}$	4.2/1.3	−5.1/−1.0
	V2	3612	980/106	26.7/0.87	$4.44 \times 10^{-2}/9.79 \times 10^{-3}$	7/2.3	−4.2/−0.6
	V3	3706	948/49	2.67/0.66	$6.01 \times 10^{-2}/1.17 \times 10^{-2}$	6.2/2.3	−2.2/−0.4
	V4	3346	261/19	0.39/0.09	$1.06 \times 10^{-2}/4.10 \times 10^{-3}$	10/2.7	−1.9/−0.4
	V5	3451	360/17	0.79/0.14	$1.68 \times 10^{-2}/4.61 \times 10^{-3}$	4.3/0.9	−8.4/−1.1
2018	V1	4737	1419/14	17/0.56	$7.63 \times 10^{-3}/4.84 \times 10^{-4}$	14.4/2.9	−4.6/−1.5
	V2	4201	299/8.99	5.92/0.23	$1.27 \times 10^{-2}/1.28 \times 10^{-3}$	9.6/2.4	−2.1/−0.4
	V3	4265	114/5.76	3.43/0.28	$3.36 \times 10^{-3}/8.58 \times 10^{-4}$	9.1/3.1	−0.6/−0.2
2019	V4	4496	1190/217	3.44/0.17	$7.3 \times 10^{-3}/1.4 \times 10^{-3}$	3.7/1.4	−3.9/−0.9
	V8	4708	693/139	0.08/9.2	$0.01/1.2 \times 10^{-3}$	3.3/1.4	−5.7/−1.1

On 21 May 2018, there existed some areas with high super-cooled liquid water content but low ice particle number concentration, which had strong precipitation enhancement potentiality. The CDP was generally unimodal, but bimodal at the height of 1750 m. The HVPS showed a bimodal distribution at the heights of 4800 m and 3300 m, and the peak particle diameter increases with the decrease of height. According to the vertical radar reflectivity, the convective cell embedded in stratiform clouds was detected in the V1 section, and the downdraft in the V1 section was stronger than that in other stages. In the precipitation particle spectrum of the V1 segment, it is observed that the peak particle diameter increases in the isothermal layer, and there is collision of unmelted particles at this time. The temperature of the V1 section was higher than those of the V2 and V3 sections, and the 0 °C layer of the V1 section was higher than those of the V2 and V3 sections (Table 3, Figures S4–S6 in Supplementary Materials).

Two vertical detections on 24 August 2019 also showed that the spectrum width of precipitation particles below the 0 °C layer was significantly smaller than that in the negative temperature layer. In the melting process of ice precipitation particles, the particle spectrum was mainly unimodal, with bimodal distribution in the V4 section (Figure S7 in Supplementary Materials) at 4300 m, 3400 m, and 2850 m, and in the V8 section (Figure S8 in Supplementary Materials) at 4300 m, 3900 m, and 3550 m. The region with larger peak diameter corresponded to wider particle spectrum. This indicated that the melting rates of ice particles with different particle sizes are different, and the melting rates of ice particles with larger particle sizes were relatively faster.

3.2. The Image of Ice Particles

On 22 May 2017, the temperature of the detection process was low. It could be seen that plate crystals existed in the high negative temperature layer (−13 °C), column crystals existed in the negative temperature layer near zero (−3 °C), and rime-attached ice phase particles appeared in other areas (Figure 11). Close to the 0 °C layer, there still existed the riming growth of ice particles. Below the 0 °C layer, some large drops existed. At the temperature of −3.4 °C, it could be seen that the particles have smooth edges, which indicated that the ice particles have begun to melt in the negative temperature layer above the 0 °C layer.

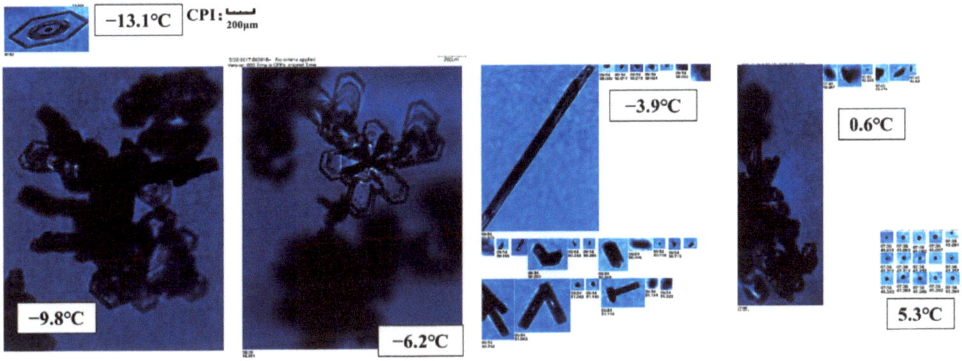

Figure 11. Particles image of the flight on 22 May 2017.

On 21 May 2018, aggregates formed by collision and riming of ice particles were observed in the negative temperature layer close to the 0 °C layer. There were large drops below the 0 °C layer. At the V2 detection stage, it was found that the ice phase particles at the same temperature had different shapes, while at different heights. At the same temperature (−1.6 °C), the particles at the height of 4975 m were dominated with aggregates of columns, co-exited with riming process, while most of the particles at 4583 m were already melted based on their regular edge. In the negative temperature layer close to the 0 °C layer, as the particles began to melt, absorbing latent heat, so an isothermal layer was formed near the 0 °C layer (Figure 12).

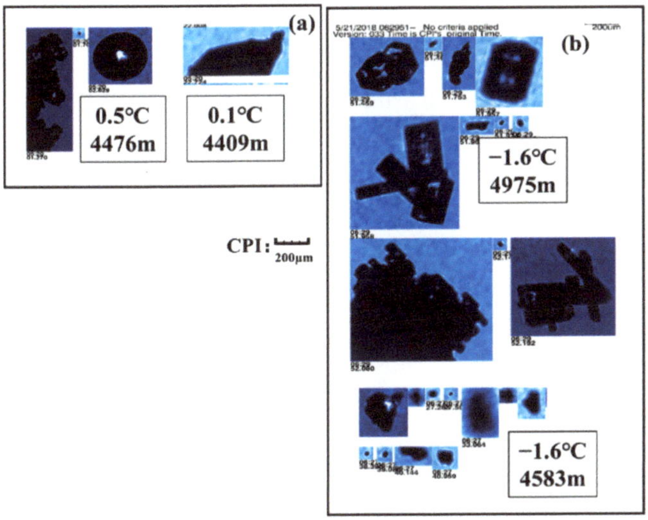

Figure 12. Particles image of the flight on 21 May 2018, (**a**) V1 stage, (**b**) V2 stage.

On 24 August 2019, the riming of particles in the negative temperature layer was strong. Large droplets were observed near the 0 °C layer. When the temperature reached 6.5 °C, particles with irregular edges were still observed (Figure 13).

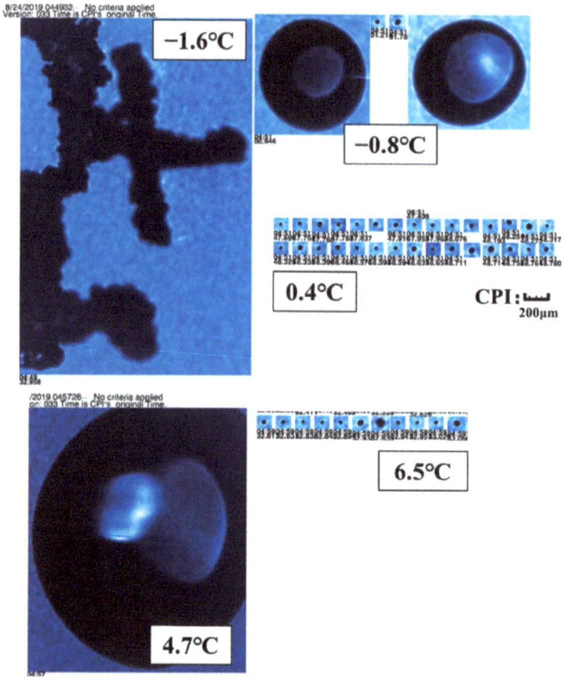

Figure 13. Particles image of the flight V4 on 24 August 2019.

3.3. Particle Size Distribution (PSD)

According to Gunn and Marshall (1958), based on ground observation, the size spectrum distribution of ice particles conformed to the form of negative exponential:

$$n(D) = N_0 \cdot e^{-\lambda D} \tag{1}$$

In which D is the diameter of the ice phase particle; and N_0 and λ are the intercept and slope.

In terms of vertical distribution (Figures 14–16), N_0 above the 0 °C layer was larger than that below the 0 °C layer in each stage, and N_0 above the 0 °C layer showed a decreasing trend with the increasing temperature. At each stage, λ above the 0 °C layer was less than that below the 0 °C layer. In the negative temperature layer close to the 0 °C layer, λ showed an increasing trend with increasing temperature. The maximum particle size (D_{max}) observed by the HVPS was greater above the 0 °C layer than that below the 0 °C layer. In general, N_t above the 0 °C layer was greater than that below the 0 °C layer. In the negative temperature layer close to the 0 °C layer, N_t decreased with increasing temperature.

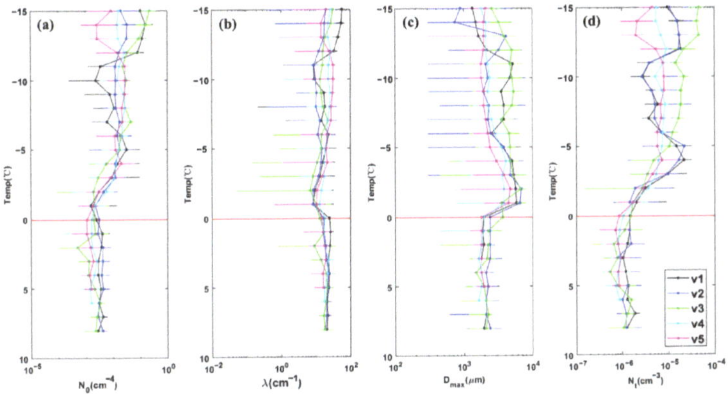

Figure 14. The parameters of 22 May 2017, (**a**) intercept (N_0), (**b**) slope (λ), (**c**) the maximum particle size (D_{max}), (**d**) the total particle concentration (N_t). The red line is the 0 °C layer.

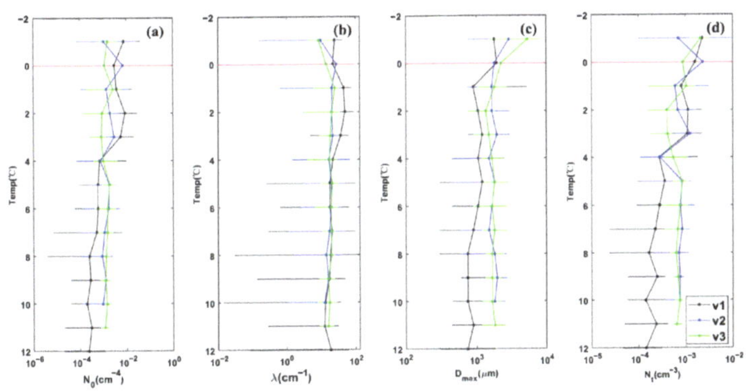

Figure 15. The parameters of 21 May 2018, (**a**) intercept (N_0), (**b**) slope (λ), (**c**) the maximum particle size (D_{max}), (**d**) the total particle concentration (N_t) The red line is the 0 °C layer.

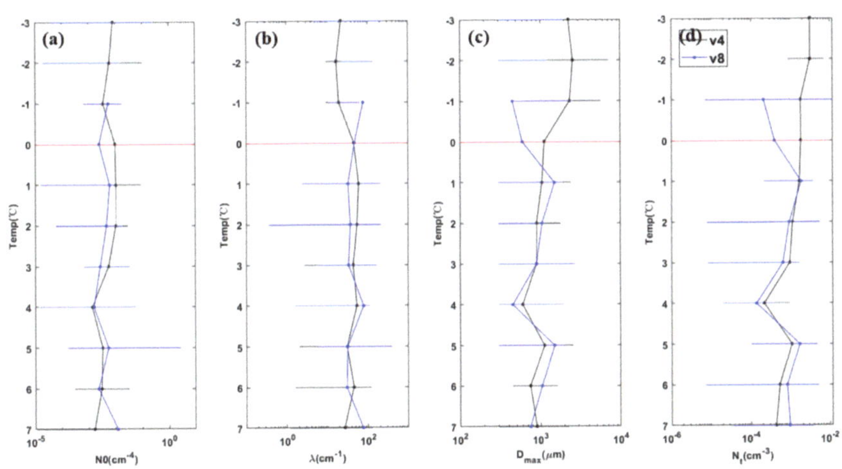

Figure 16. The parameters of 24 August 2019, (**a**) intercept (N_0), (**b**) slope (λ), (**c**) the maximum particle size (D_{max}), (**d**) the total particle concentration (N_t) The red line is the 0 °C layer.

On 22 May 2017, the N_0, λ, and N_t of the V1 and V2 sections in the convective cell, were larger than other areas. In the convective cell between −13.5 °C and −11 °C, N_t, N_0, and λ decreased with the increase of temperature, and D_{max} increased with the increase of temperature or had no obvious change, which was caused by aggregation. In the negative temperature layer from −10 °C to −5 °C, N_t, N_0, and λ in the convective cell are smaller than those in the stratiform cloud region, and D_{max} are larger than those in the stratiform cloud region. The observed data indicate that there are more small-sized particles in the upper stratiform cloud region at this height. In the cloud, N_t reaches its maximum value in the range of −5 °C to −4 °C, and the Hallett-Mossop ice crystal multiplication mechanism exists in this stage, which further proves that the riming process in the convective cell is more active. From −3.5 °C to the 0 °C layer, N_t and N_0 decreased with temperature increasing, and D_{max} decreased and λ increased with increasing temperature starting from −1 °C, so it could be judged that ice phase particles have started to melt above the 0 °C layer (Figure 14).

In the three detection processes on 21 May 2018, from the negative temperature layer near the 0 °C layer to the 0 °C layer of the V1 and V2 sections, N_t increased and D_{max} decreased with the increase of temperature. From the negative temperature layer near the 0 °C layer to the 0 °C layer, the N_t of V1 section decreased with the increase of temperature, and D_{max} increased, which was caused by aggregation. In the two processes, there may be ice phase particle sublimation in the negative temperature layer near the zero layer. The large particles were broken into small particles, resulting in the increase of the total particle concentration and decrease of the maximum particle size. The V1 segment experienced the convection cell during the measurement, and its N_0, λ, and N_t were larger than other detection areas in the negative temperature layer. From 1 °C to 4 °C, N_0 *and* λ of the V1 segment was larger than other detections, and D_{max} was less than other detections above and below the 0 °C layer. Below the 4 °C level, N_t was less than other detections (Figure 15).

On 24 August 2019, in V4 section from −2 °C to 0 °C, N_0, λ, and D_{max} decreased, and N_t did not change significantly. In this phase, particles melted and sublimated at the same time. In V8 section, N_t increased from 0.5 °C to 0 °C layer, D_{max} increases, *with* N_0 and λ decreased, sublimation and aggregation appeared at the same time in this stage. The relative humidity of V8 segment is smaller than that of V4 section. In V4 section, the particles melted in the negative temperature layer close to 0 °C layer, and the total particle concentration and maximum particle diameter decreased accordingly. In V8 section, the sublimation process led to the increase of the total particle concentration, and the decrease of the maximum particle diameter. (Figure 16).

It was found that the particle spectrum parameters were significantly related to temperature and maximum particle size. On the whole, the intercept (N_0) was positively correlated with temperature (T), except that the relationship of the V3 section in 2018 is not obvious. The exponential function based on e could better fit the relationship between the two (Table 4, Figure 17). Data from each detection have shown that the slope (λ) and intercept (N_0) are positively correlated, and the overall fitting found that logarithmic function could well fit the relationship between them. The variation rules of data above and below the 0 °C layer were different. The particles below the 0 °C layer changed faster than those above the 0 °C layer. The fitting coefficients of the two were different, and the coefficient above the 0 °C layer was lower than that below the 0 °C layer (Table 4, Figure 18). It could be found that the slope (λ) was positively correlated with the largest diameter (D_{max}). The power function could well fit the relationship between them (Table 4, Figure 19).

Table 4. The fitting of relationships between spectral parameters.

Relation		a	b	c	R^2
$N_0 = e^{a \cdot T^2 + b \cdot T + c}$		8.6×10^{-4}	−0.16129	−5.50	0.60
$\lambda = a \cdot \ln(N_0) + b$	Above the 0 °C Layer	5.61	44.59		0.47
	Below the 0 °C Layer	12.11	107.82		0.32
$D_{max} = a \cdot \lambda^b$		5.02×10^4	−0.80		0.77

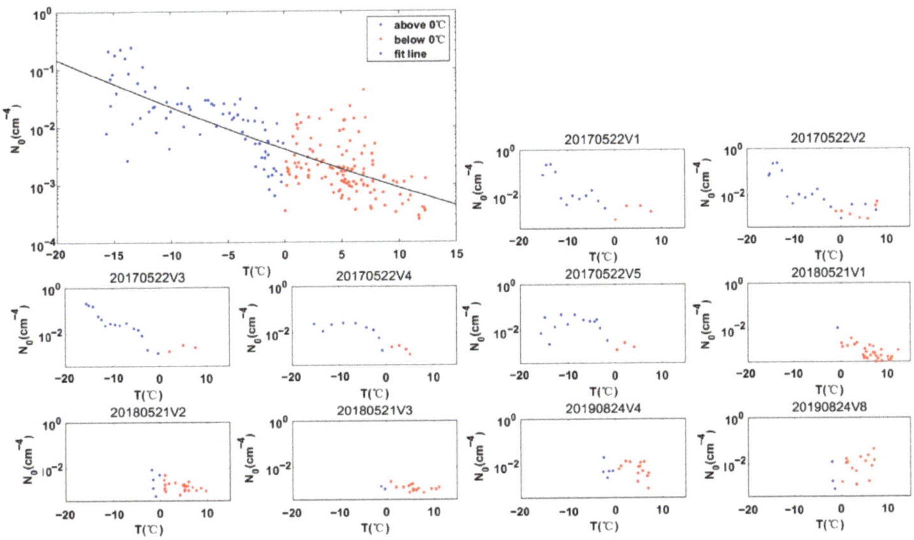

Figure 17. The relationship of N_0 and T.

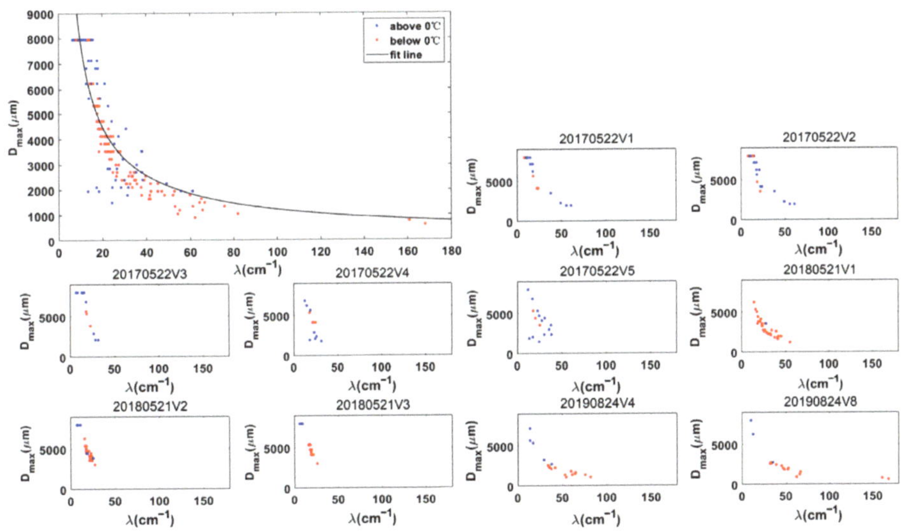

Figure 18. The relationship of N_0 and λ.

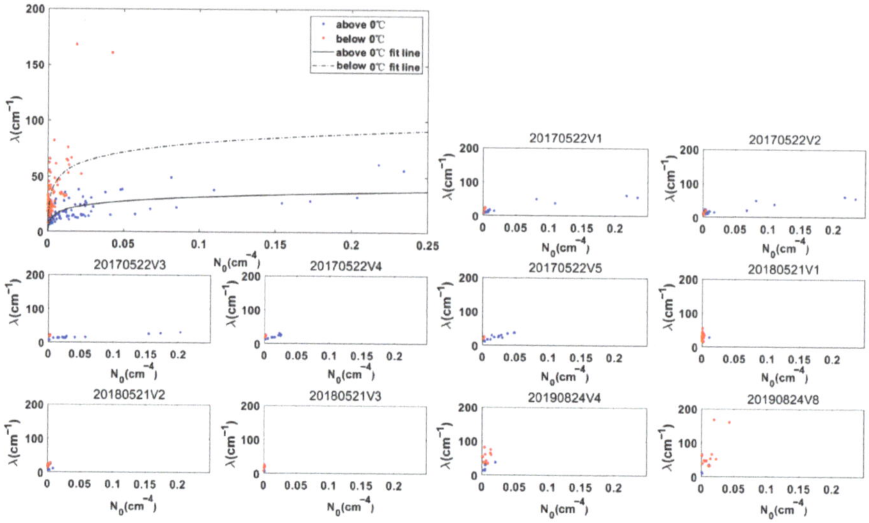

Figure 19. The relationship of λ and D_{max}.

4. Discussion

Generally, there are differences in cloud macro, micro, and precipitation characteristics in different parts of stratocumulus clouds. In this paper, based on the analysis of ten flights' cloud physical observation data, it was also found that the temperature of the cumulus cloud area at the same height was higher than that of the stratus cloud area, and the height of the 0 °C layer was also higher than that of the stratus cloud area. This was supposed to be due to stronger updrafts in the embedded cumulus leading to more condensation, releasing more latent heat. Along with stronger updrafts, ice particles grew to larger particle sizes and fell to form precipitation, which made the spectrum intercept and slope parameters in the melting level of the cumulus area higher than those of the same stratiform cloud area. At the same time, the stronger gravitational dragging at the cumulus cloud, induced stronger downdrafts.

The analysis of the detection data of different flights showed that the particle spectrum parameters and their evolution in the melting layer may be different from those in the negative temperature layer. With the beginning of melting, the spectral width and number concentration of particles decreased. However, the melting process of ice particles might be accompanied by a variety of physical processes, including sublimation, fragment, and aggregation, resulting in inconsistent evolution of particle spectrum parameters in different flights. The latent heat absorption in the melting process might lead to a decrease in the rate of temperature change with height, and an approximate isothermal layer appears around the 0 °C layer.

With the melting of ice particles, the intercept and slope of the particle spectrum below the 0 °C layers and the correlation between them have changed significantly compared with the negative temperature layer, which means that in conducting cloud microphysical parameterization and inversion research of cloud microstructure through radar and satellite observation data, it is better to adopt a parameter relationship and parameters different from the negative temperature layer below the 0 °C layers or the melting layer.

The differences in cloud microphysical characteristics in different regions require us to carry out cloud characteristics research in regions with different climatic backgrounds and underlying surface characteristics, which is very important for weather and climate prediction. The parameters and their relationships obtained in this paper have certain reference significance for North China and similar climate regions.

At present, most studies focus on the differences in precipitation mechanisms in different parts of stratocumulus clouds [24], but little attention is paid to the differences in updraft and temperature in different parts of the clouds. Compared with the studies in China, the studies on the North China stratocumulus clouds are mostly concentrated in the negative temperature layer [1], while the particle spectrum distribution under the 0 °C layer is lacking study. As shown in Table 5, the spectrum parameters of particles in different temperature ranges in stratocumulus clouds are different. The analysis of a particle spectrum parameters relationship usually focuses on temperature (T)-intercept (N_0) and intercept (N_0)-slope (λ) [22], but in China the research mostly focuses on the relationship between N_0 and λ [11]. In this paper, there is a new study on the relationships between T-N_0, N_0-λ, and λ-D_{max}, which can provide a reference for using temperature prediction particle spectrum distribution for China. The particle spectrum parameters of stratocumulus clouds found in this paper are different from other studies abroad and in China, which proves that the study of stratocumulus cloud observation in different regions is very necessary.

Table 5. Comparison of spectral parameters of different detection processes.

	Detection Temperature (°C)	The Maximum Particle Number Concentration (cm^{-3})	N_0(cm^{-4})	λ (cm^{-1})
Hou et al. (2021) [1]	−12~0	10^{-1}	10^{-6}~10^{-2}	10^{-1}~10^{1}
Heymsfiled et al. (2015) [22]	−4~4	10^{-1}	0.01~100	10~25
Feng et al. (2021) [11]	−20~0	10^{2}	0~10^{4}	10~10^{4}
Xiong et al. (2023)	−15.9~8.6	10^{-2}	10^{-5}~10^{0}	10^{-2}~10^{2}
Xiong et al. (2023)	−1.8~12.4	10^{-2}	10^{-6}~10^{0}	10^{-2}~10^{2}
Xiong et al. (2023)	−2.3~7.3	10^{-2}	10^{-5}~10^{0}	10^{-1}~10^{2}

5. Conclusions

From the observation data of ten flights of three precipitation processes, there were differences in the macro- and micro-characteristics of cumulus and stratiform areas of stratocumulus clouds. The particle spectrum parameters and their evolutions below the melting level were also different from those in the negative temperature layer. Based on the analysis, it can be found that study on the distribution and evolution of cloud vertical structure and particle spectrum parameters in different parts of the stratocumulus cloud, especially the evolution of the particle spectrum in the ice particle melting layer and below, may have great significance for the prediction of precipitation intensity. The specific conclusions of this paper are as follows:

(1) The downdraft in the cumulus cloud area was stronger, and the temperature at the same height was higher than that in the stratus cloud area, and the 0 °C layer height was correspondingly higher. In terms of particle spectrum parameters, the intercept and slope parameters of the particle spectrum of the cumulus area below the melting layer were higher than those of the stratus cloud area, for the same weather process. There were significant differences in the characteristics of vertical evolution and parameter evolution of the particle spectra in different parts of stratocumulus clouds, which showed that the microphysical characteristics of stratus clouds and cumulus clouds area were different.

(2) In different vertical detections, it was found that the ice phase particles have narrowed the particle spectrum, decreased the total concentration, and smoothed the edges of the ice particle images in the negative temperature layer near the 0 °C level, indicating that the ice particles have begun to melt in the negative temperature layer near the 0 °C level. As the ice particles began to melt, the $D_{\max}$ of the particle spectrum decreased rapidly, indicating that the larger particles might melt earlier than the middle-sized ones. In different flights, the variation trends of intercept, slope, maximum particle size, and total concentration of particle spectrum with the increase of temperature, were not consistent. The analysis of the evolution of the spectrum parameters showed that there might be sublimation, fragmentation, and aggregation in the melting process of ice phase particles.

In addition, with the latent heat released by the melting process of ice particles, it was possible to have an approximate isothermal layer with a small temperature change rate around the 0 °C level.

(3) The melting process changed the spectral parameters greatly and also changed the correlation between the intercept and slope of the particle spectrum. The slope below the 0 °C level increased with the increase of intercept, which was greater than that above the 0 °C level. Through the fitting calculation of the data of ten flights, the expression relation of N_0 increasing with temperature for North China was obtained. It was also found that, with the increase of the slope parameter of the precipitation particle spectrum, the maximum particle size of the particle spectrum shows an exponential decrease trend.

Supplementary Materials: The following supporting information can be downloaded at: https://www.mdpi.com/article/10.3390/rs15082168/s1.

Author Contributions: X.L. conceived the study; J.W. contributed to the investigation; J.X. analyzed the results and contributed to the original draft preparation; X.L. contributed to the reviewing and editing. All authors have read and agreed to the published version of the manuscript.

Funding: National Natural Science Foundation of China (Grant Nos. 41975176 and 42061134009).

Data Availability Statement: Data available upon request from all the authors.

Acknowledgments: The authors are grateful to three anonymous reviewers and associate editors for providing valuable comments and feedbacks on this work. We acknowledge the High Performance Computing Center of Nanjing University of Information Science and Technology for their support of this work.

Conflicts of Interest: The authors declare no conflict of interest.

References

1. Hou, T.J.; Lei, H.C.; He, Y.J.; Zhao, Z.; Hu, Z.X. Aircraft Measurements of the Microphysical Properties of Stratiform Clouds with Embedded Convection. *Adv. Atmos. Sci.* **2021**, *38*, 966–982. [CrossRef]
2. Zhu, S.C.; Guo, X.L. Ice crystal habits, distribution and growth process in stratiform clouds with embedded convection in North China: Aircraft measurement. *Acta Meteor. Sinica.* **2014**, *72*, 366–389.
3. Qi, P.; Guo, X.L.; Lu, G.X.; Duan, Y.; Li, B.D.; Wu, Z.H.; Dong, X.B.; Hu, X.F.; Yang, Y.S.; Fan, H.; et al. Aircraft Measurements of a Stable Stratiform Cloud with Embedded Convection in Eastern Taihang Mountain of North China: Characteristics of Embedded Convection and Melting Layer Structure. *Chin. J. Atmos. Sci.* **2019**, *43*, 1365–1384.
4. Kang, Z.M.; Li, Z.L.; Liu, W.; Dong, X.B.; Mai, R.; Sun, Y.W. Aircraft observations on physical properties of precipitation clouds in Hebei province. *J. Meteor. Environ.* **2019**, *35*, 1–7.
5. Yang, J.F.; Hu, X.F.; Lei, H.C.; Duan, Y.; Lv, F.; Zhao, L.W. Airborne Observations of Microphysical Characteristics of Stratiform Cloud Over Eastern Side of Taihang Mountains. *Chin. J. Atmos. Sci.* **2021**, *45*, 88–106.
6. Gao, Q.; Guo, X.L.; He, H.; Liu, X.E.; Huang, M.Y.; Ma, X.C. Numerical Simulation Study on the Microphysical Characteristics of Stratiform Clouds with Embedded Convections in Northern China based on Aircraft Measurements. *Chin. J. Atmos. Sci.* **2020**, *44* 899–912.
7. Hu, X.F.; Xiao, H.X.; Cui, Y.; Lv, F.; Zhao, L.W.; Ji, X.S. Microphysical characteristics of precipitating stratiform clouds in north China revealed by joint observations of an aircraft and a polarimetric radar. *J. Atmos. Sci.* **2022**, *79*, 2799–2811. [CrossRef]
8. Wei, L.; Lei, H.; Hu, W.; Huang, M.; Zhang, R.; Zhang, X.; Hou, T.; Lv, Y. An Analysis of the Microstructure of the Melting Layer of a Precipitating Stratiform Cloud at the Dissipation Stage. *Atmosphere* **2022**, *13*, 284. [CrossRef]
9. Huang, M.S.; Lei, H.C. Ice particle habit distribution in stratiform clouds at different precioitation stages. *Trans. Amos. Sci.* **2022** *45*, 247–256.
10. Zhao, Z.; Lei, H. Observed microphysical structure of nimbostratus in northeast cold vortex over China. *Atmos. Res.* **2014**, *142* 91–99. [CrossRef]
11. Feng, Q.J.; Niu, S.J.; Hou, T.J.; Fan, X.P.; Shen, D.D.; Yang, J.M. Aircraft-Based Observation of the Physical Characteristics of Snowfall Cloud in Shanxi Province. *Chin. J. Atmos. Sci.* **2021**, *45*, 1146–1160.
12. Rauscher, S.A.; Pal, J.S.; Diffenbaugh, N.S.; Benedetti, M.M. Future changes in snowmelt-driven runoff timing over the western US. *Geophys. Res. Lett.* **2008**, *35*, L16703. [CrossRef]
13. Ashfaq, M.S.; Ghosh, S.C.; Kao, S.C.; Bowling, L.C.; Mote, P.; Touma, D.; Rauscher, S.A.; Diffenbaugh, N.S. Nearterm acceleration of hydroclimatic change in the western U.S. *J. Geophys. Res. Atmos.* **2013**, *118*, 10676–10693. [CrossRef]
14. Marty, C.; Tilg, A.M.; Jonas, T. Recent evidence of largescale receding snow water equivalents in the European Alps. *J Hydrometeorol.* **2017**, *18*, 1021–1031. [CrossRef]

15. Mote, P.W.; Li, S.H.; Lettenmaier, D.P.; Xiao, M.; Engel, R. Dramatic declines in snowpack in the western US. *Npj. Clim. Atmos. Sci.* **2018**, *1*, 1–6. [CrossRef]
16. Tamang, S.K.; Ebtehaj, A.M.; Prein, A.F.; Heymsfield, A.J. Linking global changes of snowfall and wet-bulb temperature. *J. Clim.* **2020**, *33*, 39–59. [CrossRef]
17. Kain, J.S.; Goss, S.M.; Baldwin, M.E. The melting effect as a factor in precipitation-type forecasting. *Weather. Forecast.* **2000**, *15*, 700–714. [CrossRef]
18. Oraltay, R.G.; Hallett, J. Evaporation and melting of ice crystals: A laboratory study. *Atmos. Res.* **1989**, *24*, 169–189. [CrossRef]
19. Matsuo, T.; Sasyo, Y. Empirical formula for the melting rate of snowflakes. *J. Meteorol. Soc. JPN. Ser. II* **1981**, *59*, 1–9. [CrossRef]
20. Mitra, S.K.; Vohl, O.; Ahr, M.; Pruppacher, H.R. A wind tunnel and theoretical study of the melting behavior of atmospheric ice particles. IV: Experiment and theory for snow flakes. *J. Atmos. Sci.* **1990**, *47*, 584–591. [CrossRef]
21. Heymsfield, A.J.; Bansemer, A.; Field, P.R.; Durden, S.L.; Stith, J.; Dye, J.E.; Hall, W.; Grainger, T. Observations and parameterizations of particle size distributions in deep tropical cirrus and stratiform precipitating clouds: Results from in situ observations in TRMM field campaigns. *J. Atmos. Sci.* **2002**, *72*, 2902–2928. [CrossRef]
22. Heymsfield, A.J.; Bansemer, A.; Poellot, M.R.; Wood, N. Observations of ice microphysics through the melting layer. *J. Atmos. Sci.* **2015**, *72*, 2902–2928. [CrossRef]
23. Liu, S.Y.; Zhao, C.F.; Zhou, Y.Q. SPEC airborne cloud detection system and its cloud physics research progress. *Torrential Rain Disaster* **2021**, *40*, 280–286.
24. Guo, X.L.; Fu, D.H.; Guo, X.; Fang, C.G. Study on physical plane observation of cloud precipitation in China. *JAM* **2021**, *32*, 641–645.

remote sensing

MDPI

Article

Technical Evaluation of Precipitation Forecast by Blending Weather Radar Based on New Spatial Test Method

Junchao Wang [1,2,3,*], Zhibin Wang [1], Jintao Ye [1], Anwei Lai [1,2], Hedi Ma [1] and Wen Zhang [1]

1 China Meteorological Administration Basin Heavy Rainfall Key Laboratory/Hubei Key Laboratory for Heavy Rain Monitoring and Warning Research, Institute of Heavy Rain, China Meteorological Administration, Wuhan 430205, China; wangzb@whihr.com.cn (Z.W.); yejintao@whihr.com.cn (J.Y.); laianwei@whihr.com.cn (A.L.); mahedi@whihr.com.cn (H.M.); zhangwen@whihr.com.cn (W.Z.)
2 Three Gorges National Climatological Observatory, Yichang 443099, China
3 Hubei Key Laboratory of Intelligent Yangtze and Hydroelectric Science, China Yangtze Power Co., Ltd., Yichang 443000, China
* Correspondence: wjc@whihr.com.cn

Abstract: The Fourier–Merlin transform method, multi-scale optical flow method, and Weibull distribution are used to integrate the GRAPES_3 km model and Radar Extrapolation Forecast (REF) both developed independently by China. Taking GRAPES_3 km, Wuhan Rapid Update Cycle (WHRUC), and the REF as examples, the prediction performance of the Blending forecast is evaluated comprehensively by the traditional point-to-point method. A new spatial test method is introduced to evaluate the applicability and difference of high-resolution model evaluation. The area, position, shape, and intensity of the precipitation area are matched through the target object test method. The potential forecast information of the spatial field is obtained and the related results are compared and analyzed. The results show that: (1) the comprehensive application of various evaluation methods can evaluate the convective storm forecast more comprehensively. The Blending forecast effect is obviously better than those of other models by using the point-to-point scoring method, especially in the heavy precipitation forecast. The shorter the prediction time is, the better the effect is. (2) The new spatial test method can evaluate the prediction effect of convective storm characteristics, and the target recognition hit rate of the Blending forecast is highest. The scores of target area, position, shape, and median intensity of precipitation are better than those of other forecasts. The variation in the east–west direction is less than that in the north–south direction, which is basically consistent with the actual observation. The variation range of the forecast grid before and after translation is the closest to the reality. (3) The Blending forecast method combines the advantages and disadvantages of the numerical model and REF, which can not only grasp the precipitation area but also improve the prediction ability of rainfall intensity. The traditional point-to-point scoring method and the new spatial test method have the same conclusion as the convective storm forecast of the high-resolution model, which has a certain reference value, and the new spatial test method can provide more detailed evaluation information.

Keywords: mesoscale numerical prediction; radar extrapolation; blending technology; short-term and impending forecast of precipitation echo; MODE

1. Introduction

The flood and geological disasters caused by severe convective weather cause great harm to people's life and property. It is of great significance to forecast them and provide early warnings [1,2]. The Radar Extrapolation Forecast (REF) and Meso-scale Numerical Weather Prediction (NWP) have become the key technical support for the short-term Quantitative Precipitation Forecast (QPF) at present [3–5]. The problem of "spin-up" always exists in the model forecast in the first few hours, which leads to poor prediction results in the first few hours and cannot be directly applied to the short-term approaching forecast. REF and NWP have their own advantages, and their combination can improve the prediction ability of 0~6 h. Therefore, the integrated precipitation forecast based on the numerical model and radar extrapolation is developing rapidly [6–9].

The spatial inspection methods of the precipitation field can be divided into the following two categories: traditional point-to-point-based inspection technology and object-oriented inspection technology [10]. Brownlee classified weather events through a two-variable forecast test contingency table [11]. He calculated a series of scoring indexes, such as hit rate and false alarm rate. Doswell found that the real skill score TSS often tended to the hit rate when calculating the forecast score of small-probability events [12]. So, he revised the HSS score. Since then, a series of scoring indicators have been developed and the deterministic prediction of binary events mainly includes: forecast deviation, probability ratio, accuracy, etc. For the deterministic prediction of classified events, the 2*K contingency table can be used to classify the probabilities of different levels to calculate the scoring criteria. In recent years, meteorologists have conducted a lot of work around the precipitation forecast test. Wang analyzed the predictability of the radar echo by the decorrelation time method and quantitatively analyzed the error of the extrapolated forecast by means of the forecasting skill score and relative absolute error [13]. In addition, the relationship between REF error and scale, and the relative importance of the echo intensity change and echo motion field change in the prediction error were also analyzed. Although the traditional lattice comparison method can reflect certain forecasting characteristics, the test results cannot give specific reasons for the deviation. It still cannot comprehensively test the prediction ability of NWP and objective methods. On the other hand, the object-oriented test technology divides the precipitation field into discrete targets and comprehensively measures the forecasting effect according to the characteristic attributes, which can provide users with more abundant information [14,15]. Xue studied the objective performance of precipitation objects and the prediction ability of the Japanese fine grid model to cases based on the MODE of object-oriented model diagnosis and analysis [16]. The methods adopted involved more advanced spatial testing [17,18], which reflect the spatial structure and scale changes of the precipitation forecast and play a leading role in the inspection industry.

Since 2016, the national self-developed GRAPES_3 km high-resolution numerical model has been widely used in severe convective weather forecast and early warning. More scientific and technical personnel have carried out a variety of inspection and evaluation work for the GRAPES_3 km model. The test results show that the GRAPES_3 km model has a good ability to predict high-threshold and small-scale convective events. Xu evaluated the precipitation forecast of the model and pointed out that the variation in the daily frequency in summer was similar to the observation [19]. The heavy-precipitation frequency and regional distribution are in good agreement with the observation and can reflect the diurnal variation characteristics of the precipitation process. Tang compared and evaluated the prediction ability of GRAPES_3 km in several typical severe convective weather processes in North China by using the fractional technique score (FSS) and discussed the good prediction performance of this model in severe convective weather [20]. The test results show that the GRAPES_3 km model has a good ability to predict high-threshold and small-scale convective events, which are difficult to forecast. However, it has not been comprehensively evaluated by a variety of test methods [21]. Therefore, this paper combines GRAPES_3 km and REF, and compares and analyzes the prediction ability of mesoscale models with different resolutions in severe convective weather by taking the Blending results as the evaluation object, to provide a reference basis for the further development of intelligent grid forecasting technology and products of severe convective weather.

2. Materials and Methods

2.1. Materials

The observation data are based on the national radar combination reflectivity data, with a spatial resolution of 0.01° and a time resolution of 6 min. The regional range is 12.2°~54.2°N and 70°~135°E. The NWP data are based on the GRAPES_3 km model prediction and Fast Update Cyclic Assimilation model (WHRUC) in Table 1. The Blending data are based on GRAPES_3 km and REF, and the bilinear interpolation method is adopted. Due to the difference in temporal and spatial resolution between GRAPES_3 km and actual observation data, which are sparse when the spatial resolution is unified, it was necessary to take the maximum value of the area around the sparse mesh so the strong echo information of convective weather could be preserved.

Table 1. NWP model systems.

	WHRUC	GRAPES_3 km
Forecast area	28°~34°N, 108°~116.5°E	20°~50°N, 73°~139°E
Spatial resolution	0.01°	0.03°
Time resolution	08/20:00 1 h	08/20:00 1 h
Forecast time limit	12 h	36 h
Background field	WHRAP/NCEP GFS	T639
Initial value of model	3D-VAR	Downscaling cloud analysis

2.2. Methods

The REF aimed to calculate the moving speed and direction of precipitation based on the images observed by the radar at the previous time and the current time, and to speculate the position of the future time [22–24]. In view of the severe convective weather, the improved variational optical flow method was used to retrieve the wind field from the radar data and the change in the echo optical flow field was used to obtain the motion vector field. The improved variational optical flow method organically combined the local optical flow method with the global optical flow method through an energy function and added a high-order smoothing operator to solve the equation to obtain the flow field structure. Additionally, the 9-point moving average was then used to obtain the motion vector field. Finally, the semi-Lagrange method was used for extrapolation prediction.

In order to improve the ability of the short-term forecast of disastrous weather, the Institute of Heavy Rain, China Meteorological Administration established a kilometer-scale high-resolution fast updating cyclic assimilation forecast system WHRUC in 2019 [25]. The horizontal resolution of the system is 1.5 km. The update frequency of analysis is 15 min. The update frequency of prediction is 1 h and the time effectiveness of prediction is 12 h. The center of the simulation area is 113.0°E, 30.5°N. The number of horizontal grid points is 801 × 701. The vertical direction is 51 layers and the integration time step is 10 s.

GRAPES_3 km is a high-resolution numerical model developed and operated by the numerical Forecast Center of the China Meteorological Administration. In order to improve the prediction accuracy and stability of the numerical model throughout the country, the numerical Forecast Center integrates national weather radar data, formulates a quality control scheme, and assimilates SC/CD radar data. The prediction accuracy and stability of the model are improved by introducing the monotone high-order horizontal diffusion scheme, adopting the automatic modulation time step scheme, and optimizing the initial reference profile. The stability of the model is evaluated to solve the problem of instability. The physical process parameterization scheme and the prediction ability are improved under weak dynamic forcing by using in-depth analysis of the sources of prediction deviation. The spatial resolution of the model is 3 km and the current coverage is 10°~60°N. It runs four times a day. The reporting time is 02, 08, 14, and 20:00 (BT).

3. Key Technologies of Blending Forecast

The radar extrapolation prediction and intensity correction method are developed based on the RAPIDS technology of the Hong Kong Observatory, Fourier–Merlin transform, and Weibull distribution. Combined with the radar extrapolation prediction of the multi-scale optical flow variational method (multi-scale optical flow by variational analysis, MOVA), the hyperbolic tangent function is used to integrate the radar extrapolation prediction and the corrected model prediction. The Blending method is preliminarily realized and the technical flow is shown in Figure 1.

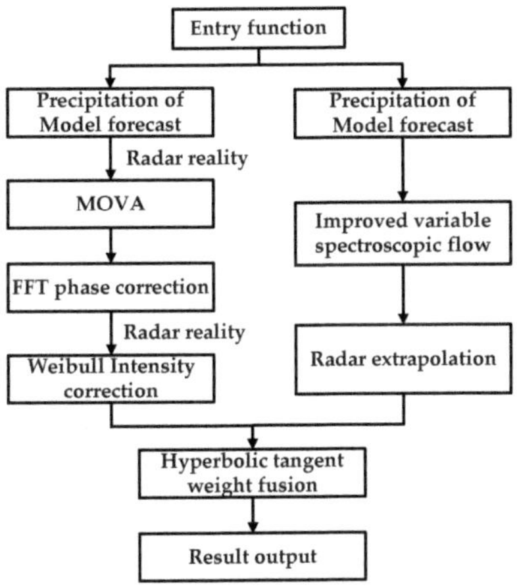

Figure 1. The Blending technology process of numerical prediction into short-term and impending extrapolation prediction of radar echo.

3.1. Phase Correction of Precipitation Forecast by Model

The Fourier–Merlin transform is the global phase correlation between the fast Fourier transform and logarithmic polar transformation. Suppose $f_1(x, y)$ is the template image and $f_2(x, y)$ is the image to be matched. There are rotations, translations, and scaling between them, which are set to $\Delta\theta$, $(\Delta x, \Delta y)$, and λ, respectively, that is:

$$f_2(x,y) = f_1[\lambda(x\cos\Delta\theta + y\sin\Delta\theta) - \Delta x, \lambda(-x\sin\Delta\theta + y\cos\Delta\theta) - \Delta y] \quad (1)$$

By the Fourier transform, the following results can be obtained:

$$F_2(u,v) = \frac{1}{\lambda^2}e^{-2\pi i(u\Delta x + v\Delta y)}F_1[\frac{1}{\lambda}(u\cos\Delta\theta + v\sin\Delta\theta), \frac{1}{\lambda}(-u\sin\Delta\theta + v\cos\Delta\theta)] \quad (2)$$

In the formula: $f_1(u, v)$ and $f_2(u, v)$ are the Fourier transform results of $f_1(x, y)$ and $f_2(x, y)$, respectively. It can be seen from Formula (2) that the relative translation between them is only in the phase spectrum. The amplitude spectra on both sides of Formula (2) are calculated, respectively, and the following results are obtained:

$$M_2(u,v) = \frac{1}{\lambda^2}M_1[\frac{1}{\lambda}(u\cos\Delta\theta + v\sin\Delta\theta), \frac{1}{\lambda}(-u\sin\Delta\theta + v\cos\Delta\theta)] \quad (3)$$

In the formula: $M_1(u, v)$ and $M_2(u, v)$ are the amplitude spectra of $F_1(u, v)$ and $F_2(u, v)$, respectively. By converting the amplitude spectrum to the logarithmic–polar coordinate space, we can obtain:

$$M_2(\lg\rho, \theta) = \frac{1}{\lambda^2}M_1(\lg\rho - \lg\lambda, \theta - \Delta\theta) \quad (4)$$

In the formula: $M_1(\lg\rho, \theta)$ and $M_2(\lg\rho, \theta)$ are logarithmic polar transformations of $M_1(u, v)$ and $M_2(u, v)$, respectively. Because $1/\lambda^2$ only affects the value, it has no effect on the calculation results of rotation, translation, and scaling parameters. So, it can be ignored. It can be seen from Formula (4) that the rotation and scaling between $f_1(x, y)$ and $f_2(x, y)$ are converted into translation in logarithmic polar coordinates, namely $(\lg\lambda, \theta)$. By using the phase correlation algorithm for the amplitude spectrum in logarithmic–polar coordinates, the translation can be calculated and then the rotation and scaling parameters can be obtained. The rotation and scaling parameters are applied to the template image $f_1(x, y)$ to obtain the image with only translation. Then, the translation can be obtained by using the phase correlation algorithm. Formula (4) is called the Fourier–Merlin transform. Figure 2 shows the flow chart of the algorithm.

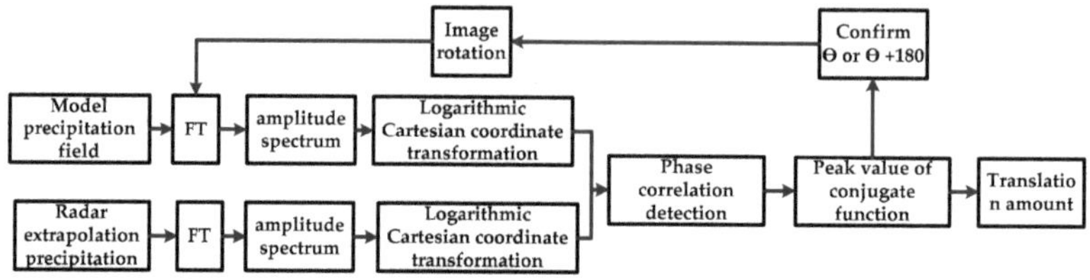

Figure 2. Flow chart of the algorithm.

3.2. Correction of Precipitation Intensity Forecast by Model

The difference between the precipitation intensity predicted by the numerical model and the actual precipitation may be caused by the physical processes such as model resolution, convective parameterization, and cloud microphysical schemes. The intensity adjustment is adjusted by gradually moving the model forecast precipitation field toward the quantitative estimation precipitation field. Statistics show that both the model forecast precipitation and the actual precipitation satisfy the Weibull distribution, and their probability density distribution functions are the same. The model forecast precipitation intensity correction $I_{f\text{-}mod}$ model is as follows:

$$I_{f-mod} = F_{\mathrm{e}}^{-1}(x_0)F_f(x_0) \tag{5}$$

In the formula: $F_f(x_0)$ and $F_{\mathrm{e}}(x_0)$ are the cumulative distribution functions of model forecast precipitation and radar extrapolation forecast precipitation at the initial time, respectively.

3.3. Blending of Radar Extrapolation Prediction and Model Correction Prediction

Both kinds of forecast results are fused by combining with the optimization of the radar extrapolation forecast method after the precipitation forecast by the numerical model is corrected and adjusted according to the time series. The weight change of the model prediction is expressed by the hyperbolic tangent function and its empirical equation is as follows:

$$W(t) = a + \frac{1}{2}(b - a) \times \{1 + \tanh[k(t - 3)]\} \ (1 \le t \le 6) \tag{6}$$

In the formula: t is time; a and b are the Blending weights of 1 h and 6 h model forecasts, respectively. The weights can be determined according to the historical statistical results of precipitation types and precipitation evolution characteristics or dynamically specified in combination with the position error and intensity error. k is the slope of $W(t)$ in the middle part of the Blending period and its value can be determined according to the weather type. The spectral spatial correlation of radar reflectivity and the change speed of the weight curve are determined by adjusting the k value. The calculation formula of the Blending forecast is as follows:

$$R_{\text{blending}}(t) = W(t) \times R_{\text{GRAPES\_3km}}(t) + [1 - W(t)] \times R_{\text{RFST}}(t) \ (1 \le t \le 6) \tag{7}$$

In the formula: $R_{\text{GRAPES\_3km}}(t)$, $R_{\text{RFST}}(t)$, and $R_{\text{blenging}}(t)$ represent t time GRAPES_3 km, REF, and Blending forecast, respectively.

4. Test Methods

4.1. Point-to-Point Comprehensive Test

The evaluation process aims to compare the regional average value of each grid point of the precipitation forecast field (resolution is 0.01° × 0.01°, about 1 km × 1 km) and the adjacent 3 × 3 grid point with the observed precipitation of this grid point. The TS (equitable threat score) and bias forecasting skills scoring methods commonly used in the world are used to test the forecasting effect of the above four precipitation cases. The prediction accuracy cannot be directly given by the prediction skill score, so the average absolute error (mean absolute error, MAE) and hit rate (probability of detection, POD) are selected to describe the prediction accuracy.

In order to quantitatively describe the correlation between forecast precipitation and the radar quantitative estimation of precipitation, the correlation coefficient between the observed area forecast precipitation and radar quantitative precipitation estimation at the same time is calculated as follows:

$$r = \frac{\sum_{i=1}^{N}(F_i - \overline{F})(O_i - \overline{O})}{\sqrt{\sum_{i=1}^{N}(F_i - \overline{F})^2}\sqrt{\sum_{i=1}^{N}(O_i - \overline{O})^2}} \quad (8)$$

where F_i is the predicted value, F is the average value of the predicted value, O_i is the observed value, and O is the average value of the observed value.

4.2. Spatial Comprehensive Test

Although the above lattice-based analysis can reflect certain forecast characteristics—it can deal with all the causes of forecast errors in the same way—it cannot distinguish the sources of forecast errors nor can it give the overall properties of the precipitation field. For this reason, the spatial diagnosis evaluation method is used to further test the results of the precipitation forecast.

4.2.1. MODE Test

MODE is a test technique based on "object". The weight coefficients of different attributes are set on the basis of defining and calculating different attributes of the precipitation object. The characteristics of the target are calculated and compared once the "target area" is correctly identified. The fuzzy logic algorithm is used to calculate the total return function of the forecast performance to judge the overall performance of the forecast. Finally, the spatial position, strength, and shape of the prediction field and the observation field are given to provide more detailed inspection information. The target-based precipitation detection method includes three basic steps: target recognition, target pairing, and target detection. First, the Unicom targets are identified according to the input observation and forecast grid data. The multiple Unicom regions are merged according to the proximity degree and these Unicom targets are marked with serial numbers in turn. Then, the target in the forecast field is matched to the observed target. The number of forecast targets in the matching will be the same as the number of observation targets. There is a many-to-many situation between the observation field and the forecast field target. In the process of matching, the forecast targets will be further merged according to the position distribution to improve the coincidence between the observation targets and the observation targets. Finally, the axis attribute, face attribute and area, center of gravity position, and shape and strength parameters of each target object are calculated. Through the target attributes of the observation field and forecast field, the similarity matrix of observation and prediction is calculated.

(1) Area score:

$$S_A = \left(2 \cdot \left|\frac{A_{mod} - A_{obs}}{A_{obs}}\right| + 1\right)^{-1} \quad (9)$$

where S_A is the area score, A_{mod} is the forecast target area, and A_{obs} is the actual target area.

(2) The score of the center of gravity:

$$S_{GC} = \begin{cases} 0 & L \geq L_{max} \\ 1 - \frac{L - L_{min}}{L_{max} - L_{min}} & L_{min} < L < L_{max} \\ 1 & L \leq L_{min} \end{cases} \tag{10}$$

where S_{GC} is the center of gravity, L_{max} is the maximum tolerance distance, L_{min} is the best distance, the range of the east–west direction is about 940 km, and the range of the north–south direction is about 660 km. For short-term prediction, the tolerance maximum deviation L_{max} is 470 km, the corresponding minimum deviation L_{min} is 47 km, the corresponding minimum deviation L_{max} is 330 km, and the corresponding minimum deviation L_{min} is 33 km.

(3) Shape score:

$$S_{Axial} = \begin{cases} 0 & D_{AxA} \geq 90^\circ \\ \frac{90 - D_{AxA}}{90 - 10} & 10^\circ \leq D_{AxA} < 90^\circ \\ 1 & D_{AxA} < 10^\circ \end{cases} \tag{11}$$

where S_{Axial} is the axial angle score and D_{AxA} is the axial angle difference between the predicted and the real object.

$$S_{Ellip} = \begin{cases} 0 & D_{Ellip} \geq 0.5 \\ \frac{0.5 - D_{Ellip}}{0.5 - 0.1} & 0.1 \leq D_{Ellip} < 0.5 \\ 1 & D_{Ellip} < 0.1 \end{cases} \tag{12}$$

where S_{Ellip} is the ellipticity difference score and D_{Eliip} is the difference between the ellipticity of the predicted object and the actual object.

(4) Intensity score of precipitation center:

The intensity score of the precipitation center is defined as follows: based on the actual precipitation level, the grade of the model is consistent with that of the real heavy precipitation center, or the absolute value of the difference between the model and the real heavy precipitation center is less than 10 mm. The score of the above two cases is 1; when the difference between the model and the actual heavy precipitation center is one grade, the score is 0.5; if the difference is more than one grade, the score is 0.

4.2.2. SAL Test

The SAL (structure, intensity, scale) method is based on the MODE algorithm, which counts the uniformity, average intensity, and centroid distribution of precipitation from the marked grid. It verifies the deviation attributes such as uniformity, average intensity, and distance between prediction and observation. During the calculation, the mesh points that are not marked as targets are first set to 0, and then the intensity error (A), distance error (L), and structure error (S) are calculated.

5. A Case of Inspection and Evaluation

5.1. Case Introductions

Since 8 August 2021, heavy rainfall occurred in the north, east, and southwest of Hubei Province, in which torrential rainfall occurred at 16 stations in Xiangyang from 20:00 on 11 August to 20:00 on 12 August 2021. The maximum daily rainfall was 519 mm in the willow forest, followed by 495 mm in Yinghe. From 8 August to 12 August 2021, there were 7 stations with an accumulated rainfall exceeding 400 mm and those with a heavier hourly rainfall had 118 mm, 105 mm, and 104 mm. From 3:00 a.m. on 12 August 2021, there was a sudden heavy rainfall at night in Liulin Town. From 11 to 12 August, the accumulated rainfall was 503 mm, the rainfall reached 373.7 mm from 4:00 to 7:00 on 12 August 2021, and the rainfall exceeded 100 mm for two consecutive hours from 5:00 to 6:00, all of which were the historical extremes since meteorological records began. The average depth of stagnant water in the rainfall center was 3.5 m and the deepest depth was 5 m.

Figure 3 shows the comparison between the extrapolation forecast, WHRUC model forecast, GRAPES_3 km forecast, and Blending forecast of 0~3 h precipitation from 04:00 (BT) on 12 August 2021. On the whole, the main areas of precipitation are basically concentrated in Jingzhou, Xianning, and Wuhan, and the location remains relatively unchanged, indicating that the predicted precipitation location is consistent with the actual situation. It is found that the Blending forecast results improve the deviation in the heavy precipitation area and precipitation intensity predicted by radar extrapolation, and the overall effect is better than that of the single numerical model forecast and radar extrapolation forecast.

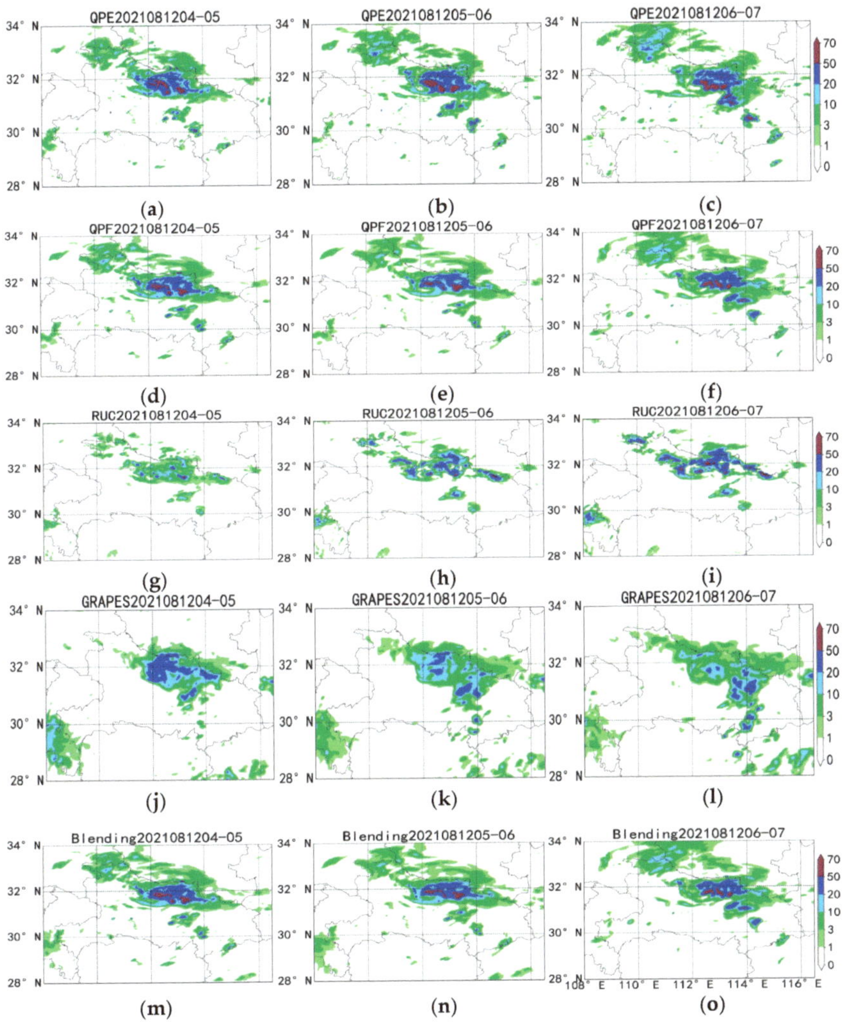

Figure 3. Comparison of 0~3 h radar-extrapolated precipitation forecast (**d–f**), WHRUC model forecast (**g–i**), GRAPES_3 km model forecast (**j–l**), and Blending forecast (**m–o**) with real time (**a–c**) from 04:00 (BT) on 12 August 2021.

5.2. Forecast Effect Test

5.2.1. Point-to-Point Comprehensive Test

Zawadzki defined the "de-correlation time" L when studying the predictability of 11 precipitation cases [26], that is $L = \int_0^\infty C(t)dt$. It is no longer predictable when the correlation value between prediction and observation is less than 0.5. It is found that L is equal to the time constant in the index if $C(t)$ conforms to the exponential law and corresponds to the time when the correlation coefficient decreases to 1/e = 0.37. The decorrelation time defined in the formula can be used to measure the predictability of precipitation.

The correlation coefficients calculated by the formula for each precipitation process are averaged in order to better evaluate the prediction results of 0~3 h (Figure 4). Figure 5 shows the variation in the correlation coefficients of the four methods with the prediction time. For the four methods with the forecast precipitation process, the variation in the correlation coefficient with the forecast time effect is basically decreasing exponentially. It can be seen from the straight line of stroke 1/e in Figure 5 that the decorrelation time of the QPF and Blending forecast precipitation process is much more than 3 h. The decorrelation time of the GRAPES_3 km forecast precipitation process is about 3 h and the decorrelation time of the WHRUC forecast precipitation process is less than 3 h. It can also be seen from Figure 5 that large-scale precipitation systems correspond to longer persistence, while for storms with a faster evolution and smaller scale, their persistence is shorter.

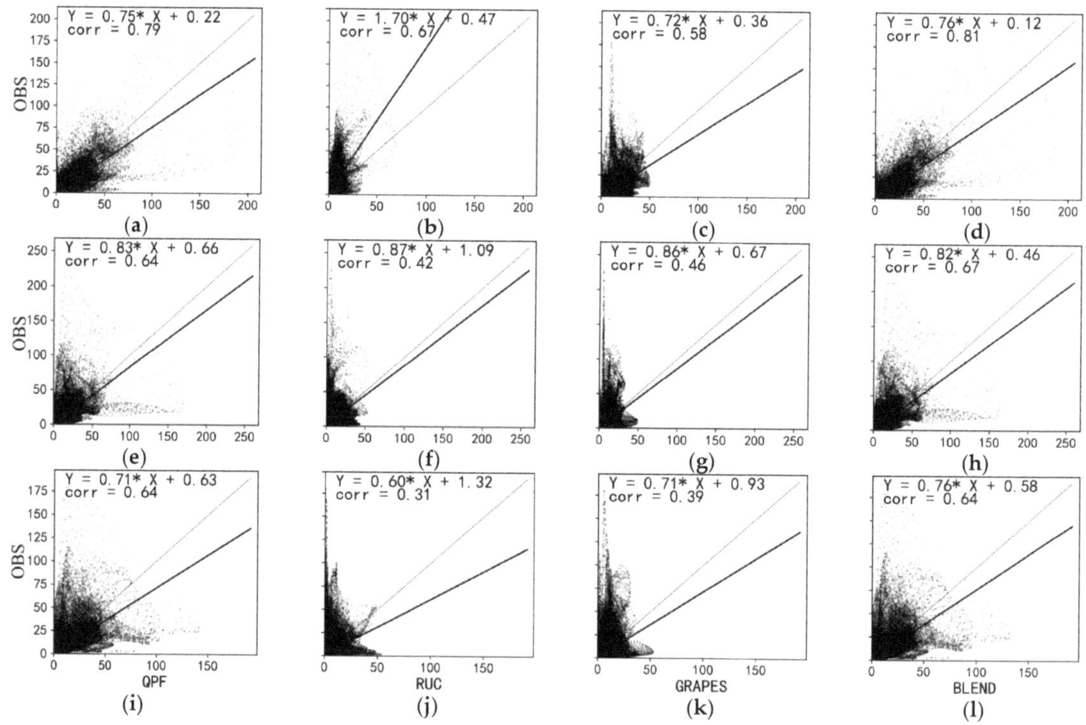

Figure 4. Comparison of 0–3 h radar-extrapolated precipitation forecast (**a**,**e**,**i**), WHRUC model forecast (**b**,**f**,**j**), GRAPES_3 km model forecast (**c**,**g**,**k**), and Blending forecast (**d**,**h**,**l**) with real time from 04:00 (BT) on 12 August 2021.

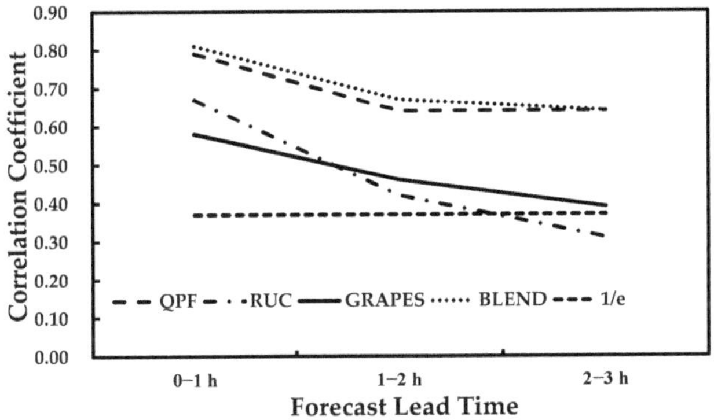

Figure 5. The variation in the correlation coefficient of the four methods with the prediction time.

Figure 6a,c,e show the comprehensive performance of the threshold values of 1 mm, 5 mm, 10 mm, 20 mm, and 50 mm for the four prediction methods. The abscissa is the success rate and the ordinate is the hit rate. The auxiliary lines of equal bias and equal TS curves are drawn. The test results are displayed in the chart in the form of dots so that you can directly browse the test indexes such as success rate, HIT rate, BIAS, and TS. It can be seen that the accuracy of WHRUC prediction is obviously lower than those of other forecasts and the overall BIAS value is smaller. The prediction effect of QPF and Blending is the best. The BIAS value of the 1 mm precipitation threshold is too large. Additionally, the prediction accuracy is close to 0.6, which is slightly lower than that of QPF. With the increase in threshold, the prediction accuracy of the Blending forecast is obviously improved. The prediction deviation is similar to that of QPF. There is little difference between the prediction deviation of 10 mm and 20 mm and that of 5 mm. The prediction accuracy is obviously improved. As can be seen from the chart, when the threshold of WHRUC and GRAPES_3 km exceeds 50 mm, the sample of the heavy precipitation forecast is almost zero. BIAS and TS scores are the lowest. While the TS score of the QPF and Blending forecast is close to 0.3, the BIAS value is close to 1. The accuracy of each forecast shows a downward trend with the increase in time. The WHRUC forecast BIAS value gradually increases and the overall predicted value is close to the actual value. The other predicted BIAS value gradually decreases and the overall forecast value shows a trend from high to low. It can be seen from the chart that the effect of the Blending forecast with different thresholds in 1~2 h is better than those of other forecasts. The effect of the 2~3 h forecast is similar to that of QPF, which is better than those of the other forecasts especially in the heavy precipitation forecast.

Figure 6b,d,f show Taylor diagrams. They show the standard deviation, rmse, and correlation coefficient of both observed and predicted data, respectively. The graph is a polar coordinate system where the radius r represents the standard deviation of the observation or forecast data itself. The value of the correlation between prediction and observation can be determined according to the blue ray (dotted line) in the graph and the scale value on the outermost circle arc. The green arc (dashed line) represents the concentric circle around the observation data and its radius represents the rmse of the forecast data. The standard deviation reflects the discrete degree of the precipitation data set and rmse indicates the degree to which the predicted precipitation deviates from the observed value. It can be seen that the standard deviation of the Blending forecast is closest to the actual situation in the forecast of 0~1 h. With the increase in the forecast time, the standard deviation of the Blending forecast increases at first and then decreases. The discretization degree of the precipitation data set is closer to reality than those of other models. The correlation coefficient of the Blending forecast decreases with the increase in prediction time, which is higher than those of other models. The rmse value of the Blending forecast in 0~1 h is smallest compared to those of other models and the deviation in the predicted precipitation from the observed value is smallest. With the increase in failure, the rmse value is basically the same.

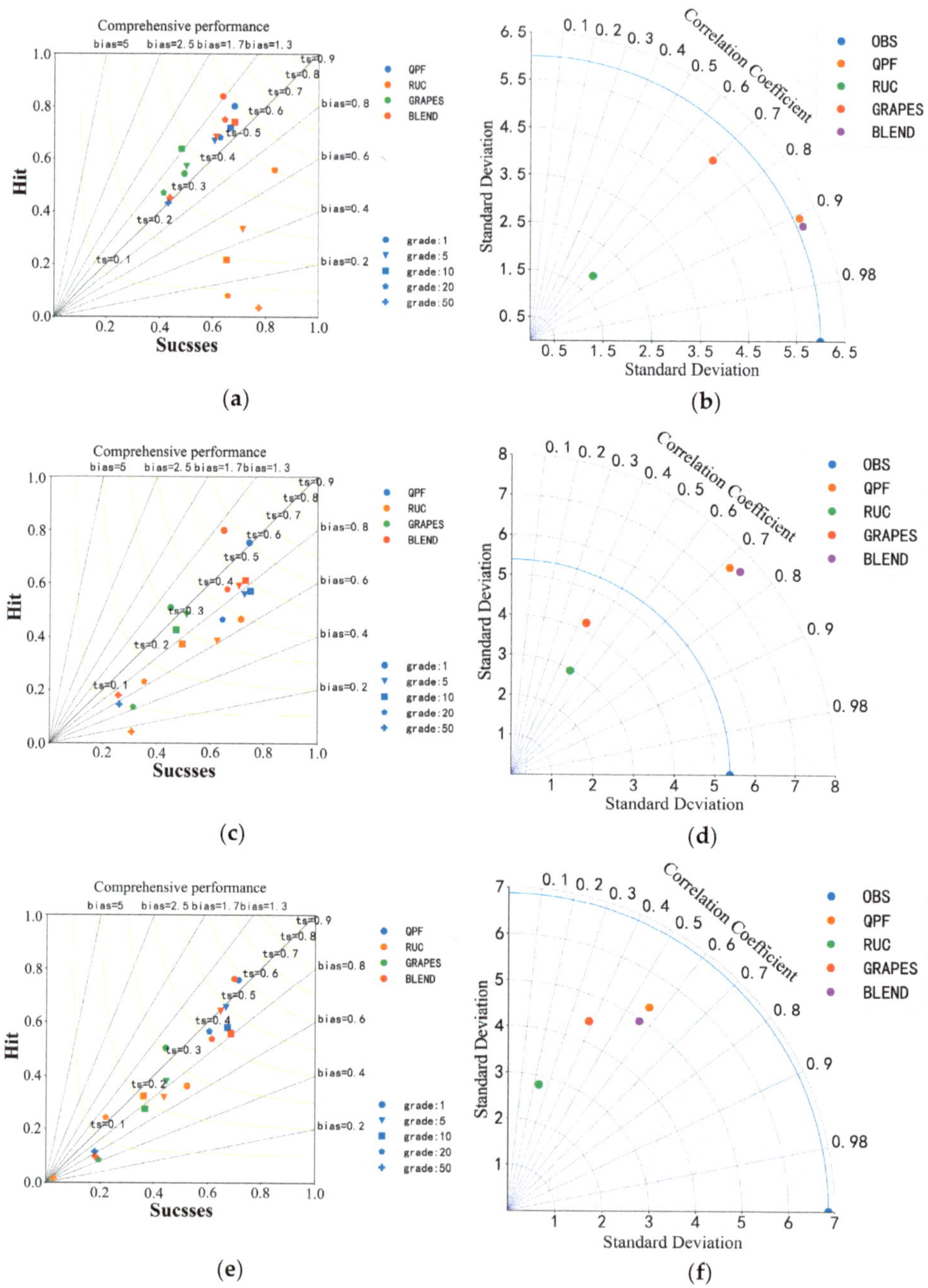

Figure 6. Comparative Analysis of Comprehensive performance (0~1 h (**a**), 1~2 h (**c**), 2~3 h (**e**)) and Taylor Chart (0~1 h (**b**), 1~2 h (**d**), 2~3 h (**f**)) of different prescription modes.

The statistical observation and forecast represent the number of samples of various categories. The frequency statistics in Figure 7a–c are drawn in the form of a histogram. The abscissa is the value category of the precipitation sample and the ordinate is the sample proportion. The observed and predicted values are sorted from small to large, respectively, and the two groups of data after sorting are drawn into a frequency matching relation in Figure 7d–f, with the abscissa as the predicted value and the ordinate as the observed value. From the frequency matching mapping diagram, it can be seen that the prediction results of WHRUC and GRAPES_3 km are smaller than the actual observations. The observed values of 0~1 h prediction results are lower than 100 mm. GRAPES_3 km is closer to the reality than WHRUC is. The QPF and Blending forecast 0–1 h are larger than the actual observation, the Blending forecast result is closer to reality, 1~2 h and 2~3 h are smaller than the actual observation, and the QPF prediction effect is closer to reality. Generally, the prediction effect of the QPF and Blending is better than that of WHRUC and GRAPES_3 km.

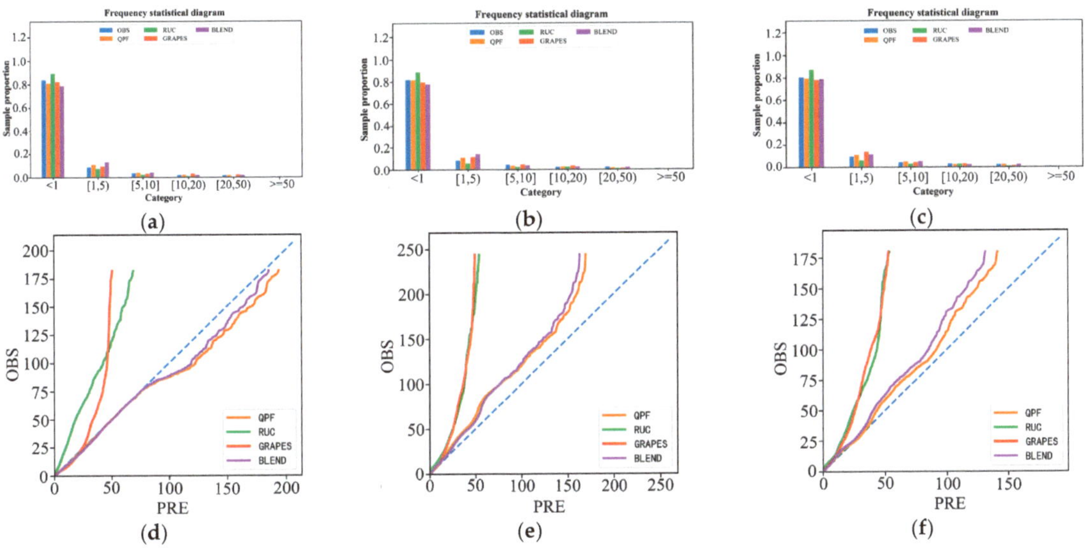

Figure 7. Comparison of frequency statistics (**a**–**c**) and matching mapping (**d**–**f**) of different time-effect prediction methods.

5.2.2. Spatial Contrast Test

Figure 8 shows the spatial distribution of errors of the four forecasting methods. The point-to-point errors between the predicted results and the actual results of all grid points are calculated. For a grid point, the samples with different starting times and prediction times are tested together. The size of the error on the site is represented by the color of the site and the absolute value of the error is represented by the area of scattered dots. The larger the area, the greater the error. The site with large errors is highlighted by setting the site size. It can be seen that the precipitation intensity predicted by WHRUC is obviously smaller than those of other methods and the precipitation error is obviously larger.

Grid data are used for target recognition and the recognition steps include the following: (1) select a disk convolution kernel with a radius of smooth = 5, and convolution-smooth the observation field and the prediction field; (2) set threshold to 5 and set the grid value of the smoothed value to less than threshold, 0; (3) identify the targets in the observation and prediction field by the connected domain extraction algorithm; (4) set minsize = 100 and delete targets whose area (number of grid points) in the forecast or observation field is less than minsize; (5) judge the closeness between two targets and take the maximum value of multiple closeness. Merge the corresponding two targets if it is greater than near_rate; (6) repeat step 5 until the closeness between all pairwise targets is less than the termination of the near_rate algorithm.

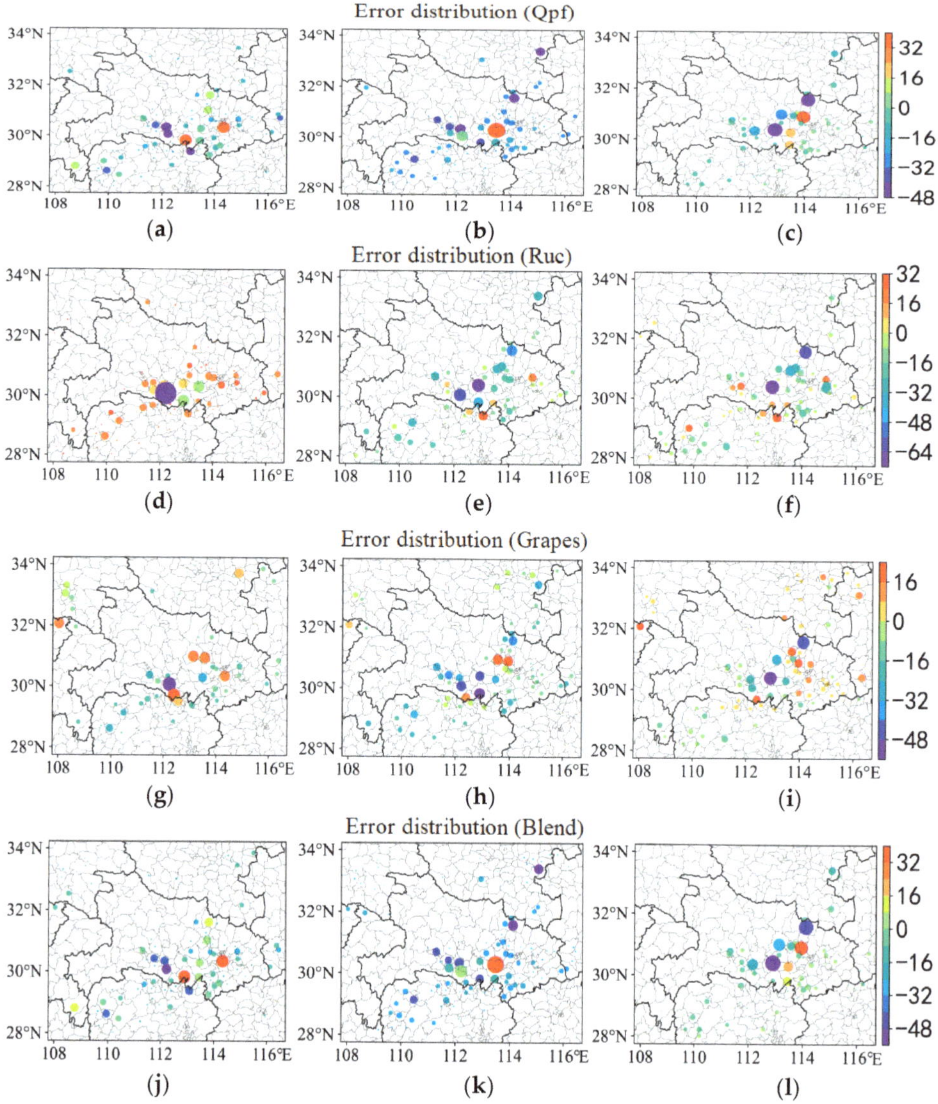

Figure 8. From 04:00 (Beijing time) on 12 August 2021, the error spatial distribution maps of the 0–3 h radar-extrapolated precipitation forecast (**a–c**), WHRUC model forecast (**d–f**), GRAPES_3 km model forecast (**g–i**), and Blending forecast (**j–l**).

Figure 9 shows the hourly target recognition results of different prediction methods for 0~3 h. It can be seen that the ellipses identified and matched by MODE can basically reflect the scale and shape of convective storms. It can be used as a basis for model evaluation. Three target objects, which are completely consistent with the actual precipitation area, are identified by the target object method. Table 2 shows the number of targets matched (hit), missed, and empty.

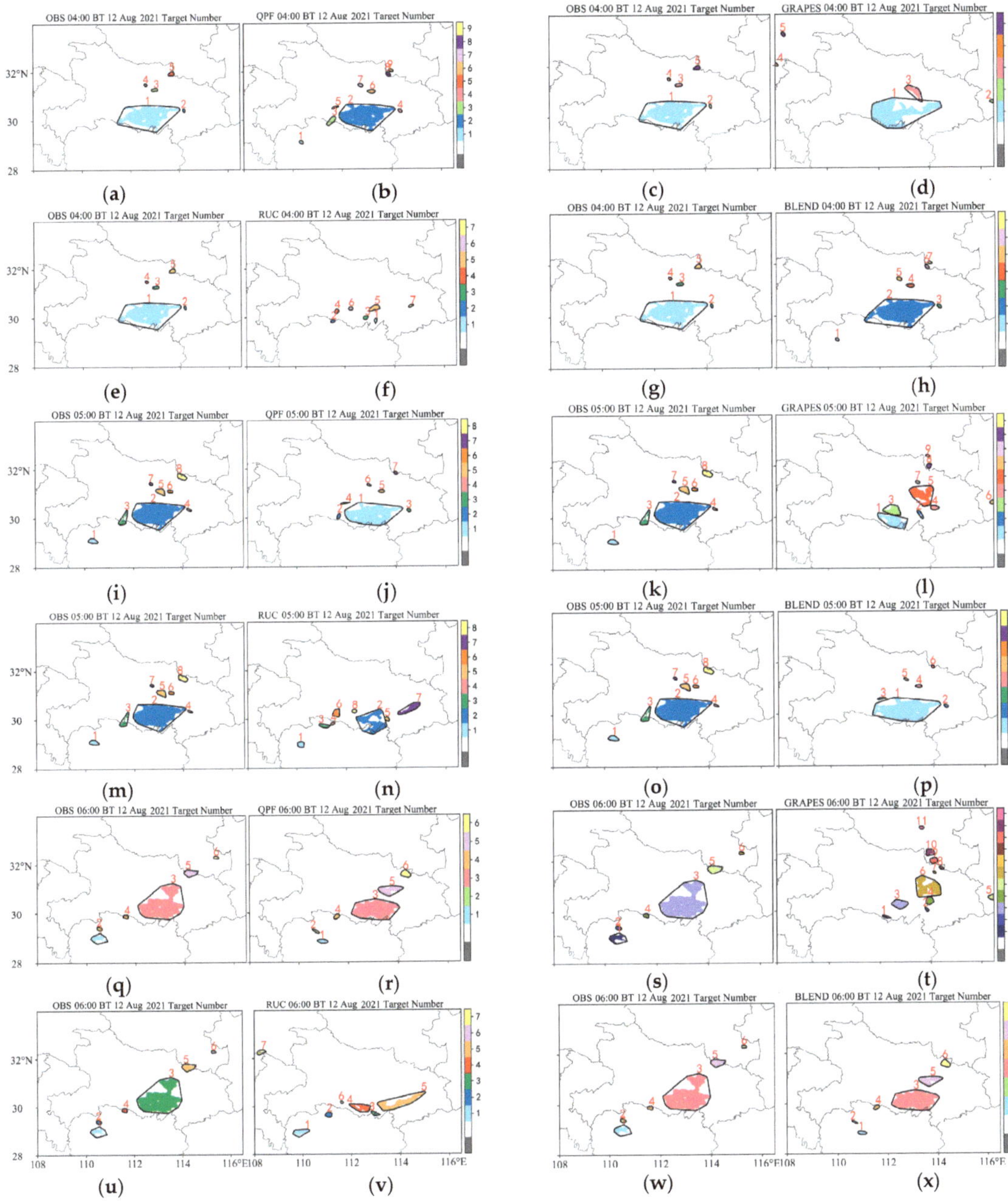

Figure 9. From 04:00 (BT) on 12 August 2021, the live observation (**a,c,e,g,i,k,m,o,q,s,u,w**), the 0~3 h radar-extrapolated precipitation forecast (**b,j,r**), the WHRUC model forecast (**f,n,v**), the GRAPES_3 km model forecast (**d,l,t**), and the Blending forecast (**h,p,x**) target recognition results.

Table 2. Statistics of target matching results.

		Hits	Misses	Fales	Correct Negatives
0–1 h	WHRUC	2	3	2	96
	GRAPES_3 km	3	2	2	90
	BLEND	3	1	1	52
1–2 h	WHRUC	1	3	3	98
	GRAPES_3 km	2	3	2	91
	BLEND	3	1	2	70
2–3 h	WHRUC	2	3	3	105
	GRAPES_3 km	2	4	2	109
	BLEND	5	1	1	89

Two classification test indexes are calculated for inspection, scoring, and evaluation (Figure 10a) based on the attribute information of the identified target, and the specific prediction of the model is obtained. Figure 10b shows the intensity error (A), distance error (L), and structure error (S) of the SAL test. It can be seen that the prediction ability of GRAPES_3 km for the intensity of the 0~1 h rain areas is better than those of location and shape. The prediction ability of rainfall intensity is larger with the passage of time. The prediction ability of the WHRUC model is worse compared to those of other models and the prediction of precipitation intensity is obviously smaller. The rain area and rainfall intensity forecast of the Blending forecast are close to the actual situation. The rain intensity error also fluctuates due to the influence of the GRAPES_3 km model. On the whole, the Blending forecast integrates the advantages and disadvantages of the model and extrapolation forecast. It can grasp the precipitation area forecast of the system and the forecast ability of rainfall intensity improves.

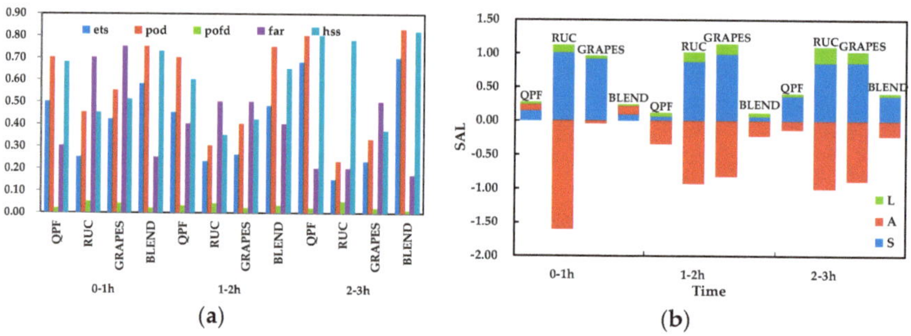

Figure 10. Target matching evaluation (**a**) and SAL test of different forecasting methods (**b**).

According to the scoring results in Table 3, the Blending forecast has the highest score for the actual goal of 1 and the area scores of 0~3 h are 0.88, 0.79, and 0.88, respectively, which are higher than those of the GRAPES_3 km model. The Blending forecast is obviously superior in the area score of identifying targets in terms of area ratio and overlap area ratio. The position score of the two models in 0~3 h is 1 and there is basically no deviation from the predicted center of gravity of target 1. The centroid distance of the 1~2 h Blending forecast is 0.1, which is less than the centroid distance predicted by GRAPES_3 km at the same time. The hourly axial angle differences of the two prediction methods are 1.2, 7.5, and 0, respectively, while those of the Blending forecasts are 3.1, 6.9, and 1.6, respectively, which are all less than 10°. The score of ellipticity is 1, except that predicted by GRAPES_3 km for 2~3 h and the difference between the ellipticity and real object is less than 0.1. The intensity score is calculated based on the precipitation intensity of the 50% quantile and the values are both 1. The precipitation intensity predicted by the two methods is consistent with the actual precipitation grade. From the intensity difference, it can be seen that GRAPES_3 km is small and the Blending forecast is larger. However, the deviation is obviously less than that of GRAPES_3 km. Generally, the score of the Blending forecast is significantly higher than that of GRAPES_3 km for the forecast of goal 1. The target object test method not only gives the evaluation of area, location, shape, and the extreme value of the precipitation center, but also analyzes the forecast performance of heavy precipitation in terms of precipitation area and precipitation intensity, which provides the scientific calculation and test results for forecasters and can mine more valuable information from failed forecast cases.

Table 3. Comprehensive test of target attributes.

	0–1 h				1–2 h				2–3 h			
	GRAPES _3 km		BLEND		GRAPES _3 km		BLEND		GRAPES _3 km		BLEND	
	Obs	Pre	Obs	Pre	Obs	Pre	Obs	Pre	Obs	Pre	Obs	Pre
Spindle length	2.8	2.6	2.6	2.8	2.4	2.4	3.0	3.1	1.9	2.7	1.9	2.1
Spindle inclination angle	1.6	1.3	2.0	2.0	1.6	1.5	1.6	1.6	1.4	1.3	1.4	1.4
Rectangular window(x0)	17.8	18.9	9.9	6.9	17.7	25.2	19.6	12.7	16.1	16.1	16.1	17.7
Rectangular window(y0)	111.1	111.4	110.1	109.4	111.5	112.0	111.1	111.0	111.8	111.8	111.8	112.0
Rectangular window(x1)	29.9	29.8	30.1	30.0	30.0	29.8	29.0	29.9	29.9	30.0	30.0	30.0
Rectangular window(y1)	114.4	114.5	115.6	117.2	114.3	114.5	109.9	114.2	115.2	115.1	114.7	114.6
Centroid(x)	30.9	30.9	31.0	30.9	30.9	31.0	30.9	30.8	30.9	30.9	30.9	30.8
Centroid(y)	112.9	112.9	112.7	112.8	112.9	113.1	112.9	113.0	113.1	113.2	113.1	113.2
Area	30.2	30.3	30.2	30.2	30.3	30.4	30.3	30.3	30.4	30.3	30.4	30.4
Median intensity	1.5	1.3	1.4	1.5	1.4	1.0	1.5	1.7	1.4	1.2	1.4	1.5
Strength difference	28.2	23.1	24.7	26.1	28.8	20.2	28.0	29.8	26.9	25.0	26.9	28.3
Centroid distance	−5.1		1.4		−8.6		1.8		−1.9		1.4	
Angle difference	0.1		0.1		0.2		0.1		0.1		0.1	
Area ratio	1.2		3.1		7.5		6.9		0		1.6	
Overlap area ratio	0.9		0.9		0.7		0.9		0.9		1.0	
Area score	0.7		0.8		0.4		0.8		0.7		0.8	
Location score	0.79		0.88		0.64		0.79		0.78		0.88	
Median intensity	1		1		1		1		1		1	
Axial Angle score	1		1		1		1		1		1	
Ellipticity score	1		1		1		1		0.61		1	
Intensity score	1		1		1		1		1		1	
Total goal score	0.94		0.96		0.89		0.94		0.90		0.96	

In a plane field, the values of two points in different positions are different and the difference between them usually increases with the increase in distance. The variogram is used to detect whether the above increasing trend in the prediction field is consistent with that in the observation field. Figure 11 shows the variation chart with a grid number of 10 in east–west and south–north trends. It can be seen from the above variation diagram that the larger the number of translation grids, the lower the coincidence of grid field before and after translation. The variation predicted by QPF and BLEND is smaller than that in the south–north direction, which is basically consistent with the actual observation. This is consistent with the fact that the north–south gradient of the temperature field is greater than that of the east–west gradient. WHRUC and GRAPES_3 km predict that the variation in the east–west direction is similar to that in the north–south direction. In addition, it is impossible to directly identify the slight deviation between the forecast and the actual situation when the observation and prediction are close. It is necessary to further test the variation range of the variation with distance.

Figure 12 shows the amplitude of change before and after the grid translation of different prediction methods. The maximum number of grid points of horizontal and vertical translation is the upper half of the graph, which is the average value (thick real line) of the variation value of elements in a certain interval. Further statistics are made of the standard deviation of the variation value in the same distance interval and the average value is added (minus) to the standard deviation in two dotted lines on the top (bottom) of the graph, which is used to represent the approximate range of the variation value. The lower part is the statistics sample number, which is used to calculate the variation. The result corresponding to a moving mode is recorded as one. In the above example, the maximum distance of the horizontal and vertical translation is set to 100. The translation step is set to 5 and the types of horizontal and vertical movement are both 41. The total movement mode is 41 × 41 = 1681. It can be seen from the figure that the change range of the BLEND forecast grid before and after translation is closer to reality.

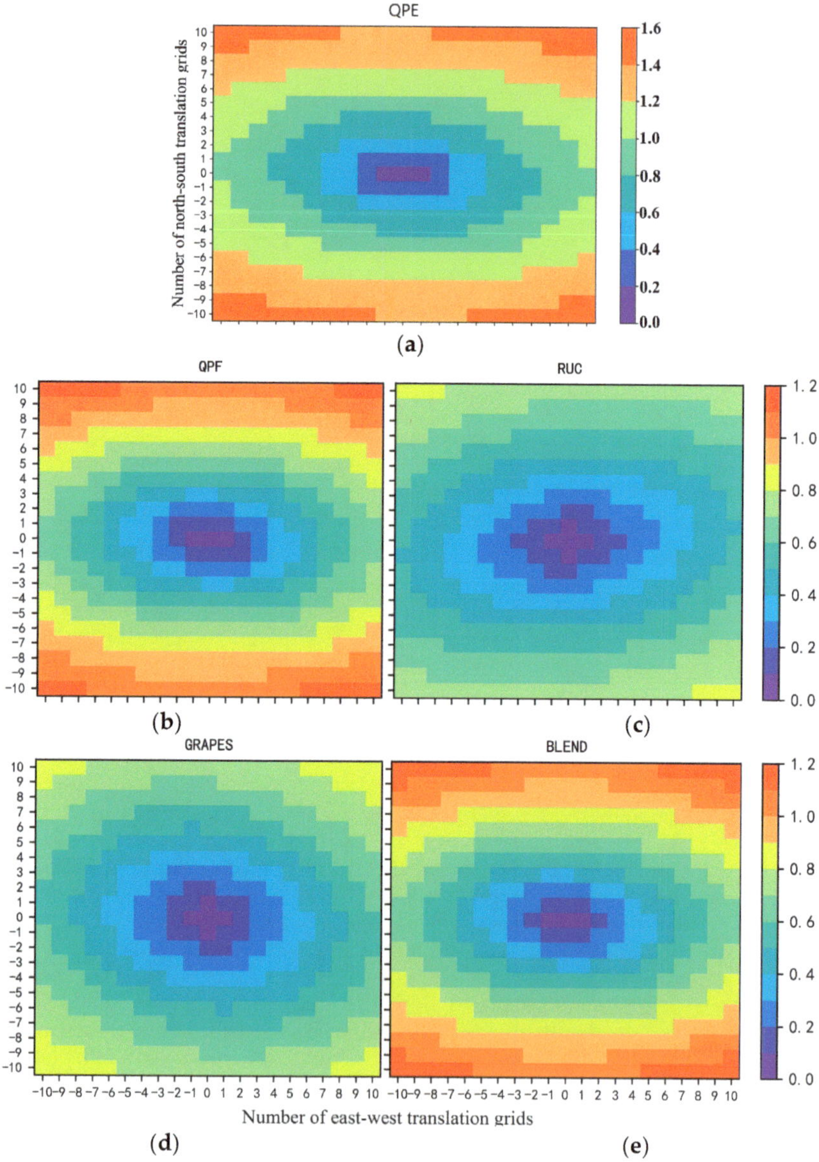

Figure 11. Comparative analysis of precipitation variation maps predicted by live product (**a**), radar-extrapolated precipitation forecast (**b**), WHRUC model forecast (**c**), GRAPES_3 km model forecast (**d**), and Blending forecast (**e**).

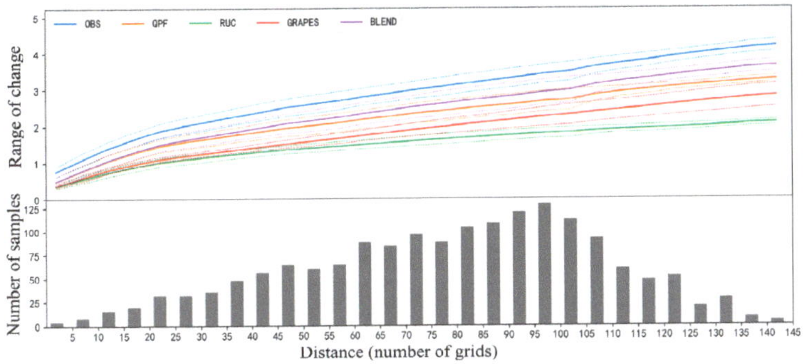

Figure 12. Comparative analysis of the variation range of precipitation variation with distance predicted by different forecasting methods.

6. Conclusions and Prospects

In this paper, the prediction abilities of the Radar Extrapolation Forecast (REF), WHRUC, GRAPES_3 km, and Blending are compared and analyzed by using traditional point-to-point evaluation and the new spatial test method. The conclusions are as follows:

(1) The comprehensive application of various evaluation methods can evaluate the convective storm forecast more comprehensively. By using the point-to-point scoring method, for the forecasts with different time limits, the shorter the time effect is, the better the prediction effect is. The Blending forecast effect of 1~2 h is obviously better than that of other models, and that of 2~3 h is similar to that of QPF and better than those of the others, especially in the heavy precipitation forecast.

(2) The new spatial test method can evaluate the prediction effect of convective storm features. The precipitation intensity of Blending is larger than those of other models. The error is smallest and the target recognition hit rate is highest. The scores of target area, position, shape, and median intensity of precipitation are better than those of other forecasts.

(3) The variation in the Blending-identified target in the east–west direction is less than that in the north–south direction, which is basically consistent with the actual observation, and the variation range of the forecast grid before and after translation is the closest to the reality.

It can be seen that Blending is obviously better than the single forecast, especially in the heavy precipitation echo forecast, and plays a positive role in the convective forecast. Blending technically has an important operational reference value for the 0~3 h quantitative precipitation forecast. The Blending method combines the advantages and disadvantages of NWP and EP, which can not only grasp the precipitation area forecast of the system but also improve the prediction ability of rainfall intensity. The traditional point-to-point scoring method and the new spatial test method have the same conclusion to the convective storm forecast of the high-resolution model, which has a certain reference value, and the new spatial test method can provide more detailed evaluation information.

Author Contributions: J.W., Z.W. and H.M. contributed to the conception and design of the study. Z.W. and J.Y. organized the database. J.W. and J.Y. performed the statistical analysis. Z.W. and A.L. wrote the first draft of the manuscript. W.Z., H.M. and J.W. wrote sections of the manuscript. All authors have read and agreed to the published version of the manuscript.

Funding: This research was funded by the Special Project of Innovation and Development of China Meteorological Administration (National and Provincial Co-ordination, CXFZ2022J018, CXFZ2022-J019), Hubei Provincial Natural Science Foundation (2022CFD129, 2018CFB706), the Open Research Fund of Hubei Key Laboratory of Intelligent Yangtze and Hydroelectric Science, China Yangtze Power Co., Ltd. (242202000907) and the Science and Basic Research Fund of WHIHR (202306, 202314).

Institutional Review Board Statement: The study did not require ethical approval.

Informed Consent Statement: Informed consent was obtained from all subjects involved in the study.

Data Availability Statement: The data is unavailable due to privacy.

Conflicts of Interest: The authors declare no conflict of interest.

References

1. Rinehart, R.E.; Garvey, E.T. Three-dimensional storm motion detection by conventional weather radar. *Nature* **1978**, *273*, 287–289. [CrossRef]
2. Zheng, Y.G.; Zhang, X.L. Review on Severe Convective Weather Short-Term Forecasting and Nowcasting. *Meteorol. Mon.* **2010**, *36*, 33–42.
3. Wang, J.; Zhang, J.G. Methods and Techniques of Multiscale Composite Rainfall Nowcasting. *Meteorol. Sci. Technol.* **2008**, *36*, 524–528.
4. Han, L.; Wang, H.Q. Application of Optical Flow Method to Nowcasting Convective Weather. *Acta Sci. Nat. Univ. Pekin.* **2008**, *44*, 751–755.
5. Kong, R.; Wang, J.J. Applying Scale Decomposition Method to Verification of Quantitative Precipitation Nowcasts. *J. Appl. Meteor. Sci.* **2010**, *21*, 535–544.
6. Cheng, C.L.; Chen, M.X. Short-term quantitative precipitation forecast experiments based on blending of nowcasting with numerical weather prediction. *Acta Meteorol. Sin.* **2013**, *71*, 397–415.
7. Wang, D.; Wang, G.L. Comparisons Analysis on Short-Term Precipitation between the Radar-Based Extrapolation and the Meso-Scale Numerical Model Weather Prediction. *Plateau Meteorol.* **2014**, *33*, 811–822.
8. Cao, C.Y.; Chen, Y.Z. The optical flow method and its application to nowcasting. *Acta Meteorol. Sin.* **2015**, *73*, 471–480.
9. Guo, H.Y.; Chen, M.X. High resolution nowcasting experiment of severe convections based on deep learning. *Acta Meteorol. Sin.* **2019**, *77*, 715–727.
10. Pan, L.J.; Zhang, H.F. Neighborhood-Based Precipitation Forecasting Capability Analysis of High-Resolution Models. *J. Trop. Meteorol.* **2015**, *31*, 632–642.
11. Brownlee, K.A. A Wiley Publication in Applied Statistics. In *Statistical Theory and Methodology in Science and Engineering*, 2nd ed. John Wiley & Sons Wiley: New York, NY, USA, 1965; pp. 26–30.
12. Doswell, C.A.; Jones, R.D. On summary measures of skill in rare event forecasting based on contingency tables. *Weather. Forecast.* **1990**, *5*, 576–585. [CrossRef]
13. Wang, G.L.; Zhao, C.G. Error Analysis of Radar Echo Extrapolation. *Plateau Meteorol.* **2013**, *32*, 874–883.
14. Long, F.C.; Wang, G.R. The Application Analysis of MODE Method to the Rainfall Forecast Test. *Meteorol. Mon.* **2011**, *37*, 1498–1503.
15. Mao, M.; Dai, J.H. Object-Based Verification and Evaluation for Different Types of Severe Convection Forecasting Products. *Meteorol. Mon.* **2016**, *42*, 389–397.
16. Xue, C.F.; Pan, L.J. Diagnostic Analysis of Precipitation Forecasting from Japan Thin-Grid Model based on Mode. *Plateau Meteorol.* **2016**, *35*, 406–418.
17. Liu, C.H.; Niu, R.Y. Object-Based Preciptitation Verfication Method and Its Application. *Meteorol. Mon.* **2013**, *39*, 681–690.
18. Qu, Q.N.; Sheng, C.Y. Comparison of the multi-model forecasts for severe precipitation based on the object verification. *Meteorol. Mon.* **2019**, *45*, 908–919.
19. Xu, C.L.; Wang, J.J. Evaluation on QPF of GRAPES-Meso4.0 model at convection-permitting resolution. *Acta Meteorol. Sin.* **2017**, *75*, 851–876.
20. Tang, W.Y.; Zeng, Y.G. FSS-based evaluation on convective weather forecasts in North China from high resolution models. *J. Appl. Meteorol. Sci.* **2018**, *29*, 513–523.
21. Zhang, X.W.; Tang, W.Y. Comprehensive evaluations of GRAPES_3 km numerical model in forecasting convective storms using various verification methods. *Meteorol. Mon.* **2020**, *46*, 367–380.
22. Horn, B.K.; Schunck, B.G. Determing optical flow. *Artif. Intell.* **1981**, *17*, 185–204. [CrossRef]
23. Aubert, G.; Deriche, R. Computing optical flow via variational techniques. *SIAM J. Appl. Math.* **1999**, *60*, 156–182. [CrossRef]
24. Wang, Z.B.; Xiao, Y.J. Implementation of Improved Variational Optical Flow Parallel Algorithm. *Comput. Appl. Softw.* **2019**, *36*, 105–110.
25. Lai, A.W.; Ma, H.D. A Squall Line Case Study of Assimilating the Radar Data, Retrieval of Water Vapor and In-Cloud Potential Temperature from Reflectivity in a 3DVAR Framework. *Meteorol. Mon.* **2021**, *47*, 932–952.
26. Zawadzki, I.; Torlaschi, E.; Sauvageau, R. The Relationship between Mesoscale Thermodynamic Variables and Convective Precipitation. *J. Atmos. Sci.* **1981**, *38*, 1535–1540. [CrossRef]

remote sensing

MDPI

Technical Note

Multiscale Representation of Radar Echo Data Retrieved through Deep Learning from Numerical Model Simulations and Satellite Images

Mingming Zhu [1,2], Qi Liao [3], Lin Wu [1,4,*], Si Zhang [1,2], Zifa Wang [1,2,5], Xiaole Pan [1], Qizhong Wu [6], Yangang Wang [7] and Debin Su [3]

1 State Key Laboratory of Atmospheric Boundary Layer Physics and Atmospheric Chemistry, Institute of Atmospheric Physics, Chinese Academy of Sciences, Beijing 100029, China; zhumingming@mail.iap.ac.cn (M.Z.); zhangsi@mail.iap.ac.cn (S.Z.); zifawang@mail.iap.ac.cn (Z.W.); panxiaole@mail.iap.ac.cn (X.P.)
2 College of Earth and Planetary Sciences, University of Chinese Academy of Sciences, Beijing 100049, China
3 College of Electronic Engineering, Chengdu University of Information Technology, Chengdu 610225, China; liaoqi@cuit.edu.cn (Q.L.); sudebin@cuit.edu.cn (D.S.)
4 Carbon Neutrality Research Center, Institute of Atmospheric Physics, Chinese Academy of Sciences, Beijing 100029, China
5 Center for Excellence in Urban Atmospheric Environment, Institute of Urban Environment, Chinese Academy of Sciences, Xiamen 361021, China
6 College of Global Change and Earth System Science, Beijing Normal University, Beijing 100875, China; wqizhong@bnu.edu.cn
7 Computer Network Information Center, Chinese Academy of Sciences, Beijing 100083, China; wangyg@sccas.cn
* Correspondence: wlin@mail.iap.ac.cn

Abstract: Radar reflectivity data snapshot fine-grained atmospheric variations that cannot be represented well by numerical weather prediction models or satellites, which poses a limit for nowcasts based on model–data fusion techniques. Here, we reveal a multiscale representation (MSR) of the atmosphere by reconstructing the radar echoes from the Weather Research and Forecasting (WRF) model simulations and the Himawari-8 satellite products using U-Net deep networks. Our reconstructions generated the echoes well in terms of patterns, locations, and intensities with a root mean square error (RMSE) of 5.38 dBZ. We find stratified features in this MSR, with small-scale patterns such as echo intensities sensitive to the WRF-simulated dynamic and thermodynamic variables and with larger-scale information about shapes and locations mainly captured from satellite images. Such MSRs with physical interpretations may inspire innovative model–data fusion methods that could overcome the conventional limits of nowcasting.

Keywords: deep learning; multiscale representation; model–data fusion

1. Introduction

Meteorological forecasts at the convective scale are crucial to mitigate environmental hazards such as storms and floods that cause huge socioeconomic damages, but face fundamental challenges in representing the convective weather regime in numerical weather prediction (NWP) models, which is a less-known "gray zone" compared to the relatively well-resolved synoptic-scale systems [1,2]. Radar is an invaluable instrument to scan the convective atmosphere in nearly real time. The extrapolations of these sequential radar echo data can provide state-of-the-art nowcasts of precipitation patterns and severe weather events within a few hours based on the persistence principle [3–5]. However, these radar echo data are snapshots of the complex atmosphere with fine-grained details but not readily related to the dynamics of the atmosphere and the information from remote sensing satellites. This lack of representation of the dynamical and global atmospheric information poses a limit for convective nowcasting. For instance, when combining extrapolation-based methods with the dynamical information from NWP model simulations [6,7], the forecast skill in general decreases rapidly in the first forecast hour and remains low beyond a few forecast hours [8,9], despite advances in convective-permitting modeling [10], radar and satellite data assimilation [9,11,12] and high-resolution observations [13].

The representation gap, particularly when model–data fusion is to be conducted, is difficult to bridge [14,15]. Representing processes related to turbulence, convection, and topography is challenging for numerical models in gray zones [16]. Parameterizing and resolving the atmospheric motions in grid spacing for deep convection (1–10 km) and turbulence (0.1–1 km) requires ponderation and in-depth explorations [17–19]. The precipitation and storms observed using radar and satellites are among the most difficult to simulate, which is a consequence of the intertwined consecutive physical processes of NWPs with multiplicative error propagations [20]. Convective-scale NWPs also have fundamental theoretical challenges such as the mathematical characteristics of the underlying partial differential equations as well as the predictability and probability issues related to the nonlinear dynamics of the convective systems with effective dimensions much higher than those of the balanced synoptic systems [1]. It has long been recognized that the convective atmospheric motions are of multiscale interactions [19,21–23]. Currently, the first-principle multiscale formulation is beyond the traditional NWP modeling paradigm. In addition, a representation gap exists between satellite images and radar data. Geostationary satellites observe cloud evolution from a global perspective [24], but they are limited to detecting cloud tops and hardly probe the internal structure of clouds.

Deep learning (DL) has recently emerged as a general data-driven technology to represent the spatiotemporal features that cross multiple scales and are not captured well by geophysical models [25]. DL techniques can explore the rich patterns in radar data with deep networks of neurons and improve the precipitation nowcasting skill [26]. Numerous DL applications for the convective atmosphere have been proposed, ranging from radar- or satellite-based nowcasting [27–30] to reconstructions of radar data from satellites [31–33]. However, there are few applications aiming at bridging the representation gap for the fusion between radar data, satellite images, and NWP simulations. Accordingly, it remains largely unexplored how the deep networks represent the convective atmosphere. In addition, these deep networks are usually considered black boxes with limited physical interpretations. Here, we attempt to retrieve the deep network representations by reconstructing the radar reflectivity data from NWP simulations and satellite observations and then probe the structure of the obtained representations by diagnosing their relations with physical quantities such as NWP variables and satellite images. This attempt aims to reveal the potential of data-driven DL models to bridge the representation gaps between multiscale multi-source data. Hopefully, this potential of multiscale representation with investigations of physical interpretations could make the DL models more transparent and inspire innovative model–data fusion methods that could overcome the conventional limits of nowcasting.

2. Data and Methods

The study area is the Beijing–Tianjin–Hebei (BTH) region ([36°N, 113°E] × [43°N, 120°E]), which is vulnerable to floods caused by heavy summer precipitation. The study period is from June to September for the years 2015 and 2016, when precipitation is more frequent over this region.

2.1. Radar Echo Data

Radar reflectivity data are collected from six Doppler radars that sufficiently cover the BTH region from the Chinese new generation weather radar network (ChIna New generation doppler weather RADar, CINRAD), with four radars of type CINRAD/SA located in Beijing, Tanggu, Shijiazhuang, Qinhuangdao, and two radars of type CINRAD/CB located in Zhangjiakou and Chengde, respectively (Figure S1a). The SA and CB Doppler radars are S-band radars and C-band radars, respectively. The collected radar echo data are interpolated from plan position indicators (PPIs) to constant altitude PPIs (CAPPIs) with a vertical linear interpolation method [34,35]. Then, the radar data in the polar coordinate are mapped onto 0.01° × 0.01° Cartesian grids using the nearest neighbor interpolation method. After that, the data from different radars around the same time (e.g., within 3 min) are combined, and maximum values are preserved for overlapped grid cells. The resulting radar echo data have a temporal resolution of 6 min. The combined radar data are then smoothed and filtered using a convolution threshold method [36,37]. We calculate the maximum radar data based on CAPPIs at 1500 m, 2000 m, 2500 m, 3000 m, and 3500 m above sea level, a typical range of elevations around the level of free convection where convection develops actively.

2.2. Numerical Model Simulations

The Weather Research and Forecasting (WRF) model [38] is used as a convection-permitting modeling system over the BTH region with three nested domains at horizontal resolutions of 9, 3, and 1 km, respectively (Figure S1b). By switching off the cumulus parameterization of the inner two nested domains, convections are explicitly resolved in this setting. Detailed configuration is listed in Table 1. We run a 36 h simulation at a temporal resolution of 30 min, beginning at 12:00 UTC each day with the first 12 h of simulations as spin-ups. The remaining 24 h of simulations of the innermost domain are used to provide the meteorological input for deep networks. The initial and boundary conditions for the simulations are provided by the NCAR/NCEP 1° × 1° reanalysis data. We compare the simulations with weather station observations (Table S1) and verify that the performance on most selected meteorological factors is close to those in other state-of-the-art WRF studies [39,40]. We select 14 daily simulated variables commonly used in convective nowcasting from three categories (i.e., dynamic variables, thermodynamic variables, and moisture-related variables) to build the dataset for learning, such as the three components of wind velocity (U, V, W), K index (K), water vapor mixing ratio (WVMR), and relative humidity (RH). Five of the fourteen variables are three-dimensional and extracted from the pressure levels of 850 hPa, 700 hPa, and 500 hPa, generally around the elevations of the radar echo data. The other nine variables are two-dimensional. Therefore, the input data from WRF simulations have 24 channels ($5 \times 3 + 9 = 24$). Detailed information about all the selected variables can be found in Table S2. These variables are mapped onto the same grid of radar data using a linear interpolation method.

Table 1. Parameterization schemes used in WRF simulations.

Process	Parameterization Scheme
Microphysics	WSM3 [41]
Longwave radiation	RRTM [42]
Shortwave radiation	Dudhia [43]
Surface layer	Revised MM5 Monin–Obukhov [44]
Surface physics	Unified Noah land surface [45]
Planetary boundary layer	YSU [46]
Cumulus	Modified Tiedtke [47] (only for the outermost domain)

2.3. Geostationary Satellite Images

Five infrared bands (5th, 8th, 13th, 15th, and 16th) of data are collected from the Himawari-8 geostationary satellite products. These satellite images can provide global information on cloud properties such as phases and heights (Table 2) with high spatiotemporal resolution (2 km and 10 min) [48]. We also extract the deep convective cloud classification (CCC) data from the Himawari-8 cloud type products by assigning 1 to the grids of the deep convective cloud and 0 to the grids of other cloud types. Therefore, we obtain 6-channel input data from the Himawari-8 satellite products. All the satellite products are remapped in a way similar to the mapping of the WRF variables.

Table 2. Descriptions of the selected Himawari-8 satellite images.

Band Number	Central Wavelength (μm)	Concerned Physical Properties
5	1.6	Cloud phases
8	6.2	Middle and upper tropospheric humidity
13	10.4	Clouds and cloud top
15	12.4	Clouds and total water
16	13.3	Cloud heights

Note. Himawari-8 satellite images listed above were supplied by the P-Tree System, Japan Aerospace Exploration Agency (JAXA).

2.4. Data Preprocessing

We have obtained 30-channel input data from the WRF simulations and the Himawari-8 satellite products, and 1-channel labels from the radar echo data on the common grid of $0.01^\circ \times 0.01^\circ$ over the BTH region (i.e., 700×700 horizontally). We first match the input data with labels for the same time and form a dataset of 2647 samples. We then use the min–max normalization to scale each channel of data in the dataset to be in the range of [0, 1], where the maximum value for the radar echo data is set at 70 dBZ, so that the effect of outliers can be suppressed. Moreover, we fill in missing or invalid values with 0 for the normalized dataset.

2.5. Deep Network Model

We adopt a U-Net for the representation learning of the radar echo data (Figure 1). The U-Net deep network is a convolutional neural network (CNN) variant originating from biomedical image segmentation [49] and is here repurposed for a regression task as in many previous studies [31,33,50,51]. It preserves the hierarchical convolutional structure of a CNN in its left contracting path, and uses upsampling operations in successive layers to form a right expansive path. Consequently, the network has a 'U' shape, hence its name. The U-Net is an encoder–decoder network architecture that allows the end-to-end learning of multiscale features and outputs with desired dimensions (i.e., 700×700 in this study). In general, early layers in the contrasting path learn small-scale features such as textures and edges, whereas deep layers learn large-scale features such as semantic information. The U-Net is equipped with so-called skip connections that perform identity mappings of low-level features from the contrasting path (encoder) to the expansive path (decoder) at corresponding levels. The U-Net combines large-scale information with small-scale information brought by the skip connections for reconstructing the data from the learnt multiscale features. Such a network architecture and reconstruction process are appealing for our study on how radar data are represented by deep networks.

Concretely, the U-Net depicted in Figure 1 has eight blocks (Block-As in gray) in the encoder and eight blocks (Block-Bs in blue) in the decoder, followed by an individual convolutional block (Block-C in orange). Each Block-A consists of a convolutional layer followed by a batch normalization layer [52] and a LeakyReLU activation layer. The 1st, 2nd, 4th, 6th, and 7th Block-As convolve the data with 4×4 convolution filters with 2×2 strides to reduce resolutions, enabling the subsequent layers to detect patterns in expanded areas. The 3rd Block-A employs 3×3 convolution filters with 2×2 strides to produce the output of a certain dimension (i.e., 88×88 horizontally). The remaining Block-As contain 3×3 convolution filters with 1×1 strides. All convolutional operations are carried out with a zero padding of size 1. The respective numbers of convolution filters in the eight Block-As are 36, 72, 144, 288, 288, 288, 288, and 288. Each Block-B in the decoder section consists of a bilinear upsampling layer, a 3×3 convolutional layer, a batch normalization layer, and a LeakyReLU activation layer, except for the 8th Block-B, which does not have the LeakyReLU activation layer. The respective numbers of convolution filters in eight Block-Bs are 288, 288, 288, 288, 144, 72, 36, and 1. Finally, an individual Block-C is added, which is composed of a 1×1 convolutional layer, a 3×3 convolutional layer, a 1×1 convolutional layer, and a ReLU-6 activation layer, which is considered to facilitate the learning of sparse features [53]. Therefore, the 30-channel input data are mapped into one-channel radar echo data reconstructions.

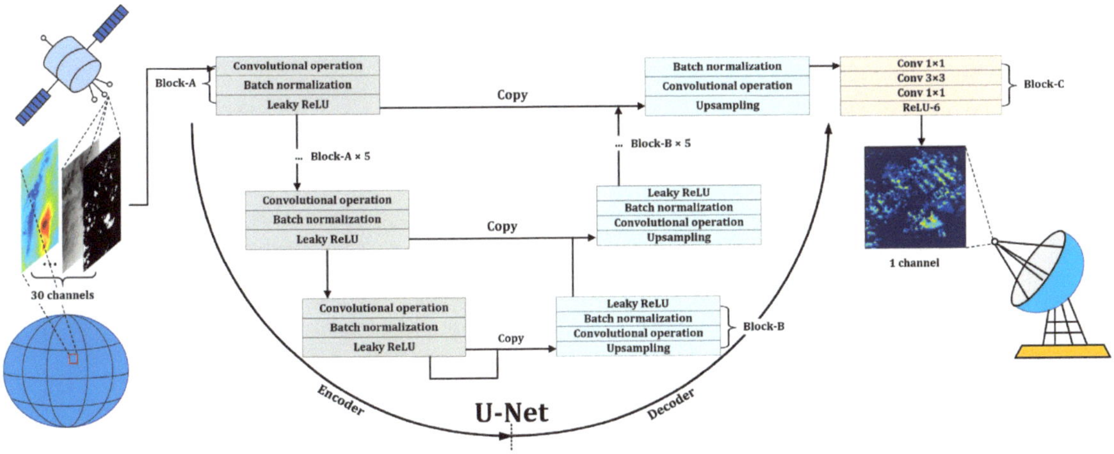

Figure 1. The U-Net model architecture.

2.6. Training

Since we do not optimize the hyperparameters of the U-Net, we divide the dataset (2647 samples) in time order only into a training set (the first ~90%, 2387 samples) and a test set (the remaining 260 samples). We employ two types of loss functions for training (that is, the objective function penalizing the discrepancy between the radar echo observations and the reconstructions generated by the U-Net). The first type is the mean square error (MSE), and we denote the resulting network of this type as UNet-MSE. The second type of loss function is the echo-weighted mean square error (EWMSE), with larger weights assigned to grid cells of higher echo intensity [28]. The resulting network of this type is denoted as UNet-EW. The calculations of the MSE and the EWMSE can be found in Text S1. The U-Net models are trained using the stochastic gradient descent (SGD) with the momentum algorithm [54] (momentum = 0.9) with batch size 8. The initial learning rate is set as 1×10^{-5}. The U-Net model is trained until the loss function shows no reduction on the test set for 100 subsequent epochs. We stop the trainings of the UNet-MSE and the UNet-EW at the 96th and the 92nd epoch, respectively. Since we obtain satisfactory trained models and the reconstruction accuracy is not our ultimate goal, we do not use data augmentation for potential improvement.

2.7. Evaluations and Interpretations

The performance of the U-Net reconstructions is evaluated by five indices, namely the root mean square error (RMSE), the mean error (ME), the critical success index (CSI), the probability of detection (POD), and the false alarm rate (FAR). RMSE and ME account for the difference between the reconstructed and observed echo data. CSI and POD measure the precision of reconstructions, whereas FAR measures the degree of overestimation. All indices range from 0% to 100%. Ideal CSI and POD values are supposed to approach 100%, whereas ideal FAR values are the opposite. Moreover, structural similarity (SSIM) is calculated in the following analysis, which quantifies the similarity between the visible structures of two images. SSIM is a value between -1 (perfect anti-correlation) and +1 (perfect similarity) and a value of 0 indicates no similarity. The calculations of these indices are detailed in Text S2.

To investigate the physical interpretations of the learnt deep network models, we propose a sensitivity analysis method inspired by Ankenbrand et al. [55]. We intentionally perturb the input data and then diagnose the relationships between the multiscale features and the input physical quantities by checking the consequences of perturbations on the reconstructions. Two types of perturbations are considered. First, we flip the input data from left to right and scale them by multiplying a coefficient. We denote this sensitivity analysis experiment as SA-a. Second, we evaluate the role of each input quantity by nullifying it while keeping other input quantities unchanged. This sensitivity analysis experiment is denoted as SA-b.

3. Results

3.1. Echo Reconstructions

Figure 2 shows the performance of reflectivity data reconstructions with the deep networks. The UNet-MSE and UNet-EW networks reconstruct the echoes well, with RMSEs of 4.76 and 5.38 dBZ, respectively. Because most radar echoes are of low intensity, training with unweighted MSE loss functions will bias towards echoes of low intensity. The UNet-MSE networks systematically underestimate the echoes, especially those with high intensity. However, echoes of higher intensity are the most valuable radar signals that directly observe the precipitation and convective processes. Adding more weights on intense echoes for training is beneficial for reconstructing richer sparse high intensity patterns. The reconstructions in this case thus either overestimate or underestimate the intense echoes, which explains the slight increases in the RMSE and FAR values of the UNet-EW networks over the UNet-MSE networks. The richer patterns reconstructed by the UNet-EW networks have better CSI and POD scores, and we will further diagnose their network structures.

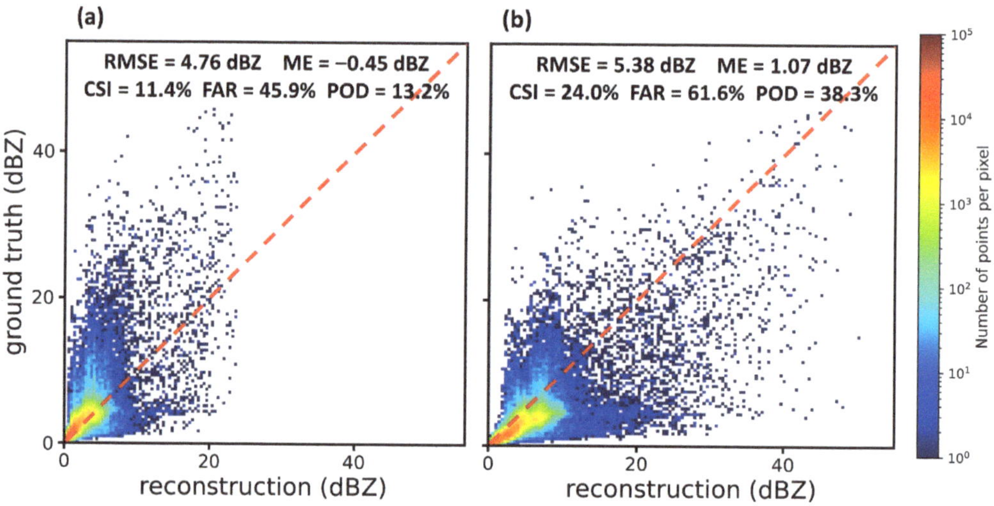

Figure 2. Performance of reflectivity data reconstructions with the (**a**) UNet-MSE and (**b**) UNet-EW networks. The observed and reconstructed echoes are averaged on a 700 × 700 grid over the BTH region for the test set.

Figure 3 details a case of reconstruction at 09:30 UTC on 10 September 2016 from the test set. The spatial distribution of clouds from satellite images (Figure 3a) outlines the overall shape and location of echoes (Figure 3e). By contrast, the WRF-simulated reflectivity data (Figure 3b) miss a majority of the observed echo distribution. Indeed, the precipitation and convective processes are among the most difficult to simulate with NWPs. Nevertheless, our simulations are qualified to represent the general atmospheric conditions (Table S1), so they can provide dynamic and thermodynamic information for echo reconstruction. Note that we do not include the WRF-simulated reflectivity data in the input data of the U-Net deep networks. Based on the multi-source information, the reconstructions by deep networks well produce echo patterns, locations, and intensities (Figure 3c,d). Compared with the UNet-MSE network, the UNet-EW network (Figure 3d) exhibits more precise details such as clearer edges and high values. Data-driven deep learning techniques effectively reduce the gap of representation from the NWP simulations and satellite images to the radar echoes and tend to encode the missing physics in their network weights. Other cases of reconstructions show similar performances (Figures S3–S9).

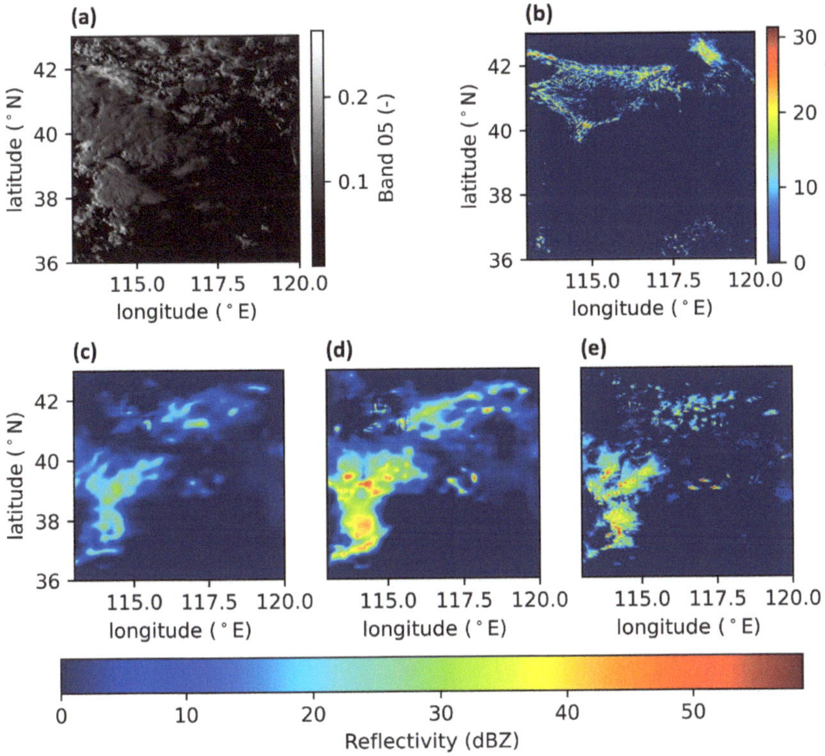

Figure 3. A case of reconstruction at 09:30 UTC on 10 September 2016 from the test set. (**a**) The satellite image of the 5th band. (**b**) The WRF-simulated reflectivity data calculated according to [56]. (**c**) The echo reconstruction by the UNet-MSE network. (**d**) The echo reconstruction by the UNet-EW network. (**e**) The observed radar reflectivity data. The rest of the input data can be found in Figure S2.

3.2. Multiscale Representation

The multiple layers of features encoded in the contracting path of U-Net are visualized in a naïve way [57]. In this way, a multiscale representation (MSR) of the radar data can be revealed and analyzed. Figure 4 exemplifies such an MSR for the reconstruction case in Figure 3. The multiscale features of the radar echo data appear to be automatically stratified in the multi-layer hierarchical structure of the U-Net, and are especially manifested in the deeper part, indicating the greater capacity of deep networks than shallow networks to represent multiscale high-dimensional relationships among multi-source data (e.g., [58,59]). Apart from the first layer with mostly texture-like information, the locations of echo signals (upper left triangles in Figure 4) are recognized in all subsequent layers and largely correspond to the cloud locations from satellite images. The location-aware ability of U-Net is one of the key factors for its success in image segmentation. The small-scale features in the shallow layers (first–third) still demonstrate no clear echo patterns. These small-scale features such as the echo intensities (hexagons in Figure 4) as well as larger-scale features like the echo shapes (ellipses in Figure 4) emerge in the middle layers (fourth–sixth). The shapes are more visibly related to the satellite images, whereas the sources of intensities need further investigation (see Section 3.3). The features in the deepest layers (seventh and eighth) are rather global descriptions of echoes and may have some semantic meanings. MSRs for other reconstruction cases manifest similar stratifications (see Figures S10–S16 for more examples).

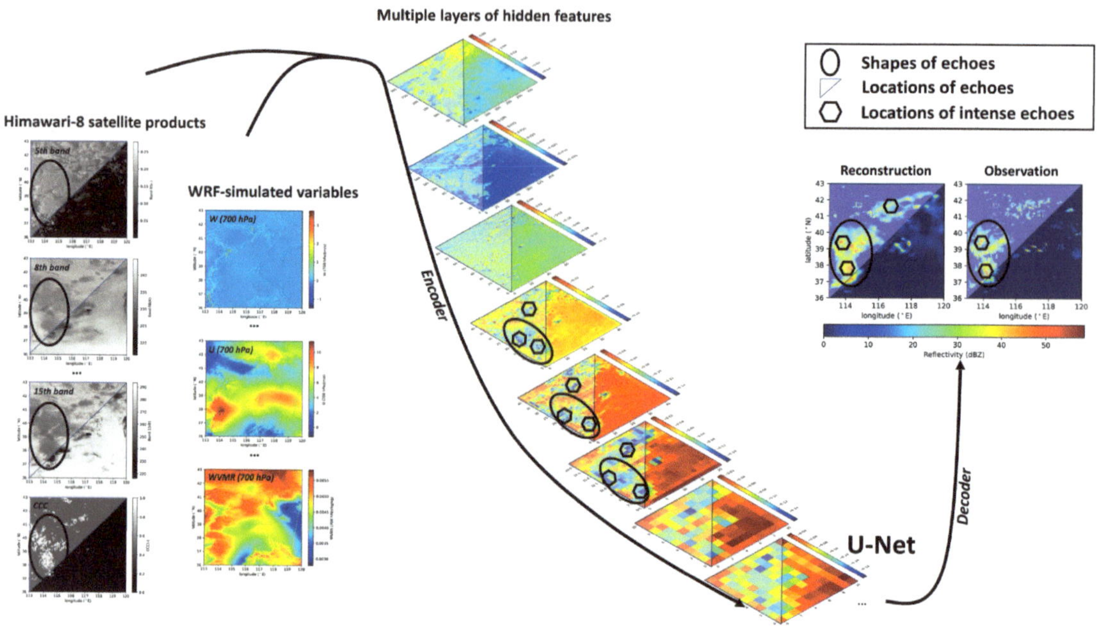

Figure 4. A multiscale representation of the radar echo data at 09:30 UTC on 10 September 2016. The UNet-EW deep network is used for the reconstruction of echo data. Only parts of the input data for this reconstruction are presented. Hidden features at each hidden layer are averaged along the channel dimension.

3.3. *Physical Interpretations of the MSR*

We further investigate the physical interpretations of the MSR through sensitivity analysis. The SA-a experiments assess the overall contributions of the WRF simulations and satellite images to reconstructions by flipping or attenuating these different sets of input data. Figure 5 shows the resulting diverse reconstructions, where the CSI and SSIM of reconstructions of this case and over the test set are calculated. When all input data are flipped, the reconstruction flips accordingly (Figure 5a), but does not yield a perfect flipping of the original reconstruction (Figure 5i). This indicates that factors other than the input data (e.g., the topographic conditions) may also play a role in the spatial distribution of the reconstructions. The echo shapes and locations are much more influenced by the satellite images (Figure 5b, SSIM = −0.07) than by the WRF simulations (Figure 5c,g,h with higher SSIM). Moreover, this influence is mainly related to the spatial distribution of the satellite images rather than changes in magnitude (Figure 5e,f). Concerning the echo intensities, the representation is much more sensitive to the WRF simulations (Figure 5g,h, especially in Figure 5h CSI = 0.26 and CSI-Test = 0.17) than to the satellite images (Figure 5e,f). This sensitivity appears to depend more on the value than on the spatial distribution of the simulations (Figure 5c). Other cases of reconstructions have similar interpretations (Figures S17–S23).

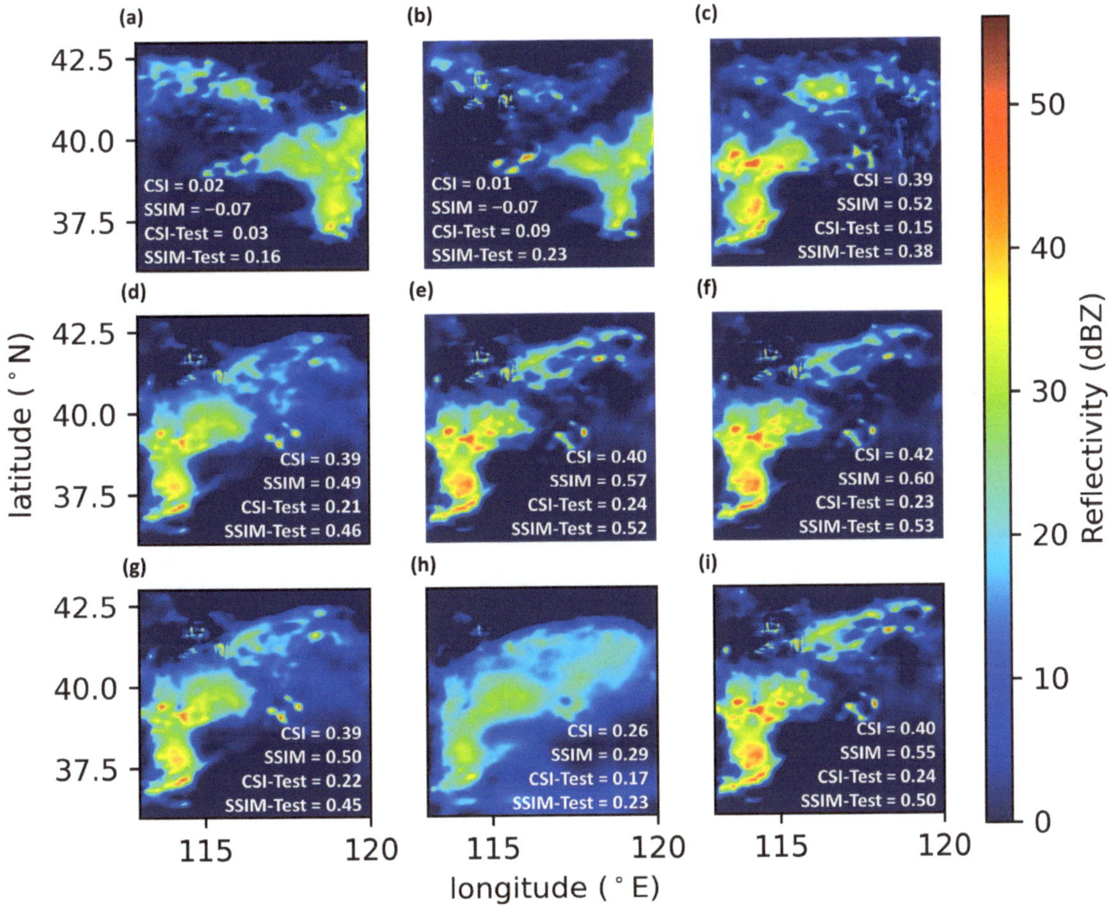

Figure 5. Reconstructions of the radar echo data at 09:30 UTC on 10 September 2016 by the UNet-EW deep network for the SA-a experiments. The perturbations are conducted by (**a**) flipping all the input data, (**b**) flipping only the satellite images, (**c**) flipping only the WRF simulations, (**d**) multiplying all the input data (except for the binary CCC data) by 0.9, (**e**) multiplying only the satellite images (except for the binary CCC data) by 0.9 and (**f**) 0.8, and (**g**) multiplying only the WRF simulations by 0.9 and (**h**) 0.8. Also shown is the (**i**) original reconstruction without perturbation. CSI/SSIM and CSI-/SSIM-Test are CSI/SSIM of reconstructions of this case and over the test set, respectively, compared with radar echo observations.

Figure 6 shows the reconstructions with CSI for the SA-b experiment (see Figures S24–S30 for more examples). The MSR here is predominantly sensitive to the satellite images from the 8th and 13th bands of Himawari-8 as well as the WRF-simulated W, K, and RH. These bands of satellite images provide information on middle and upper tropospheric humidity and cloud-top properties, respectively. The three dominant WRF variables describe the vertical motion of the air, the atmospheric instability, and the water vapor content in the atmosphere. They correspond well with the three key ingredients for deep convection initiation and evolution, namely lift, instability, and moisture [60]. Therefore, for the current study, the MSR may encapsulate these complex atmospheric physics in their learnt multiscale features.

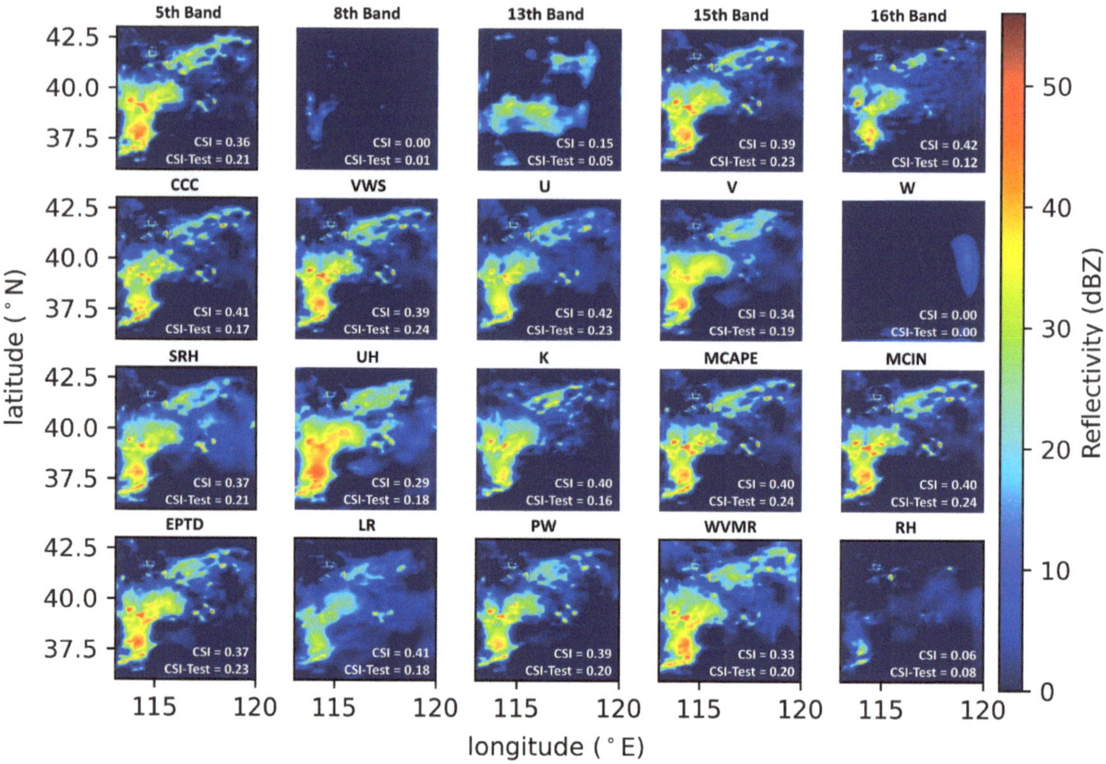

Figure 6. Reconstructions of the radar echo data at 09:30 UTC on 10 September 2016 by the UNet-EW deep network for the SA-b experiments. Each reconstruction is obtained by setting the listed physical quantity to zero while keeping other quantities unchanged. CSI and CSI-Test are CSI of reconstructions of this case and over the test set, respectively, compared with radar echo observations.

4. Summary and Discussions

We have addressed the difficulty of model–data fusion for convective nowcasting with a representation problem. Deep learning techniques were applied to represent the fine-grained radar reflectivity data by reconstructing them from the WRF model simulations and the Himawari-8 satellite products. We learnt a multiscale representation (MSR) of the radar data using the U-Net deep networks. The MSR manifested a stratification, and the richest features were in the middle layers. Sensitivity analyses on the retrieved representation showed that small-scale patterns like echo intensities were more sensitive to the magnitude of numerical model simulations, whereas larger-scale information about the shapes and locations was mainly from the spatial distribution of satellite images.

The retrieved multiscale representation takes advantage of the ability of deep learning techniques [61] to find complex relationships in data, which are otherwise difficult to model or formulate in traditional approaches, especially when the underlying physics is complex and multiscale or even unknown. The deep network representation can organize the learnt features at increasing levels of abstraction from local fine-grained details to global semantic information [62–64]. Such multiscale representation could inspire innovative methods that make use of the features in the mapping from numerical model simulations to radar data as well as in convective nowcasting, where machine learning has been demonstrated to be a useful tool [65]. Note that multiscale representations can also be obtained using traditional methods such as wavelets in a more compact manner [64,66]. However, in general, these traditional methods require domain expertise for feature extraction and should be combined with deep learning techniques for automatic end-to-end feature extractions [67,68].

This study has a number of limitations. First, although radar data provide routine verifications for convection-permitting models [10,69], they are not perfect and can suffer from various errors [70,71]. Hence, the retrieved multiscale representation may occasionally fit noise rather than signals. Second, our WRF simulations were configured at a convection-permitting resolution, but we did not tailor and optimize the WRF configurations and examine in-depth model–radar comparisons (e.g., [69]), which is not the main objective of this representation study. Third, we did not test other deep learning techniques such as CNN variants other than U-Nets or generative deep networks. These deep networks, once successfully trained, are known to have similar performances [72]. Finally, convective processes can vary by regions, so the learnt multiscale representation needs evaluation across regions, such as considering the role of terrain during training or applying the obtained representation to different regions. Hence, there is still room for qualitative and quantitative improvement in the accuracy of the reconstructed echo reflectivity. Despite these limitations, our finding on the functioning and potential of the deep network representation for convective atmosphere should not be affected.

For now, the proposed representation of radar echo data may not meet the requirements of practical convective nowcasting. We consider our attempt as a step towards a deep network framework of multiscale representation with physical interpretations that relate the features in hidden network layers with the convective atmosphere. Our sensitivity analyses for physical interpretations have been experimental. It is possible to apply more formal methods—notably explainable artificial intelligence techniques [62,73–75]—to analyze the cause-and-effect relationships among radar data, satellite images, and convective dynamics. The multiscale features with physical interpretations are seldom investigated in radar data assimilation practices [9,12]. Data assimilation based on multiscale features should merge with artificial intelligence techniques [76,77] and should be inevitably "deep" in some sense, either in models [78,79], in data (this study), or in assimilation algorithms [80]. Such deep assimilating techniques may be essential to overcome the conventional limits of convective nowcasting.

Supplementary Materials: The supporting information can be downloaded at: https://www.mdpi.com/article/10.3390/rs15143466/s1, Figure S1: (a): The selected six radar stations in the Beijing-Tianjin-Hebei region. (b): WRF simulation area with three nested domains; Figure S2: Visualizations of input data of the reconstruction sample at 09:30 UTC on 10 September 2016; Figure S3: A case of reconstruction at 00:30 UTC on 4 September 2016 from the test set; Figure S4: A case of reconstruction at 05:30 UTC on 4 September 2016 from the test set; Figure S5: A case of reconstruction at 08:30 UTC on 4 September 2016 from the test set; Figure S6: A case of reconstruction at 08:00 UTC on 10 September 2016 from the test set; Figure S7: A case of reconstruction at 03:30 UTC on 11 September 2016 from the test set; Figure S8: A case of reconstruction at 06:30 UTC on 13 September 2016 from the test set; Figure S9: A case of reconstruction at 07:30 UTC on 23 September 2016 from the test set; Figure S10: A multiscale representation of the radar echo data at 00:30 UTC on 4 September 2016; Figure S11: A multiscale representation of the radar echo data at 05:30 UTC on 4 September 2016; Figure S12: A multiscale representation of the radar echo data at 08:30 UTC on 4 September 2016; Figure S13: A multiscale representation of the radar echo data at 08:00 UTC on 10 September 2016; Figure S14: A multiscale representation of the radar echo data at 03:30 UTC on 11 September 2016; Figure S15: A multiscale representation of the radar echo data at 06:30 UTC on 13 September 2016; Figure S16: A multiscale representation of the radar echo data at 07:30 UTC on 23 September 2016; Figure S17: Reconstructions of the radar echo data at 00:30 UTC on 4 September 2016 by the UNet-EW deep network for the SA-a experiments; Figure S18: Reconstructions of the radar echo data at 05:30 UTC on 4 September 2016 by the UNet-EW deep network for the SA-a experiments; Figure S19: Reconstructions of the radar echo data at 08:30 UTC on 4 September 2016 by the UNet-EW deep network for the SA-a experiments; Figure S20: Reconstructions of the radar echo data at 08:00 UTC on 10 September 2016 by the UNet-EW deep network for the SA-a experiments; Figure S21: Reconstructions of the radar echo data at 03:30 UTC on 11 September 2016 by the UNet-EW deep network for the SA-a experiments; Figure S22: Reconstructions of the radar echo data at 03:30 UTC on 11 September 2016 by the UNet-EW deep network for the SA-a experiments; Figure S23: Reconstructions of the radar echo data at 07:30 UTC on 23 September 2016 by the UNet-EW deep network for the SA-a experiments; Figure S24: Reconstructions of the radar echo data at 00:30 UTC on 4 September 2016 by the UNet-EW deep network for the SA-b experiments; Figure S25: Reconstructions of the radar echo data at 05:30 UTC on 4 September 2016 by the UNet-EW deep network for the SA-b experiments; Figure S26: Reconstructions of the radar echo data at 08:30 UTC on 4 September 2016 by the UNet-EW deep network for the SA-b experiments; Figure S27: Reconstructions of the radar echo data at 08:00 UTC on 10 September 2016 by the UNet-EW deep network for the SA-b experiments; Figure S28: Reconstructions of the radar echo data at 03:30 UTC on 11 September 2016 by the UNet-EW deep network for the SA-b experiments; Figure S29: Reconstructions of the radar echo data at 06:30 UTC on 13 September 2016 by the UNet-EW deep network for the SA-b experiments; Figure S30: Reconstructions of the radar echo data at 07:30 UTC on 23 September 2016 by the UNet-EW deep network for the SA-b experiments; Table S1: Average evaluation indices comparing hourly WRF simulations and observations; Table S2: Descriptions of the selected and computed physical quantities from WRF simulations; Text S1: The calculations of the MSE and the EWMSE; Text S2: The calculations of the evaluation indices.

Author Contributions: Conceptualization, L.W. and M.Z.; methodology, M.Z. and Q.L.; software, Q.W. and Y.W.; validation, M.Z., Q.L. and S.Z.; formal analysis, M.Z.; investigation, M.Z. and L.W.; resources, L.W. and Q.W.; data curation, D.S.; writing—original draft preparation, M.Z. and L.W.; writing—review and editing, M.Z., L.W. and S.Z.; visualization, M.Z.; supervision, L.W., Z.W. and X.P.; project administration, L.W. and Z.W.; funding acquisition, L.W. and Z.W. All authors have read and agreed to the published version of the manuscript.

Funding: This research was funded by the Informatization Plan of Chinese Academy of Sciences (Grant No. CAS-WX2021SF-0107), the National Key Basic Research Program of China (Grant No. 2014CB441401), the major science and technology project of Inner Mongolia Autonomous Region (Grand No. 2020ZD0013), and the Pioneer Hundred Talents Program of the Chinese Academy of Sciences.

Data Availability Statement: The test dataset (processed Himawari-8 satellite images, WRF simulations, and radar echo data), the codes supporting the analysis, and the trained U-Net model parameter files are available in a Zenodo archive (https://doi.org/10.5281/zenodo.7798803, accessed on 1 June 2023).

Acknowledgments: The authors thank the editors and reviewers for their valuable contributions and support throughout this research.

Conflicts of Interest: The authors declare no conflict of interest.

References

1. Yano, J.-I.; Ziemianski, M.Z.; Cullen, M.; Termonia, P.; Onvlee, J.; Bengtsson, L.; Carrassi, A.; Davy, R.; Deluca, A.; Gray, S.L.; et al. Scientific challenges of convective-scale numerical weather prediction. *Bull. Am. Meteorol. Soc.* **2018**, *99*, 699–710. [CrossRef]
2. Wyngaard, J.C. Toward numerical modeling in the "terra incognita". *J. Atmos. Sci.* **2004**, *61*, 1816–1826. [CrossRef]
3. Dixon, M.; Wiener, G. TITAN: Thunderstorm identification, tracking, analysis, and nowcasting—A radar-based methodology. *J. Atmos. Ocean. Technol.* **1993**, *10*, 785–797. [CrossRef]
4. Pulkkinen, S.; Nerini, D.; Hortal, A.A.P.; Velasco-Forero, C.; Seed, A.; Germann, U.; Foresti, L. Pysteps: An open-source Python library for probabilistic precipitation nowcasting (v1.0). *Geosci. Model Dev.* **2019**, *12*, 4185–4219. [CrossRef]
5. James, P.M.; Reichert, B.K.; Heizenreder, D. NowCastMIX: Automatic integrated warnings for severe convection on nowcasting time scales at the German weather service. *Weather Forecast.* **2018**, *33*, 1413–1433. [CrossRef]
6. Foresti, L.; Sideris, I.V.; Panziera, L.; Nerini, D.; Germann, U. A 10-year radar-based analysis of orographic precipitation growth and decay patterns over the Swiss Alpine region. *Q. J. R. Meteorol. Soc.* **2018**, *144*, 2277–2301. [CrossRef]
7. Sideris, I.V.; Foresti, L.; Nerini, D.; Germann, U. NowPrecip: Localized precipitation nowcasting in the complex terrain of Switzerland. *Q. J. R. Meteorol. Soc.* **2020**, *146*, 1768–1800. [CrossRef]
8. Sun, J.; Xue, M.; Wilson, J.W.; Zawadzki, I.; Ballard, S.P.; Onvlee-Hooimeyer, J.; Joe, P.; Barker, D.M.; Li, P.-W.; Golding, B.; et al. Use of NWP for nowcasting convective precipitation. *Bull. Am. Meteorol. Soc.* **2014**, *95*, 409–426. [CrossRef]
9. Fabry, F.; Meunier, V. Why are radar data so difficult to assimilate skillfully? *Mon. Weather Rev.* **2020**, *148*, 2819–2836. [CrossRef]
10. Clark, P.; Roberts, N.; Lean, H.; Ballard, S.P.; Charlton-Perez, C. Convection-permitting models: A step-change in rainfall forecasting. *Meteorol. Appl.* **2016**, *23*, 165–181. [CrossRef]

11. Gustafsson, N.; Janjic, T.; Schraff, C.; Leuenberger, D.; Weissmann, M.; Reich, H.; Brousseau, P.; Montmerle, T.; Wattrelot, E.; Bucanek, A.; et al. Survey of data assimilation methods for convective-scale numerical weather prediction at operational centres. *Q. J. R. Meteorol. Soc.* **2018**, *144*, 1218–1256. [CrossRef]
12. Bannister, R.N.; Chipilski, H.G.; Martinez-Alvarado, O. Techniques and challenges in the assimilation of atmospheric water observations for numerical weather prediction towards convective scales. *Q. J. R. Meteorol. Soc.* **2020**, *146*, 1–48. [CrossRef]
13. Bluestein, H.B.; Carr, F.H.; Goodman, S.J. Atmospheric observations of weather and climate. *Atmosphere-Ocean* **2022**, *60*, 149–187. [CrossRef]
14. Vignon, E.; Besic, N.; Jullien, N.; Gehring, J.; Berne, A. Microphysics of snowfall over coastal east Antarctica simulated by Polar WRF and observed by radar. *J. Geophys. Res.-Atmos.* **2019**, *124*, 11452–11476. [CrossRef]
15. Troemel, S.; Simmer, C.; Blahak, U.; Blanke, A.; Doktorowski, S.; Ewald, F.; Frech, M.; Gergely, M.; Hagen, M.; Janjic, T.; et al. Overview: Fusion of radar polarimetry and numerical atmospheric modelling towards an improved understanding of cloud and precipitation processes. *Atmos. Chem. Phys.* **2021**, *21*, 17291–17314. [CrossRef]
16. Chow, F.K.; Schar, C.; Ban, N.; Lundquist, K.A.; Schlemmer, L.; Shi, X. Crossing multiple gray zones in the transition from mesoscale to microscale simulation over complex terrain. *Atmosphere* **2019**, *10*, 274. [CrossRef]
17. Jeworrek, J.; West, G.; Stull, R. Evaluation of cumulus and microphysics parameterizations in WRF across the convective gray zone. *Weather Forecast.* **2019**, *34*, 1097–1115. [CrossRef]
18. Kirshbaum, D.J. Numerical simulations of orographic convection across multiple gray zones. *J. Atmos. Sci.* **2020**, *77*, 3301–3320. [CrossRef]
19. Honnert, R.; Efstathiou, G.A.; Beare, R.J.; Ito, J.; Lock, A.; Neggers, R.; Plant, R.S.; Shin, H.H.; Tomassini, L.; Zhou, B. The atmospheric boundary layer and the "gray zone" of turbulence: A critical review. *J. Geophys. Res.-Atmos.* **2020**, *125*, e2019JD030317. [CrossRef]
20. Tapiador, F.J.; Roca, R.; Del Genio, A.; Dewitte, B.; Petersen, W.; Zhang, F. Is precipitation a good metric for model performance? *Bull. Am. Meteorol. Soc.* **2019**, *100*, 223–234. [CrossRef]
21. Koo, C.C. Similarity analysis of some meso-and micro-scale atmospheric motions. *Acta Meteorol. Sin.* **1964**, *4*, 519–522. [CrossRef]
22. Tao, S.Y.; Ding, Y.H.; Sun, S.Q.; Cai, Z.Y.; Zhang, M.L.; Fang, Z.Y.; Li, M.T.; Zhou, X.P.; ZHao, S.X.; Dian, S.T.; et al. *The Heavy Rainfalls in China*; Science Press: Beijing, China, 1980; p. 225.
23. Surcel, M.; Zawadzki, I.; Yau, M.K. A study on the scale dependence of the predictability of precipitation patterns. *J. Atmos. Sci.* **2015**, *72*, 216–235. [CrossRef]
24. Mecikalski, J.R.; Bedka, K.M. Forecasting convective initiation by monitoring the evolution of moving cumulus in daytime GOES imagery. *Mon. Weather Rev.* **2006**, *134*, 49–78. [CrossRef]
25. Reichstein, M.; Camps-Valls, G.; Stevens, B.; Jung, M.; Denzler, J.; Carvalhais, N.; Prabhat, F. Deep learning and process understanding for data-driven Earth system science. *Nature* **2019**, *566*, 195–204. [CrossRef] [PubMed]
26. Ravuri, S.; Lenc, K.; Willson, M.; Kangin, D.; Lam, R.; Mirowski, P.; Fitzsimons, M.; Athanassiadou, M.; Kashem, S.; Madge, S.; et al. Skilful precipitation nowcasting using deep generative models of radar. *Nature* **2021**, *597*, 672–677. [CrossRef]
27. Shi, X.; Chen, Z.; Wang, H.; Yeung, D.-Y.; Wong, W.-K.; Woo, W.-C. Convolutional LSTM network: A machine learning approach for precipitation nowcasting. In *Advances in Neural Information Processing Systems 28*; Cortes, C., Lawrence, N.D., Lee, D.D., Sugiyama, M., Garnett, R., Eds.; Advances in Neural Information Processing Systems; Neural Information Processing Systems: La Jolla, CA, USA, 2015; Volume 28.
28. Shi, X.; Gao, Z.; Lausen, L.; Wang, H.; Yeung, D.-Y.; Wong, W.-K.; Woo, W.-C. Deep learning for precipitation nowcasting: A benchmark and a new model. In *Advances in Neural Information Processing Systems 30*; Guyon, I., Luxburg, U.V., Bengio, S., Wallach, H., Fergus, R., Vishwanathan, S., Garnett, R., Eds.; Advances in Neural Information Processing Systems; Neural Information Processing Systems: La Jolla, CA, USA, 2017; Volume 30.
29. Wang, Y.; Long, M.; Wang, J.; Gao, Z.; Yu, P.S. PredRNN: Recurrent neural networks for predictive learning using spatiotemporal LSTMs. In Proceedings of the 31st Annual Conference on Neural Information Processing Systems (NIPS), Long Beach, CA, USA, 4–9 December 2017.
30. Lebedev, V.; Ivashkin, V.; Rudenko, I.; Ganshin, A.; Molchanov, A.; Ovcharenko, S.; Grokhovetskiy, R.; Bushmarinov, I.; Solomentsev, D. Precipitation nowcasting with satellite imagery. In Proceedings of the 25th ACM SIGKDD International Conference on Knowledge Discovery & Data Mining, Anchorage, AK, USA, 4–8 August 2019; pp. 2680–2688.
31. Hilburn, K.A.; Ebert-Uphoff, I.; Miller, S.D. Development and interpretation of a neural-network-based synthetic radar reflectivity estimator using GOES-R satellite observations. *J. Appl. Meteorol. Climatol.* **2021**, *60*, 3–21. [CrossRef]
32. Veillette, M.S.; Hassey, E.P.; Mattioli, C.J.; Iskenderian, H.; Lamey, P.M. Creating synthetic radar imagery using convolutional neural networks. *J. Atmos. Ocean. Technol.* **2018**, *35*, 2323–2338. [CrossRef]
33. Duan, M.; Xia, J.; Yan, Z.; Han, L.; Zhang, L.; Xia, H.; Yu, S. Reconstruction of the radar reflectivity of convective storms based on deep learning and Himawari-8 observations. *Remote Sens.* **2021**, *13*, 3330. [CrossRef]
34. Mohr, C.G.; Vaughan, R.L. Economical procedure for Cartesian interpolation and display of reflectivity factor data in 3-dimensional space. *J. Appl. Meteorol.* **1979**, *18*, 661–670. [CrossRef]
35. Zhang, J.; Howard, K.; Xia, W.W.; Gourley, J.J. Comparison of objective analysis schemes for the WSR-88D radar data. In Proceedings of the 31st Conference on Radar Meteorology, Seattle, WA, USA, 6–12 August 2003.

36. Davis, C.; Brown, B.; Bullock, R. Object-based verification of precipitation forecasts. Part I: Methodology and application to mesoscale rain areas. *Mon. Weather Rev.* **2006**, *134*, 1772–1784. [CrossRef]
37. Davis, C.; Brown, B.; Bullock, R. Object-based verification of precipitation forecasts. Part II: Application to convective rain systems. *Mon. Weather Rev.* **2006**, *134*, 1785–1795. [CrossRef]
38. Skamarock, W.C.; Klemp, J.B.; Dudhia, J.; Gill, D.O.; Barker, D.M.; Wang, W.; Powers, J.G. A description of the Advanced Research WRF version 3. *NCAR Tech. Note* **2008**, *475*, 113. [CrossRef]
39. Blanco-Ward, D.; Rocha, A.; Viceto, C.; Ribeiro, A.C.; Feliciano, M.; Paoletti, E.; Miranda, A.I. Validation of meteorological and ground-level ozone WRF-CHIMERE simulations in a mountainous grapevine growing area for phytotoxic risk assessment. *Atmos Environ.* **2021**, *259*, 118507. [CrossRef]
40. Giordano, C.; Vernin, J.; Ramio, H.V.; Munoz-Tunon, C.; Varela, A.M.; Trinquet, H. Atmospheric and seeing forecast: WRF model validation with in situ measurements at ORM. *Mon. Not. R. Astron. Soc.* **2013**, *430*, 3102–3111. [CrossRef]
41. Hong, S.Y.; Dudhia, J.; Chen, S.H. A revised approach to ice microphysical processes for the bulk parameterization of clouds and precipitation. *Mon. Weather Rev.* **2004**, *132*, 103–120. [CrossRef]
42. Mlawer, E.J.; Taubman, S.J.; Brown, P.D.; Iacono, M.J.; Clough, S.A. Radiative transfer for inhomogeneous atmospheres: RRTM, a validated correlated-k model for the longwave. *J. Geophys. Res.-Atmos.* **1997**, *102*, 16663–16682. [CrossRef]
43. Dudhia, J. Numerical study of convection observed during the winter monsoon experiment using a mesoscale two-dimensional model. *J. Atmos. Sci.* **1989**, *46*, 3077–3107. [CrossRef]
44. Jimenez, P.A.; Dudhia, J. Improving the representation of resolved and unresolved topographic effects on surface wind in the WRF model. *J. Appl. Meteorol. Climatol.* **2012**, *51*, 300–316. [CrossRef]
45. Chen, F.; Dudhia, J. Coupling an advanced land surface-hydrology model with the Penn State-NCAR MM5 modeling system. Part I: Model implementation and sensitivity. *Mon. Weather Rev.* **2001**, *129*, 569–585. [CrossRef]
46. Hong, S.-Y.; Noh, Y.; Dudhia, J. A new vertical diffusion package with an explicit treatment of entrainment processes. *Mon Weather Rev.* **2006**, *134*, 2318–2341. [CrossRef]
47. Zhang, C.; Wang, Y.; Hamilton, K. Improved representation of boundary layer clouds over the southeast Pacific in ARW-WRF using a modified Tiedtke cumulus parameterization scheme. *Mon. Weather Rev.* **2011**, *139*, 3489–3513. [CrossRef]
48. Bessho, K.; Date, K.; Hayashi, M.; Ikeda, A.; Imai, T.; Inoue, H.; Kumagai, Y.; Miyakawa, T.; Murata, H.; Ohno, T.; et al. An introduction to Himawari-8/9—Japan's new-generation geostationary meteorological satellites. *J. Meteorol. Soc. Japan. Ser. II* **2016**, *94*, 151–183. [CrossRef]
49. Ronneberger, O.; Fischer, P.; Brox, T. U-Net: Convolutional networks for biomedical image segmentation. In *Medical Image Computing and Computer-Assisted Intervention, Pt Iii*; Navab, N., Hornegger, J., Wells, W.M., Frangi, A.F., Eds.; Lecture Notes in Computer Science; Springer: Berlin/Heidelberg, Germany, 2015; Volume 9351, pp. 234–241.
50. Trebing, K.; Stanczyk, T.; Mehrkanoon, S. SmaAt-UNet: Precipitation nowcasting using a small attention-UNet architecture. *Pattern Recognit. Lett.* **2021**, *145*, 178–186. [CrossRef]
51. Li, Y.; Liu, Y.; Sun, R.; Guo, F.; Xu, X.; Xu, H. Convective storm VIL and lightning nowcasting using satellite and weather radar measurements based on multi-task learning models. *Adv. Atmos. Sci.* **2023**, *40*, 887–899. [CrossRef]
52. Ioffe, S.; Szegedy, C. Batch normalization: Accelerating deep network training by reducing internal covariate shift. In Proceedings of the 32nd International Conference on Machine Learning, Lille, France, 7–9 July 2015.
53. Krizhevsky, A.; Hinton, G. Convolutional Deep Belief Networks on cifar-10. 2010. Available online: http://www.cs.toronto.edu/~kriz/conv-cifar10-aug2010.pdf. (accessed on 26 May 2023).
54. Ruder, S. An overview of gradient descent optimization algorithms. *arXiv* **2016**, arXiv:1609.04747.
55. Ankenbrand, M.J.; Shainberg, L.; Hock, M.; Lohr, D.; Schreiber, L.M. Sensitivity analysis for interpretation of machine learning based segmentation models in cardiac MRI. *Bmc Med. Imaging* **2021**, *21*, 27. [CrossRef] [PubMed]
56. Ladwig, W. *Wrf-Python*, Version 1.2.3; Github: San Francisco, CA, USA, 2017. [CrossRef]
57. Toda, Y.; Okura, F. How convolutional neural networks diagnose plant disease. *Plant Phenomics* **2019**, *2019*, 9237136. [CrossRef]
58. Poggio, T.; Mhaskar, H.; Rosasco, L.; Miranda, B.; Liao, Q. Why and when can deep-but not shallow-networks avoid the curse of dimensionality: A review. *Int. J. Autom. Comput.* **2017**, *14*, 503–519. [CrossRef]
59. Mhaskar, H.; Liao, Q.; Poggio, T. When and why Are deep networks better than shallow ones? In Proceedings of the 31st AAAI Conference on Artificial Intelligence, San Francisco, CA, USA, 4–9 February 2017.
60. Doswell, C.A.; Brooks, H.E.; Maddox, R.A. Flash flood forecasting: An ingredients-based methodology. *Weather Forecast.* **1996** *11*, 560–581. [CrossRef]
61. Bengio, Y.; Courville, A.; Vincent, P. Representation learning: A review and new perspectives. *IEEE Trans. Pattern Anal. Mach. Intell.* **2013**, *35*, 1798–1828. [CrossRef]
62. McGovern, A.; Lagerquist, R.; Gagne, D.J., II; Jergensen, G.E.; Elmore, K.L.; Homeyer, C.R.; Smith, T. Making the black box more transparent: Understanding the physical implications of machine learning. *Bull. Am. Meteorol. Soc.* **2019**, *100*, 2175–2199 [CrossRef]
63. Bai, J.; Gong, B.; Zhao, Y.; Lei, F.; Yan, C.; Gao, Y. Multi-scale representation learning on hypergraph for 3d shape retrieval and recognition. *IEEE Trans. Image Process.* **2021**, *30*, 5327–5338. [CrossRef]
64. Jiao, L.; Gao, J.; Liu, X.; Liu, F.; Yang, S.; Hou, B. Multi-scale representation learning for image classification: A survey. *IEEE Trans Artif. Intell.* **2021**, *4*, 23–43. [CrossRef]

5. Foresti, L.; Sideris, I.V.; Nerini, D.; Beusch, L.; Germann, U. Using a 10-year radar archive for nowcasting precipitation growth and decay: A probabilistic machine learning approach. *Weather Forecast.* **2019**, *34*, 1547–1569. [CrossRef]
6. Mallat, S. Group invariant scattering. *Commun. Pure Appl. Math.* **2012**, *65*, 1331–1398. [CrossRef]
7. Michau, G.; Frusque, G.; Fink, O. Fully learnable deep wavelet transform for unsupervised monitoring of high-frequency time series. *Proc. Natl. Acad. Sci. USA* **2022**, *119*, e2106598119. [CrossRef]
8. Ramzi, Z.; Michalewicz, K.; Starck, J.-L.; Moreau, T.; Ciuciu, P. Wavelets in the deep learning era. *J. Math. Imaging Vis.* **2023**, *65*, 240–251. [CrossRef]
9. Wilson, J.; Megenhardt, D.; Pinto, J. NWP and radar extrapolation: Comparisons and explanation of errors. *Mon. Weather Rev.* **2020**, *148*, 4783–4798. [CrossRef]
0. Fabry, F. *Radar Meteorology: Principles and Practice*; Cambridge University Press: Cambridge, UK, 2018.
1. Sokol, Z.; Szturc, J.; Orellana-Alvear, J.; Popova, J.; Jurczyk, A.; Celleri, R. The role of weather radar in rainfall estimation and its application in meteorological and hydrological modelling-a review. *Remote Sens.* **2021**, *13*, 351. [CrossRef]
2. Greff, K.; Srivastava, R.K.; Koutnik, J.; Steunebrink, B.R.; Schmidhuber, J. LSTM: A search space odyssey. *IEEE Trans. Neural Netw. Learn. Syst.* **2017**, *28*, 2222–2232. [CrossRef]
3. Gagne, D.J., II; Haupt, S.E.; Nychka, D.W.; Thompson, G. Interpretable deep learning for spatial analysis of severe hailstorms. *Mon. Weather Rev.* **2019**, *147*, 2827–2845. [CrossRef]
4. Toms, B.A.; Barnes, E.A.; Ebert-Uphoff, I. Physically interpretable neural networks for the geosciences: Applications to Earth system variability. *J. Adv. Model. Earth Syst.* **2020**, *12*, e2019MS002002. [CrossRef]
5. Davenport, F.V.; Diffenbaugh, N.S. Using machine learning to analyze physical causes of climate change: A case study of U.S. midwest extreme precipitation. *Geophys. Res. Lett.* **2021**, *48*, e2021GL093787. [CrossRef]
6. Bocquet, M.; Brajard, J.; Carrassi, A.; Bertino, L. Bayesian inference of chaotic dynamics by merging data assimilation, machine learning and expectation-maximization. *Found. Data Sci.* **2020**, *2*, 55–80. [CrossRef]
7. Geer, A.J. Learning earth system models from observations: Machine learning or data assimilation? *Philos. Trans. R. Soc. A-Math. Phys. Eng. Sci.* **2021**, *379*. [CrossRef]
8. Penny, S.G.; Smith, T.A.; Chen, T.C.; Platt, J.A.; Lin, H.Y.; Goodliff, M.; Abarbanel, H.D.I. Integrating recurrent neural networks with data assimilation for scalable data-driven state estimation. *J. Adv. Model. Earth Syst.* **2022**, *14*, e2021MS002843. [CrossRef]
9. Bocquet, M.; Farchi, A.; Malartic, Q. Online learning of both state and dynamics using ensemble Kalman filters. *Found. Data Sci.* **2021**, *3*, 305–330. [CrossRef]
0. Arcucci, R.; Zhu, J.; Hu, S.; Guo, Y.-K. Deep data assimilation: Integrating deep learning with data assimilation. *Appl. Sci.* **2021**, *11*, 1114. [CrossRef]

MDPI AG
Grosspeteranlage 5
4052 Basel
Switzerland
Tel.: +41 61 683 77 34

Remote Sensing Editorial Office
E-mail: remotesensing@mdpi.com
www.mdpi.com/journal/remotesensing

www.ingramcontent.com/pod-product-compliance
Lightning Source LLC
Chambersburg PA
CBHW041423210926
51850CB00013B/3337